Biology and Ecology of Fishes

Biology and Ecology of Fishes

Third Edition

JAMES S. DIANA

TOMAS O. HÖÖK

Registered Office(s)
John Wiley & Sons, Inc., 111 River Street, Hoboken, NJ 07030, USA
John Wiley & Sons Ltd, The Atrium, Southern Gate, Chichester, West Sussex, PO19 8SQ, UK

For details of our global editorial offices, customer services, and more information about Wiley products visit us at www.wiley.com.

Wiley also publishes its books in a variety of electronic formats and by print-on-demand. Some content that appears in standard print versions of this book may not be available in other formats.

Library of Congress Cataloging-in-Publication Data
Names: Diana, James S., 1951– author. | Höök, Tomas O., author.
Title: Biology and ecology of fishes / James S. Diana, Tomas O. Höök.
Description: Third edition. | Revised edition of: Biology and ecology of fishes /
  James S. Diana. 2nd edition. 2003. | Includes bibliographical references and
  index.
Identifiers: LCCN 2022055936 (print) | LCCN 2022055937 (ebook) | ISBN
  9781119505778 (hardback) | ISBN 9781119505761 (adobe pdf) | ISBN
  9781119505747 (epub)
Subjects: LCSH: Fishes–Ecology. | Fishes–Feeding and feeds.
  | Fishes–Reproduction. | Fish populations. | Freshwater fishes–Ecology.
  | Freshwater fishes–Physiology.
Classification: LCC QL639.8 .D54 2023 (print) | LCC QL639.8 (ebook) | DDC
  597.17–dc23/eng/20221219
LC record available at https://lccn.loc.gov/2022055936
LC ebook record available at https://lccn.loc.gov/2022055937

Cover Design: Wiley
Cover Image: © RLS Photo/Shutterstock; Justas in the wilderness/ Shutterstock; FedBul/Shutterstock; Yannick Tylle/Getty Images

Set in 10/12pt STIX Two Text by Straive, Pondicherry, India
Printed and bound by CPI Group (UK) Ltd, Croydon, CR0 4YY

C9781119505778_060623

Contents

Preface

Studies of fishes have historically focused on one of two perspectives. The first—a more basic, academic approach emerging from universities—has dealt with systematics, anatomy, physiology, and theoretical ecology. The second—a more practical and applied approach emanating from natural resource organizations—has emphasized fishery biology, population ecology, and fishery management. The distinction between applied and basic perspectives has become less marked over time as researchers, managers, educators, and students have realized the importance of both lines of inquiry and have crossed this boundary. However, this dual perspective has been less common in the published literature—particularly textbooks—in the field of fish ecology.

This is the third edition of a book attempting to consider both biological and management needs in this field. It is written as a textbook for upper division undergraduate or graduate courses in fish ecology or fishery biology. It should be broad enough to include many areas of interest to practicing fish ecologists and fishery managers, but the writing style, selection of materials, and coverage of concepts are more appropriate for a textbook than a research or reference text.

In writing this book, we decided that keeping the interest of readers is more important than including all details or literature. We have attempted to develop text that is accessible for a reader and focuses on describing concepts and studies. A case study approach is more interesting to us than an exhaustive review approach, and we rely on case studies throughout the text. Following this approach, we also decided to limit the number of references to primary literature. In many cases, our understanding and generally accepted facts replace documented reviews of papers to support a particular point. We did not want to develop text that requires readers to pause after every sentence to comprehend its content and reflect upon multiple references. We believe this approach leads to a more interesting and readable work. However, it does not blend well with highly specialized graduate courses, so this book should be considered for the first courses students take in fish ecology, biology, or management—not for more specialized courses.

This book emphasizes how fishes respond to environmental conditions at the individual, population, and community levels of biological organization. This subject is commonly considered fish ecology although emphasis from some sections is often found in books on fish biology. Broadly, the first three chapters are intended as an introduction on aquatic systems, fish diversity, morphology, and functionality. These chapters are very much intended to orient the reader, and we emphasize that other texts are much more suitable for in-depth instruction related to aquatic systems and ichthyology. Chapters 4–8 emphasize growth of individual fish from a bioenergetics perspective. The subsequent 16 chapters provide a population perspective, addressing population processes (Chapters 9–16), feeding and predation mechanisms (Chapters 17–20), and reproduction and life histories (Chapters 21–24). The book concludes with chapters related to community and food web structuring (Chapters 25–31), followed by consideration of human influences on fish and fisheries, including fisheries, aquaculture, and climate change (Chapters 32–36). Processes such as growth, population structure, and behavior are included in the book as they are the underpinnings for a foundation in fishery management. We have expanded material on fishery management in this edition of the book, but all areas of concern to fishery managers and fishery scientists are not included. Again, the emphasis of the book is concepts, not methods. On occasion, methods—including quantitative methods—are described to better educate the reader in these areas but not in great detail or in great breadth. The above perspectives attempt to put this text into its proper place in the field of ichthyology and indicate to users the main emphasis and direction.

In selecting case studies for the book, we were aware that there is a historical lack of diversity (broadly defined) in terms of scientists conducting fish ecology research. To this point, many of the historical studies we describe were authored by white males. While still apparent, this bias is less dramatic in more recent studies, and we have attempted to select recent citations from a diversity of authors.

Finally, we would like to emphasize that this book should provide good coverage of current and historic themes in fish ecology and prepare readers to understand and analyze the concepts of importance in the field. An approach covering concepts allows the reader to cope more effectively with future changes in the field and remain current in spite of changing methods or emphases. It is our hope that this book will assist readers to further develop their insight and ability to understand and critically evaluate ecological literature.

Acknowledgments

Many people have aided us during the preparation of all three editions of this book. We owe them a debt of gratitude. This third edition expands authorship to include Tomas Höök as coauthor and Emmanuel Frimpong as a contributing author. Frimpong contributed to the development of overall topics for the book, and in particular did much of the writing for the chapters on morphology and evolution, positive interactions, and movements and habitat occupancy. Barbara Diana has done much of the word processing, edited and reviewed the writing, and helped us remain organized and on track to finish this work. It would never have been completed without her help. All three editions of the book have her personal touch and improved dramatically because of her effort.

In developing the third edition, we have added several chapters and combined and revised extant chapters. In so doing, we have added a great deal of text and several new figures and tables. However, this edition very much builds directly from the two previous editions, and in many cases text, figures, and tables are retained from the earlier editions.

This book initially arose from a course taught at the University of Michigan. Besides Diana, Drs. Paul Webb, Gerald Smith, Ed Rutherford, Richard Clark, James Breck, and Paul Seelbach have all taught portions of this course at various times and have influenced our thinking in many ways. Material was also informed from Höök's experience teaching fisheries and fish ecology courses at Purdue University. We are grateful to all of the students we have interacted with in these courses, as they have helped shape how we present this information.

In developing the first two editions, much of the actual writing was accomplished during two sabbatical leaves, funded by the University of Michigan. Diana produced the first edition during a leave at the Institute for Fisheries Research, Michigan Department of Natural Resources in Ann Arbor. Diana produced the second edition during a sabbatical leave at Griffith University, Brisbane, Australia. In producing this third edition, Diana and Höök did not take leaves from their academic positions, and hence, drafting this edition stretched across multiple years. At this time, they were partially supported by the University of Michigan, Purdue University, Michigan Sea Grant, and Illinois-Indiana Sea Grant.

Many colleagues reviewed draft portions of this book and provided useful comments. For this edition, we particularly thank Jim Breck, who not only reviewed two of the chapters for content, but then provided a review of the entire book, leading to a number of important suggestions. For this edition, we thank the many people who reviewed chapters and provided critical feedback, including Zoe Almeida, Karen Alofs, Nancy Auer, Mark Bevelheimer, Russell Borski, Jim Breck, David "Bo" Bunnell, Leandro Castello, Paris Collingsworth, Alison Coulter, David Coulter, Derek Crane, Jacob Daley, Solomon David, Jason Doll, Damilola Eyitayo, Troy Farmer, Zach Feiner, Michelle Fonda, Lee Fuiman, Tracy Galarowicz, Joel Hoffman, Dana Infante, Brian Irwin, Yan Jiao, Yoichiro Kanno, Conor Keitzer, Holly Kindsvatar, Steve Kohler, Hernan Lopez-Fernandez, Stu Ludsin, Chuck Madenjian, Ellen Marsden, Christine Mayer, Steve Midway, Don Orth, Brandon Peoples, Steve Pothoven, Mark Pyron, Frank Rahel, James Roberts, Jamie Roberts, Brian Roth, Lars Rudstam, Carl Ruetz, Justin VanDeHey, Hui-Yu Wang, Paul Webb, Earl Werner, Rusty Wright, and Mitch Zischke. For the second edition, we particularly thank Jim Breck, Chuck Madenjian, Jeff Schaeffer, and Kevin Wehrly for reviewing the new chapters and providing constructive feedback. For the first edition, Michael Benedetti, Jim Breck, Scott DeBoe, Dan Dettweiler, Gary Fahnensteil, Neal Foster, Roger Haro, Liz Hay-Chmielewski, Leon Hinz, Pat Hudson, Dave Jude, Carl Latta, Jeff Schaeffer, Jim Schneider, Paul Seelbach, Kelley Smith, Paul Webb, and Troy Zorn all provided useful feedback. In addition, Joanne Cooper did an excellent review and copy edit of the final manuscript.

In covering the varied aspects of fish ecology, the distinctions became unclear to us regarding what we had learned from our own reading and research compared to what we had learned from personal interactions. Much of the knowledge we possess today was driven into us by former professors and teachers, often after extreme efforts on their parts. Jim Diana in particular thanks Drs. Dave Lane, Cliff Hill, Don Nelson, Bill Mackay, and Dave Beatty for helping him understand the basics of fish ecology. Tomas Höök in particular thanks Paul Webb, Jim Breck, and Ed Rutherford.

Finally, we both thank our families, colleagues, and students for being understanding of our time and effort (and occasional frustration) in developing this book.

PART 1

Introduction

CHAPTER 1

Introduction to Aquatic Ecosystems

The organisms that scientists consider fishes represent the most diverse groups of all vertebrates as there are more species of fishes than of all other vertebrates combined. Fishes occupy diverse habitats such as normal surface waters, great ocean depths exceeding 8,000 m, heated desert pools and caverns that may exceed 40 °C, caves deep in the Earth, under the ice in the Arctic and Antarctic seas, and a variety of other extreme habitats. Adaptation to these habitats has resulted in extreme diversity in fish physiology and anatomy, coupled with differences in foraging patterns and other behaviors. This variability makes it hard to generalize the responses of fishes to local conditions but makes their adaptation to environments very interesting.

There has been much recent debate about what exactly is a fish. Historically, fishes have been considered to include five classes of vertebrates, although the taxonomic status of fossil groups is unsure. Currently existing fishes represent three classes: Agnatha – hagfish and lamprey; Chondrichthyes – cartilaginous fishes including sharks and rays; and Osteichthyes – bony fishes including the most current species. A recent novel by Miller (2021) has examined not only the history of fish biology and systematics but also the evidence that fishes do not represent a single evolutionary line, which means they have not evolved from a single common ancestor. Genetic evidence has recently shown closer relationships between some of the groups of fishes and reptiles, amphibians, and mammals, questioning whether what we call fishes actually are an evolved group or are just an animal life form. Ecologists have included all aquatic living vertebrates with gills, scales, and fins as fishes (although some species have secondarily lost some of these traits), and we will follow that pattern in this book. Certainly cladists (evolutionary biologists that study species relationships) can better explain the evolution of different classes of vertebrates, which may change this higher level of taxonomy. For the purpose of this book, we will continue to consider the three classes of vertebrates listed above as fishes as they share similarities in their life form.

Freshwater organisms are among the most endangered of all species in the world. Analysis by The Nature Conservancy shows that 70% of all freshwater mussels, 50% of crayfishes, and 40% of freshwater fishes are at risk of extinction in the United States (Master et al. 1998). In comparison, approximately 18% of reptiles, 15% of mammals, and 14% of birds are in a similar status. This is particularly daunting when we realize that the highest diversity of vertebrates is found in the classes known as fishes, where there exist at least 32,000 species. This is more than all the species of birds, mammals, reptiles, and amphibians, combined. Such a high fraction of the fauna being endangered among freshwater fishes is due to the various challenges we

Biology and Ecology of Fishes, Third Edition. James S. Diana and Tomas O. Höök.
© 2023 John Wiley & Sons Ltd. Published 2023 by John Wiley & Sons Ltd.

place on freshwater ecosystems, including the use of water for irrigation, industry, and human consumption, as well as the discharge of chemicals into water for disposal. In addition to direct use of water, humans alter habitat by building dams, channelizing streams for ship passage, and building canals. All of these changes in aquatic ecosystems have resulted in major reductions in the fish fauna. There have been a number of evaluations of factors causing animal extinction in various ecosystems. All of these divide the causes of extinction into three main groups of approximately the same magnitude: introduction of exotic species, overexploitation, and habitat disruption. Since fishes are the only major group of organisms remaining that are hunted as food on a global scale, much damage is due to overexploitation, as well as habitat disruption and exotic species. It is no wonder why freshwater organisms, in particular freshwater fishes, are under such threat.

Ecology has a variety of popular, or lay, definitions, but the science of ecology has been well defined and accepted by most scientists. The definition has evolved over time, depending largely on the level of our understanding of ecological interactions. Krebs (2009) provided the best definition: ecology is the study of the interactions that determine the distribution and abundance of organisms. These can be categorized as interactions with physical, chemical, or biological factors in the environment. The purpose of this textbook is to overview the means by which fish distributions and abundances are influenced by physical, chemical, and biological factors.

This book is divided into six main topics that focus on the three major disciplines of ecology: physiological, behavioral, and community ecology. These three disciplinary areas of ecology have boundaries that are intentionally unclear, so some concepts will be presented several times throughout the book.

To appreciate the ecology of fishes, it is important to first understand the habitat in which fish exist – the aquatic system. This chapter reviews living in the water, the characteristics of fish, and aquatic ecosystems, emphasizing several systems that are more familiar. A key theme throughout this book (highlighted in this chapter) is that the environments in which fish exist differ in two important dimensions: the distribution of temperature in time and space, and the distribution of food in time and space.

Properties of Water

Water has a number of physical properties that are challenging to organisms living in the water and yet promote life within the water because of the long-term stability of water conditions. Water is one of the few compounds that is liquid at ambient temperatures and has high viscosity and surface tension. This means that movement through the water is difficult, and diffusion across the water surface level is limiting. Animals moving within the water must overcome this high viscosity in order to shoulder their way through this dense and difficult medium. The maximum density of water occurs at 3.9 °C, which is unusual for liquids because the freezing point of water is 0 °C. The fact that water does not freeze at its maximum density allows water to exist under the ice in winter conditions and is key to sustaining life in many aquatic ecosystems. Water also has an extremely high heat capacity. It requires 1 kcal of energy to increase 1 kg of water by 1 °C. In fact, this demonstrates the importance of water to humans because many of our characteristics in physics are based on water, such as the Celsius scale of temperature and the caloric scale of energy. Because of this high specific heat, water does not change temperature very easily and remains relatively consistent over time. As a result, living in the water is actually living in a moderate thermal condition, where it is neither extremely cold nor extremely hot. In fact, fishes utilize this thermal characteristic to specialize within even narrower ranges within the typical temperatures of surface waters.

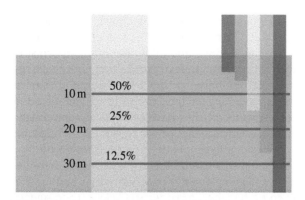

FIGURE 1-1 Schematic of penetration by different wavelengths of light into freshwater.

In addition to the characteristics above, water has several other characteristics that are important to life in the water. There is very low gas concentration in water. The atmosphere contains 21% oxygen and is relatively light and reasonable to ventilate. In contrast, water at saturation contains maximum level of 14.6 mg of oxygen for every liter of water. If we equate the two, 1 kg of air would contain 0.23 kg of oxygen, while 1 kg of water contains only 0.0000146 kg of oxygen, over 5 orders of magnitude less than the same mass of air. Clearly, this low oxygen concentration results in difficulties passing enough water across respiratory surfaces to allow animals living in water to attain high metabolic rates. This is one of the major specializations in fish – that of extracting oxygen from water at low oxygen concentration.

Water is known as the universal solvent, which means almost all materials can dissolve in water. This is both a benefit and a difficulty for aquatic life as many materials from the land and industrial processes dissolve in the water and influence physiology of fish breathing water containing that material. This universal solvent property is important in the discharge of waste as assimilation by natural ecosystems is one of the ways humans dispose of sewage. At the same time, waste dissolved in water can cause significant damage to aquatic species living in the region receiving wastes.

Finally, light is absorbed rapidly with depth in water and differentially depending on the wavelength of the light. Maximum light transmission in distilled water is approximately 100 m, and only the blue wavelengths of light penetrate to this depth. In contrast, infrared wavelengths, which include heat, only penetrate to a very shallow depth – usually less than 1 m – and there is differential distribution of wavelengths between those two extremes (Figure 1-1). These penetration depths are achieved only in very clear water; once water contains dissolved materials, there is considerably less light penetration as well as different penetration for various wavelengths. Given the oceans have a mean depth of 4,000 m, most of the ocean is below the level of light penetration, and organisms living there have difficulty using eyesight as a means of orientation.

What is a Fish?

Given the constraints placed upon fish by living in aquatic habitats, the characteristics of fish vary dramatically to allow locomotion, respiration, and feeding in this medium. A fish is an aquatic vertebrate, that is, fish spend most of their life cycle in water and belong to the phylum Chordata (chordates), subphylum Vertebrata. There are fishes that can remain out of water for considerable periods of time, utilizing modified lungs to breathe from air under moist conditions. All chordates have a notochord, which is a flexible nerve cord that extends the length of the body. In most fishes, this notochord is surrounded by bony or calcified tissue to protect it from damage and allow for more vigorous muscle action. Fishes have 5–7 gill arches, which vary in complexity and range, from each gill arch having a separate opening to existence of one opening under the operculum for more advanced fishes. There are many fishes that have lungs, which evolved into swim bladders in more advanced fishes. Some of the more advanced fishes have secondarily developed lungs as a means of living in poorly oxygenated environments or for respiration while transporting themselves across land. Fishes have fins to help propel themselves through the water, including medial fins along the middle of the body, a caudal fin at the posterior extreme, and paired pectoral and pelvic fins. Most fishes also have scales to protect their bodies; these have evolved over time from being absent in early fishes, to very stout scales, to more flexible and lightweight scales in advanced fishes. Finally, fishes have a two-chambered heart, from which the blood is circulated through the gills to the body and then returns to the heart. All of these characteristics vary among groups of fishes but are common characteristics of the three classes of vertebrates considered to be fishes.

The details above indicate that there are many challenges to being a fish. Water is a dense and viscous medium; thus, swimming through the water utilizes much muscle power and fin activity. The low concentration of dissolved gases means the gill structure and function must be efficient in order to allow for high rates of metabolism. The high thermal capacity of water means that most fishes are ectothermic and poikilothermic, or what we refer to as cold blooded. Their body temperature will vary with the temperature of water, and there is generally no difference between their body temperature and water temperature. In addition, the conditions are generally dark and difficult to see in aquatic systems. Some advantages of being a fish include buoyancy that can be maintained by simple changes in swimming motion, inflation of swim bladder, retention of lipids in the body, and other such characteristics. Since fishes live in the water, it is never too hot or cold, and few fishes have developed a system of insulating their body from extremely cold external conditions. Water transmits chemicals and sound from long distances, and thus, senses of smell, hearing, and taste are acute and can occur over long distances, allowing fish to utilize these senses very efficiently. In addition, fishes can sense movement in the water and water currents and are capable of adjusting to these through the lateral line sense. These senses are also effective in some species of fish for finding prey or escaping predation.

Lentic Systems

Much of our knowledge of aquatic ecology comes from work in temperate freshwater ecosystems. This is not due to any special importance of these ecosystems, but more to their proximity to scientists and institutions interested in aquatic ecology. This section on fish habitats will begin with details on temperate lakes. The emphasis on aquatic systems has broadened to include tropical and Arctic freshwaters, as well as marine ecosystems, but the preponderance of ecological studies still deals with organisms living in freshwaters of the temperate zone.

Most people are familiar with standing water ecosystems, that is, lakes and ponds. They are termed lentic because the movement of water is relatively unimportant in the ecology of the system. The kinds of habitats available in lentic systems are related to lake types, which are determined by the internal processes in lakes and in watersheds where lakes are located. The largest differences within temperate lakes occur during midsummer when productivity and thermal processes vary most dramatically in regions of a lake.

Vertical Stratification of Temperature

One interesting seasonal process in temperate lakes occurs because of the physical relationship between temperature and density of water. Water that is colder or warmer than 4 °C floats on top of this cool and heavy water. As water warms, the warmer water floats on the surface of colder water. This process of vertical segregation of water by temperature is termed vertical stratification. The classic stratification pattern of an inland lake varies with season (Figure 1-2). During winter, ice covers the lake, and immediately below the ice, temperatures between 0 and 4 °C occur. Water at 4 °C is at the bottom of the lake because of its density, and the amount of water between 0 and 4 °C depends on the rate of cooling that occurred in the lake prior to freezing.

In spring, warming by the Sun allows the ice to melt on the lake, and wind blowing across the surface of the lake mixes water from top to bottom in most lakes. During this time, lakes are isothermal, having the same temperature from top to bottom. Also during this time, oxygen is mixed throughout lakes, and nutrients and other chemicals that might be tied up in bottom sediments or in deep water are redistributed to the surface.

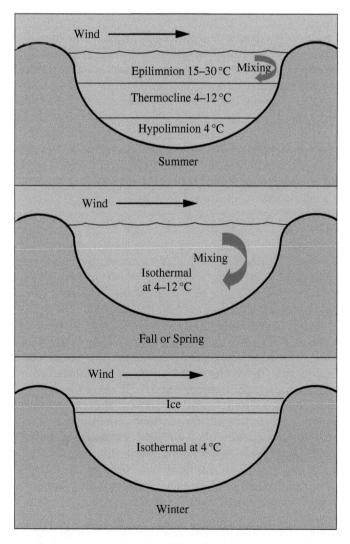

FIGURE 1-2   Diagram of a lake and its zonation by depth during climatic seasons.

Two other processes affect temperature stratification of lakes: the absorption of light within a lake and circulation of lakes by wind. Wind mixing occurs as wind blows across the surface of lakes and causes turbulence. The amount of wind mixing depends on the length (fetch) of lake exposed to wind, the wind intensity, and whether the nearby topography is hilly or flat. Wind mixing dramatically varies among lakes due to local and regional geographic effects. During warming in spring, all heat is absorbed at the surface and is spread throughout the lake by wind mixing. During spring, this mixing often occurs top to bottom and causes isothermal conditions. However, as warming at the surface continues into summer, surface water becomes much warmer and lighter than deep water, and the lake stratifies into three zones. The epilimnion, which is the surface area that is mixed completely by winds, has the warmest temperature in the lake. The metalimnion, or thermocline, has temperatures that dramatically plunge from near surface conditions at the top of the thermocline to near 4 °C at the bottom. In fact, the thermocline is defined as the region in which temperature declines 1 °C for every meter in depth. The hypolimnion, or deep water, has a constant temperature of 4 °C. As stratification proceeds, wind cannot continue mixing between the density layers in a lake; therefore, all wind mixing occurs only at the surface. The epilimnion remains completely mixed, the metalimnion partly mixed, and the hypolimnion receives no mixing throughout the course of a stratified summer.

As the year progresses to fall and air temperature drops, the temperature of surface waters declines until another isothermal condition occurs, when wind can blow on the lake and cause it to mix again. This period is called fall turnover; the lake is entirely mixed at this time, and water conditions are isothermal from top to bottom.

Obviously, the extent and significance of the seasonal cycle for any lake largely depends on climate and geography. Winter conditions may not exist in semitropical lakes, and summer stratification may not occur in Arctic lakes. Additionally, whether the zonation of lakes into three habitats occurs may depend on the depth and surface area of the lake. Relatively shallow lakes may not completely stratify. However, this classic stratification pattern occurs in many lake systems, modified by local conditions.

Lake Productivity

In addition to temperature, primary productivity dramatically varies among lakes, from an extreme of an oligotrophic or lowly productive lake to a eutrophic or highly productive lake. Primary production is the elaboration of new plant material due to photosynthesis, and it can be measured by carbon fixation or oxygen evolution in the water column.

Productivity cycles in oligotrophic lakes are partly limited by morphological characteristics of the lake and partly by internal nutrient processes. Oligotrophic lakes are generally deep and V-shaped. There is limited littoral (nearshore) habitat, and volume of the hypolimnion is most commonly much greater than volume of the epilimnion. Most oligotrophic lakes occur in areas in which nutrient runoff from the watershed is relatively low and sediments in the lake are minimal, so there is low nutrient availability. The main nutrients necessary for primary production – phosphorous and nitrogen – are in very limited supply. There is a very low rate of primary production in oligotrophic lakes, and similarly low fish production because of the limited nutrient availability. Since respiration by plants and animals is relatively low, oxygen is usually available at all depths, even during summer stratification. Oligotrophic lakes comprise the deep blue mountain lakes many people consider classic lakes, yet in reality, they have limited primary and secondary productivity. Secondary productivity is the elaboration of new animal tissue due to consumption of food.

The other extreme in lake productivity is a eutrophic lake. Eutrophic lakes have much larger littoral areas than oligotrophic lakes. These lakes are usually shallow, bowl-shaped basins. The epilimnions of eutrophic lakes are much larger than the hypolimnions; in fact, many eutrophic lakes do not have hypolimnions. Lake sediments are muddy and occasionally have covered lake bottoms that were formerly deep hypolimnetic waters. These sediments are high in nutrients. Eutrophic lakes have much higher primary production, particularly in phytoplankton, because they have higher nutrient concentrations than do oligotrophic lakes. High nutrient availability and abundance of organisms lead to larger oxygen demand, and oxygen may become limiting in the hypolimnion of a eutrophic lake during summer or winter stratification. Oxygen becomes limiting because these lakes have low light penetration into the hypolimnion, where much decomposition and respiration occurs.

The identification of lakes as either eutrophic or oligotrophic is rather arbitrary. Obviously, there is a continuum of lakes across these two extremes, and many systems of classification are used to distinguish different categories of lakes. For example, the trophic state index (Carlson 1977) evaluates the lake's relative status on a scale of 1–100, where 1 would be highly eutrophic and 100 highly oligotrophic. This index uses phosphorus concentration, chlorophyll *a* content, or Secchi disk depth as indicators of productivity. Phosphorus is commonly the most limiting nutrient in a lake, while chlorophyll *a* is a measure of phytoplankton biomass, and Secchi disk depth is related to water transparency, and therefore inversely related to chlorophyll *a* content.

The characteristic of most importance to ecology is that the interaction between temperature stratification and oxygen availability allows for much habitat segregation with depth in a lake system. For example, an oligotrophic lake may have warmwater fishes, such as bass and bluegill, in the epilimnion, coolwater fishes, such as perch and pike, in the thermocline, and coldwater fishes, such as trout, in the hypolimnion. These fishes can survive through summer near their optimal temperatures for growth by vertically segregating because oxygen is available at all depths in the lake. This segregation is very common in the Laurentian Great Lakes. The stratification of habitat vertically allows coexistence of different thermal guilds of fishes in the same water body. It may be limited in smaller inland lakes by low productivity and food availability of oligotrophic systems.

In comparison, a eutrophic lake might not produce this "two-story" or "three-story" type of fish community. A eutrophic lake with an anoxic hypolimnion will contain no trout because trout cannot survive high summer temperatures at the surface or low oxygen levels at depth. Similarly, eutrophic lakes with anoxic or poorly oxygenated thermoclines may not have coolwater fishes. Due to the high productivity of eutrophic lakes, energy is available for a much higher standing crop of fish if there are adequate physical conditions.

The combination of summer stratification and oxygenated conditions allows for a more diverse fish fauna in a given lake. At any time, lakes can be categorized on the basis of their oligotrophic or eutrophic conditions, but a lake proceeds from an oligotrophic to a eutrophic state over time. An oligotrophic, or relatively new lake, has low nutrient availability, low primary production, low decomposition, and low sedimentation, and remains in that state for some time. Eventually, erosion from the watershed and internal processes in the lake increase sedimentation and nutrient concentrations, which boost primary production and further sedimentation, and eventually result in filling in of the lake. Over time, oligotrophic lakes become eutrophic lakes by these erosion and internal processes.

This book emphasizes the relationship of fishes with temperature and food, key factors driving metabolism of fishes. Brown et al. (2004) have formalized the interactions between metabolism and size into the metabolic theory of ecology, where these defining characteristics of organisms set the stage on which animals perform, and can reflect processes in populations and communities as well. The primary production and trophic (food web) relationships of various habitats within a lake vary significantly. In the epilimnion, the basis for most productivity is phytoplankton. Phytoplankton, or small floating algae, dominate the plant community of most lakes and are limited by sunlight, temperature, and nutrient availability. Phytoplankton produce tissue through photosynthesis, and they are utilized by bacteria or herbivorous zooplankton and fishes. The production of herbivores must be lower than the productivity of phytoplankton because of the inefficiency of biological processes. Carnivores, feeding on herbivorous animals, are even lower in productivity, and this sort of a trophic pyramid follows the typical pattern of declining productivity with increasing trophic level.

Phytoplankton have very rapid birth and death rates and very high individual rates of productivity per unit biomass; therefore, the biomass of phytoplankton in a lake at any given time is relatively low. Similarly, bacteria with rapid life cycles also have relatively low biomass per unit of production. Herbivores and carnivores, on the other hand, have higher amounts of biomass maintained per unit of productivity. The trophic pyramid of a lake epilimnion based on biomass may be inverted, indicating a lower biomass of phytoplankton compared to the biomass of herbivores and carnivores. This sort of inverted pyramid indicates that food availability at the lower trophic levels is very much related to primary production within the phytoplankton rather than to the standing crop of phytoplankton present.

In the hypolimnion of a stratified lake, photosynthesis becomes unimportant because light does not penetrate with sufficient intensity into the depths to allow much primary productivity. The basis for nutrient availability is detritus in the hypolimnion, and the primary consumers are the detritivores. Detritus is dead and decaying animal and plant tissue, and detritivores are animals that feed upon such material. Within a hypolimnion, there is a much higher base in detritus than there are detritivores or carnivores, whether measured on a productivity or biomass basis.

In comparison with the epilimnion of an oligotrophic lake, a highly eutrophic lake has much macrophyte production within the littoral zone as well as phytoplankton production in the pelagic zone (the open water zone not influenced by the lake bottom). Macrophytes are rooted vascular plants. Compared to phytoplankton, macrophytes are relatively poor food sources for most herbivores. In fact, many herbivorous invertebrates that live on macrophytes actually feed upon periphyton or algae attached to the surface of these macrophytes.

In summary, combinations of physical processes, nutrient availability, and oxygen availability influence the vertical profile of the lake and the distribution of plants and animals within the lake. This distribution affects the source and sink of energy within a lake and therefore the rate of productivity as well.

Lotic Systems

In comparison with freshwater lakes, flowing freshwater systems generally show very limited, if any, stratification. Lotic, or flowing water systems, commonly have longitudinal zonation rather than vertical zonation. Physical and chemical conditions within streams are constant with depth because of the movement and mixing of water. The longitudinal zonation of streams occurs along a continuum from a headwater area to a very mature area near the mouth of most rivers.

Stream Types

Water inputs to headwater streams include runoff from snow melt or precipitation as well as ground water. Runoff may vary dramatically with season due to the snow melt or irregularity of rain. Often these water sources, through snow melt or ground water, are relatively cold, and water may warm during day due to insolation and cool during night due to additional runoff. This may produce large diel (24-hour) fluctuations in temperature at a given site within a headwater stream. Since the runoff for headwater streams occurs over a limited catchment area, there are few nutrients within the stream, and most of these nutrients are from external sources, such as erosion. Headwater streams tend to segregate longitudinally into riffles and pools. Riffles are quick-running areas with rapid currents and usually rocky substrates. Pools are deeper, slower-moving waters with more sedimentation of smaller particle sizes. Riffles and pools may be separated by runs, which are geomorphic units sharing some of the characteristics of both riffles and pools – deep with quick flow but not as turbulent as riffles.

In contrast with headwater areas, a more mature stream receives most of its water from other tributary streams rather than direct runoff from the nearby watershed. These tributary streams contain water that has been exposed to terrestrial conditions for some period of time and are often warmer and more stable in temperature on a diel basis, although they can fluctuate seasonally with air temperature. As water continues to flow into the river systems, more of the nutrients spiral or cycle internally, rather than being added from external sources, and many nutrients are retained in sediments. Mature streams, rather than having clearly defined riffles and pools, tend to have meandering zones, with slower water on the inside bend of a meander and faster water on the outside. This causes segregation of substrate types and flows across the stream rather than in a longitudinal pattern. This meandering pattern of a more mature stream is typical of most flood plains.

Most streams have oxygen conditions near saturation as oxygen is continually diffused into the moving water and mixed by physical motion. Since motion is important, plankton communities (which drift in water) are also uncommon in streams and rivers. Primary production for a youthful stream is mainly based on periphyton and is combined with detritus from external sources as the primary inputs of energy to the trophic pyramid. Many temperate streams rely on detritus from annual leaf fall as the major input of energy and have large communities of shredding insects that feed upon leaves that accumulate in the steams from deciduous trees, as well as on periphyton or attached algae that grow on the leaf substrate. External detritus and periphyton form the basis of the food chain, which usually flows through benthic (bottom dwelling) insects, or other benthic invertebrates, and ultimately to carnivores. A more mature stream has periphyton or detritus, produced by internal processes rather than external loadings, forming the basis of primary production.

Fish Communities in Streams

Much as stream types zone longitudinally in physical and biological conditions from headwater to mouth, fish communities in streams also change longitudinally. The zonation of streams has been illustrated by data from European and North American fish communities. Headwater areas of streams are the trout zones, where coldwater fishes such as trout predominate. Mid-elevations of streams, where stream velocity has slowed and temperatures are warmer and more constant on a diel basis, include fishes such as the grayling and minnows. The lower reaches of streams, which are more lake-like, are considered the bream zones, where species such as sunfishes and catfishes predominate. Thus, stream fish communities and ecosystems vary with elevation from headwater to mouth. Obviously, there are areas of overlap among these different communities at the edges of the zones, but these categories are often fairly distinct.

Marine Systems

The marine environment is extremely diverse compared to the freshwater environment because oceans continuously span a variety of climatic as well as physical areas. Over 70% of the world's surface is covered by oceans, so obviously the largest available habitat for animals and the largest area for fish habitation is the ocean. The open pelagic zone of the ocean has some characteristics similar to a freshwater lake. It commonly stratifies with temperature although the stratification may be seasonal; it has oxygen and primary production limited by light penetration, and it has primary production due to phytoplankton. The pelagic zone differs from a freshwater lake in other ways, mostly in depth. The average depth of the ocean is about 4,000 m, so most of the ocean water is at a depth exceeding the penetration of light. The epipelagic zone, or the shallow area of the ocean, is a limited habitat in terms of area (at least among oceanic habitats) but is the source of virtually all primary production, as well as much of the fish production in the oceans.

In addition to the pelagic zone of the oceans, the nearshore environment also differs significantly from that of lakes. The largest difference in the nearshore environment is in the importance of tides and intertidal zones to the production and distribution of plants and animals. Tides are caused by the force of gravity from the moon and occur differently in various regions of the world but result in diurnal changes in water level. As the water level changes, some attached plants and animals are exposed to air during a portion of the day and to water during another portion of the day. Tidal heights also vary at different times of the month, so within extreme zones there may be exposure to air or water for only a short period each month. Many attached algae grow in the intertidal zone and are the basis for much food production for the native fishes and invertebrates.

Two other marine ecosystems addressed several times in this book that receive specific emphasis are coral reefs and upwelling areas because they differ from typical lake systems in important ecological processes. A brief overview of each is important at this point because fishes occupying these conditions are mentioned sporadically throughout this text.

Coral Reefs

Coral reefs are ecosystems limited geographically to a zone of 30° north or south of the equator. They occur in tropical, relatively low productivity areas and can grow to a depth of 50 m. The physical basis for coral reefs is the coral organism itself, which secretes a hard calcareous skeleton that becomes the structure of the reef after coral polyps themselves die. The calcareous material is secreted in relationship to growth of the coral polyps, so there is much structural complexity. A wide variety of fishes occurs on coral reefs, and there are more fish species per unit area associated with coral reefs than with any other ecosystem. These fishes also have developed a wide range of feeding specializations.

The coral reef ecosystems, while existing in waters low in nutrient availability, actually have a high degree of primary production because symbiotic zooxanthellae coexist with the corals. The symbiotic algae can fix nitrogen for photosynthesis, and surplus energy produced by them is available for coral nutrition. These symbionts, plus the attraction of wider-ranging animals to the reef, result in high productivity and biomass of organisms occurring on coral reefs. Coral reefs are rather isolated habitats spatially and usually do not cover large geographic areas.

Upwelling in the Ocean

Upwelling systems occur when the cold bottom waters of the ocean move upward and mix with surface waters. This generally occurs where currents move offshore, taking surface waters offshore as well. Upwelling areas bring up cold, nutrient-rich waters and are characterized by high primary productivity due to nutrient enrichment and high fish yield due to food availability. Deep oceanic waters are nutrient rich due to decomposition in that zone but generally do not return these nutrients to shallow waters during mixing. High pelagic fish production occurs due to upwelling of nutrients. Many of the major fisheries of the world occur in these upwelling zones, including the coasts of Peru and California. Upwelling zones are also important features of large lakes but are transient in location due to the general lack of consistent current patterns.

Summary

Ecology is the study of the interactions that determine the distribution and abundance of organisms. Several major factors in aquatic environments vary, including temperature, oxygen, and food. These factors vary with habitat type and may affect fish distribution and abundance. Some habitats in lakes, streams, or coral reefs may have large fluctuations in conditions daily or seasonally, while others experience relative consistency. The absolute level of primary productivity, as well as seasonal timing of production, strongly affects the distribution and abundance of animals. The remainder of this book addresses some of the physiological effects of these factors on fish distribution and abundance as well as the means by which fishes adjust, both behaviorally and in terms of population changes. It is important to keep the physical nature of the habitats in mind throughout the following chapters.

Literature Cited

Brown, J.H., J.F. Gillooly, A.P. Allen, V.M. Savage, and G.B. West. 2004. Toward a metabolic theory of ecology. Ecology 85:1771–1789.

Carlson, R.E. 1977. A trophic state index for lakes. Limnology and Oceanography 22:361–369.

Krebs, C.J. 2009. Ecology: The Experimental Analysis of Distribution and Abundance, 6th edition. Benjamin Cummings, San Francisco, California.

Master, L.L., S.R. Flack, and B.A. Stein, editors. 1998. Rivers of Life: Critical Watersheds for Protecting Freshwater. The Nature Conservancy, Arlington, Virginia.

Miller, L. 2021. Why Fish Don't Exist: A Story of Loss, Love, and the Hidden Order of Life. Simon and Schuster, New York.

CHAPTER 2

Fish Diversity

Globally, fishes display a high degree of behavioral, physiological, and anatomical diversity. This diversity is most evident when considering interspecific (among species) differences, but for many species, intraspecific (within species) diversity is also high. There are greater than 30,000 known species of fishes (32,500 estimated by Nelson 2006). These species are distributed throughout the world's waters, from the poles to the equator, from huge oceans to small pools, and from dry deserts to dark caves. In addition, there are new fish species described each year at a rate of about one new species per day (estimated from Fricke et al. 2021), with 3890 named between 2006 and 2015 (Nelson et al. 2016).

The number of species of fish is very high with more species of fish than all other vertebrates combined (Table 2-1). Such a large number of fish species also contribute to the high number of fish species considered threatened or endangered. In early 2022, the U.S. Fish and Wildlife Service listed 140 fishes as endangered under the Endangered Species Act. While most of these fishes were listed at the species level, several evolutionary significant units (ESUs) were also listed. An ESU is a population of a certain species, which is judged to be sufficiently distinct and evolutionarily important to conserve independent of broader species considerations. The International Union for Conservation of Nature (IUCN) maintains the *IUCN Red List of Threatened Species* and categorizes threatened species as vulnerable, endangered, critically endangered, extinct in the wild, and extinct. Here, too, fish species constitute larger numbers than other vertebrate groups in most categories (IUCN 2021). However, as fishes are more speciose (consisting of many species) than other vertebrate taxa, the proportion of fish species listed as threatened is lower than for some other taxa, particularly amphibians, which have a much larger fraction that are threatened.

The number of known fish species is roughly equivalent between marine and fresh water. For example, Vega and Wiens (2012) reported that for ray-finned fishes (actinopterygian; including over 95% of fish species), there were 15 150 known species in fresh water and 14 740 known species in marine systems. If one considers all species of fishes – not just actinopterygian – the species numbers may be slightly greater for marine systems (the exact number of species by habitat differ by source), but nonetheless the total species numbers are similar. Some species can be classified as diadromous – inhabiting both freshwater and saltwater environments during their life. While species are split fairly evenly between marine and fresh water, there are likely many more individual fish in marine environments than in fresh water as most marine fish populations have many more individuals than freshwater species. There is 1.3–1.4 billion km³ of salt water globally compared to less than 0.04 billion km³ of fresh water. Also, the vast majority of the Earth's fresh water is in the form of ice or groundwater. Less than 0.5% of global fresh water is in lakes, streams, and wetlands, with the majority of this surface water – >85% – in lakes, specifically a few large lakes in North America, Africa, and Asia. The ratio of

Biology and Ecology of Fishes, Third Edition. James S. Diana and Tomas O. Höök.
© 2023 John Wiley & Sons Ltd. Published 2023 by John Wiley & Sons Ltd.

TABLE 2-1 The Estimated Number of Described Species and Number of Threatened Species by Vertebrate Group (Based on Report by Iucn Red List of Species from December 2021), and The Number of Species Listed as Endangered Under the U.S. Endangered Species Act in Early 2022.

	Number of Globally Identified Species	Number of IUCN Threatened Species	Number of US Species Listed as Threatened or Endangered
Mammals	6,578	1,333	80
Birds	11,162	1,445	106
Reptiles	11,690	1,839	48
Amphibians	8,395	2,488	38
Fishes	36,058	3,322	140

non-frozen surface salt water to fresh water inhabitable by fishes is approximately 8,000:1, so it is not surprising there are many more individual fish in the marine environment. However, on a per unit volume, there are undoubtedly many more individual fish in fresh water. Assuming an equivalent number of fish species in fresh water and salt water, on a volumetric basis the ratio of fish species in fresh water versus salt water is 8,000:1.

Speciation – the emergence of new species – can occur through multiple processes, and there are many reasons why there is such a disproportionally high number of freshwater fish species compared to saltwater species. However, two important contributing factors include the great variation of habitat conditions across freshwater environments and the relative isolation of many fish populations in freshwater systems.

Fish Trait Diversity (Interspecific)

While individual species are delineated based upon their genetic makeup and ability to produce viable offspring, they are often characterized by observable and measurable traits. The breadth of trait variation among species of fishes is incredible. Species of fish vary not only in terms of externally obvious traits – such as pigmentation, size and position of fins, and overall body shape – but also in terms of internal morphology, such as relative shape and length of the digestive tract, brain lobes, and proportional composition of skeletal muscle. They also vary physiologically (e.g., salinity and temperature tolerance), behaviorally (e.g., schooling and foraging strategies), and reproductively (e.g., internal or external fertilization, genetic or environmental determination of sex, and sex reversal). For example, the whale shark can grow to over 18 m, while the stout infant fish may reach a maximum size around 8 mm. Many species of fishes cannot survive outside of water for even short periods of time and can only tolerate very narrow ranges of salinity. However, other species, such as African lungfish, can survive outside of water for months, and species, such as bull shark, can tolerate a very wide range of salinities. Species generally express traits and forms that allow them to function well in the specific environments they inhabit, therefore allowing fish to maintain relatively high fitness – the ability to pass on their genetic material to subsequent generations. Measuring true fitness (contribution to future generations) of individual fish is exceedingly difficult, so fitness traits have instead been related to vital rates, such as survival, growth, and reproduction. The belief is individuals that experience higher survival, grow faster or have higher reproductive output will pass on relatively more genetic material to subsequent generations and therefore experience greater fitness.

Fishes are the geologically oldest vertebrate and since initial development of fishes several processes have contributed to speciation of fishes. Allopatric speciation occurs when two or more groups of the same species become geographically isolated and over time diverge to the point they can no longer exchange genetic material. During allopatric speciation, dissimilar selective pressures in the distinct environments select for different phenotypes and genotypes, and processes, such as genetic drift – random, non-selective change, leading to differences in relative frequency of genotypes – and mutations can lead to genetic differences between the now isolated species. Entire genome duplication events led some fishes, including the family Salmonidae, to become tetraploid – having four sets of chromosomes – and facilitated rapid discrimination and speciation. In contrast to allopatric speciation, sympatric speciation occurs without geographic isolation of groups. Under sympatric speciation, groups of fish evolve to specialize on distinct resources, such as food or habitat. These specializations are often reinforced by selective reproduction among distinct specialists, making these fish unable to reproduce with sympatric groups and causing speciation over time.

One of the most common cited examples of sympatric speciation is the haplochromine cichlids of the African rift lakes. Cichlids in these lakes are thought to have experienced adaptive radiation – the relative rapid speciation of multiple species from a common ancestor. Many cichlid species are endemic to individual lakes, meaning they only exist in that one lake. This suggests that allopatric and sympatric speciation has contributed to the distinct cichlid fauna across the African rift lakes. What is perhaps more impressive is the large number of cichlid species in individual lakes. Lake Victoria and Lake Malawi each have over 500 species of cichlids, although the taxonomy of these species is not entirely clear, and many species have become rare – or even extinct – due to human-induced changes.

Intraspecific Trait Variation

While trait diversity may be clearest among different species of fishes, intraspecific trait variation is another key source of variation among individuals. Different populations of the same species may display distinct mean trait values. Fish of similar size, age, and sex within a population may express dramatically different traits. The traits that characterize an individual change dramatically within the lifetime of a single fish. These trait differences are partially related to genetic differences among individuals. Individuals also differ in the environment they experience, and genetically similar individuals may express traits differently when placed in distinct environments. Phenotypic plasticity – the phenomenon of variable (plastic) expression of a trait (phenotype) when experiencing a different environment – is common among fishes.

Differences Among Populations

Populations of fish experience microevolution, and their gene frequencies change over time. Changes in gene frequencies are often influenced by selective pressures in the particular environment occupied by a population and selection for certain alleles that improve performance in an environment over time. Separate populations of the same species may display different mean traits because they have adapted to their environment through microevolution. Differences in biotic and abiotic conditions among environments may lead to expression of different traits, even in the absence of genetic differences. For example, fishes that grow in an environment with warmer temperatures and more prey may grow more rapidly and develop a deeper, wider body, because the environment leads to such performance and trait responses, not simply because of genetic differences. In many cases, it is difficult to determine the relative contribution of genetic differences or environmental influences on observed intraspecific differences in mean trait values among populations. The genetics underlying some traits are well understood, and in such cases, molecular genetic analyses may allow researchers to compare gene frequencies among populations and relate the contribution of genetic differences to trait differences. Researchers may also use common garden experiments in which organisms are held in identical environments to evaluate if trait expression differs when environmental conditions do not.

Populations of Atlantic silversides are distributed along the Atlantic coast of North America – from northern Florida to the Maritime Provinces of Canada. Populations of the species display various trait differences along this range, including differences in growth capacity. Atlantic silversides have a very short life cycle compared to many other fishes. They are essentially an annual species with almost all adults reproducing at age one and very few adults surviving to reproduce a second time. Due to this short life cycle – as well as their sensitivity to environmental conditions and strong response to selection – Atlantic silversides have been the frequent subject of research. In particular, David Conover and colleagues from Stony Brook University have explored a variety of ecological and evolutionary questions using Atlantic silversides as a model organism.

Conover and Present (1990) explored the observation that northern populations of Atlantic Silversides have the capacity to grow much faster than southern populations. While southern populations have a much longer growing season, Atlantic silversides from northern or southern locations appear to reach a similar mean size by the end of the first growing season. Since northern populations have a much shorter growing season, this would suggest that northern populations grow faster. Conover and Present (1990) tested this idea through common garden experiments with Atlantic silversides from different latitudes and showed that northern fishes were in fact able to grow faster than southern fishes. In particular, they found northern fishes grew faster at relatively high temperatures, but there were no strong differences in growth rates at cooler temperatures. This experiment suggested that northern Atlantic silversides populations have the capacity to take advantage of short periods of warm temperatures to grow faster and reach a similar size as southern populations by the end of the first growing season. This is an example countergradient trait variation, where a phenotypic trait value (in this case silverside growth and size) is expected to vary in one direction based on environmental influence (e.g., faster growth and larger size in warmer, lower latitude locations), but varies in the opposite direction based on genetic influences. Conover et al. (2009) demonstrate that countergradient trait variation is evident for various animal populations, and among fish species, in particular, there are several examples of countergradient trait variation across latitudes.

Differences Within Populations

Populations of fish differ dramatically in terms of intrapopulation genetic variation. Smaller populations tend to have less genetic variation than larger populations. Low genetic variation may also reflect processes that contributed to reduced genetic variation in the past, such as previous survival bottlenecks, strong selective events, or limited genetic variation among initially colonizing individuals. To the extent which phenotypic traits are reflective of genotypes, populations with low genetic variation may be expected to also display low trait variation. Therefore, individual fish in populations with limited genetic variation may display less variation in traits such as mouth size, digestive tract length, and feeding behavior.

Trait expression is not simply dependent on genetics as environmental conditions can affect expression of many traits. Again, phenotypic plasticity is common among fishes, and the distribution of trait values among individuals in a population may reflect both canalized and plastic processes. For many traits, intrapopulation trait values may center on the population mean value and follow a normal (bell-curved) distribution. For other traits, genetic and plastic processes may lead to bimodal or multimodal distributions of trait values.

Eurasian perch differentiate between individuals occupying more pelagic conditions and individuals occupying more littoral conditions in several lakes in Europe (Svanbäck and Eklöv 2002). Perch in the pelagia exist in a much more open environment and specialize on pelagic prey such as crustacean zooplankton. In contrast, littoral perch inhabit a zone with high structural complexity due to aquatic plants and other nearshore structures, and specialize more on larger, benthic, and terrestrial invertebrate prey, such as insects. This trophic and habitat specialization leads pelagic and littoral perch to have different carbon and nitrogen isotope signatures – biochemical measures of long-term trophic reliance – and to display distinct body shapes. Pelagic perch display more streamlined body shapes, while littoral perch have deeper bodies with more down-turned mouths.

These intrapopulation differences in Eurasian perch appear to be largely plastic. Body shape differences can be elicited by rearing Eurasian perch in different controlled environments, and body shape can even be reversed by moving a perch reared in a simulated pelagic environment to a simulated littoral environment and vice versa (Olsson and Eklöv 2005). There is no evidence that assortative mating by Eurasian perch is based on diet specialization or body shape. While the distinction of more streamlined pelagic perch from deeper bodied littoral perch may persist in many populations, there is no evidence that this will lead to speciation of two different perch species.

Interspecific and intraspecific trait variation was highlighted above. However, in some cases, these two types of trait variation blur. Before settlement by Europeans, the Laurentian Great Lakes contained various forms of ciscoes of the genus *Coregonus*. Lake Michigan alone was thought to have contained nine different species of *Coregonus*: *C. artedi* (cisco or lake herring), *C. clupeaformis* (lake whitefish), *C. hoyi* (bloater), *C. johannae*, *C. kiyi*, *C. nigripinnis*, *C. reighardi*, and *C. zenithicus*. Several of these species were extirpated through overfishing and the negative effects of invasive species and habitat degradation. In addition to lake whitefish, only two species of ciscoes (cisco and bloater) are currently present in Lake Michigan. Distinct types of *Coregonus* differed in habitat occupied, prey consumed, maximum size attained, and morphology (e.g., mouth position, and size and shape of gillrakers). Many of the same forms of *Coregonus* were present in multiple lakes. While these distinct forms were previously thought to constitute different species, more recent studies suggest that some may actually be forms of the same species (see review by Eshenroder et al. 2016).

Ciscoes in the Laurentian Great Lakes are now generally referred to as a species flock. The lake whitefish is clearly a distinct species, but differences among all other *Coregonus* forms in the lakes may not constitute species. While there is some genetic differentiation among *Coregonus* forms within individual lakes, in general, *Coregonus* forms are genetically similar to other forms that occupy the same lake, yet there is stronger genetic differentiation between similar morphs from different lakes.

Ontogeny

Another huge source of trait variation among fishes occurs through the life of an individual fish. The extremity of ontogenetic (developmental) trait changes can be quite different among fish species. Some species hatch from eggs or emerge from placental development in a more precocial (well-developed) state with ability to actively move and feed exogenously. However, many fish species transition ontogenetically from a tiny egg to a very small yolk-sac larva in a very poorly developed state (termed altricial), and then to a small exogenously feeding larva, juvenile, and ultimately a large adult. For most species of fishes, this ontogeny involves a change in body size from larva to adult, which spans many orders of magnitude. Some fishes do not grow to a large size as adults, and therefore the magnitude of body size change from larva to adults is not as extreme. A walleye may hatch as a 7-mm larva, weighing approximately 1 mg. A few years later, it may reach an adult length of 700 mm and weigh over 3,000 g. The physical forces acting on an organism in an aquatic environment are very different between these two extreme sizes. During ontogeny, the ability of fish to move, feed, respire, osmoregulate, and sense their environment all change dramatically. A young fish may be potential prey for a medium-sized forager, but that same fish may develop to prey on that medium-sized forager when it is a large adult. In addition to interspecific trait variation and intraspecific variation among individuals, intraindividual variation through ontogeny can strongly influence performance of individual fish.

Convergent Evolution

Trait variation can be characterized both inter- and intraspecifically, and while trait variation is generally expected to be greatest among species, species can also evolve to become more similar. Evolution is said to be convergent when two or more organisms of independent lineages evolve similar morphological, anatomical, or physiological traits that perform the same functions. Convergent evolution is widespread among fishes and is especially commonly associated with adaptive radiations, such as cichlids evolving in different African rift lakes. The phenomenon involves similar environmental constraints selecting for analogous traits not present in the last ancestor the species shared. Therefore, through the process of evolution, biological organization often predictably arrives at the same solution to a particular need (Morris 2003).

An intriguing example of convergent evolution is the loss of sight (and external eyes in most cases) in North American cave fishes and blind catfishes. Although only distantly related, the species occupying environments characterized by the complete absence of light have traded sight for increased sensitivity to water movement and chemical gradients. Blind fishes have been documented to navigate water around obstacles as small as thin threads. In addition to absence of eyes, species inhabiting caves lack body pigmentation, a trait cave fishes share with many deep sea fishes.

In other examples of convergent evolution, Langecker and Longley (1993) list additional traits they describe as "deep sea syndrome" of traits shared between blind catfishes and deep sea fishes, including a total regression of swim bladder, accumulation of large fat deposits, small body size, an enlarged head, a weakly ossified skeleton, and reduced muscles – all in response to extremely high hydrostatic pressures. A deep sea species that shows almost all the aforementioned adaptations to high hydrostatic pressure is blobfish. Often cited as one of the ugliest species of fish, it inhabits deep waters off the coast of Australia and Tasmania at depths of 600–1,200 m, where hydrostatic pressure is 60–120 times as high as at sea level. At this depth, a gas-filled swim bladder cannot efficiently maintain buoyancy and increased adipose (fat) deposits give the fish a somewhat lower density than water, so it floats above the sea bottom ingesting floating prey. The appearance of a melting fish when observed above water is due to the reduced pressure in addition to excessive fat in the body of the blobfish.

Convergent evolution is also widely seen in antifreeze protein production in Antarctic notothenioids, northern cods, and Arctic flounder – all phylogenetically distant species that thrive in seas at subfreezing temperatures. Chen et al. (1997) showed in a study of two species groups that not only do Antarctic notothenioids and northern cods use antifreeze glycoproteins to survive in freezing water, but also the specific glycoprotein compounds produced by these phylogenetically distant species are near identical. The ichthyofauna of the Antarctic continent is dominated by the notothenioids – a good example of adaptive radiation in the marine environment, capitalizing on the key innovation of antifreeze proteins (Matschiner et al. 2015).

Summary

Fishes are a very diverse taxonomic group with high trait variation both intra- and interspecifically. Within species, traits of individuals differ both among and within populations through a combination of genetically canalized and phenotypically plastic traits. Trait variation is even greater across species. For example, the largest species of fish – whale shark – can grow to over 18,000 kg, while species, such as the dwarf minnow, top out at less than 12 mm. The exact number of described fish species is always changing but was estimated as 32,500 in 2006 (Nelson 2006) and 36,058 in 2021 (IUCN 2021). These different species have developed through various processes, including allopatric and sympatric speciation. While speciation by definition involves differentiation, through evolution, different species have developed similarities via convergent evolutionary processes. Convergent evolution reinforces the important concept of form and function in ecology.

Literature Cited

Chen, L., A.L. DeVries, and C.-H.C. Cheng. 1997. Convergent evolution of antifreeze glycoproteins in Antarctic notothenioid fish and Arctic cod. Proceedings of the National Academy of Science, USA 94:3817–3822.

Conover, D.O., and T.M.C. Present. 1990. Countergradient variation in growth-rate — compensation for length of the growing season among Atlantic silversides from different latitudes. Oecologia 83:316–324.

Conover, D.O., T.A. Duffy, and L.A. Hice. 2009. The covariance between genetic and environmental influences across ecological gradients. Annals of the New York Academy of Sciences 1168:100–129.

Fricke, R., W.N. Eschmeyer, and R. Van der Lan, editors. 2021. Eschmeyer's Catalog of Fishes: Genera, species, references. *researcharchive.calacademy.org/research/ichthyology/catalog/fishcatmain.asp* (Accessed 13 January 2022).

Eshenroder, R.L., P. Vecsei, O.T. Gorman, et al. 2016. Ciscoes (*Coregonus*, Subgenus *Leucichthys*) of the Laurentian Great Lakes and Lake Nipigon. Great Lakes Fishery Commission Miscellaneous Publication 2016–01.

IUCN 2021. Red List Version 2021–3: Table 1a *www.iucnredlist.org/resources/summary-statistics* (Accessed 13 January 2022)

Langecker, G., and G. Longley. 1993. Morphological adaptations of the Texas blind catfishes *Trogloglanis pattersoni* and *Satan eurystomus* (Siluriformes: Ictaluridae) to their underground environment. Copeia 1993:976–986.

Matschiner, M., M. Colombo, M. Damerau, et al. 2015. The adaptive radiation of notothenioid fishes in the waters of Antarctica. Pages 35–57 *in* R. Riesch, M. Tobler, and M. Plath, editors. Extremophile Fishes. Ecology, Evolution and Physiology of Teleosts in Extreme Environments. Springer.

Morris, S.C. 2003. Life's Solution: Inevitable Humans in a Lonely Universe. Cambridge University Press, Cambridge, UK.

Nelson J.S. 2006. Fishes of the World. 4th edition. John Wiley & Sons. Hoboken, NJ, USA.

Nelson, J.S., T.C. Grande, and M.V.H. Wilson. 2016. Fishes of the World. 5th edition. John Wiley & Sons. Hoboken, NJ, USA.

Olsson, J., and P. Eklöv. 2005. Habitat structure, feeding mode and morphological reversibility: Factors influencing phenotypic plasticity in perch. Evolutionary Ecology Research 7:1109–1123.

Svanbäck R, and P. Eklöv. 2002. Effects of habitat and food resources on morphology and ontogenetic growth trajectories in perch. Oecologia 131:61–70.

Vega, G.C., and J.J. Wiens. 2012. Why are there so few fish in the sea? Proceedings of the Royal Society B 279:2323–2329.

CHAPTER 3

Morphology and Evolution of Fishes

The morphology, anatomy, and physiology of fishes are shaped by the environments they occupy over both evolutionary and short-term time scales. The phenotypes expressed by fish directly also affect how they respond to dynamic abiotic conditions and interact with potential prey, predators, competitors, and other organisms. As a result of the close link between body form and function, much of the ecology of a species – resting and sheltering habitat, movement, food and mode of feeding – can be inferred from the individual's morphology, anatomy, and movement. Nelson et al. (2016) in the most recent account of *Fishes of the World* recognized 85 orders and 536 families. These groups of fishes range from simple body forms and evolutionarily ancient species – such as hagfishes and lampreys – to extremely derived species, such as ocean sunfishes and porcupine fishes. Throughout this book, we refer to how traits of various fishes shape the biology and ecology of individuals, populations, and communities. We introduce some details on body forms of fishes in this chapter and review how these forms evolved. This chapter is intended as a limited survey of how body form relates to the functional ecology of a fish. We review the evolution of some key fish body forms and focus on some select adaptations that have influenced the ecology of fishes by changing energetic needs, feeding modes, and risk of being consumed by predators. We provide a summary of some ecologically influential morphological and anatomical attributes. Various other examples of form and function relationships are described in chapters related to feeding and predation, growth, reproduction, and movement. For a more in-depth study of fish evolution and ichthyology, readers may consider reviewing textbooks focused on these topics.

Body Plan and External Anatomy

Figure 3-1 depicts basic terms to describe the orientation and external morphology of representative modern fishes. Fishes are externally bilaterally symmetrical along the median plane, with the exception of flatfishes of the order Pleuronectiformes that undergo asymmetrical eye migration across the top of the skull over time. Other planes relevant for descriptive terms are transverse (vertical) – dividing the body into anterior (front) and posterior (rear) sections – and frontal (horizontal) – dividing the body into dorsal (top) and ventral (bottom) sections. These terms are applied to the whole body and parts of the body such as fins. Lateral is away from the medial plane.

Biology and Ecology of Fishes, Third Edition. James S. Diana and Tomas O. Höök.
© 2023 John Wiley & Sons Ltd. Published 2023 by John Wiley & Sons Ltd.

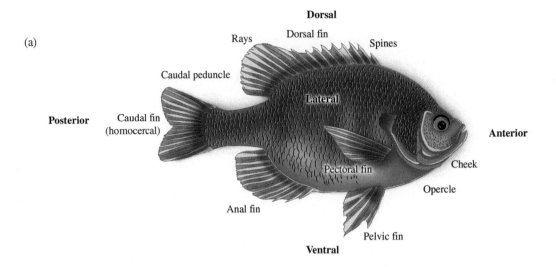

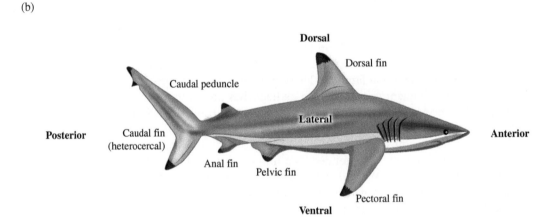

FIGURE 3-1 Basic terms to describe the orientation (**bolded**) and external morphology (not bolded) of representative modern fishes: (a) bluegill and (b) blacktip shark.

Evolution of Fishes

Modern fishes evolved from various earlier groups, several of which are no longer extant. Fishes are the oldest vertebrates in the fossil record, with the earliest fossilized fish dating back to Cambrian strata about 510 million years ago. Fishes are thought to have evolved from a tunicate (sea squirt) – a marine invertebrate whose larval stage is free swimming and has a notochord, dorsal nerve cord, and gill slits. While the filter-feeding adult remains fixed to hard substrates and looks nothing like a fish, the body plan of the larvae is strikingly similar to that of some of the earliest fishes, including lampreys. The conventional view of scientists is that characteristics of the free-swimming larval stage of tunicates were retained as an adult – a process called neoteny.

As a group, fishes have changed a lot since the early days of the Cambrian Period. During that time, most living fishes were characteristically bottom dwellers, heavily armored, and lacked paired fins and jaws, known as ostracoderms – a paraphyletic group (meaning they arose

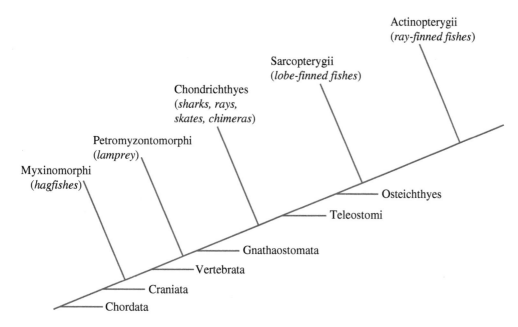

FIGURE 3-2 Phylogeny of major groups of extant fishes. *Source: Adapted from Nelson et al. (2016).*

from the same evolutionary ancestor as modern fishes) that is no longer extant. Morphological additions and innovations to the ostracoderm body plan form the historical basis for the classification of fishes (Figure 3-2). Major innovations – such as the evolution of flexible paired fins – underlay radiations of successive groups into various aquatic and semiterrestrial habitats.

Evolution of Jaws and Branchiostegal Rays

The earliest lineage of fishes were known as agnathans (a- = lacking; -gnath = jaw). Their lack of jaws is important because it constrained feeding, which likely had indirect effects on habitat use and body size. Using knowledge of the morphology and associated ecology of the limited set of bottom-dwelling agnathans that exist today (hagfishes and lampreys), ancestral agnathans were likely not highly selective and fed on algal matter and detritus at or near the ocean floor. The strong association with the ocean floor was related to these ancestral fishes lacking a swim bladder and well-developed fins that could provide thrust, not only lift. These constraints likely limited their diets to detritus (decomposing organic matter typically associated with the bottom) or prey items of small size or limited mobility. The earliest agnathans were intermediate in size (the largest individuals were thought to be 60 cm in total length), which is not too different from the agnathans of today (maximum sizes: hagfishes range 18–127 cm; lampreys range 13–100 cm).

Jaws evolved about 525–430 million years ago and were an important anatomical advancement promoting fish diversity by allowing fishes to consume a broader set of prey. Jaws appear to have evolved from a modification to the pharyngeal arches, the supporting skeleton of the gills in primitive species. Jawed species, such as placoderms (an extinct class), are considered the earliest jawed fishes. Placoderms constituted the least derived class of gnathostomes (gnath- = jaw; -stome = mouth), a superclass that includes all jawed vertebrates, including most fish species.

A second important adaptation that altered the ecology of fishes and promoted fish diversity was the evolution of branchiostegal rays – long, curved bony rays that support the gill membrane (also called the branchiostegal membrane). The timing of their origin is unknown; they are first found in Devonian ray-finned fishes (actinopterygia), 360–408 MYBP. These rays are important in facilitating water pumping from the mouth across the gills, exiting around the operculum, Low pressure in the oro-branchial chamber sucks water into the mouth, and the branchiostegal membrane creates a seal to prevent water entering around the operculum. With the evolution of branchiostegal rays preventing backflow into the oro-branchial chamber, the mouth can function like a straw, allowing fishes to suck water – and any prey located within it – into the mouth cavity with greater force. Some species of fishes have such strong suction abilities that they no longer need to physically grasp prey with their jaws.

The importance of a sealed gill membrane and the evolution of branchiostegal rays are highlighted by the fact that they evolved independently at least 11 times in the actinopterygians and perhaps more often (Farina et al. 2015). This adaptation is especially important in a wide range of bony fishes and is effective for feeding on plankton (planktivores), bottom-dwelling macroinvertebrates (benthivores), and other fishes (piscivores). In fact, many common sport-fishes, such as largemouth bass, striped bass, and walleye, rely on the sucking force of branchiostegal rays to "inhale" the lure anglers use as bait.

Another adaptation that can help explain the diversity in fishes is the ability to separate the premaxilla – the most anterior bone at the end of the upper jaw from the maxilla. As with branchiostegal ray complex, this is common in more derived species and has evolved independently at least five times in actinoperygians (Westneat 2004). In less derived ray-finned fishes – such as the brook trout – the premaxilla is small and non-mobile. It has few teeth, most of which are on the maxilla. By comparison, in more derived species – such as the large-mouth bass or slingjaw wrasse – the premaxilla is expanded in length, highly mobile, and carries many teeth.

Separation of the premaxilla is among several modifications of the jaw affecting feeding that have evolved among fishes, notably the bony fishes. The various modifications increased the range of food items that members of a clade could access, therefore affecting their radiation into various ecological niches. The lengthening of the premaxilla, combined with its ability to flex outward, increased gape to consume relatively larger prey items. Increased premaxilla length and mobility contributed to the ability to expand the volume of the mouth for suction feeding, and premaxilla separation allowed for increased biting force (by more than a third; Holtzman et al. 2008).

Evolution of Traits Related to Buoyancy and Movement

A key difference in most modern bony fishes (Class Osteichythes) and all ancestral fishes is the absence of a swim bladder (also referred to as a gas or air bladder) in ancestral fishes. Extant cartilaginous fishes (Class Chondrichthyes), which include ~1,000 species of sharks, rays, skates, and chimeras, Class Petromyzontimorphi (lamprey), and Class Myxinimorphi (hagfishes) lack a swim bladder. The presence of a swim bladder is not a requirement for fish species to be successful, and many species of fishes – including many less derived fishes – have thrived without a swim bladder. There are benefits and challenges of either having or lacking a swim bladder related to buoyancy, stability, and ability to rapidly change depth and ambient pressure. Species that lack a swim bladder are generally required to move more to generate lift and maintain vertical position in the water column. Swim bladders may decrease the stability of fish in the water column. Swim bladders may also limit the depth range occupied by individual fish as changes in pressure – especially pressure changes associated with rapid shift in depth occupancy – can lead to rapid and even lethal expansion or contraction of the swim bladder. Many benthic species of fish lack a swim bladder as their stability and ability to remain on the bottom is enhanced without this structure.

The swim bladder – like a lung – forms as an outpocket of the gut during larval development and has been shown to serve two primary purposes: it facilitates buoyancy regulation in most fishes, and in a smaller set of species – i.e., the lobe-finned fishes, Subclass Sarcopterygii – the swim bladder is used for breathing when out of water. No species of fish has both a swim bladder and lungs. However, the swim bladder can play a role in hearing and producing sounds, as well as in buoyancy regulation.

In some fishes – typically more ancestral ray-finned fishes – a lifelong connection is maintained between the swim bladder and gut. In these species – which include carp, catfish, and trout – this connection allows the swim bladder to be filled by gulping air from the water surface or emptied by expelling air through the mouth. Species exhibiting these characteristics are commonly referred to as physostomes (physo- = bladder; -stome = mouth). The connection between the swim (gas) bladder and gut is lost during ontogeny in more derived ray-finned fishes. Species with these characteristics – referred to as physoclists (-clist = closed) – include the vast majority of modern teleosts, and their swim bladder is regulated via the gas gland and oval window.

The buoyancy properties of a swim bladder allow individuals to occupy pelagic areas of the water column in a more efficient manner. An ability to efficiently move in all three dimensions of the water column probably facilitated survival by helping exploit new foraging arenas and detect prey within them, as well as helping in predator detection and evasion. The hearing capabilities provided by the swim bladder likely augmented these benefits. Through this combination of improved foraging abilities, reduced risk of predation, and heightened access to new habitats, the swim bladder promoted fish diversity by allowing additional niches to be filled.

Evolution of the swim bladder in more derived species may be associated with key adaptations involving the caudal fin and paired fins. As fish evolved, there was (1) a shift from a heterocercal to homocercal caudal fin (Lauder 1989), and (2) movement of the pectoral fins from abdominal to thoracic and pelvic fins from a posterior to anterior position on the body (Murata et al. 2010). These adaptations are believed to have promoted fish diversity by increasing the range of niches in the water column in which individuals could search for and capture food, evade predators, and provide opportunities for energy conservation during migrations.

Early fishes (e.g., ostracoderms, placoderms, and chondrichthyes) were thought to be bottom-dwelling species. An effective mechanism is required to leave the bottom in order to utilize the water column. This mechanism was provided by an asymmetrical heterocercal caudal fin (hetero- = different; -cercal = tail) (Figure 3-1). In this type of fin structure, typical of most sharks, the upper (dorsal) lobe of the fin is longer than the lower (ventral) lobe, therefore providing lift while swimming. In more derived ray-finned fishes with swim bladders, homocercal (homo- = same) caudal fins are more prevalent. This type of fin is externally symmetrical, with equal dorsal and ventral lobes ranging from being lunate (shaped like a crescent moon) to round (shaped like a half moon).

In addition to a shift from a heterocercal to homocercal caudal fin in ray-finned species, the placement of paired (pelvic and pectoral) fins has evolved (Figure 3-1). The position of these fins – used primarily for stability and fine-scale maneuverability (braking, tilting, and diving) – has shifted rostrally (toward the snout) in more derived species. When in a more rostral location, the pelvic fins confer greater maneuverability than when located farther toward the caudal fin. For example, more primitive species – such as sturgeons and sharks that are often benthic dwelling – may be armored or lack a swim bladder, and generally do not forage in the tight, interstitial spaces associated with macrophytes and reefs; the pelvic fins are generally located between the abdomen and the anus. By contrast, among many, more derived teleosts – including those inhabiting structurally complex habitats (e.g., many centrarchids, cichlids, and pomacentrids) – the pelvic fins are located toward the rostrum or snout, typically in the thoracic region near the pectoral fins.

Ancestral species were heavily armored with structures such as bony plates. In contrast, most derived species have flexible scales – or lack scales – and use structures such as fin spines for protection. This shift in the nature of external armor appears parallel with evolution of the swim bladder and enhanced maneuverability of fishes. In many extinct ancestral fishes – including both ostracoderms and placoderms – bony plates and large scales were used for protection from predation. Many modern-day teleosts can easily seek refuge and maneuver in any number of places, unlike their ancestors. The evolution of fin rays has also allowed for evolution of spines that offer a more energy-efficient form of protection than heavy body armor and large scales.

Fish Body Structure

Body Shape and Fin Placement

The wide range of habitats occupied by fishes and the behaviors they exhibit are represented in the form of the body, plus the types and placements of the paired (pelvic and pectoral) and median, unpaired (dorsal, caudal, and anal) fins. A fish's body is often reflective of how it moves through the environment it occupies, but body shapes and fin positions may also reflect adaptations to overcome risk of predation (e.g., deep body shape to escape gape-limited predators) and facilitate successful reproduction (e.g., narrow caudal peduncle and large caudal fin to fan and oxygenate incubating eggs) or various other behaviors. Figure 3-3 demonstrates

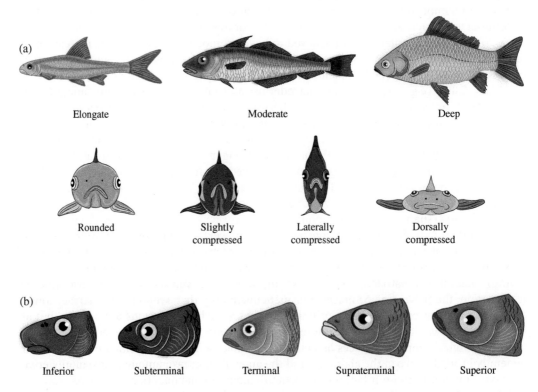

FIGURE 3-3 (a) General fish body shapes and terminology for their description. (b) General mouth positions and orientations.

the frontal and lateral views of generalized body shapes of bony fishes and the terminology used to describe them. Different body shapes confer different advantages for fishes, based on physical properties. The relative body depth of a fish – from elongate to moderate to deep – may influence swimming ability. More elongate fishes are generally able to swim straightforward in a rapid manner. More deep bodied fishes may be well suited for rapid turning, but may not be as successful in rapid, straight-line swimming. Fishes with deep body depths often also express strong lateral compression, which may further facilitate turning in structurally complex environments. While many fishes that occupy more open environments have a rather rounded body shape, there are also many open-water cruising species with compressed bodies. In more open environments, somewhat laterally compressed bodies (fusiform) may provide hydrodynamic benefits, while also limiting total mass and allowing for more efficient fin placement. Finally, a flattened or dorsoventrally compressed shape may facilitate fishes that rest on the bottom of a system, as such a body shape is effective for benthic fishes resisting displacement in water currents.

In addition to general body shape, various specific body attributes are associated with the behavior of fishes. For example, the depth of the caudal peduncle varies greatly among fish species. A very narrow caudal peduncle should reduce drag and require less muscle mass and energy expenditure to move the caudal fin, but it also limits the total thrust that can be generated with a single beat of the caudal fin. Therefore, cruising species that more or less constantly engage their caudal fin often have a narrow caudal peduncle. Species of sticklebacks also have a very narrow caudal peduncle. These species do not generally cruise through open water; however, they do use their caudal fin to continuously fan their eggs to allow for oxygenation while the eggs incubate in their nest. In contrast to species that repeatedly engage their caudal fin, ambush predators may hover for long periods of time and then accelerate to catch their prey. They tend to have thick caudal peduncles, containing relatively large amount of muscle and allowing for high thrust generation over short distances, using large amounts of energy over a very short period of time.

Shape of the body is often complemented with fin shape and placement. Ambush foragers often have large anal and dorsal fins located toward the posterior of their body. Similar to a thick caudal peduncle, the placement of these fins creates large surface area toward the rear of the body, allowing a fish to generate very high, transient thrust forces for high acceleration. Cruising foragers that rely on body and caudal motions are generally characterized by medial fins more anterior on the body and are often characterized by a caudal fin with a high aspect ratio, A:

$$A = h^2/s \qquad \text{(Equation 3-1)}$$

where h is the height of the caudal fin and s is this surface area. A high aspect ratio allows for relatively high lift and low drag. These physical attributes imply that a cruising fish with high caudal fin aspect ratio can propel itself through the water with less energy requirement. For example, consider the caudal fins shapes of northern pike and bluefin tuna (Figure 3-4). Ambush foraging northern pike have caudal fins with a relatively low height, large surface area, and low aspect ratio, while cruising bluefin tuna have very high aspect ratio caudal fins.

Paired fin characteristics also vary among fishes and may reflect movement behavior. For example, pelagic cruising fish may have relatively long and narrow pectoral fins that extend out from the body – similar to wings on an airplane – and can serve to provide lift and allow for more stable and efficient swimming. In contrast, many fishes that maneuver in spatially complex environments have large, flexible pectoral fins allowing for small radius turns.

The associations between fin shape, position, and movement behavior of fishes are simply generalities. There are several examples of species that do not adhere to these generalities.

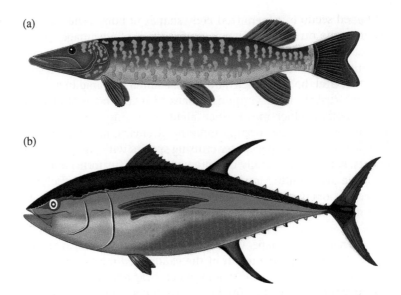

(a)

(b)

FIGURE 3-4   Sketches of the body form for (a) ambush, northern pike and (b) cruising, bluefin tuna, foragers. Note that the pike has medial fins located toward the posterior of its body with a thick caudal peduncle. In contrast, the bluefin tuna has a very narrow caudal peduncle and a caudal fin with a high aspect ratio. The tuna also has narrow pectoral fins that provide additional lift when cruising.

Many species also have highly modified and specialized fins. In some species, the pelvic fins act as suction discs, while in others, they act as limbs for walking or keeping balance at rest. In many groups (e.g., order Siluriformes), the first pectoral fin is ossified and serrated (denticulate) to form a formidable defensive spine. Fins may even be missing in some species; for example, the family Schilbeidae do not have dorsal fins.

Skeletal and Muscular Systems

The skeleton is made of all cartilage (in cartilaginous fishes) or mostly bone and some cartilage in most bony fishes with some variation in between, such as the partially ossified cartilaginous vertebrae of ratfish and the primarily cartilaginous skeletons of sturgeon. The skeleton is associated with the muscles, tendons, and ligaments that collectively provide support and tension throughout the body. The skeleton of fishes is usually composed of axial and appendicular components. The axial skeleton consists of the skull and vertebral column, the caudal, dorsal, and anal fins, and their supporting hard structures. The appendicular skeleton consists of the pectoral and pelvic girdles and their associated fins. Fish vertebrae have centra that align to provide a tube around and protection for the spinal cord, except in lampreys, where centra are absent, and in hagfishes, where the vertebrae are represented by a sheath of cartilage around the notochord. The vertebrae also have short neural and hemal spines and attached short ribs in the abdominal area. The vertebral column provides the structural support for muscle attachments and swimming movements, in addition to providing structure and a conduit for major blood vessels and nerves. As in other vertebrates, the bones of the skeletal system incorporate other vital functions, such as production of red and white blood cells and mineral storage.

Skeletal muscle of almost all fish is made up of the large muscle masses of the body (trunk) and caudal peduncle, as well as smaller muscles associated with the head (eyes, jaws, branchial arches, and gill covers) and fins. Muscles are divided along the body length vertically by complex-shaped dorsoventral sheets of connective tissue called myosepta, resulting in a series of blocks or w-shaped sections called myomeres (also myotomes). The number of myomeres closely correspond to the number of vertebrae in a fish. The muscles in the upper and lower halves of the body are also separated by a horizontal septum, creating the epaxial (upper) muscles and hypaxial (lower) muscles. The myomeres are visible as early as the larval stage where the fish is transparent, and the count and appearance of myomeres is often diagnostic in species identification at the larval stage. The body-caudal peduncle muscle may include a lateral band of muscle along or slightly below the midline of the fish.

Fish muscles are informally referred to as red, white, or pink. Originally, this classification was based on color, attributed to the amount of myoglobin in the tissue. A functional classification is preferable, reflecting metabolic, force, and power characteristic. Therefore, "red" muscle is better referred to as SO muscle – or slow oxidative muscle – in which force production is relatively small, as is speed of shortening and power output. However, a high level of vascularization and muscle hemoglobin (myoglobin) allows for a more continuous supply of oxygen to the muscles and their sustained operation. This is in contrast to more white, or fast glycolytic, muscles. Without a continuous supply of oxygen, fast glycolytic muscles can only be engaged intensely for a short time. These different attributes of slow oxidative versus fast glycolytic muscles imply that they generally control different types of movements within individual fish – continuous, low intensity movement versus short bursts of intense movement, respectively. Different species also vary greatly in terms of the relative amount of slow oxidative versus fast glycolytic muscle they contain, generally reflecting their movement patterns.

Consider three species of piscivorous fish: northern pike, largemouth bass, and bluefin tuna. Northern pike is an ambush predator that generally hovers in the water column and then accelerates rapidly over a short distance for a very short time to capture prey. Not surprisingly, northern pike are characterized by a large amount of fast glycolytic muscle. In contrast, bluefin tuna is a cruising predator that is constantly in motion and swims huge distances across oceans. Bluefin tuna have large amounts of slow oxidative muscle, whose sustained power output is further enhanced by trapping heat to warm this muscle. Largemouth bass are generally more intermediate and flexible in terms of foraging strategy. They may hover and ambush prey but also cruise over intermediate differences and are characterized by a more balanced mix of muscle types.

Head

The skull is a collection of bones of the head housing and protecting the brain and most sensory organs, as well as gills, the mouth and jaws, and teeth. The size of a fish's head and the position, orientation, and characteristics of its external features are often reflective of a fish's behavior, including foraging, sensing of predators and prey, and reproduction. Features of a fish's head are also frequently employed for species identification. Jaws of all fishes are adapted for feeding and, in many species, perform other vital functions as transport of objects (e.g., pebbles for nest building), digging, defense (e.g., biting and carrying brood), and courtship. Different feeding modes and other behaviors are reflected in fishes expressing a variety of mouth adaptations. Some species have a protrusible or protractible mouth for sucking and sifting through debris, while others may have lips strongly developed to perform suction feeding or anchor functions in habitats with strong currents. As a specific example, the stonerollers of North America have thick cartilaginous inferior horseshoe-shaped lips especially adapted for scraping epilithic algae (i.e., algae that grows on the surface of a structure) as their predominant diet.

The arrangements of the paired lower (mandibular) and upper (maxillary) jaw bones determine the positioning of the mouth. In some species, the jaws are of equal length and may be relatively short or elongated to form a beak. Upper and lower jaws of equal length result in a terminal mouth ordinarily associated with adaptation to feeding in the water column. Unequal jaw lengths will either lead to supraterminal to superior (top positioned) or subterminal to inferior (bottom positioned) mouth, corresponding to predominantly surface feeding or bottom (benthic) feeding, respectively (Figure 3-3).

Teeth are a common feature of a fish and, depending on species, may be found on any of the bones of the mouth and pharynx, the rim of jaw bones, the palate, bones in the throat, or places in between. In parasitic lampreys, teeth are small plates of keratin on the tongue and the hard lining of the mouth. Large pharyngeal teeth, common in cyprinids and catostomids and especially herbivorous species, are adapted for tearing and shredding food and generally useful as an identifying feature of closely related species. Some fishes lack teeth. The teeth of sharks and some rays are located only on the jaws. Teeth are named by their shape and the number of cusps they have. Therefore, there are monocuspid teeth that may be straight, conical, or caniniform, cutting or curved, as well as bicuspid and molariform polycuspid teeth with cusps forming a crown (Paugy et al. 2003).

Barbels are fleshy projections of varying sizes or shapes on or around the snout and jaws found commonly in catfishes and minnows and other families such as sturgeons, paddlefishes, and drums. Barbels bear taste buds and are also tactile organs that are particularly useful in nocturnal or turbid conditions where sight or vision has limited utility. The short barbels of minnows are usually situated in the posterior extreme of the upper jaw, whereas catfishes may have multiple barbels, including nasal, maxillary, and mandibular or chin barbels.

Another feature that is not prominent but extremely important in fish is the nares on the snout. Nares are small holes that look like nostrils and open into a sensory chamber. The fish senses smell as water moves over the sensory pads in the chamber from the pressure of water flowing over the head from currents or swimming or by the micropumping action of cilia through the olfactory chamber. The sense of smell is important to fishes, for example, olfactory cues are used to an astonishing precision in navigation to natal grounds in migratory salmon and likely most migratory species. Smell is also used to locate food, to sense chemical or hormonal cues in reproduction, and to receive signals of wounded individuals in a group and move away from danger.

The mouth is separated into an anterior buccal region and a posterior opercular region by gill arches. The gill arch is formed by the ceratobranchial, epibranchial (upper), and hypobranchial (lower) bones. These bear gill filaments and gill rakers. The latter serve various functions, including in feeding. For example, gill rakers can be used as a sieve for filtering particulate food and tend to be finely spaced in planktivores compared to other species. Gill filaments carry highly vascularized lamellae that provide a large surface area for effective absorption of oxygen as water passes over the gills. There are various modifications of the gill area that are handy for identification of species in certain groups, for example, the genus *Synodontis* of the order Siluriformes where the branchiostegal membranes may or may not be fused with the isthmus of the throat (Paugy et al. 2003).

Gills are ventilated by a unidirectional flow of water, generated by muscles of the head in branchial pumping or by ram ventilation involving opening the mouth and opercula while swimming forward. However, inflow to the gills occurs through a single mouth: there is variation among fishes in the excurrent flow from the opercular cavity. Bony fishes have a single gill cover – the operculum. This is widely open in most species but significantly reduced in some cases, such as the Mormyrids. In Chondrichthyes, each gill slit opens separately to the outside (Paugy et al. 2003). In some primitive fishes (Protopteridae and Polypteridae), juveniles have external gills that are reabsorbed later.

Eyes are a very important feature of the fish head but are lacking in some species that dwell in caves and deep aquifers, where low or no light renders vision obsolete. In general, the position and size of eyes indicates the extent to which a fish relies on sight, for example, for finding prey and noticing and escaping predators. Most pelagic species have lateral eyes, while species that live deeper in water or benthic species tend to have more dorsal eyes. Both eyes are located on the side of the head and oriented upward in flatfishes (Pleuronectifomes). The eyes may visibly protrude in a manner similar to amphibians (e.g., Periophthalmus). An adipose eyelid – a nictitating fold or nictitating membrane that partially covers the eye – may be present in some species, but most species have no eyelids or eye protection. In a number of species of deep-sea fishes, the eyes are disproportionately large or tubular to concentrate light.

Summary

Fish diversity and the ability of fishes to inhabit extreme and divergent habitats are the results of evolution. Fishes are the oldest vertebrates in the fossil record, and some fish species have evolved greatly from the ancestral forms. Also, extant less-derived species of fishes continue to be very successful in their exploitation of diverse resources and environments. In short, the large number of known species of fishes (over 30,000) is evidence that there is no single approach for success as a fish, and points to the highly adaptable morphology, anatomy, and physiology of these animals.

Literature Cited

Farina, S.C., T.J. Near, and W.E. Bemis. 2015. Evolution of the branchiostegal membrane and restricted gill openings in Actinopterygian fishes. Journal of Morphology 276:681–694.

Holtzman, R., S.W. Day, R.S. Mehta, and P.C. Wainwright. 2008 Jaw protrusion enhances forces exerted on prey by suction feeding fishes. Journal of the Royal Society Interface 5:1445–1457.

Lauder, G.V. 1989. Caudal fin locomotion in ray-finned fishes: Historical and functional analyses. American Zoologist 29:85–102.

Murata, Y., M. Tamuraa, Y. Aitaa, et al. 2010. Allometric growth of the trunk leads to the rostral shift of the pelvic fin in teleost fishes. Developmental Biology 347:236–245.

Nelson, J.S., T.C. Grande, and M.V.H. Wilson. 2016. Fishes of the World, 5th edition. Wiley & Sons, Inc., Hoboken, NJ, USA.

Paugy, D., C. Leveque, and G.G. Teugels. 2003. The Fresh and Brackish Water Fishes of West Africa, Vol. I. Institute for Research and Development, Paris.

Westneat, M.W. 2004. Evolution of levers and linkages in the feeding mechanisms of fishes. Integrative and Comparative Biology 44:378–389.

PART 2

Bioenergetics and Growth

CHAPTER 4

Balanced Energy Equation

Ecology is the study of interactions that determine the distribution and abundance of organisms. In examining concepts of importance in physiological ecology, one of the first characteristics that determine whether animals survive and thrive in an area is whether they can find sufficient food to meet all the energy requirements of daily life. Animals function as closed systems and follow the laws of thermodynamics, so all energy ingested must be accounted for in energy use or growth. This partitioning of energy into different uses is termed a balanced energy equation.

Energy Flow in Animals

The components of a balanced energy equation will be developed in subsequent chapters, but it is important to first get an overall concept of the balanced energy equation. The equation describes how an animal makes a living, that is, how it uses accumulated food energy for processes, such as metabolism, maintenance of the body, growth, and reproduction. Obviously, for any animal to survive it has to consume at least enough energy to account for its needs. For reproductive fitness, it must ingest enough to have some surplus energy for growth and breeding. Survival of an individual does not necessarily require breeding or growing, but continuance of a species or population certainly necessitates successful reproduction.

The first law of thermodynamics (conservation of energy) states that energy cannot be created or destroyed, although it can be changed. Energy available for photosynthetic organisms comes from energy in the Sun, and this energy is fixed into structural material. Energy available for an animal comes from its food. The equation of energy consumption and use must mathematically balance because an animal cannot create extra energy and cannot destroy it; it can only change energy from one form to another.

The second law of thermodynamics indicates that in the process of changing energy from one form to another, much energy will be released as waste in the form of heat (entropy). The metabolic basis for homeothermy (temperature regulation) is to utilize energy to generate heat, and the heat generated by metabolism helps maintain body temperature. Homeotherms often generate just enough heat to maintain body temperature, but in some situations, they produce excess heat. During exercise, animals may use energy stored in materials, such as lipids, and convert it into adenosine triphosphate, which is used for muscle activity. In converting energy from stored materials into action, heat is generated and body temperature increases. If surplus heat is generated, sweating may dissipate it. This equation fits the second law of thermodynamics: waste material (heat) can still be accounted for and is dissipated into the environment.

Biology and Ecology of Fishes, Third Edition. James S. Diana and Tomas O. Höök.

In describing the balanced energy equation, rates of change from one form of energy to another are evaluated. For example, the daily rate of food consumption or the hourly rate of metabolism are rates in time, and it is important that time intervals and units are also similar across these various components.

Basic Energy Budgets

The most basic energy budget equation for an immature fish has some proportion (p) of the food consumed (C) available for metabolism (M) or growth (G) (Webb 1978):

$$pC = M + G \hspace{5cm} \text{(Equation 4-1)}$$

The value of p includes assimilation efficiency, which is the efficiency by which energy is absorbed from food eaten. Not all of the total energy ingested can be assimilated because quite a bit ends up as indigestible feces. Assimilation efficiency indicates how efficiently ingested material is broken down into substances, such as glucose, free fatty acids, and triglycerides, that can be absorbed across the gut. Assimilation efficiency varies tremendously, mainly based on diet. In addition, nitrogenous waste is included in the above measure of p, so p in Equation 4-1 is actually the fraction of digestible energy. During digestion of food or activity, animals break down proteins. Nitrogen is generated during this process (usually as ammonia) and must be excreted because it is toxic.

If a fish such as grass carp eats a diet very high in plant material, especially cellulose, this food is not very digestible, and most of the energy is lost as feces. If the same animal eats plant material rich in digestible sugars (e.g., soft fruit), more of the food energy is digestible and less ends up lost as fecal material. Similar comparisons could be made for fishes eating arthropods (with much indigestible chitin) or earthworms. Clearly, the quality of food ingested strongly influences assimilation, and p is mainly dependent on what is ingested. Once food is assimilated, available energy can either be used in maintaining the body or in growth.

The use of energy for normal bodily functions is termed catabolism. It is the breakdown of stored energy for activity, maintenance, replacement of damaged tissue, ion and water balance, or any other regulation. This net loss of energy can be contrasted with the process termed anabolism. Anabolism is the storage of energy by growth of body tissues, energy depots, or reproductive tissues.

Through natural selection, animals making appropriate energetic choices that increase survival are favored, while animals making inappropriate choices are selected against, so the frequency of animals using appropriate strategies should increase in future generations. Therefore, the balanced energy equation is strongly influenced by natural selection. Patterns of energy allocation should indicate how those animals have adapted to environmental conditions in different populations. The effect of natural selection on energy budgets of different populations is well illustrated by northern pike in a variety of lakes (Diana 1983).

The average northern pike first breeds at age two or three in an unexploited lake. However, northern pike from a heavily fished lake in central Michigan (Houghton Lake) do not survive very long because they are removed by anglers. They have little chance of living for three years, and they reproduce very early – on average at age one. If pike in Houghton Lake bred at age three or four, few would survive to breed. Natural selection – due to humans removing some fish prior to their first breeding – has favored fish that breed early and resulted in a reduced age at first breeding for this exploited pike population. Even though the typical pattern for pike is to breed at age three, there are some individuals in every population that breed first at an earlier or later age as well. The fish that survived in Houghton Lake had genotypes that put more energy into reproductive growth at an early age. Early maturation should also result in energetic differences among populations – particularly in ultimate size – because when an animal begins breeding earlier, it uses energy otherwise available for growth in order to breed and cannot grow as rapidly. In this example, all energy budget parameters could be changed by natural selection. This influence has the same effect whether it is natural or human induced, as in the case of fishing.

Detailed Energy Budgets

The balanced energy equation defined in Equation 4-1 is interesting ecologically because it can compare energy allocations to survival and growth, but it is difficult to deal with experimentally because none of the parameters – except ration – is fundamental. For example, metabolism, as included in Equation 4-1, can be partitioned into several components; each of which varies under a variety of conditions. The equation must be broken down further to understand how different processes affect the energy budget. A more detailed form of the energy budget, which will be used subsequently, is

$$C = R + A + H + N + F + G + \text{Re} \qquad \text{(Equation 4-2)}$$

where R is standard metabolism, A is activity costs, H is heat increment (also termed SDA), N is nitrogen excretion, F is defecation, and Re is the cost of reproduction (Webb 1978; Deslariers et al. 2017).

    Standard metabolism for a fish is similar to basal metabolism for a mammal. Basal metabolism is measured in mammals by forcing them to remain inactive without feeding at a comfortable temperature and measuring metabolic rate. In the case of fish, the minimum energy usage is measured under ambient temperatures. Minimum usage occurs when fish are undergoing no digestive activity, there is no swimming activity, and they are unstressed. It is an attempt to measure the metabolic rate of a docile fish. There are many methods for doing this because fish generally are not docile. Fish become active when placed in an aquarium, and when they are active, measurements of their metabolic rates are higher than standard.

    Another reason why this level is defined as standard rather than basal metabolism is related to the effect of temperature. Basal metabolism is the minimum metabolic rate of an animal, measured at temperatures within the thermal neutral zone. This thermal neutral zone for a mammal occurs when ambient temperature causes it to neither shiver to generate heat nor sweat to lose heat. Fishes are poikilotherms. They do not generate or lose heat to maintain body temperature (with few exceptions); rather their body temperatures are determined by ambient temperature. Therefore, their standard (or minimum) metabolic rate changes with temperature. Basal metabolism cannot be evaluated as in mammals because, for fishes, any increase in temperature causes an increase in the metabolic rate. However, the minimum metabolic rate at the new temperature is still a standard rate. It is a different standard because it is at a different temperature.

Application of Energy Budgets

The energy budgets of fishes have interested scientists for many years and for some very good reasons. For over 2,000 years, humans have grown fishes under "controlled" conditions and wanted to understand how to obtain more fish flesh for food from this process. The balanced energy equation predicts that fishes that fed more food will grow faster. However, this may not occur in nature if temperature is inappropriate or if the fishes are stressed and active. All of these factors influence fish growth and food production from aquaculture. While humans have been interested in this process for many years, the energy budget has not been described to a detailed level until fairly recently.

Winberg Equation

Winberg (1956) tried to estimate a balanced energy equation. He did not break down the budget to the level of detail in Equation 4-2; it was more similar to Equation 4-1. He evaluated a series of experiments and estimated that average assimilation efficiency was about 80%, most fishes use about twice as much energy when they are active as they do for standard metabolism, and any leftover energy results in growth. The Winberg equation is

$$0.8C = 2R + G \qquad \text{(Equation 4-3)}$$

Equation 4-3 roughly fits for the fishes included in Winberg's analysis and many other species. However, it does contain inaccuracies.

The first inaccuracy – and the most difficult part of the energy budget to estimate – is the cost of activity. When fish swim at maximum sustained speeds, their metabolic rates may be as much as 25 times higher than the standard metabolic rate. The duration of activity and the speed at which a fish swims strongly influence how much energy is necessary for activity. Obviously, the swimming speed and duration vary under a multitude of conditions. Most energy budget estimates were made under laboratory conditions, and apparently, $2R$ as a measure of total metabolism is a reasonable estimate of activity in those experiments. This cost may vary greatly in nature with predators and competitors. Still today, we have very little data to estimate active metabolism in the field with much precision. So, while $2R$ is not a very good estimate for a wide range of behavior of fishes in nature, there is little general information with which to replace it.

The second inaccuracy, as discussed earlier, is that assimilation efficiency varies tremendously with diet. In the case of most piscivorous fishes (consume other fishes), 0.8 is a reasonable approximation. In the case of herbivorous fishes consuming vegetation, it is more likely 0.3–0.5.

Finally, one other problem in the Winberg equation is that temperature influences all of these processes. Temperature is the most significant factor influencing the behavior, distribution, and survival of fishes. Fishes must live in the water, and it is difficult (due to the heat capacity of water) to maintain a body temperature different from water temperature. Some fishes have managed to do this, but only with extreme physiological and anatomical adaptation. If water temperature is high, maintenance metabolism may take all the energy available in the ration. If temperature is low, fishes may not be able to eat at all because temperature may limit digestion. Temperature influences the distribution of energy as well. Strong seasonal patterns in the growth of fishes are very common because of the influence of temperature on energetic parameters as well as the influence of temperature on food abundance.

Brett Equations

J.R. Brett from the Pacific Biological Station, part of the Department of Fisheries and Oceans in Canada, accomplished more to forward our understanding of fish energetics than any scientist to date. Brett and Groves (1979) summarized available energy budgets of fishes and estimated the breakdown of the energy budget into each parameter (Figure 4-1). For non-herbivores on an excess consumption (C), about 80% is digestible energy and 20% ends up as feces. Of this

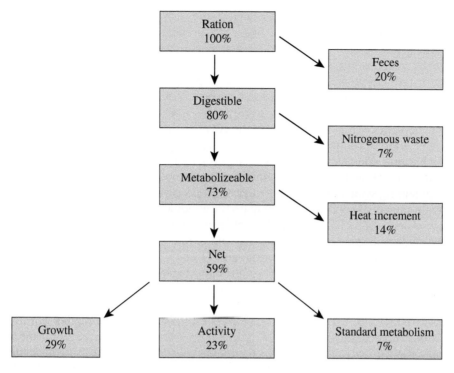

FIGURE 4-1   The energy budget for an average carnivorous fish. *Source: Adapted from Brett and Groves (1979).*

digestible energy, however, the amount available for metabolism is lowered by another 7% because there is a commensurate loss due to nitrogen excretion. So, 73% of the energy ingested by a non-herbivore is available for metabolism. About 59% is net energy because an additional 14% is lost to the metabolic cost of digestion (*H*). Net energy – the amount actually available for use – is only 59% of what was originally eaten. This emphasizes the inefficiencies defined in the laws of thermodynamics. Approximately 29% of the initially consumed energy is deposited in growth, and 30% is used in metabolism. Activity costs for the 30% in metabolism are around 23% (this would be *A*), and standard metabolism is about 7% of the consumed energy. These activity costs, which consume a large portion of energy ingested, are difficult to estimate. Methods for such analyses are covered later in chapters on bioenergetics models as well as fish activity.

This average budgeting provides details on parameters that are costly in the energy budget. It is surprising that standard metabolism is only as important as excretion, in terms of an overall energy cost for a fish at high ration. It is also obvious that energy loss due to activity can be quite large. Reproduction, which is not even included in this budget, remains another problem in energy budget estimates. Most frequently, immature fishes are used in measuring an energy budget in the lab. Growth is observed over short time intervals, and reproductive growth or behavior is not measured. The energetic cost of reproduction cannot be estimated by these controlled experiments, yet species such as northern pike may grow as much ovarian tissue in a given year as body tissue.

The patterns of energy allocation to each budget component tremendously depend on conditions. Growth only occurs when sufficient food is eaten to account for all other processes and leaves some "surplus" energy. An animal will not grow if it consumes less than required for all of these functions. Growth is the first component sacrificed under energy stress. It is not a process regulated by itself, but one dependent on surplus energy availability. This is an important point emphasizing growth in ecological studies of fishes. Growth is a sensitive indicator; anything that affects abundance of food, such as competition, influences growth because other energetic processes must be dealt with first. For any stress that requires increased activity or other regulatory costs, the simplest place to measure the effect is in reduced body growth.

The previous energy budget was for a non-herbivore, and there is a big effect of diet on an energy budget. Fewer herbivorous fishes have been studied extensively. Herbivorous mammals cannot digest the energy in cellulose very well even though cellulose is the major structural tissue of plants. Symbiotic bacteria in the gut of some herbivorous mammals break down cellulose into simpler compounds that can be digested by the mammal. Maintaining optimal symbiotic bacteria populations in the gut requires fairly consistent conditions, particularly temperature and pH. Fishes cannot maintain constant body temperatures and may have difficulty maintaining symbiotic bacteria to digest cellulose. As a result, few fishes – particularly in temperate climates – consume macrophytes (large aquatic plants).

A good example of herbivore energy budgets is the grass carp. Grass carp eats large vegetation. The energy budget of a grass carp includes only 19% of ration as metabolizable energy, while 81% is defecated and excreted. Of the remainder, metabolism (activity, heat increment, and standard metabolism combined) requires about 16% of the budget and growth about 3%. Therefore, a grass carp must eat large volumes of vegetation to survive. A Winberg equation for a grass carp would use $0.19 \times R$ rather than his standard of 0.8.

Grass carp may also eat other foods. If a grass carp is fed a meal of tubifex worms (which are highly digestible for most fishes), 60% of the energy budget goes into feces or urine and 40% is metabolizable. Of that, only 23% shows up in metabolism or about the same total amount as when carp consumes vegetation. Growth consumes 17% of the energy budget, again indicating that more net energy available leads to more growth. Grass carp can assimilate energy much more efficiently when fed food that is more nutritious. Still, 60% is egested compared to 20% for carnivores.

Virtually every fish has highly adapted digestive anatomy to suit its diet. For a grass carp, its gut, mouth parts, and digestive enzymes are evolved to break down plants. The grass carp has not evolved to take advantage of tubifex worms and is not very efficient at it. Adaptation to plant diet includes elongation of the hind gut to provide a larger surface area for absorbing energy from cellulose. Herbivores may also not have stomachs at all because the stomach is involved in acid production for breaking down materials such as proteins or lipids, which are not common in an herbivorous diet. Predatory fishes, on the other hand, commonly swallow prey whole and have a large stomach for acidic breakdown of this unprocessed food. Protein and lipids are easily transported across the gut, so predators have short hind guts. Those anatomical adaptations, as well as differing mouth parts, have important influences on digestive ability and therefore energy budgets.

As an aside, in Chinese polyculture, the grass carp is a key species. The fish itself is not highly efficient in converting food to fish flesh. Pond managers add leaves, grass, and other vegetable waste to ponds. The grass carp eats these materials and does not digest them very well, but it breaks them down while excreting and egesting nutrients into the water. This fertilizes the water and causes phytoplankton production to increase. The other fishes utilize natural food pathways that increase in response to this bio-fertilization. The grass carp is a pivotal animal because it takes large waste vegetation and turns it into nutrients that can be used for pond fertilizers. It turns that vegetation into soluble phosphorus and nitrogen, which stimulate phytoplankton growth, as well as into particulate material for zooplankton and bacterial growth. This is obvious in the energy budget; 81% of ingested vegetation is lost as feces or urine.

Surprisingly, after the pioneering work of Brett, there has been little effort to change the basic ideas he developed. There have been a number of evaluations of energetic parameters for different fish species – often conducted to fit bioenergetics models that can be used to predict fish growth or consumption in natural waters – but few overall assessments of energetic performance of fishes. Jobling (1994) published a book detailing his and other work on fish bioenergetics, which has many specific examples and details of fish energy budgets. More recently, scientists have become interested in evaluating energy budgets based on the protein and lipid content of food consumed and body tissues produced (Lupatsch et al. 2003). Such applications are important in the aquaculture field where different food components have largely different costs and values in terms of growth. However, the details of the overall fish energy budgets as defined by Brett have not been updated.

Summary

Diet and digestibility strongly influence an energy budget. If a fish from one location is transplanted to another site, the fish will have a different energy budget, based on altered availability of food alone, unless the same food resources are at the new site. Activity, temperature, and fish size also affect this budget. The energy budgets of fishes are interesting for a variety of reasons but require site-specific studies. The budget of largemouth bass at one lake will not be the same at another lake because the ecosystems of lakes have different characteristics. Fishes should selectively adapt to the conditions of the ecosystems in which they live.

Literature Cited

Brett, J.R., and T.D.D. Groves. 1979. Physiological energetics. Pages 279–352 *in* W.S. Hoar, D.J. Randall, and J.R. Brett, editors. Fish Physiology, Vol. 8. Academic Press, New York.

Deslariers, D., S.R. Chipps, J.E. Breck, J.A. Rice, and C.P. Madenjian. 2017. Fish Bioenergetics 4.0: An R-based modeling application. Fisheries 42(11):586–596.

Diana, J.S. 1983. Growth, maturation, and production of northern pike in three Michigan lakes. Transactions of the American Fisheries Society 112:38–46.

Jobling, M. 1994. Fish Bioenergetics. Chapman and Hall, London.

Lupatsch, I., G.W. Kissil, and D. Sklan. 2003. Comparison of energy and protein efficiency among three fish species gilthead sea bream (*Sparus aurata*), European sea bass (*Dicentrarchus labrax*) and white grouper (*Epinephelus aeneus*): Energy expenditure for protein and lipid deposition. Aquaculture 225:175–189.

Webb, P.W. 1978. Partitioning of energy into metabolism and growth. Pages 184–214 *in* S.D. Gerking, editor. Ecology of Freshwater Fish Production. John Wiley and Sons, New York.

Winberg, G.G. 1956. Rate of metabolism and food requirements of fishes. Fisheries Research Board of Canada, Translation Series 194.

CHAPTER 5

Metabolism and Other Energy Uses

Chapter 4 reviewed the balanced energy equation with the intent of deriving the relative importance of each component. The purpose of this chapter is to give an overview of fish metabolism. Many factors affect metabolic rate, particularly activity level, temperature, fish mass, and feeding level. In addition, measures of fish respiration vary in methodology and accuracy. Therefore, this chapter first reviews respiration, metabolism, and the means to measure these rates in fishes, and then develops the relationships among activity level, temperature, mass, and fish metabolic rates. Since feeding level also influences metabolism, the relationship between food intake, digestion, and metabolic rate is also evaluated here.

Respiration and Metabolism

The basis for metabolism is the conversion of glucose and oxygen into carbon dioxide, water, and energy:

$$C_6H_{12}O_6 + 6O_2 \rightarrow 6CO_2 + 6H_2O + \text{energy} \qquad \text{(Equation 5-1)}$$

In this process, stored energy is released, and this energy is converted into high-energy bonds in adenosine triphosphate (ATP). ATP temporarily stores this energy, which is used in tissues or organs requiring energy for catabolism, or is deposited in energy storage sites. Heat is also generated, mainly because metabolism is not 100% efficient. Equation 5-1 can be short-circuited in the absence of enough oxygen to undergo aerobic metabolism. During anaerobic metabolism, secondary byproducts (such as lactic acid) accumulate, and anaerobic respiration does not liberate as much energy as aerobic respiration. For 1 mole (180 g) of glucose, anaerobic respiration can produce 2 moles of ATP molecules (83.6 kJ), while aerobic respiration can produce 38 moles of ATP molecules (1588 kJ). Since a mole of glucose has 2867 kJ of energy stored, aerobic cellular metabolism is 55% efficient and anaerobic cellular metabolism is only 3% efficient, while much energy is lost as heat. Anaerobic metabolism also results in buildup of an oxygen debt, which has to be paid back later.

When a fish undergoes heavy exercise beyond its aerobic capacity, lactic acid builds up in muscles during this anaerobic activity until eventually the fish must stop swimming because

Biology and Ecology of Fishes, Third Edition. James S. Diana and Tomas O. Höök.
© 2023 John Wiley & Sons Ltd. Published 2023 by John Wiley & Sons Ltd.

of exhaustion. Lactic acid at high levels is toxic to muscle action, and the fish must rest and ventilate for some time to pay back that oxygen debt. All of the lactic acid is cycled back through aerobic metabolism into carbon dioxide and water. Therefore, there are two types of cellular metabolism: aerobic, which can be continued indefinitely as long as there is enough oxygen and glucose, and anaerobic, which cannot be continued indefinitely but can result in higher levels of burst performance. These two types of metabolism are inversely related: aerobic metabolism is rate limited but not limited in capacity, while anaerobic metabolism is capacity limited but unlimited in rate (or nearly so).

There is often confusion between the terms respiration and metabolism. Respiration may be considered as the consumption of oxygen but more commonly is related to the actual process of breathing. For a fish, this involves passing water across the gills and extracting oxygen from it. Metabolism, on the other hand, is the use of energy. Metabolic rate is often expressed in units of joules or calories of energy consumed per kilogram of fish per hour. The kilocalorie, or the amount of energy required to raise 1 kg of water by 1 °C, is commonly used in dietary analyses, and many people are familiar with this unit. A joule, defined as the amount of work performed by exerting 1 N of force over a distance of 1 m, is the standard international unit for energy. The two can be mathematically converted: 1 cal = 4.18605 J or 1 kcal = 4.18605 kJ. We will use joules as our measure of energy throughout this book. Respiration and metabolism are very closely related, and the relationship between them is developed in the subsequent sections. When scientists discuss respiratory mechanisms, they are generally describing the process of ventilating the gills and extracting oxygen from the water, not necessarily the process of using that oxygen to liberate energy.

Calorimetry

Several variables from Equation 5-1 can be measured to estimate metabolic rate of an animal. These include the rate of oxygen consumption, glucose removal, carbon dioxide production, water production, or heat production. In fact, mammalian metabolic rates are often measured in heat production or CO_2 production. When fishes respire, (1) the oxygen content of water should decrease, (2) the carbon dioxide content should increase, and (3) the temperature should increase. These changes may differ depending on aerobic or anaerobic processes in the fish. However, this gives two generalized methods to measure respiration: direct or indirect calorimetry. Calorimetry is the measurement of energy used in a process or stored in a substance. Direct calorimetry measures metabolism by directly measuring the heat production of an animal.

Some of the first experiments in calorimetry used mice in a container with a layer of ice surrounding the container and measured how fast the ice melted as a correlate of how much heat the mouse generated. Metabolic rate is low in fishes, and the amount of heat produced is low (Box 5-1). Fishes also live in water, a medium with high heat capacity. It takes much heat to warm water (4.2 kJ for each kg of water to increase 1 °C). These two problems combine to make it difficult to accurately measure the heat generated by a fish in water. Direct calorimetry with a fish would require extremely sensitive thermometers, a very low volume of water, and a fish with a fairly high metabolic rate (Box 5-1). This might still be possible with actively swimming fishes, but in general, these measures are a lot less sensitive than measurements taken by oxygen consumption.

The common method to measure metabolism of a fish is by indirect calorimetry, which is measuring the other components of Equation 5-1 and relating them to energy production. Either oxygen consumption or carbon dioxide production could give a measure of aerobic metabolic rate. This measure would have to be converted into the rate of energy production to estimate metabolism in energy terms. When an animal uses glucose as an energy source, 1 g

Box 5-1 Metabolic Performance in Fishes

Fishes generally have low metabolic rates compared to birds and mammals. For example, a 10-g fish at 25 °C has a metabolic rate of about 1.93 $J \cdot g^{-1} \cdot h^{-1}$. This compares to a rate of 41.67 $J \cdot g^{-1} \cdot h^{-1}$ for a mammal, which is 21.6 times higher. For a 10-g fish in 1 l of water, heat generated in one hour (19.3 J) would only warm the water from 25 to 25.0046 °C.

For the same system, assuming an RQ of 1, oxygen consumption would be 0.14 $mg \cdot g^{-1} \cdot h^{-1}$, and CO_2 production would be the same. For a one-hour experiment, dissolved oxygen would decline from saturation (8.24 mg/l) to 6.84 mg/l, or 17%. If alkalinity were 100 mg/l and pH 8.3, free CO_2 would be undetectable in solution and all carbon would be in the form of bicarbonate. Bicarbonate concentration would change from 100 to 101.4 mg/l, or 1.4%, under these conditions.

of oxygen is consumed for every gram of carbon dioxide produced. That is not true; however, when an animal catabolizes other energy substrates for metabolism, major energy sources can be categorized as carbohydrates (like glucose), lipids, and protein. These are very broad categories. A specific fatty acid or protein may not be very similar to any other fatty acid or protein as they all vary considerably in chemical structure. The ratio of elements in their chemical structures determines how much carbon dioxide will be evolved in respiration. All of these energy substrates contain carbon, but carbon is not their only component. Respiratory quotient (RQ) is the ratio of grams of carbon dioxide produced for each gram of oxygen consumed. For carbohydrates, the accepted RQ is 1 (at least for glucose); for lipids, it is approximately 0.7; and for proteins, it is 0.9.

Besides the variable respiratory quotient of carbon dioxide, other factors make measuring carbon dioxide difficult. One of the major factors is the dissociation of carbon dioxide in water. Carbon dioxide in water, under normal pH, becomes carbonic acid (H_2CO_3), which becomes dissociated into H^+ and HCO_3^-. Depending on the pH of the water, carbon dioxide in water can be predominantly carbon dioxide, carbonate, bicarbonate, or a combination of these. To estimate carbon dioxide given off by metabolism, one cannot simply measure carbon dioxide production in the water. The dissociation of carbon dioxide into those other anions is also important. Alkalinity, which is a correlate of carbon dioxide and its derivatives in water, is often measured in hundreds of milligrams per liter. The transformation of carbon dioxide into related forms, and the large amount of carbon contained in many water sources, combines to make the estimation of carbon dioxide production by fishes imprecise (Box 5-1).

For all these reasons, oxygen consumption is most commonly measured for metabolism of fishes. Oxygen does not dissociate into other compounds in water, and, with the exception of anaerobic metabolism, oxygen consumption is directly related to energy production by an animal, unlike carbon dioxide, which has a variable respiratory quotient. Saturated levels for oxygen at most temperatures of water are less than 20 mg of oxygen per liter of water; this means changes in oxygen content over time can be measured more precisely.

Most metabolism measures for fishes are taken in oxygen consumption per unit of time and then converted back to heat production. For metabolism, energy stores already in the body are being catabolized for energy needs. Depending on recent food consumption or the trophic nature of the animal, some animals preferentially store lipid or protein and collectively utilize these different energy substrates for metabolism. This may depend on their histories; animals in good energetic condition may have surplus lipid, which can be used as a reserve. Animals in poor condition may have used their entire surplus lipid, and they may have to catabolize muscle protein. Many factors affect which substrates might be used.

Respirometers

Before expanding on energy storage, an evaluation of methods to measure metabolism is in order. Metabolism can be measured in a very simple or complex system, depending on the level of metabolism in question. A fish can be placed in an aquarium filled with water, and the aquarium can be sealed by placing a glass lid over the water surface so air cannot diffuse in or out (Figure 5-1a). The oxygen content of the water can be measured at the start of the experiment, the fish can be allowed to metabolize for some time, and the oxygen content can be measured again (Box 5-2). The difficulty with this measurement is that a fish in an aquarium does what it pleases. It swims around if it wants to; its activity cannot be controlled. Activity is a very important determinant of total metabolism, and, to do an accurate measure of metabolism, activity must be controlled. This kind of a static system measures what is termed routine metabolism because there is no control on the level of metabolism measured. Activity of the fish

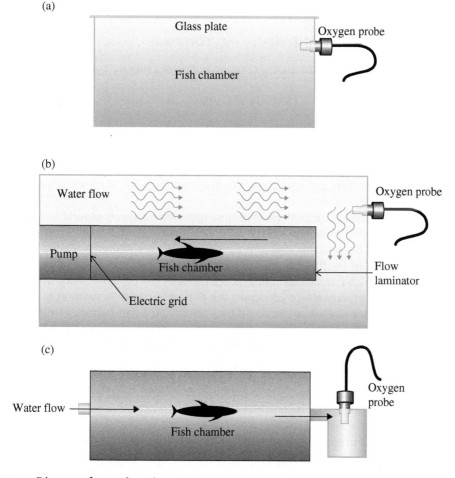

FIGURE 5-1   Diagram of several respirometers:

a. Closed respirometer (glass aquarium sealed with a glass plate).

b. Enclosed swimming respirometer (Blazka et al. 1960), constructed of PVC pipe; water flows through the larger outer pipe into the smaller fish chamber and recirculates.

c. Flow through respirometer also of PVC pipe.

Box 5-2 Estimating Metabolic Rates for Fishes

For a static respirometer (Figure 5-1a), metabolic rate can be calculated as

$$MR\left(mg\,O_2/h\right)=\left[\left(DO_{initial}-DO_{final}\right)\times volume\right]/time$$

In an example using an initial DO of 7.1 and a final DO of 6.1 over a 10-hour experiment in a 1-l aquarium, metabolic rate would be [(7.1 − 6.1) × 1000]/10 or 100 mg O_2 per hour.

For a swimming respirometer with no water input, metabolic rate would also be calculated as above. An example of using an initial DO of 7.1 and a final DO of 6.1, over 10 hours in a 10-l aquarium with water velocity at 35 cm/s, would give a metabolic rate of

$$\left[\left(7.1-6.1\right)\times 10\,000\right]/10\,or\,1000\,mg\,O_2/h$$

at a swimming speed of 35 cm/s.

For a flow-through respirometer, metabolic rate can be calculated as

$$MR=\left(DO_{in}-DO_{out}\right)\times flow\,rate$$

An example using a DO_{in} of 7.1, a DO_{out} of 6.1, and a flow of 2 l/h would give a metabolic rate of

$$\left(7.1-6.1\right)\times 2=2\,mg\,O_2/h)$$

is controlled only by the fish, and the proportion of metabolism due to standard metabolism (S) or activity cannot be separated.

The problems involved in interpreting routine metabolism are considerable. For example, consider a comparison of how fishes react to certain chemicals in water. A fish in an aquarium at low pH might become stressed and actively try to escape the poor conditions. What is actually measured in routine metabolism at normal pH as compared to low pH might not be the effect of pH on metabolism but rather the effect of pH on activity. To account for activity, other respirometers either constrain the fish or cause it to swim at a set speed. A swimming respirometer has water circulated by a pump and a fish swims against the current (Figure 5-2). To force the fish to remain swimming, an electrical current is put into a grid, which shocks the fish on contact with the grid. The fish learns not to touch the grid after a short period and will maintain its swimming speed as long as possible. The oxygen content of the water is measured over time as the fish swims against the current. Swimming speed must equal the flow rate of the water, and metabolic rate is the change in oxygen content of that water over time (Box 5-2). The result of one measurement is the metabolic rate for a fish swimming at a known speed.

Fish can be forced to swim at many different speeds with this apparatus, and activity can be controlled. For the pH example, the metabolic rate at a lower pH and a swimming speed of 30 cm/s could be compared to normal pH and a speed of 30 cm/s. With this comparison, the effect of pH on metabolism is not due to increased activity but to increased cost related to regulating homeostasis at altered pH. This swimming respirometer could be used to measure metabolism at many different swimming speeds, and standard metabolism could also be estimated using the relationship between swimming speed and metabolism. A second swimming respirometer is shown in Figure 5-1b.

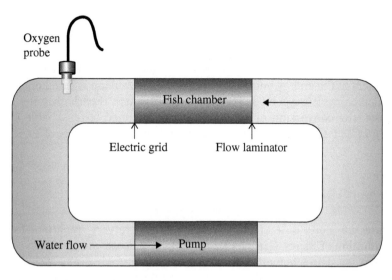

FIGURE 5-2 Schematic diagram of a swimming respirometer (Brett 1964) used to determine metabolic rate at different swimming speeds. The respirometer is composed of PVC pipe, usually transparent and 10–20 cm in diameter.

Another respirometer design is a flow-through system. After some time in the aquarium discussed earlier, the fish will have respired so much oxygen from the water that oxygen becomes depleted. As oxygen levels drop in a static respirometer, the fish may change its activity or ventilation to survive under low oxygen conditions. Flow-through systems used for long-term measurements add new water continuously and measure the rate of water addition as well as oxygen content of inflow and outflow water. The oxygen content in the respirometer remains constant over time. A flow-through system is often just a Plexiglas tube. Water input rate is usually slow, so the flow rate is not sufficient to force fish to swim against a current. Activity may be constrained in these kinds of flow-through systems by making the respirometer fit the fish (Figure 5-1c). If a fish is placed in a tube only slightly larger than the fish itself, it cannot swim and may remain quiescent. In these cases, standard metabolism can be measured by constraining activity. Some species of fish will not remain quiescent but will struggle no matter how small the tube is; other species are content to remain in a tube and not move around. Thus, in a flow-through respirometer, activity might sometimes be controlled by the size of the tube. In this system, metabolic rate (mg O_2/h) is calculated using the oxygen content of inflow and outflow water and the flow rate of water (l/h; Box 5-2). Of course, a swimming respirometer could also be modified to have flow-through water for longer-term metabolic measures.

Recently, there has been much interest in intermittent flow respirometers. These systems essentially create a series of short-term static measures while maintaining higher water quality. Since water is regularly replaced, long-term measurements can be made to allow fish to adjust to the respirometer and minimize stress. The design of these systems can vary (Svendsen et al. 2016), and measures of standard metabolism can be made by measuring oxygen consumption (MO_2) over the time series, then using the lowest 10–30% of MO_2 measures as an estimate of Chabot et al. (2016) consider these measures to be the best estimates of S, especially for fishes that will not swim continuously over long time periods.

As respirometers vary in design and characteristics of water flow, no single system is ideal for all species of fish. Fishes that swim more or less continuously, like salmon, respond well to swimming respirometers, while typically inactive fishes, like pike, or irregular swimming fishes, like various sunfishes, respond better to flow through or static respirometers.

Metabolism, which is the rate of energy expenditure, is usually measured in mg O_2/h. However, joules are preferable units because food consumption, fecal production, and all other components of the energy budget can be expressed in joules but not in oxygen terms. Conversion of metabolism from oxygen consumption to energy use allows a comparison of metabolism to other parameters of the energy budget equation. This conversion is based on an oxycalorific coefficient, which converts oxygen consumption to energy use. If carbohydrates are the main energy substrate, the conversion of 14.76 J/mg O_2 consumed is reasonable (Brafield and Solomon 1972): 13.38 for proteins, 13.59 for lipids, and an overall average of 13.54.

Energy Substrates

Energy substrates are quite different in many ways and are very important in metabolism and food consumption of fishes. Carbohydrates, such as glucose, are combinations of carbon, hydrogen, and oxygen. The schematic formula is $(CH_2O)_n$. Carbohydrates include compounds, such as starch and cellulose, as well as a variety of other structural components. The more structural a carbohydrate is, the larger the molecule and the less digestible it is. The lower the n in the previous schematic formula, as in sugars, the more digestible carbohydrates may be. In aquatic animals, carbohydrates are not generally very important energy sources; they are not a large component of the diet and are a very small component of any energy stored in the body itself. Carbohydrates, containing nothing more than carbon, hydrogen, and oxygen, are converted into simple molecules such as carbon dioxide and water through respiration. Proteins are more complex, containing amino acids of some sort as well as C, H, and O. They contain nitrogen and possibly other elements. When a protein is metabolized, nitrogen or some other nitrogenous compound (ammonia, NH_3, in most fishes or urea, N_2H_4CO, in some fishes and mammals) must be excreted as well as carbon dioxide and water. Lipids, like carbohydrates, are a fairly long chain of C, H, and O but may also include other elements.

The energy released by completely oxidizing a unit of material is called the heat of combustion or calorific equivalent. Lipid material combusted in a bomb calorimeter would generate a certain amount of energy, which is its heat of combustion. For 1 g of substrate, 39.5 kJ of energy is produced for lipids, 23.7 kJ for proteins, and 17.1 kJ for carbohydrates. There are large differences in the amount of energy each substrate contains.

Heat of combustion measures the energy content of something completely oxidized. However, a fish may catabolize only some of the bonds in these compounds and extract energy from those bonds. Each substrate is differentially extractable because it contains energy that may or may not be utilized by the animal. A complex carbohydrate, such as plant cellulose, may not be digested very well. Metabolizable energy is the energy that can actually be metabolized by the animal per gram of material consumed. Remember, metabolizable energy excludes production of feces and urine in the balanced energy equation. For fishes, metabolizable energy of lipids decreases to about 33.44 kJ/g, of protein to 17.98, and of carbohydrates to 6.69. An average carbohydrate is only about 40% digestible, while a lipid is about 85% digestible.

When an animal catabolizes protein completely, it also has to process the nitrogen in the protein. This nitrogen is toxic to animals and must be excreted. Fishes excrete nitrogen as ammonia through their gills. This is not a very costly process, except the energy in those ammonia bonds cannot be extracted from the protein. Consequently, about 3.97 kJ of energy is excreted as ammonia not catabolized, per gram of protein.

If an animal stores surplus energy to use later, the most effective way to store it is as lipid because lipid has more energy per gram. Proteins are not particularly useful energy stores since energy is lost in excretion, and their energy content per gram is lower than that of lipids. Most animals avoid catabolizing protein in time of energy need, which is termed protein sparing.

Most proteins in fishes are deposited in cells of muscle tissues, which, if broken down, also cause the animal to become debilitated. Therefore, there is not only an extra cost of energy in excreting ammonia, but the animal also becomes physically less capable if it utilizes muscle protein. There is no negative effect on performance if the animal utilizes lipid stores for energy.

Metabolic Rates

Methods of measuring metabolic rates were considered earlier in this chapter. To measure the standard metabolic rate of fish in a swimming respirometer, oxygen consumption can be measured at several swimming speeds for a fish with no food in its gut. The results approximate Figure 5-3, measuring metabolism during sustained levels of activity. Over time, fishes undergoing significant anaerobic activity build up such oxygen deficits that they fatigue and collapse against the back of the respirometer because they cannot sustain such levels of activity. Therefore, the maximum swimming speed a fish can maintain without eventually collapsing is considered to produce its maximum or active metabolic rate. However, there is evidence that some fishes build up partial oxygen debts even during routine swimming (Webb 1978). At swimming speeds below the maximum sustained level, there is a linear relation between swimming speed and log of metabolic rate as shown in the following equation:

$$Q_{02} = a\,e^{gs} \qquad\qquad \text{(Equation 5-2)}$$

where Q_{02} is metabolic rate (mg $O_2 \cdot kg^{-1} \cdot h^{-1}$), a is a constant, s is swimming speed, and g is a constant. The constant a is actually standard metabolic rate for the conditions of that experiment (weight and temperature). A straight line can be extrapolated to the point where swimming

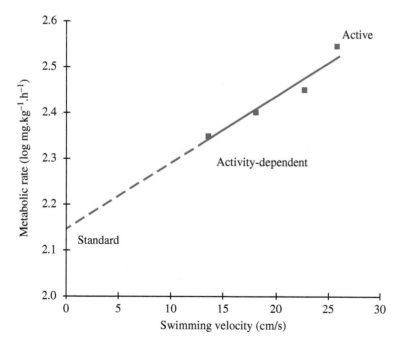

FIGURE 5-3 The relationship between metabolic rate and swimming speed of a 150-g largemouth bass at 20 °C. *Source: Adapted from Beamish (1970).*

speed is zero, which is the standard metabolic rate. The upper extreme is the maximum metabolic rate (considered here to be active metabolic rate), i.e. the highest aerobic metabolic performance a fish can sustain for a specified time interval. Activity-dependent metabolic rate (Rice et al. 1983) occurs at levels of swimming between 0 and the maximum sustained speed.

Temperature Effects on Metabolic Rate

Now that we have reviewed methodology, compared direct and indirect calorimetry, and examined the kinds of respirometers that might be used, it is reasonable to proceed from the methodological system to biotic and abiotic factors influencing metabolism. The first factor to examine is temperature because it has a very significant effect. Brown et al. (2004) proposed the metabolic theory of ecology as possibly the conceptual foundation for much of ecology. The theory deals with the relationships among metabolism, body size, and temperature. Since most individual fish processes are affected by these factors, survival, growth, and reproduction are all dependent on metabolic rate and temperature influences. Brown et al. (2004) then extended these ideas beyond organisms to populations, communities, and even the biosphere – indicating the importance they place on metabolism. In the case of fishes, they are ectotherms and poikilotherms: ectotherm indicates that body temperature is controlled by external temperature, while poikilotherm indicates that body temperature varies. At any given temperature, a fish's body temperature will be almost precisely the temperature of water. A fish living in a stratified lake might move from deep cold water up to warm surface water. In this case, there might be a brief period in which the fish's body temperature is below ambient water temperature, but this would only last a few minutes.

A few fishes generate heat and store it – mainly some of the large sharks and tunas. They are actually endotherms, able to raise their body temperatures above ambient water temperatures. They can control temperature but not as precisely as mammals and birds, which control their temperatures accurately (homeotherms). Many sharks continually swim to survive, using ram ventilation (opening their mouths while swimming and forcing water past their gills). These fishes must maintain high metabolic rates and are unable to ventilate without swimming. These high metabolic rates, coupled with large size, make endothermy possible.

Most fishes, however, are ectothermic poikilotherms, whose body temperatures fluctuate in accordance with ambient water temperatures. Metabolic rate and other physical or chemical processes are related to water temperature for these fishes because temperature influences the rate of enzyme activity, mobility of gases, diffusion, and osmosis. As body temperature changes, reactions (physical and physiological) change as well.

Standard Metabolism

Standard metabolism (S) for a sockeye salmon (Brett 1964) increases as a positive exponential function of temperature (Figure 5-4) until a lethal temperature is reached (about 25 °C). Salmon can survive down to a temperature near 0 °C but cannot exist below zero because body fluids freeze. This exponential relationship is fairly consistent for many species, and the relationship between metabolic rate and temperature is proportional to e raised to the power of mT. It can be calculated as

$$S = a\,e^{mT}$$

(Equation 5-3)

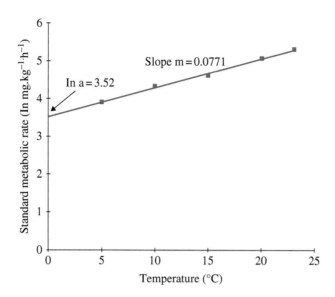

FIGURE 5-4   The relationship between standard metabolism (natural log scale) and temperature for a 10-g sockeye salmon. *Source: Adapted from Brett and Glass (1973).*

or, because $\ln(S) = \ln a + mT$, a plot of $\ln S$ against T should produce a straight line with slope of m (Figure 5-4). Values for m vary among species to some degree but are usually between 0.07 and 0.11. The constant a can vary dramatically among species, which sets the base level of metabolism (standard metabolism at 0 °C).

Metabolism can be expressed in a variety of ways. Metabolic rate is the amount (mg) of oxygen consumed or kJ of energy used per unit of time. Using the oxycalorific coefficient allows calculation of one from the other. However, those values (mg O_2/h) become a problem when fishes such as guppies are compared to sharks. A small guppy has a much lower total metabolic rate than a large shark. To take body weight into account, the weight-specific metabolic rate (mg $O_2 \cdot$kg body wt$^{-1} \cdot$ h^{-1}) is calculated. To compare metabolism and temperature for the same individual fish, it does not matter whether weight-specific or total metabolic rate is used. Obviously, for fishes of different masses, these units affect the comparison.

The influence of temperature change on the rate of reaction has been frequently studied in chemistry, and chemists use Q_{10} to predict these changes. The Q_{10} for any reaction is the proportional increase in the rate of reaction for a 10 °C increase in temperature. Q_{10} is equal to R_2/R_1 for reactions at two temperatures exactly 10 °C apart. In nonenzymatic chemical reactions, the rate of reaction roughly doubles for every 10 °C increase in temperature. For example, salt dissolves faster in warmer water; warming the water increases the rate of dissolution. The rate of dissolution should increase the same proportion for every increase in temperature of 10 °C. The general formula for calculating Q_{10} is

$$\log\left(Q_{10}\right) = \left[10\left(\log R_2/R_1\right)\right]\Big/\left(T_2 - T_1\right) \qquad \text{(Equation 5-4)}$$

where R is the rate of reaction and T is temperature (°C). If the metabolic rate of a fish was measured at 10 °C and the water temperature was increased to 20 °C for another metabolic rate measure, these data could be used to calculate Q_{10}. From Equation 5-3, $S = a\,e^{mT}$ and $Q_{10} = Q_{S20}/Q_{S10} = (a\,e^{20m})/(a\,e^{10m}) = e^{10m}$. Q_{10} is constant over a wide range of temperatures for many physical and chemical reactions. However, most biological processes are often enzymatically controlled, and enzymes may have optimal temperatures over which they operate. Certain enzymes might produce substrates more rapidly at 10 °C than at 20 °C; in fact, the processes animals use to acclimate physiologically to temperature produce enzymes with different thermal capabilities.

Acclimation

Physiological adaptations to temperature are important to fish ecology. The first of three changes fishes may undergo to adapt to temperature is acclimation, defined here as the physiological change of an individual animal in response to seasonal temperature change. Traditionally, physiologists define acclimation as the adaptations shown by animals in the laboratory to variations in temperature alone. Our definition of acclimation overlaps with acclimatization (which is discussed later in this chapter) in their terminology.

Fishes are limited in their acclimation to temperature. A good example of this acclimation is in gobies, which are small fishes that live in tide pools. Tide pools have variable temperature conditions due to tides. At high tide, water washes into pools and cools them; at low tide, the pools are stranded in the Sun, and the water may warm drastically, especially during the day. High tide or night return to bring in cool water. Tide pool gobies regularly undergo large changes in temperature depending on tide and weather, and they also acclimate to seasonal temperature cycles.

An example of this acclimation is laboratory work by da Silva et al. (2019) from University of Queensland, Australia. They began this study with the expectation that goby would not acclimate to seasonal temperature because of their regular exposure to variable temperature conditions. They exposed fish to summer conditions by maintaining them at 30 °C during the night and increased temperature to 35 °C during day, typical of the diel cycle. For winter acclimated fish, they used 12 °C at night and 17 °C at day. They held fish under these conditions for six weeks, then tested their routine metabolism, active metabolism, and burst swimming speeds at a variety of test temperatures between 12 and 37 °C. They found a strong acclimation response to seasonal temperature conditions (Figure 5-5).

The curve in Figure 5-4 (standard metabolism increasing with temperature) represents an animal that was acclimated to the same temperature at which it was tested for each point on the curve. Conditions were unlike those that were applied to animals in Figure 5-5. When gobies were acclimated to cold temperature (12–17 °C), they had increased maximum metabolic rates and maintained higher scopes for metabolism at 15–20 °C than gobies acclimated to 32–37 °C. Scope for metabolism is the difference between minimum and maximum metabolic rates and is discussed in detail in the subsequent sections. It is largely recognized as the metabolic potential for swimming and other performance. Just the opposite occurred at warm temperatures with warm acclimated fish showing higher active metabolism and metabolic scope at 36 °C compared to cold acclimated fish. A poikilotherm living in cold temperatures has a high metabolic rate and scope for that temperature. Acclimation to high temperature alters the metabolic rate to a produce a larger scope, in this case by increasing maximum metabolic rate.

The goby acclimated to compensate for temperature but could not completely compensate (note that the two temperatures do not have equal minimum or maximum metabolic rates or scopes for metabolism in Figure 5-5). Acclimation is the process of changing internal physiology to adjust to seasonal changes in temperature. Fishes generally acclimate to temperature within one to two weeks. The adaptation to cold temperature would increase the metabolic rate to an intermediate level; the adaptation to warm temperature would increase the metabolic scope to a high level. This means an individual fish is quite different physiologically during winter than it is during summer.

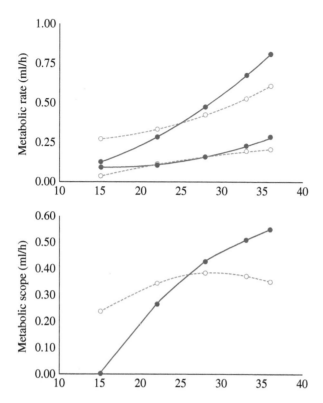

FIGURE 5-5   Metabolic rates of gobies acclimated to cool (dotted lines and open circles) and warm (solid line and filled circles) conditions and tested at different temperatures. The upper graph shows minimum and maximum metabolic rates for each group, while the lower graph shows metabolic scope for each group. *Source: Adapted from da Silva et al. (2019).*

Acclimatization

The second process is the change that occurs in a given species over a range of habitats in which it lives. We call this process acclimatization. Once again, this differs from the traditional definition of acclimatization, which is the adaptation of an individual to seasonal changes in nature (including temperature, photoperiod, and day length). Acclimatization evaluates temperature adaptations among populations of a given species that live in different climates. For example, one might compare largemouth bass in Florida with largemouth bass in Michigan. The two populations are spatially isolated and cannot interbreed. Largemouth bass in Florida never experience winter ice, and they experience maximum temperatures much warmer than largemouth bass in Michigan. Largemouth bass have experienced these climatic differences for approximately 10 000 years when they became separated after the glaciers retreated. There is a fairly long history of these populations being reproductively isolated in those two locations and quite a long time during which temperature adaptation has occurred. Venables et al. (1977) examined routine metabolic rates of largemouth bass from three populations existing under different climates in Texas, Ontario (Canada), and California (Figure 5-6). The population they studied from California was not a fair test because largemouth bass are not native to California, and only recently have they experienced that climate. California data were eliminated from Figure 5-6.

Seasonal temperature acclimation by fishes may involve adjustment to more than simply differences in temperature. Chipps et al. (2000) found that muskellunge demonstrated seasonal acclimation as described earlier, but also that routine metabolic rates in spring and fall

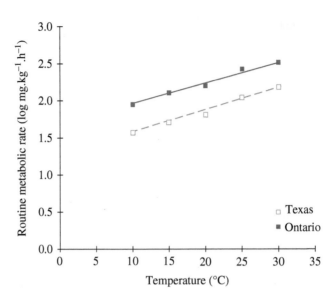

FIGURE 5-6   Routine metabolic rates (log scale) of acclimated fish at each temperature for 40-g large-mouth bass from Texas and Ontario. *Source: Adapted from Venables et al. (1977).*

were higher than in winter at the same acclimation temperature. They believe the differences were due to gonadogenesis, which occurs over fall (gonad growth) and spring (final ovulation and spermiation) for muskellunge. They also suggest that such variation in seasonal metabolism be considered in bioenergetics models.

The two largemouth bass populations showed distinctly different relationships between metabolism and temperature. Acclimated bass in Texas showed lower metabolic rates at every temperature than did bass from Ontario. The seasonal acclimation process would reduce metabolic rate at high temperature to conserve energy. Acclimatization demonstrates exactly that same pattern in populations. The warmer the climate, the lower the metabolic rate at every given temperature. These populations of largemouth bass studied by Venables et al. (1977) could still physically interbreed as they are the same species, but geographic isolation prevents interbreeding. The process of acclimatization involves evolution but not enough for speciation as not enough differences have developed among populations for acclimatization to interfere with breeding.

Metabolic Compensation

The third type of temperature adaptation – metabolic compensation – evaluates different species. Metabolic compensation is the process by which different species have evolved to largely different climatic regimes. Tropical, temperate, and Arctic fishes are the targets of analysis, not populations several hundred miles north or south. There are very large differences in climate among these zones. Metabolic compensation has been summarized by Brett and Groves (1979) using comparative data from several studies.

In temperate climates, fishes live near 0 °C every winter and experience temperatures as high as 30 °C every summer, so they must be able to tolerate both extremes. Temperate fishes are, therefore, robust in their temperature capabilities. Arctic and Antarctic fishes are almost always exposed to temperatures near 0 °C and are never exposed to temperatures as high as 25 °C. Over evolutionary time, their metabolic rates appear to have risen accordingly at cold temperatures, and they cannot tolerate warm temperatures. Arctic fishes have ambient metabolic rates comparable to temperate fishes even though they live at a much colder temperature.

Arctic fishes have evolved considerably to maintain high metabolic rates at cold temperatures, which is exactly what an individual fish does during acclimation.

Maximum Metabolism

Now that the relationship between temperature and standard (or routine) metabolism has been evaluated in depth, let us examine the relationship between the total (maximum) metabolism and temperature for two species: sockeye salmon and largemouth bass (Figure 5-7). The sockeye salmon is a coldwater, temperate fish with lethal temperature near 25 °C; the largemouth bass is a warmwater, temperate fish with lethal temperature near 35 °C. As stated earlier, standard metabolism increases with temperature for the sockeye salmon, and maximum metabolism is as much as nine times higher than standard metabolism. Maximum metabolism increases with temperature to an intermediate temperature and then declines at higher temperatures.

A largemouth bass (a lake-dwelling species) at 35 °C has about the same level of standard metabolism as a sockeye salmon (an ocean- and river-dwelling species) at 22 °C. The largemouth bass does not maintain as high a maximum metabolism as the sockeye salmon, but it shows the same general relationship to temperature. There is still an intermediate temperature at which maximum metabolism is highest and then metabolism declines with temperature. The difference between standard metabolism and maximum metabolism is termed the scope for metabolism (or scope for activity). This scope is the range of metabolic rates at which the animal can perform. A fish must maintain at least a standard metabolic rate to survive and cannot exceed the maximum metabolic rate at least not on a sustained basis. The life processes it has to maintain, including food digestion, osmoregulation, and locomotion, must remain between those two metabolic levels. Maximum scope does not occur at coldest or warmest temperature but rather at an intermediate temperature. For a sockeye salmon, the largest scope for metabolism occurs at about 15 °C. A salmon has the largest variation of metabolic activity at this temperature, where it might be able to swim fastest.

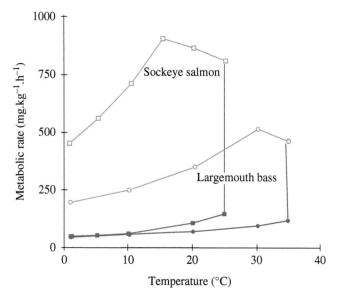

FIGURE 5-7   Standard and active metabolic rates for 10-g sockeye salmon and 150-g largemouth bass acclimated to various temperatures. *Source: Adapted from Brett and Groves (1979).*

Figure 5-7 presents several important characteristics. First, the curve for coldwater fishes – which is terminated at moderate temperatures – has fairly high levels of metabolic performance (scope) at cold temperatures. Fishes can be separated into thermal guilds, which are groups of fishes that are similar due to their temperature capabilities (Magnuson et al. 1979). The sockeye salmon is an example of a temperate stenotherm or a temperate coldwater fish. Trout are another example of temperate stenotherms. The largemouth bass is a warmwater fish or temperate eurytherm. It grows best at warm temperatures, and, while it can tolerate cold temperatures, its performance is very limited at 5 °C. An intermediate guild between these extremes includes temperate mesotherms (coolwater fish). Examples of this guild are northern pike, yellow perch, and walleye. They do not survive at 35 °C, but they do well at cold temperatures as well as moderately warm ones. These thermal guilds and their ecological significance are discussed in depth in subsequent chapters.

The second important characteristic of Figure 5-7 is the change in maximum metabolism at high temperatures. Both standard and maximum metabolism increase with temperature up to a point. Yet at high temperatures, maximum metabolic rates actually decline, indicating that heat becomes stressful to the fish. As temperature increases, the ability of water to hold oxygen declines, yet the metabolic rate of the fish increases. Therefore, as temperatures increase, fishes not only have to pump more water across their gills to maintain higher metabolic rates, but the water they pump across their gills has a lower concentration of oxygen. Fishes pay a double metabolic penalty for increases in temperature, by increases in metabolism and cost of ventilation.

Metabolism and Swimming

The relationship between swimming speed and metabolic rate was evaluated earlier to demonstrate that as swimming speed increases, metabolic rate increases as well. Plotting the log of metabolic rate against swimming speed yields a straight line, for which the intercept with the y-axis estimates standard metabolic rate, which occurs when the swimming speed is zero (see Figure 5-3). Within the range of active metabolic rates that are aerobically sustainable, there is a consistent relationship between swimming speed and metabolic rate. That covers only a part of the capacity of fish – that part which is aerobic.

Fishes can generate extreme burst speeds performed by anaerobic processes. To understand these metabolic rates, it is best to evaluate the time to utilize a similar total amount of energy. For example, Brett and Groves (1979) evaluated how much time it took a fish to consume 10 mg of oxygen: the longer it took to consume 10 mg of oxygen, the lower the metabolic rate. At a fish's maximum speed, 10 mg of oxygen was consumed in 20 seconds (Figure 5-8). Maximum sustained swimming speed (maximum metabolic rate) consumed 10 mg of oxygen in approximately 15 minutes. Swimming at cruising speed (about 60% of the maximum sustained speed) was maintained for 90 minutes with 10 mg of oxygen, and the standard metabolic rate was maintained for 180 minutes. This indicates that moving at burst speed is very costly. It takes as much energy for 20 seconds of a burst as for 15 minutes of full sustained swimming, or for three hours of inactivity. A burst speed for 20 seconds may be all the fish can perform without paying back an oxygen debt. While it takes 15 minutes of maximum standard swimming to consume 10 mg of oxygen, fishes can continue to swim at that speed indefinitely.

To summarize, swimming is costly. Metabolism varies tremendously between standard and burst rates. It is apparent from Figure 5-8 that as little as eight bursts in the course of a day could double metabolism over the standard metabolic rate. The Winberg (1956) concept of average metabolic rate ($2S$) obviously depends on activity, which may vary with the time of year and the conditions under which an animal lives. Analysis of the activity budget of fishes in the field is necessary to understand metabolic levels for fishes in natural systems. The real difficulty in such analyses would be accounting for short term but very costly bouts of burst swimming.

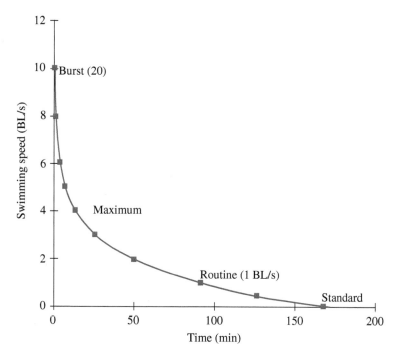

FIGURE 5-8 The relationship between time to consume 10 mg of oxygen and swimming speed for 50-g sockeye salmon at 15 °C. *Source: Adapted from Brett and Groves (1979). BL: body lengths.*

Metabolism and Size

The metabolic rate of an animal absolutely increases as the animal gets larger. However, it is also true that weight-specific metabolic rates (mg $O_2 \cdot kg^{-1} \cdot h^{-1}$) are higher in smaller animals. Standard metabolism (in mg O_2 per hour) increases with size, not in a linear fashion but curvilinearly (Figure 5-9). Standard metabolism in mg $O_2 \cdot kg^{-1} \cdot h^{-1}$ declines with mass.

Much has been hypothesized about this relationship in animals between weight-specific metabolism and size. It can be mathematically summarized as

$$MR = aW^b \qquad \text{(Equation 5-5)}$$

where a is a constant, W is mass, and b is an exponent. Experts believe that there is considerable significance to the exponent of 0.67 in homeotherms. The relationship between the surface area and the volume of a cube is 0.67, and an animal body may approximate that relationship. Since a homeothermic animal may lose heat based on surface area exposed to the environment but generate it based on volume or mass (Phillipson 1981), this heat loss may be at the root of the metabolic rate:mass relationship in homeotherms. In a similar manner, the surface area of lungs or gills may influence the oxygen-obtaining capacity per unit mass, while metabolism is based on mass, again affording the above relationship even in ectotherms.

In fishes, there is a variety of weight exponents. Brett and Groves (1979) estimated the average weight exponent at 0.86, with values from 0.5 to 1. Clarke and Johnston (1999) also summarized fish metabolic data and estimated the average weight exponent to be 0.79 with much variation. They also found an average Q_{10} value between these species to be 1.83, lower than they found within a species (2.4). This variance in weight exponent is common in other ectotherms as well (Phillipson 1981). The surface area:volume relationship may not have such

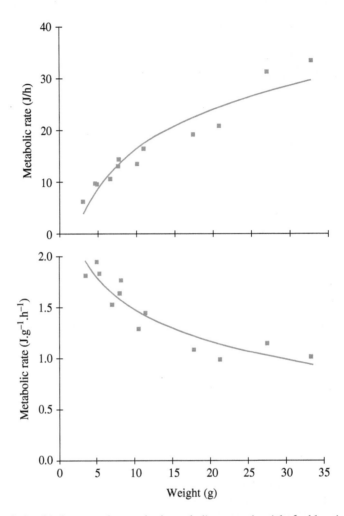

FIGURE 5-9   The relationship between the standard metabolic rate and weight for blennies at 10 °C: (upper graph) total metabolic rate and (lower graph) weight-specific metabolic rate. *Source: Data from Wallace (1973).*

importance in fishes. One reason is that fishes do not generate and maintain body temperature, so heat loss is unimportant. Phillipson (1981) believes that the membrane surfaces, as well as respiratory gas transport systems, vary among ectotherms and cause this variable exponent. The important point here is that metabolism (per unit mass) declines with increase in mass, and the average exponent is somewhere around 0.79–0.86, but that fish species show large differences in that exponent. In other words, to estimate an energy budget and changes in metabolism with size, the exponent b must be measured for that species.

Cost of Digesting Food

All of the measures listed previously were for fishes that were not digesting food. The metabolic rate of an animal is elevated over a period of time after the animal has eaten, and then declines once the food is completely digested. A species of Hawaiian reef fish (aholehole) fed a ration of 4.7% of its body weight showed a large increase in metabolism after feeding (Muir and Niimi 1972; Figure 5-10). Independent metabolic measures on this species show that the lowest

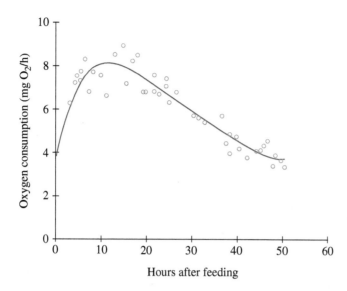

FIGURE 5-10 The oxygen consumption of an aholehole after consuming a 4.7% ration at 23 °C. *Source: Adapted from Muir and Niimi (1972).*

level of metabolism in Figure 5-10 was routine metabolic rate and that metabolism increased after feeding, not because of swimming but because of the cost of digesting food. That cost is the specific dynamic action (H), also termed the heat increment (Brett and Groves 1979).

What are the mechanisms for this increase in metabolic rate? They are partly physiological and partly experimental artifacts. Obviously, the costs of digesting, which include enzyme and acid production, gut mobility, and muscle contraction related to passing food through the gut, are true costs of digesting food and a part of the measured specific dynamic action. Secondarily, fishes fed in a tank may become very active in searching for food. If activity is not controlled, the measure of H would include the cost of activity related to feeding. Fishes may also increase their metabolic rates in anticipation of eating (due to excitation). All of these are components of the increased metabolism measured in a feeding experiment. Those costs also exist in nature, but they may not be the same as measurements in the lab. Therefore, this measurement of increased metabolism after feeding is called apparent specific dynamic action, and it includes true specific dynamic action, as well as other artifacts, due to experimental techniques. Brett and Groves (1979) estimated H at 3.7 ± 1.2 times standard metabolism (mean $\pm$ SD). A substantial increase in metabolism occurs to digest the food or, in these experiments, handle the food.

Details of the relationship between H and ration were evaluated for sockeye salmon by Brett (1976). His results (Figure 5-11) include data for a variety of rations at three temperatures: 20, 15, and 10 °C. When a fish has no food in the gut, metabolic rate is essentially standard metabolism if the fish is inactive, and since standard metabolism increases with temperature, the metabolic rate at zero ration also increases with temperature. As ration is increased, total metabolic rate also increases, indicating that specific dynamic action is higher with larger rations. However, the increase in H with ration is not similar for all three temperatures, but rather the lines in Figure 5-11 diverge as temperature increases. As temperature increases, total metabolic rate increases, and the rate at which metabolism changes with meal size also increases. Specific dynamic action, which is the difference between standard metabolism (zero ration) and the metabolic rate at any ration, increases with meal size.

A similar study on carp presented slightly different results. Huisman (1974) tested carp at 17 and 23 °C (Figure 5-11). Increasing temperature resulted in increased metabolic rate, both standard metabolism and specific dynamic action. Once again, carp ate a larger meal at warmer

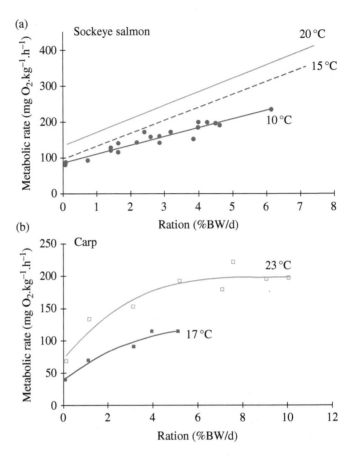

FIGURE 5-11 Metabolic rate at a variety of rations and temperatures for (a) sockeye salmon and (b) common carp. H is the difference between the metabolic rate at some condition and the rate at zero ration for that same temperature. *Source: Adapted from Brett (1976) and Huisman (1974).*

temperatures. Maximum ration at 17 °C was only about 6% of body weight (%BW) per day, whereas at 23 °C, it increased to 10% per day. The main difference between carp and salmon is that data for carp did not follow a straight line. Increased meal size for carp resulted in H costs that elevated at a reduced rate. Comparing Figure 5-11a and b, larger meals resulted in higher costs of specific dynamic action for both species. This may be related to different diets or digestive systems of these two fishes. However, the shape of these curves was not consistent across these species of fishes.

How much energy should be allocated to H in developing an energy budget for fish in nature? At present, a common estimate is that 12–16% of the food energy consumed is lost in specific dynamic action. Making H a multiple of ration mathematically produces the same relationship shown in the slope in Figure 5-11a – making H greater with increased ration. The actual slope of Figure 5-11a, when both axes are converted to kcal, is 0.20. The range between 12 and 16% is also used because of differences in proximate composition of food. Fishes that eat high-protein diets generally have a higher specific dynamic action (Knights 1985). However, Tirsgaard et al. (2015) evaluated the heat increment of Atlantic cod at different temperatures and found that not only H was affected by temperature, but the duration of elevated metabolism, as well as the amount of metabolic elevation over S, varied with temperature. They determined that H coefficients (the percent of H energy divided by meal energy) varied from 6 to 16% and was lower at cold temperatures. Other studies have shown similar variability in the proportion of a meal that is used up in the energy needs for digestion, making generalization of a value for H difficult.

Defecation and Excretion

Animals consuming food are only able to absorb energy from a portion of that food, while the remainder is lost as feces (F). The simplest way to measure egestion or defecation is to measure the amount of food consumed and divide that into the amount of fecal material produced (Box 5-3). For example, if an animal eats 100 g of food and egests 20 g of feces, the egestion rate is 20%. This is not an unbiased measure of egestion efficiency for several reasons. The material egested is considerably lower in quality than material eaten. This means that mass is not a good unit of measure for both components. A lot of energy in the food should have been absorbed, so feeding and egestion rates in kJ would be much better. The calorific coefficient for feces might be 4 kJ/g dry weight, whereas the energy content of food might be 16 kJ/g dry weight. Using these values, the original 20% egestion rate would change to 5% because energy content of feces is much lower than food. Fecal material is largely ash, bone, or cellulose that is not digestible.

Another problem measuring defecation is that fishes live in water, and as soon as a fish defecates, part of the feces can dissolve. For many species of fish, it may be necessary to filter feces from the water. Dissolution of feces during defecation can be a major problem. Since a fish may not defecate all at once but rather slowly over time, bacteria will also decompose feces before the fish finishes defecating. It is difficult to use the simple measurements outlined earlier to estimate assimilation efficiency for most fishes.

Box 5-3 Calculations of Fecal Losses

1. From food consumption and feces collection:

$$\textbf{Egestion} = \textbf{amount feces} \times \textbf{100 / amount food}$$

where amount = g or kJ (better).

Example:

100 g food eaten and 20 g feces:

$$E = 20/100 \times 100 = 20\%$$

1670 kJ (16.72 kJ/g) food eaten and 83.6 kJ (4.18 kJ/g) feces:

$$E = 83.6/1670 \times 100 = 5\%$$

2. From Cr_2O_3 or (^{14}C)

$$\textbf{Egestion} = \textbf{100}\left(\textbf{\%Cr}_2\textbf{O}_3 \textbf{ in food / \%Cr}_2\textbf{O}_3 \textbf{ in feces}\right) \times \left(\textbf{kJ feces / kJ food}\right)$$

Example: 2% Cr_2O_3 in diet, 4% in feces; 16.7 kJ in food and 4.18 in feces:

$$E = \left(100 \times 2/4\right) \times \left(4.18/16.7\right) = 12.5\%$$

Assimilation efficiency $\left(A\right)$ is then $1 - E = 87.5\%$

See Talbot (1985) for more details.

Other methods have been developed to measure defecation rate. The most common method is to use an indigestible indicator (something fish cannot digest). The most common indigestible indicator is chromic oxide (Cr_2O_3). This indigestible compound is easy to measure, and the ratio of Cr_2O_3 in the food compared to the feces can be used to estimate assimilation efficiency (see Box 5-3). A prepared diet with a known indicator concentration is used. After digestion, feces are collected to determine the ratio of food to marker in the feces. With total collection of feces, the ratio of Cr_2O_3 in the food and in the feces gives the egestion rate. Since total feces collections are difficult, a ratio of nutrient content of food and feces (in % or kJ) can be used for incomplete collections.

The chromic oxide method simplifies measurements because not all of the feces have to be collected. Similar ratios could be developed using ^{14}C as the tracer. These ratios are particularly useful for species like tilapia, which might feed and defecate continuously in nature.

Factors Influencing Assimilation Efficiency

As described earlier, the nature of food (whether it is highly digestible or indigestible) and the nature of the feeding type both influence how much energy will be assimilated. The largest factor influencing assimilation efficiency is the kind of food eaten. It overwhelms fish size, meal size, or temperature, but some of these variables also affect assimilation efficiency. Elliott (1976) fed *Gammarus* (an amphipod) to brown trout and evaluated the assimilation efficiency of trout at different temperatures. Assimilation efficiency varied from 69 to 88%. Elliott also varied meal size as a percent of satiation ration from 10 to 100%. Remember, satiation ration changes with temperature, so we are not comparing the same food amount at each temperature. Satiation may be 0.3% body weight at 5 °C, 2% body weight at 10 °C, and 4% body weight at 20 °C. In evaluating changing rations with temperature, it is difficult to conceptualize a controlled level of ration since ration changes with both temperature and fish mass. Elliott used a percentage of the maximum level a fish could eat at any temperature.

Elliott found that trout generally did not assimilate as efficiently when they passed more food through their guts. The total differences due to meal size were about 10%. Temperature also had an influence. At any fraction of a satiation ration, warmer temperature gave greater assimilation efficiency. These differences due to temperature were about 5%. Both factors (temperature and meal size) had an influence on assimilation efficiency. Both are much less important than the nature of food consumed.

Excretory Losses

In addition to assimilation, fishes also lose a fraction of ingested ration due to excretion. Excretion (N) is the cost of removing excess nitrogen, which results from a breakdown of internal tissue (endogenous), and is caused by a breakdown of amino acids in food (exogenous). Both of these produce ammonia as a waste product, which must be excreted, usually through the gills. Many fishes offset the excretion of ammonia through gills by ion exchange during osmotic regulation. These fishes exchange ammonia ions for sodium or other anions (in a freshwater fish) to maintain an electrolyte balance as they excrete ammonia. Part of the cost of excretion is balanced by a decreased cost of osmoregulation.

It is relatively simple to measure nitrogen excretion since ammonia diffuses into the water. A fish can be placed in the water for a given amount of time over which ammonia production is determined. Once again, there are two sources of such ammonia increases: endogenous and exogenous. Fishes put on starvation diets catabolize body stores to satisfy their normal metabolic requirements. Fishes on maintenance diets excrete some endogenous ammonia, plus some ammonia from their diets. At any feeding level, sorting out endogenous and exogenous sources of nitrogen may be a problem. However, the actual measurement is simple; all that is necessary is to measure the amount of ammonia produced and then convert that to energy.

The second component of Elliott's work was to measure excretion. Metabolizable fraction ($[1 - (F + N)]/R$) is the amount of energy in the food that can be metabolized. Elliott (1976) found that at a reduced ration, trout had higher assimilation efficiency and higher metabolizable fraction. However, unlike digestible fraction, which increased with temperature, metabolizable fraction declined with temperature. The difference between these two values is due to N, the excretion component. The cost of excretion must have increased considerably

---

Box 5-4 Sample Bioenergetic Model Calculations

An energetic model for largemouth bass includes details of Figure 4-1 plus information on growth and water temperature. For example, Rice et al. (1983) proposed the following parameters for bass:

$$a = 0.3478 \, \text{mg O}_2 \cdot \text{g}^{-1} \cdot \text{h}^{-1}$$
$$b = -0.325$$
$$g = 0.0196$$
$$m = 0.0313 \, \text{s/cm}$$

For these values, a mass of 100 g, a swimming speed of 30 cm/s, and a temperature of 25 °C, total metabolism for that bass becomes

$$OC = 0.3478 \left(100^{-0.325}\right) \times 2.71828^{(0.0313 \times 25)+(0.0196 \times 30)}$$
$$OC = 0.2908 \, \text{mg O}_2 \cdot \text{g}^{-1} \cdot \text{h}^{-1}$$
$$OC = 29.08 \, \text{mg O}_2 / \text{h}$$

Since 1 mg OC consumed = 13.54 kJ of energy utilized, this becomes

$$MR = 29.08 \, \text{mg} \times 13.54 \, \text{J/mg} = 393.74 \, \text{J/h}$$
$$MR = 393.74 \times 24 = 9449.8 \, \text{J} = 9.45 \, \text{kJ/day}$$

Using this to estimate a daily ration requires three more components: assimilation efficiency (or digestibility of food), cost of digestion (heat increment), and urinary energy loss. Excretion loss is estimated 0.142, urinary loss 0.079, and specific dynamic action 0.104 for each unit of ration consumed (Rice et al. 1983). Therefore, a maintenance ration for this bass (no growth) would require the following ration:

$$C = (R+G) \big/ \left[1 - (F+U+H)\right]$$
$$C(\text{kJ/d}) = 9.45 \big/ \left[1 - (0.142 + 0.079 + 0.104)\right]$$
$$C(\text{kJ/d}) = 9.45 \big/ 0.675 = 14$$

If the fish actually grew 2 g (4.18 kJ/g) per day, the ration would be

$$C = (R+G) \big/ \left[1 - (F+U+H)\right]$$
$$C = (9.45 + 8.36) \big/ 0.675$$
$$C = 17.81 / 0.675 = 26.3 \, \text{kJ/day}$$

It is obvious that many interesting manipulations can be done with an energy budget equation. An energetic model is really just a computer-operated equation with ability to generate such information over months or years of simulation, using seasonal temperature regimes and growth rates as input. See Rice et al. (1983) for more details.

with temperature and with meal size. At most rations, trout had an intermediate temperature at which the metabolizable fraction of their diet was highest. That optimum temperature increased with increasing ration size.

Summary Equation and Energetics Model

Several factors, particularly fish mass, temperature, and swimming speed, have been shown in this chapter to affect metabolic rate. These can be summarized in an overall metabolism equation:

$$OC = a \cdot W^b \, e^{mT+gs} \tag{Equation 5-6}$$

where OC (oxygen consumption) is in mg $O_2 \cdot kg^{-1} \cdot h^{-1}$, a is standard metabolism for a fish at 0 °C, b is the weight exponent, m is the metabolism-temperature coefficient, T is temperature (°C), g is the metabolism-swimming coefficient, and s is swimming speed.

Such an equation was first proposed for bioenergetic models by Kitchell et al. (1977) and Rice et al. (1983) and has been expanded on by Deslauriers et al. (2017). The logic of such models is that energy costs can be estimated more easily than ration (energy consumed by feeding) in the field, so the model can be used with measured fish growth rates to predict ration (see Box 5-4). Such predictions can be powerful management tools if the models are accurate. The history and application of bioenergetic models are covered in Chapter 8.

Literature Cited

Beamish, F.W.H. 1970. Oxygen consumption of largemouth bass, *Micropterus salmoides*, in relation to swimming speed and temperature. Canadian Journal of Zoology 48:1221–1228.

Blazka, P., M. Volt, and M. Cepela. 1960. A new type of respirometer for the determination of the metabolism of fish in an active state. Physiology Bohemoslovia 9:553–558.

Brafield, A.E., and D.J. Solomon. 1972. Oxy-calorific coefficients for animals respiring nitrogenous substrates. Comparative Biochemistry and Physiology 43A:837–841.

Brett, J.R. 1964. The respiratory metabolism and swimming performance of young sockeye salmon. Journal of the Fisheries Research Board of Canada 21:1183–1226.

Brett, J.R. 1976. Feeding metabolic rates of sockeye salmon, *Oncorhynchus nerka*, in relation to ration level and temperature. Environment Canada, Fisheries and Marine Service Technical Report No. 675, 18 pages.

Brett, J.R., and N.R. Glass. 1973. Metabolic rates and critical swimming speeds of sockeye salmon, *Oncorhynchus nerka*, in relation to size and temperature. Journal of the Fisheries Research Board of Canada 30:378–387.

Brett, J.R., and T.D.D. Groves. 1979. Physiological energetics. Pages 279–352 *in* W.S. Hoar, D.J. Randall, and J.R. Brett, editors. Fish Physiology, Vol. 8. Academic Press, New York.

Brown, J.H., J.F. Gillooly, A.P. Allen, V.M. Savage, and G.B. West. 2004. Toward a metabolic theory of ecology. Ecology 85:1771–1789.

Chabot, D., J.F. Steffensen, and A.P. Farrell. 2016. The determination of standard metabolic rate in fishes. Journal of Fish Biology 88:81–121.

Chipps, S.R., D.F. Clapp, and D.H. Wahl. 2000. Variation in routine metabolism of juvenile muskellunge: Evidence for seasonal metabolic compensation in fishes. Journal of Fish Biology 56:311–318.

Clarke, A., and N.M. Johnston. 1999. Scaling of metabolic rate of body mass and temperature in teleost fish. Journal of American Ecology 68:893–905.

Deslauriers, D., S.R. Chipps, J.E. Breck, J.A. Rice, and C.P. Madenjian. 2017. Fish Bioenergetics 4.0: An R-based modeling application. Fisheries 42:586–596.

Elliott, J.M. 1976. Energy losses in the waste products of brown trout (*Salmo trutta* L.). Journal of Animal Ecology 45:561–580.

Huisman, E.A. 1974. A study on optimal rearing conditions for carp (*Cyprinus carpio* L.). Unpublished Ph.D. Thesis, Agricultural University of Wageningen.

Kitchell, J.F., D.J. Stewart, and D. Weininger. 1977. Applications of a bioenergetics model to yellow perch (*Perca flavescens*) and walleye (*Stizostedion vitreum vitreum*). Journal of the Fisheries Research Board of Canada 34:1922–1935.

Knights, B. 1985. Energetics and fish farming. Pages 309–340 *in* P. Tytler and P. Calow, editors. Fish Energetics: New Perspectives. The Johns Hopkins University Press, Baltimore, Maryland.

Magnuson, J.J., L.B. Crowder, and P.A. Medvick. 1979. Temperature as an ecological resource. American Zoologist 19:331–343.

Muir, B.S., and A.J. Niimi. 1972. Oxygen consumption of the euryhaline aholehole (*Kuhlia sandvicensis*) with reference to salinity, swimming, and food consumption. Journal of the Fisheries Research Board of Canada 29:67–77.

Phillipson, J. 1981. Bioenergetic options and phylogeny. Pages 20–48 *in* C.R. Townsend and P. Calow, editors. Physiological Ecology: An Evolutionary Approach to Resource Use. Sinauer Associates, Sunderland, Massachusetts.

Rice, J.A., J.E. Breck, S.M. Bartell, and J.F. Kitchell. 1983. Evaluating the constraints of temperature, activity and consumption on growth of largemouth bass. Environmental Biology of Fishes 9:277–288.

Svendsen, M.B.S., P.G. Bushnell, and J.F. Steffensen. 2016. Design and setup of intermittent-flow respirometry system for aquatic organisms. Journal of Fish Biology 88:26–50.

Talbot, C. 1985. Laboratory methods in fish feeding and nutritional studies. Pages 125–154 *in* P. Tytler and P. Calow, editors. Fish Energetics: New Perspectives. Croom Helm Australia Pty Ltd., Sydney.

Tirsgaard, B., J.C. Svendsen, and J.F. Steffensen. 2015. Effects of temperature on specific dynamic action in Atlantic cod *Gadus morhua*. Fish Physiology and Biochemistry 41:41–50.

Venables, B.J., W.D. Pearson, and L.C. Fitzpatrick. 1977. Thermal and metabolic relations of largemouth bass, *Micropterus salmoides*, from a heated reservoir and a hatchery in north central Texas. Comparative Biochemistry and Physiology 57A:93–98.

Wallace, J.C. 1973. Observations on the relationship between the food consumption and metabolic rate of *Blennius pholis* L. Comparative Biochemistry and Physiology 45A:293–306.

Webb, P.W. 1978. Partitioning of energy into metabolism and growth. Pages 184–214 *in* S.D. Gerking, editor. Ecology of Freshwater Fish Production. John Wiley and Sons, New York.

Winberg, W.W. 1956. Rate of metabolism and food requirements of fishes. Fisheries Research Board of Canada, Translation Series 194.

CHAPTER 6

Patterns of Growth and Reproduction

The growth of fishes is an important indicator of environmental and physiological conditions. For example, fishes that grow fast may have better survival rates because once they become large, they are less vulnerable to predation. Therefore, growth can have a direct effect on survival. Growth also influences reproductive fitness as larger fishes of a species produce more eggs and potentially can produce more young. Growth of fishes is extremely flexible because it is often the last process in the balanced energy equation performed by fish. Metabolism, osmoregulation, and activity must all be accomplished first, and whatever energy is left is available for growth. Small differences in the energy budget can therefore result in large differences in growth. Many problems in fishery management deal with size-related concerns – including populations with many small fishes, and possibly insufficient surplus energy for both metabolism and growth. Since growth rate is so strongly related to the environment through temperature and food availability, it differs in almost every single ecosystem.

Energy accumulated for growth is the difference between the quantity of food eaten and the amount required for catabolism. Growth is not a process in itself, but it can be regulated by adjusting the rate of food intake or energy use. This regulation takes place on several time scales. From the balanced energy equation, the daily pattern of growth is estimated by determining total food consumption and growth for that day. It may be possible to extrapolate that over time, but for long time intervals such an extrapolation becomes difficult. From an ecological or evolutionary point of view, the interest is in ultimate body size, as well as ontogenetic changes in size which influence feeding, mortality, and reproduction. From an aquaculture perspective, the capacity to grow may be of interest, as well as how the environment – through temperature and food availability – influences growth. All of these factors influence the time scale and methodology involved in growth analyses.

The growth of fishes may differ considerably in natural environments from what might be predicted in the lab. Since temperature, food availability, and other behavioral characteristics influence energy accumulation and use, every population has unique factors influencing its growth. Obviously, one energy use is reproduction, particularly since body and gonad growth seem to occur in competition with each other, that is, available surplus energy can be used in either body or in gonad growth. Biologists believe a fixed amount of surplus energy can be consumed, and the use in one process eliminates that energy for other uses. Reproduction is one of the costs that can seldom be evaluated in the laboratory since studies in the lab are often short term, use immature animals, and rarely provide an environment suitable for inducing gonad

Biology and Ecology of Fishes, Third Edition. James S. Diana and Tomas O. Höök.
© 2023 John Wiley & Sons Ltd. Published 2023 by John Wiley & Sons Ltd.

growth and spawning behavior. To fully investigate field patterns of growth and reproduction, we will move from lab experiments on factors influencing fish growth to natural fish populations, evaluating the major conditions influencing body and gonad growth.

Factors Influencing Growth

Growth and Ration

The most important factor influencing growth is ration, and this is the first relationship to be examined in terms of processes affecting growth. Brett et al. (1969) first demonstrated this relationship for sockeye salmon held in laboratory conditions. The growth–ration curve (Figure 6-1) evaluates changes in growth at each ration. If an animal does not eat, it loses weight (negative growth), so zero ration results in negative growth. A fish requires at least some level of ration to achieve zero growth. There is also an internal limit to ingestion, and, as ration increases from zero to that limit, growth increases at a declining rate until it eventually reaches an asymptote. Three levels of ration are prominent in Figure 6-1. The first, maintenance ration (Rmaint), is the amount of food required just to maintain body weight. This is comparable to standard metabolism in ration terms. Energy use at standard metabolism is actually less than at maintenance ration because at maintenance ration a fish may swim and has heat increment costs, nitrogen excretion, and assimilation efficiencies. Those processes will increase metabolism above standard.

The second ration of significance, satiation or maximum ration (Rmax), is the most a fish can eat. Usually, maximum ration causes maximum growth. A final ration of interest is optimum ration (R_{opt}), which occurs at the point a tangent line, going through zero, intersects the growth–ration curve. The slope of this curve, or the rate of change in growth per unit ration, is maximum at R_{opt}. Optimum ration produces the highest gross conversion efficiency or the most efficient growth. Figure 6-1 clearly illustrates that slight changes in ration can result in large differences in growth. At low ration, slight increases in ration increase growth dramatically; while at high ration, similar increases in ration result in smaller differences in growth.

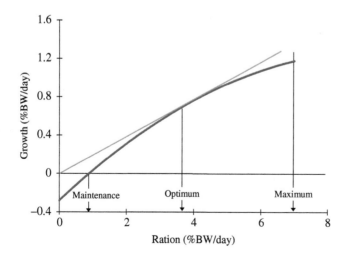

FIGURE 6-1   Relationship between growth and ration for fingerling sockeye salmon at 10 °C.
Source: Adapted from Brett et al. (1969).

Growth–ration curves are not identical for all species. Carp exhibit what may be called a glutton effect in that at maximum ration, these fishes cannot digest the meal as efficiently, and growth declines. Carp, therefore, show highest growth at less than maximum ration.

Compensatory Growth

While ration has a strong influence on growth, the frequency and pattern of food delivery also has a strong effect on consumption and ultimate growth. Malcolm Jobling from the University of Tromso in Norway has provided a good description and evaluation of compensatory growth studies in fishes (Jobling 1994). He defined compensatory growth as the capacity of an organism to grow at an accelerated rate, following a period of food shortage or reproductive weight loss. For fish populations, this process has mainly been demonstrated in laboratory settings using food deprivation over various time periods. For example, Jobling et al. (1993) evaluated compensatory growth in Arctic charr by feeding them in groups that ranged from continual feeding (control group) to feeding for three weeks and depriving for three weeks. They compared growth among four treatments at the end of four six-week periods and found that control fish grew fastest, while those deprived for various time periods grew about the same, although one week deprivation produced slightly higher growth than the other experimental groups (Figure 6-2). Since fish on cycles were only fed half of the days for an experimental period, all three experimental treatments had the same number of feeding days but differences in the duration of deprivation between feeding periods. In comparison, the control group had eaten as many total days after 84 days of experiment as the experimental groups did after 168 days. When compared over an equal number of feeding days (84), all deprived groups had no significant difference in growth, but all grew significantly more rapidly than the control group.

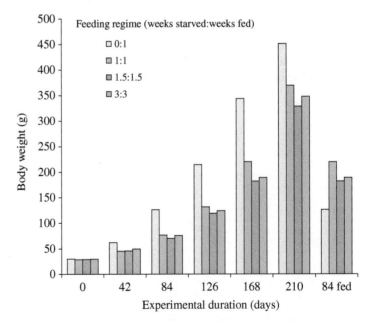

FIGURE 6-2 Mean weight of immature Arctic charr after various time durations (in days) of feeding under four cycles: continuous feeding (0:1), or alternating feeding and deprivation over one-week periods (1:1), over 1.5-week periods (1.5:1.5), and over three-week periods (3:3). Also shown are results for all groups fed an additional 42 days after completion of the feeding treatments (210 days) and sizes of fish after 84 days of actual feeding (84 fed). *Source: Drawn from data in Jobling et al. (1993).*

This indicates that the experimental groups had actually eaten more food or become more efficient at growth than the control fish. Data from other studies indicate that both processes are involved (Jobling 1994). Since experimental fish had also endured extended periods of weight loss when no food was provided, their actual growth during feeding days was even higher than it appears in Figure 6-2. When all fish were allowed to feed continuously for an additional six-week period, all experimental groups reached a similar size but were still smaller than the control group, indicating partial compensation.

Compensatory growth has been found in a number of fish species from marine and freshwater conditions. While evidence has mainly been accrued from lab experiments, the importance of such a phenomenon for recovery from energy stress would be of obvious benefit to natural populations as well. Fish under compensatory scenarios in some experiments have resulted in higher growth than control fish (Hayward et al. 1997), which would make such alternating schedules of feeding useful to aquaculture, especially if they involve not only more rapid growth but also higher conversion efficiency.

Depensatory Growth

A common behavioral result of controlled feeding trials is the phenomenon called depensatory growth. This process involves competition for food among the fish, resulting in the largest fish getting more of the food applied and the smaller ones getting less, so over time, the difference in size between the largest and smallest individual increases (Brett and Groves 1979). Depensatory growth is commonly observed in aquaculture situations and is likely to occur in natural settings where there is strong competition for food. Its occurrence in feeding experiments may also in part be due to space limitation in experimental containers. This means the presence of depensatory growth in the lab does not necessarily indicate the same process would occur in the field for that species. Experiments demonstrating depensatory growth have shown that differential food access occurs between dominant and subordinate fish but not higher conversion efficiency by the dominant fish. Depensatory growth is an obvious detriment to aquaculture systems and is commonly dealt with by grading fish in tanks so all individuals are as near in size as possible.

Determinate and Indeterminate Growth

Fishes are considered indeterminate in their growth rates (Figure 6-3). This assumes that if a fish is given more food, it will grow more rapidly. Indeterminate growth has the hypothesis that fishes show little internal control of their growth rate. A fish might increase 25-fold in size between the end of the first year and the final year of life. There is quite a considerable difference in scope for growth and how this scope changes with age over the lifetime of a fish.

What exactly is indeterminate growth? It is related not only to growth but also to reproduction. Once animals reach a certain size, they mature. Body growth rate decreases at breeding because surplus energy is used to produce gametes, migrate, care for young, build nests, etc. Maximum ration, maintenance requirements, and growth decline with increased body size, and growth decreases even further with the onset of maturation (Figure 6-4). Fishes tend to mature at a small size and continue to grow after first maturation. The curve of size against age (Figure 6-3) is similar for determinate and indeterminate growth, and the main differences are the ages and sizes at maturation and the rates at which each type achieves maximum size. Determinate and indeterminate growth are two extremes that can occur in a continuous pattern of growth. In terms of reproduction, this trade-off between body and reproductive growth is important in evolution.

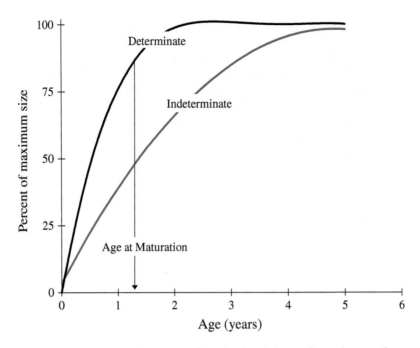

FIGURE 6-3 Schematic lifetime growth patterns for animals exhibiting determinate and indeterminate growth.

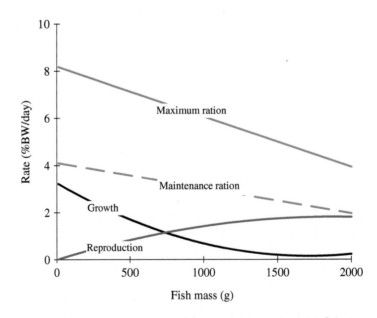

FIGURE 6-4 Hypothetical changes in maximum and maintenance ration with fish mass, and the resultant growth and reproductive allocations for an iteroparous fish.

Temperature and Growth

For poikilothermic animals, temperature influences a variety of processes that affect growth. Many of these can be summarized by examining changes in maintenance and maximum ration with temperature (Figure 6-5). Maintenance ration increases continually with temperature,

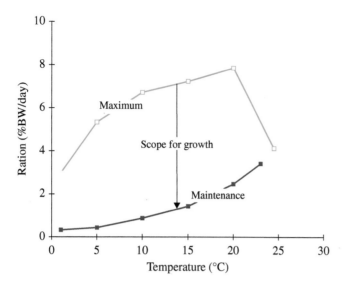

FIGURE 6-5 Maintenance and maximum ration at each temperature for fingerling sockeye salmon.
Source: Adapted from Brett et al. (1969).

while maximum ration reaches an optimum at intermediate temperature and then declines. Since the difference between these two rations is related to growth, the scope for growth (potential to grow given consumption and maintenance needs) is maximum at an intermediate temperature. The optimum temperature for growth is generally lower than the optimum temperature for consumption because of the increasing costs of maintenance at higher temperatures. At reduced ration, there is a different effect of temperature. The fixed ration could be more than the fish can consume at low temperature, while it is below Rmax at high temperature. At reduced ration, optimum temperature for growth is lower than at maximum ration. This makes it possible for a fish to move into cooler water to adjust to limited ration. That would result in a lower maintenance requirement and therefore better growth on that fixed ration. A fish might trade off among the best temperatures for metabolism, consumption, or growth by moving into different temperature locations in a water body in an attempt to maximize growth under local conditions.

The overall effects of temperature on growth are quantified by the changes in growth–ration curves for sockeye salmon at three temperatures (Figure 6-6). It is obvious that maintenance ration and maximum ration increase with temperature. At reduced rations (less than 2% BW/day in Figure 6-6) growth is best at 5 °C. At all rations, growth is best at 10 °C compared to 20 °C because of high metabolic costs at high temperature. A larger ration can be consumed at 20 °C, but growth is maximum at 10 °C, which is near the optimum temperature for growth (12 °C) for these fish.

Body Growth and Energy Storage

One major problem dealing with natural cycles of energy storage and use is the difficulty of evaluating energy storage independent of body growth. Fishes store energy in lipids at the same time they grow in total mass. Growth in mass includes gains in both length and weight, which may occur simultaneously or may not be synchronous. Fishes continue to grow in length and weight throughout their adult lives, and they may store energy in body tissue, in the liver, or in lipid depots while they grow. Relative increases or decreases in mass per unit length alone are insufficient to indicate energy cycles of growth – absolute changes must be measured.

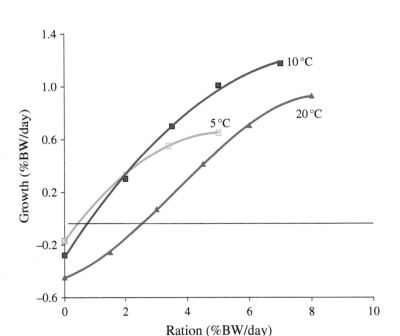

FIGURE 6-6   Relationship between growth and ration at 5, 10, and 20 °C for fingerling sockeye salmon. *Source: Adapted from Brett et al. (1969).*

This becomes important when evaluating field energetic studies because some studies use a "standard fish" concept or calculate the changes in mass for a standard-length fish (e.g., 50 cm). That 50-cm fish in February may be 75 cm the following June, so accurate measures of energy storage must include changes in length. Relative patterns of growth underestimate total growth in long-term natural studies. However, the sampling involved in determining growth for each age class is much more intensive than the amount necessary to calculate growth per unit length.

Shul'man (1974) wrote a pioneering book in this area, entitled *Life Cycles of Fish*, which evaluated patterns of lipid and protein synthesis and use. He defined two extremes in energy storage patterns for fishes. The first category is lean fishes, which do not store much lipid in body tissue. Examples of fishes from this category include northern pike, cod, and centrarchids. They do not store large lipid reserves in their bodies although they might store lipid in their livers. Some of these fishes, such as walleye, may also store visceral fat deposits, which are lipid bodies attached to the intestine. Shul'man's second category was fatty fishes, which store surplus energy predominantly as lipids deposited in body tissue not in specific organs like the liver or viscera. Some examples of this group include anchovy, salmon, and alewife. Salmon, for example, have stores of lipids in areas of their ventral musculature as part of their muscle tissue. These two extremes of lean and fatty fishes influence patterns of energy storage and use during times of energy deficit.

Sockeye Salmon

One of the first studies on energetics of fishes in the field was done by David Idler and his colleagues in British Columbia (Idler and Clemens 1959). They examined the migration of salmon and the cost of that migration in streams. When salmon leave the ocean, they cease feeding for the rest of their lives. Their intestines and stomachs degenerate to the point where they are not functional. The fish do not shrink in length but utilize internal energy stores for the entire migration, late gonad maturation, and spawning. In the case of salmon, calculating changes for a standard 50-cm fish would reasonably estimate changes in energy content that occur during this migration since the fish shrinks only in weight.

The other important behavior of salmon is that they migrate in groups. For example, a stock enters the Fraser River at the mouth at a certain time, and, within a few days, that stock has gone through the river mouth and is moving upstream. Eventually, individuals of that group will spawn at the same time many kilometers upstream. One could sample from that group at several points along its migratory route and be almost certain that the samples represent the same group of fish, which are changing over time because of the cost of migration.

Idler and his colleagues sampled fish at the mouth of the Fraser River, 400 km upstream, and at their spawning grounds, 1,200 km upstream. They evaluated changes in the carcasses, internal organs (liver and viscera), and gonads (ovaries and testes) among those samples. They determined protein, lipid, energy, and water content for each tissue. The major store of energy in any fish is the carcass, where this material is deposited in muscle and other structural components. Over the course of salmon migration, the fish depleted 91–96% of the lipid in their eviscerated carcasses. Females used 41% of carcass protein, while males depleted 30%. Lipid use accounted for 3787 kJ of energy with 1797 kJ from protein. These fish utilized nearly all the energy stored in their bodies, and this use depleted not only surplus energy depots but muscle and other functional tissues. There were major changes in the internal organs as well; the liver completely degenerated, which accounted for only 33 kJ. The viscera were depleted of 82% of the stored lipid and protein, which accounted for 167 kJ.

Males gained 50 kJ of energy in their testes, while females gained 1045 kJ in their ovaries. The salmon were migrating and degenerating their bodies at the same time they were producing gametes to use in spawning. The difference between the two sexes in protein use from the body (41% for females and 30% for males) was related to the extra energy cost of at least 836 kJ for females to produce ovaries during the migration. The sources of energy to fuel late maturation and migration of salmon were 3% from viscera, 1% from liver, 65% from body fat, and 31% from body protein.

Fishes cannot survive when they undergo losses of body components of this magnitude. Salmon have a semelparous life-history strategy, that is, they allocate all surplus energy into breeding and then die after they breed once. The evolution of reproductive strategies will be evaluated in subsequent chapters, but the concept of semelparity is that all available energy is utilized for one breeding bout. It might be unlikely that these salmon could ever have survived at a very high rate to migrate back downstream, grow gonads, and spawn again, even if all available surplus energy was not used in breeding. The evolutionary trade-off was therefore to utilize more energy in breeding, which may give the fish an extra amount of offspring production. For Pacific salmon, even fish migrating a short distance to spawning grounds show a semelparous life history, and they often put even more energy into gonad growth because of the more limited cost of migration. In following that pattern of energy use, the fish still utilize all available energy during the spawning cycle.

This pattern of energy use has importance to fishery management because when humans build a dam in the middle of a salmon spawning river, time spent during migration may be increased. If the pattern of breeding is to put all surplus energy into migration and gonad growth and just allow a sufficient surplus for courtship and breeding, then anything that delays or alters migration may cause the salmon to die before breeding.

Northern Pike

Diana (1983) evaluated age- and sex-related differences in an energy budget for a natural population of northern pike. One improvement in this study was the measurement of the activity of fish in the field by telemetry. The study attempted to couple activity measurements with the costs of standard and active metabolism. Pike are sit-and-wait predators, which were shown to remain inactive over 90% of the time, so activity costs were very low. The fish were therefore assumed to live at standard metabolism plus the costs of specific dynamic action, which were also measured, as their metabolic rate levels.

Excretion and assimilation efficiency were measured for pike consuming live fishes as prey and used in the equation. Growth was measured for each sex and age class using a cohort analysis and included body and gonad growth. Ration was estimated in the field using measured digestion rates and collected stomach contents on a regular basis. Ration methods will be covered in Chapter 19. The only method to independently test an energy budget is to measure growth, estimate metabolism and other efficiencies, and then calculate the necessary ration to fuel that budget. A comparison between this predicted ration and the field estimate of ration will give the degree of fit in the budget and point out any potential errors. Any component could be in error, but the only method to evaluate that error is by comparison of measured and predicted rations.

Balance in the budget was not good for northern pike from Lac Ste. Anne (Figure 6-7). This imbalance was assumed to be due to difficulties estimating ration in the field. Ration is one of the most difficult variables to measure for several reasons. One is that collection of fishes may result in biases for stomach contents. For example, gillnets may collect actively feeding fishes and, therefore, underestimate ration because sated fishes may not be collected. Also, fishes in gillnets may regurgitate their food, again causing an underestimate of their rations. Even if the ration estimated is reasonable for the day of estimate, it is difficult to extrapolate it from that point in time over the month or year. Undoubtedly, food consumption by fishes varies, depending on weather and other factors that change daily. Generally, ecologists may estimate ration on a monthly basis and assume that ration remains at the same level for the entire month, or they may extrapolate daily ration, using a smoothed curve through estimated points over time. Since natural growth is the accumulation of what is actually eaten over the entire month, it takes into account daily changes in food consumption. However, ration as a point estimate in time may not account well for daily variation in consumption.

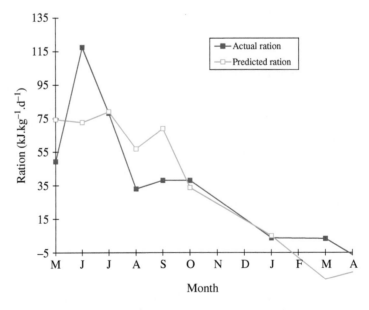

FIGURE 6-7   Comparison between predicted (open squares) and actual (solid squares) rations for three-year-old female pike from Lac Ste. Anne, Alberta. *Source: Adapted from Diana (1983).*

If we assume ration estimates for pike were in error but the remainder of the energy budget was accurate, the growth estimate, coupled with metabolic rate and the heat increment, can be summed to predict the energy budget. That estimate of predicted ration is more likely correct than the measured ration. The seasonal pattern of body and gonad growth was measured in three-year-old pike for several reasons; there were many three-year-old fish collected and most had already matured once and would mature again that year, so their pattern of gonad growth may be more distinct than the pattern in younger fish that had not previously matured. The life cycle of pike includes breeding in early spring, and spawning activity commences even before ice melts on lakes. The fish move to submerged vegetation in the shoreline area and spawn and then return to feed within a few weeks. Some fish may also move to tributary streams and marshes to spawn.

The pattern of body growth for males (Figure 6-8a) included 911 kJ of growth during summer and 489 kJ more during winter. Male pike lost about 598 kJ of stored energy from their bodies over the course of spawning. Pike, like salmon, stop feeding when they breed, undergo migrations, become very active, and use stored energy extensively. All of these activities are costs of reproduction not costs of survival over winter.

Female pike accumulated 1881 kJ of energy during summer – about twice as much as males (Figure 6-8b). But females did not grow somatic tissue at all during winter, and they lost about the same amount of stored energy as males during spawning. The average male accumulated about 67 kJ of energy in his testes, while the average female accumulated about 1672 kJ of energy in her ovaries. Female cost for gonad growth was over 20 times higher than cost for males. Females grew more gonad tissue during summer and winter; during the course of winter, 1371 kJ of energy went into gonad growth, which was more than the amount of

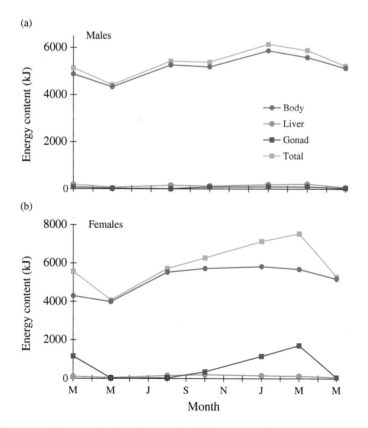

FIGURE 6-8 The seasonal accumulation of energy in the body, liver, and gonads of three-year-old male (a) and female (b) pike from Lac Ste. Anne, Alberta. *Source: Adapted from Diana and Mackay (1979).*

body growth that occurred in males. There was a difference in seasonal energy accumulation between the sexes of pike. Males grew at a slower rate throughout the entire year, while females grew at a much faster rate, allocating surplus energy into body growth in summer and into gonad growth in winter. Once spawning occurred, pike excruded the energy in gonads and resorbed whatever remained unspawned. The energy content in gonadal tissue was lost at the end of each year following spawning, while the body tissue accumulated into the next year.

There are several important points about this analysis. First, if females grow about twice as fast as males during each season, they are either more efficient at converting energy or they eat more food. There is no other way to account for these differences. In this particular case, the measured ration for females was about 1.6 times the level for males. Second, the seasonal pattern of growth is one of continual growth of the body by males, while females undergo body growth in summer and gonad growth in winter. Finally, spawning costs are also high in pike, but loss of energy during spawning is much less than for salmon.

Age-related costs of growth, reproduction, and maintenance – the major components of the energy budget – also differ for male and female pike (Table 6-1). Age-0 fish are immature and put no energy in gonad growth. There is no difference in the energy budget between males and females at this age because of the lack of reproduction. Table 6-1 indicates both absolute and relative amounts of energy into each component. Over the first year of life, 2332 kJ of energy were stored in the body of an immature northern pike and 3244 kJ were used in maintenance costs. Therefore, 58% of the energy budget went to maintenance, 42% to growth, and none to reproduction. The conversion efficiency was about 40%, which is quite high. Once pike reached age 1, maturation began and continued during their second year of life. Not every individual matured by age 2, but Table 6-1 shows estimates for individuals that did mature. Once pike matured, their growth drastically declined, both in absolute numbers and in percentages. Only 8% of the energy budget in males, and 5% in females, went to growth for age-1 fish. Reproduction accounted for only 4% of the energy budget for males and 14% for females. About 80% of the energy was used in maintenance requirements. Maintenance was by far the most costly process for mature fish.

Energy available in excess of maintenance was allocated differently between growth and reproduction depending on sex. Females put higher fractions of their surplus energy budgets into reproduction than males. They also accumulated lower fractions in growth, but they showed growth that was comparable to that of males in absolute amounts. Females ate more food, so the percentages were based on higher total energy budgets. For example, a three-year-old female ate 14,420 kJ of food, while a similar male ate only 10,868.

TABLE 6-1 The Energy Budget for Each Age and Sex of Northern Pike from Lac Ste. Anne, Alberta. Energy Allocations Equal Total Kilojoules and Percent of the Total in Parentheses.

Sex	Age	Growth	Reproduction	Maintenance
Immature	0	2,332 (42%)	0	3,243 (58%)
Male	1	573 (8%)	322 (4%)	6,713 (88%)
	2	995 (10%)	451 (5%)	8,327 (85%)
	3	803 (7%)	393 (4%)	9,994 (89%)
Female	1	426 (5%)	1,166 (14%)	6,947 (81%)
	2	794 (7%)	1,195 (11%)	8,699 (81%)
	3	1,200 (8%)	2,295 (16%)	10,905 (76%)

Source: From Diana (1983).

One other pattern of interest is the change with age in this allocation of energy into each component. Earlier analyses were used to predict that, as a relative fraction of the budget, maintenance would increase with age, growth would decline, and reproduction would increase (Figure 6-4). However, maintenance, growth, and reproduction all remained relatively constant with age for pike. One problem with this data set is that it disagrees with laboratory-based concepts.

There are two possible explanations for this disagreement, the most obvious being that concepts from laboratory studies might be wrong. Laboratory extrapolations were based on fish feeding at some rate based on their stomach capacity or maximum ration, while this study indicates that females feed more than males, and therefore the maximum ration argument may not apply well to both or either of the sexes. There may also be endogenous control of appetite that makes females more effective foragers than males. Pike, in natural systems, may increase their consumption as they grow older to compensate for the higher costs of total body maintenance. The second possibility is that these data only represent young ages in this population. Pike can live to be over 20 years of age, but in a fished lake, one is unlikely to collect many fish over 3 or 4 years old. It is therefore difficult to get field energetics data on very old fish because of their rarity. Possibly, older fish would show the expected relative declines in growth and increases in reproductive and maintenance costs.

Growth and Latitude

So far, the natural patterns of growth have been evaluated for specific locations. However, the growth rate of fishes, as an energetic result of consumption and metabolism, must vary with factors that influence energetic processes. In general, the average temperature in a location declines as latitude increases, and patterns of fish distributions and abundance change dramatically with latitude. Therefore, latitude should be one major factor affecting natural growth patterns in fishes. Changes in latitude result in orderly changes among thermal guilds of fishes as one moves to higher latitudes. In North America, temperate eurytherms – or warmwater fish – are replaced by coolwater and coldwater fishes as one moves northward. These species are better adapted at feeding, metabolizing, and growing under the increasingly colder conditions found at higher latitudes. Overall distributions of fishes often demonstrate such orderly replacements of species with latitude.

Just as changes can be found among species with different thermal capabilities, one would also expect that populations within a species would show differences in growth rate with latitude. Many other processes also affect growth, predominantly working through food abundance, so other factors may overwhelm latitude alone as the main factor influencing growth of a fish population. Still, one would expect latitude to have an important effect. In that regard, studies have focused on differences in fish growth caused by latitude.

Atlantic Silverside

Detailed work on latitudinal changes in growth has been done by David Conover and his colleagues at State University of New York at Stony Brook. Atlantic silversides are short-lived marine species that are common from Florida to Canada. Few fish of any population live beyond the end of their second year. Conover (1990) evaluated the growth of silversides from a variety of populations and found that body size at the end of the first growing season was similar among populations, even though the length of the growing season changed 2.5-fold. He extrapolated this to demonstrate that fish from higher latitudes must have more rapid growth rates during warm conditions to compensate for the shorter summer over which they grow. Laboratory experiments revealed a latitudinal difference in maximum growth potential (growth at a given ambient temperature at ad libitum ration) among populations, so fish from northern populations had a higher growth potential at any temperature than fish from southerly areas. Subsequent studies showed these differences to be genetically inherited in a population. Conover described these changes as countergradient variation in growth rate.

In order for such a process to occur, fishes from higher latitudes must have either more efficient growth at any temperature or must eat more food to allow for the higher growth. Billerbeck et al. (2000) evaluated the energetic performance of northern and southern stocks of silverside and found that food consumption and growth of individuals from a northern population were much higher at both cool (17 °C) and warm (28 °C) temperatures than a southern population. Most metabolic parameters did not differ much for the two populations, but maximum food consumption was much higher for northern fish. Routine swimming also differed among the stocks. However, it appears from these studies that the main factor influencing differences in growth was food consumption rate.

Countergradient variation in growth rate causes compensation for growth rates among fish populations located in different geographic areas. In a way, it is very similar to the acclimatization shown in Chapter 5, where more southerly populations had lower rates of metabolism at any temperature to reduce their rates of energy expenditure. In fact, if no part of the energy budget changed with temperature, acclimatization would result in some countergradient variation in growth rate. In silversides, metabolic acclimatization did not occur as populations had a similar metabolic rate at each temperature. However, countergradient variation in growth rate has also been shown in a number of other marine species, such as striped bass, American shad, and mummichog. Whether the details for silverside metabolism and food consumption apply to all of these species is not certain, but the resultant change in growth to compensate for latitude is clear. It does not appear to occur in all fish species as species such as yellow perch and muskellunge do not show differences in the growth rate among stocks held at different temperatures; while others, such as largemouth bass, Atlantic salmon, spotted gar, and rainbow trout, do show such differences.

Growth and Reproductive Tradeoffs

Studies of natural populations showed that growth differences exist between sexes due to differential costs of reproduction. The costs of reproduction have importance in the evolution of life-history traits as well. Animals may adapt to environmental conditions and competition with other animals by altering the timing and extent of reproduction. When fishes start to breed, energy that could have gone into body growth will be used in growth of gonads and in other reproductive costs and result in decreased body growth. Does the onset of reproduction really result in decreased growth rates in fishes? A parallel question to that would be: Is the energy budget fixed or flexible? In other words, is this natural ration really fixed so growth has to be sacrificed for reproduction to occur, or can fishes increase consumption when they have higher needs for energy? This possibility does not necessarily mean fishes would have to increase their maximum rations to adapt to energy demands in nature because they seldom live at maximum rations in the field.

Reproductive Strategies

The two extremes in reproductive strategies related to the pattern of energy storage and use are semelparity and iteroparity (survival to breed many times). The difference between these two strategies is related to the amount of energy used in reproduction as well as the likelihood of survival after such an energy drain. The general pattern for iteroparous fishes is depletion of 25–60% of stored energy in the body during reproduction. For semelparous fishes, the rate of energy depletion is even higher with 60–85% of somatic energy content utilized during reproduction.

Several families of fishes have individual species with reproductive strategies that include both extremes. One example already covered – the sockeye salmon – utilizes more than 80% of its somatic energy in reproduction and is semelparous. A close relative – the steelhead – uses much less of its stored energy (40–50%), and some of these fish survive to spawn again. The Atlantic salmon has an intermediate strategy with high loss of stored energy (up to 60%) and limited survival. Finally, iteroparous species, such as brown trout and brook trout, would use even less stored energy (20–30%) for reproduction and would be capable of surviving to spawn many times.

Another interesting example is the American shad. Individual populations of this species may be semelparous or iteroparous depending on the latitude of their spawning rivers (Leggett and Carscadden 1978). Northern populations of American shad are iteroparous and survive to breed again; southern populations are semelparous. Northern populations have somatic energy depletion at about 50% for spawning; southern populations have approximately 80% (Glebe and Leggett 1981). The proposed reason for this difference is that southern populations grow larger gonads and, in particular, put more energy into gonad growth during their spawning migrations. The reasons for this strategy are complex and will be dealt with in a later chapter. The main factor believed to be involved in the evolution of semelparity or iteroparity is the predictability of offspring survival.

Reproductive Costs and Age at Maturity

What factors influence the ages at which fishes mature? There are reasons why larger fishes might be more successful in breeding; these include better ability to carry out courtship and reproductive behavior, which might require significant energy reserves. Using an evolutionary fitness argument, increased size is related to the number of eggs produced per female (fecundity) since there is an exponential relationship between fecundity and mass for most fishes. Future growth after fishes first mature is most likely impaired because of the large costs of reproduction that reduce energy available for growth. Thus, the age at first maturation is a trade-off between the increased fecundity for older fishes and the reduced chance of survival to that age. The trade-off for the exact age at first breeding is between the chance of survival for another year (at least for temperate fishes that breed annually) versus the number of eggs that could be produced that year.

Heibo et al. (2005) evaluated 75 populations of Eurasian perch across a latitudinal gradient for growth rate, mortality, and age at maturation. They found that perch separated into two overriding groups based on maximum length (L_∞) achieved in the population. The small L_∞ group was considered stunted, while the large L_∞ group was considered piscivorous. Details on stunting will be developed in a subsequent section of this chapter. Maximum length was not correlated to latitude, while there was a negative correlation between all other growth parameters and latitude. Juvenile and adult mortality rates also declined with latitude (Figure 6-9), while age at maturity increased with latitude and reproductive investment (gonadosomatic index, or gonad weight divided by somatic weight) declined. Perch showed a clear latitudinal trend in their life history with slower growth and lower mortality resulting in a later age at maturation for more northern populations. The relationship between growth, survival, and reproductive investment strongly affected age at maturation at different latitudes.

There are other examples of the relationship between fishing and age at maturation for commercially fished populations of whitefish and salmon. Whitefish show the pattern of

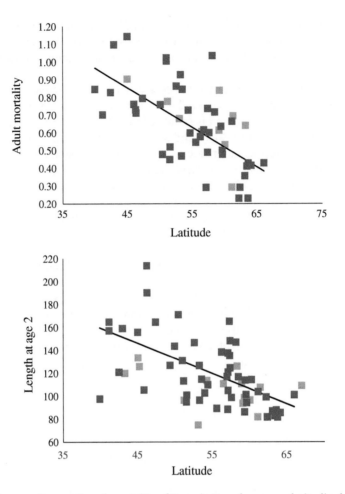

FIGURE 6-9 Differences in growth and mortality of Eurasian perch across a latitudinal gradient. Purple symbols are for piscivorous populations and blue symbols for stunted populations. *Source: Adapted from Heibo et al. (2005).*

density-dependent changes, that is, as individuals are removed from the population, the remaining individuals have more food and other resources available, so they grow better than they did before the removal. This increased growth results in an earlier age at maturation because the whitefish reach the size to mature at an earlier age (Healey 1975). Similar patterns of change have been shown for commercially exploited salmon (Ricker 1981), except salmon show earlier maturation but declining body growth at age.

One of the more interesting and controversial evaluations of the effect of fishing mortality on growth and maturation was done by Conover and Munch (2002). That study evaluated evolutionary effects of exploitation by subjecting four generations of silversides to exploitation regimes in a laboratory setting. For each treatment, they removed the largest 90% of the population, the smallest 90% of the population, or 90% of fish randomly from the population to simulate different mortality regimes. The large fish removal group showed considerable variation from the random group with a measured genetic change in growth rate to a lower size at maturation (but the same age). The rapid evolution shown in this study indicates that a very strong exploitation schedule resulted in heritable changes in growth and maturation of silversides. Hilborn (2006) questioned the validity of this study and its interpretation. Among other criticisms, he did not believe the exploitation regime (removing the upper 90% of the population) reflected any fishery, so was not related to fishery caused changes. While the exploitation schedule may have been severe, the results were similar to those shown for salmon. Clearly, high levels of size selective fishing can influence growth and maturation of fishes, and these differences become heritable traits over time.

Alternate Reproductive Patterns

Mart Gross from the University of Toronto, and David Phillip, from the Illinois Natural History Survey, have worked extensively on genetics and behavior of bluegill. The breeding pattern of bluegill involves selection and guarding of a nest – or depression in the substrate – by large male bluegill. These fish are generally of an advanced age, often 7–8 years, before first maturation. Parental males attract females to the nest to spawn. The male remains with the nest and protects the eggs and young for some time.

In addition to nest guarding (parental) males and females, there are also alternative life-history patterns in males. Male bluegill do not all guard nests because some become cuckolders. A cuckolder is a nest parasite – a male that remains near the nest when a male and female are spawning there and attempts to steal fertilization by stealth or trickery. For male bluegill, there are two cuckolder patterns. Very small bluegill may become sneakers, hiding near a nest and then sneaking into the nest quickly during a breeding bout and depositing sperm on eggs in the nest. Satellite males, which are somewhat larger than sneakers, mimic a female in coloration and behavior, and then come into a nest during a normal spawning bout and fertilize eggs while behaving like a female fish.

The pattern of life for cuckolders and normal males is exemplified in Figure 6-10. Cuckolders become reproductively mature at a very early age and forego future body growth for current reproductive growth. Gross and Charnov (1980) showed that the pattern of early reproduction clearly reduces the body growth rate for cuckolders (Figure 6-11). However, cuckolders cannot be involved in reproductive bouts unless there are nest-guarding males, so in the absence of

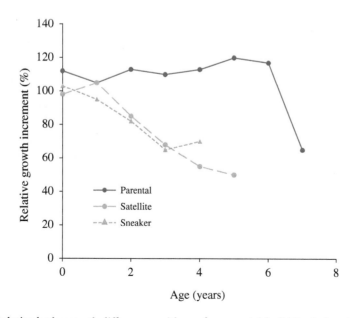

FIGURE 6-10   Relative body growth differences with age for parental (solid line), female mimic (dashed line), and sneaker male (dotted line) bluegill from Lake Opinicon. *Source: Adapted from Gross and Charnov (1980).*

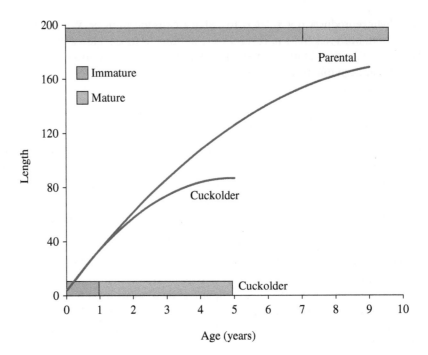

FIGURE 6-11 Schematic representation of body growth in parental and cuckolder male bluegill. Curves represent change in size with age for the two types; bars represent the time periods each are mature (open) or immature (closed).

large parental males, smaller males will build and guard nests (Jennings et al. 1997). However, Ehlinger et al. (1997) found that parentals and cuckolders have body forms that are distinguishable from each other in patterns other than body size, and the fish taking nests at earlier ages are young parental males rather than cuckolders. The frequency of cuckolders appears to be rather consistent in a population and does not generally dominate the population.

It is not entirely clear what causes a male fish to become a cuckolder. In some cases, cuckolders exhibit slower early growth than parentals but not in all cases. Once they have begun cuckolding, however, their growth rate declines dramatically and they generally cannot become parentals. The choice of early reproduction brings with it declining body growth and higher mortality, so cuckolders generally do not live long enough to become parentals. It does appear that sneakers that grow larger become satellite males but not parental males. There does appear to be some inheritance of life-history type, and parental males or cuckolders tend to produce more of their own type than of the alternate. However, this inheritance pattern is not fixed, and environment also seems to play a role in adopting a cuckolding life pattern.

Stunting

The process of stunting provides a different example of the trade-off between body and gonad growth that occurs in natural populations. Stunting occurs in some species when there is a drastic decline in growth rate. Potential reasons for this decline vary but may include overpopulation and density-dependent declines in food availability, overfishing, inappropriate temperature resources, and lack of certain sizes of prey. During stunting, the population may reach a "terminal size," in which all fish beyond a certain age are approximately of the same size. At advanced ages, there is no surplus energy available for growth. Growth ceases, but the fish remain alive. Their reproductive schedules may also vary as the stunting process occurs. Stunting is a common population response in many freshwater fishes, particularly pike, perch, bluegill, and tilapia.

Stunting is common in bluegill populations and could result from several causes, including overfishing, high density and poor food resources, high frequencies of cuckolding fish in the population, and even combined effects among these factors. Several authors have evaluated the importance of these different mechanisms. Overfishing does remove the large parental males and results in smaller males guarding nests (Drake et al. 1997). In addition to nesting fish being smaller, they also mature earlier in life and grow more slowly. So fishing not only affects the presence of large fish but also the maturation schedule of bluegill populations. Overpopulation also results in poorer growth of both parental and cuckolding males, as well as females, and generally results in smaller fish being parental types at an earlier age (Ehlinger et al. 1997). The relative frequency of cuckolders dramatically varies among lakes and appears to increase with fishing pressure and vegetation abundance and to decline with colder temperature and therefore a slower growth rate (Drake et al. 1997). However, the percentage of cuckolders does not appear to exceed 50%.

This pattern of stunting is also present in a number of fishes that do not show the reproductive complexity of bluegill. For example, both perch and pike show stunting, yet neither of these species are nest builders and guarders. Instead, they form pairs or aggregations to spawn and lay their eggs in vegetation where the eggs develop and hatch without parental influence. Cuckolding is not likely in these species as large parental males do not guard nests. There is variation in the age at first maturation, and there may be differences in ability to attract mates for small and large fish, yet there does not appear to be the variety of male types as in bluegill. However, both of these species show stunting, and stunting reduces growth rate and decreases age at first maturation. Mechanisms for stunting in these species include overfishing, temperature patterns, and overpopulation, just as in bluegill. In addition, another mechanism for stunting in perch and pike appears to be the abundance of appropriate size prey (Heibo et al. 2005). Both species make the transition of consuming different prey sizes as they grow, and the consumption of larger prey allows for more rapid growth. The absence of appropriate size prey appears to result in poorer growth and possibly stunting, depending on the size and abundance of the prey population but somewhat independent of fish density.

The changes that occur with stunting happen to fish already existing in a population, so they are physiological or behavioral, not genetic. Populations that develop stunting may have increased frequencies of fish spawning at an earlier age and smaller size. This change is not evolved, but rather results from changes in existing frequencies of phenotypes. In fact, the change appears to be physiological, that is, individual fish under poor growth conditions may mature earlier and adopt a stunted life history rather than mature later as a normal adult. This "choice" is most likely caused by patterns of the growth rate or feeding rate experienced by the fish during development. In cases where stunted fish are moved to laboratory or natural waters with normal densities and food availabilities, they again adopt a normal growth pattern. Whether they would change to later maturation is not known.

Foregone Reproduction

One final pattern to mention related to body and gonad growth is exemplified by lake sturgeon as well as by many Arctic fishes. Lake sturgeon are very large fish that eat benthic invertebrates, live for 80–90 years and grow to be 2 or more m long. Lake sturgeon apparently "measure" their current growth rates and "decide" whether or not to breed the next year. Individual sturgeon may only breed once every four or five years, and the pattern is believed to be related to storing sufficient surplus energy to breed again. In some ways, this is similar to the pattern observed with perch as sturgeon seem to evaluate the quantity of surplus energy they have accumulated before they grow gonads again. Arctic fishes also show a similar pattern in environments with cold temperatures and limited food production, where they alternate years between breeding as well (Morin et al. 1982).

For many temperate fishes, the annual maturation pattern occurs over a long time (at least several months), so the decision to breed must be made a while before breeding season. Pike, for example, grow gonad material over winter. In order to breed the next spring, they must begin gonad growth in August (Figure 6-8). Most temperate fishes probably have no means by which to predict suitability of the environment for spawning the following year. This makes it difficult for fishes to adapt the intensity of breeding to local conditions for survival of young – a very common pattern in birds and mammals. The only sensing that seems to be involved in breeding decisions is the sensing of their own surplus energy or physiological well-being.

Summary

Growth of fishes is an important component of their survival and future reproductive success. Growth dramatically varies under different conditions of fish size, temperature, and food availability. Such growth differences among natural populations may be useful in determining adaptations of fishes to their local environments. Understanding the factors influencing growth under experimental conditions is the first step in evaluating such adaptations. Field studies of fish growth have demonstrated that maintenance costs are very high in fishes, utilizing 80–90% of consumed energy in mature individuals. Fishes show seasonal patterns of energy accumulation and use that are related to food availability and temperature. They also have long seasonal patterns of gonad growth. Body and gonad growth of fishes is strongly related to reproductive strategy, evolutionary history, interactions between individuals, and food intake. Reproductive growth is not even consistent within a species but dramatically varies depending on local conditions.

Literature Cited

Billerbeck, J.M., E.T. Shultz, and D.O. Conover. 2000. Adaptive variation in energy acquisition and allocation among latitudinal populations of the Atlantic silverside. Oecologia 122:210–219.

Brett, J.R., and T.D.D. Groves. 1979. Physiological energetics. Pages 279–352 in W.S. Hoar, D.J. Randall, and J.R. Brett, editors. Fish Physiology, Vol. 8. Academic Press, New York.

Brett, J.R., J.E. Shelbourn, and C.T. Shoop. 1969. Growth rate and body composition of fingerling sockeye salmon, Oncorhynchus nerka, in relation to temperature and ration size. Journal of the Fisheries Research Board of Canada 26:2363–2394.

Conover, D.O. 1990. The relation between capacity for growth and length of growing season: Evidence for and implications of countergradient variation. Transactions of the American Fisheries Society 19:416–430.

Conover, D.O., and S.B. Munch. 2002. Sustaining fisheries yields over evolutionary time scales. Science 297:94–96.

Diana, J.S. 1983. An energy budget for northern pike (Esox lucius). Canadian Journal of Zoology 61:1968–1975.

Diana, J.S., and W.C. Mackay. 1979. Timing and magnitude of energy deposition and loss in the body, liver, and gonads of northern pike (Esox lucius). Journal of the Fisheries Research Board of Canada 36:481–487.

Drake, M.T., J.E. Claussen, D.P. Philipp, and D.L. Pereira. 1997. A comparison of bluegill reproductive strategies and growth among lakes with different fishing intensities. North American Journal of Fisheries Management 17:496–507.

Ehlinger, T.J., M.R. Gross, and D.P. Philipp. 1997. Morphological and growth rate differences between bluegill males of alternate reproductive life histories. North American Journal of Fisheries Management 17:533–542.

Glebe, B.D., and W.C. Leggett. 1981. Latitudinal differences in energy allocation and use during the freshwater migrations of American shad (*Alosa sapidissima*) and their life history consequences. Canadian Journal of Fisheries and Aquatic Sciences 38:806–820.

Gross, M.R., and E.L. Charnov. 1980. Alternative male life histories in bluegill sunfish. Proceedings of the National Academy of Sciences 77:6937–6940.

Hayward, R.S., D.B. Noltie, and N. Wang. 1997. Use of compensatory growth to double hybrid sunfish growth rates. Transactions of the American Fisheries Society 126:316–322.

Healey, M.C. 1975. Dynamics of exploited whitefish populations and their management with special reference to the Northwest Territories. Journal of the Fisheries Research Board of Canada 32:427–448.

Heibo, E., C. Magnhagen, and L.A. Vollestad. 2005. Latitudinal variation in life-history traits in Eurasian perch. Ecology 86:3377–3386.

Hilborn, R. 2006. Faith-based fisheries. Fisheries 31:554–555.

Idler, D.R., and W.A. Clemens. 1959. The Energy Expenditures of Fraser River Sockeye Salmon During the Spawning Migration to Chilko and Stuart Lakes. Report to the International Pacific Salmon Fisheries Commission, New Westminster, British Columbia.

Jennings, M.J., J.E. Claussen, and D.P. Philipp. 1997. Effect of population size structure on reproductive investment of male bluegill. North American Journal of Fisheries Management 17:516–524.

Jobling, M. 1994. Fish Energetics. Chapman and Hall, London.

Jobling, M., E.H. Jorgensen, and S.I. Siikavuopio. 1993. The influence of previous feeding regime on the compensatory growth response of maturing and immature Arctic charr, *Salvelinus alpinus*. Journal of Fish Biology 43:409–419.

Leggett, W.C., and J.E. Carscadden. 1978. Latitudinal variation in reproductive characteristics of American shad (*Alosa sapidissima*): Evidence for population specific life history strategies in fish. Journal of the Fisheries Research Board of Canada 35:1469–1478.

Morin, R., J.J. Dodson, and G. Power. 1982. Life history variations of anadromous cisco (*Coregonus artedii*), lake whitefish (*C. clupeaformis*), and round whitefish (*Prosopium cylindracium*) populations of eastern James-Hudson Bay. Canadian Journal of Fisheries and Aquatic Sciences 39:958–967.

Ricker, W.E. 1981. Changes in the average size and average age of Pacific salmon. Canadian Journal of Fisheries and Aquatic Sciences 38:1636–1656.

Shul'man, G.E. 1974. Life Cycles of Fish. John Wiley and Sons, New York.

CHAPTER 7

Estimating Growth and Condition of Fish

Growth of fish dramatically varies among individuals, populations, and species. Growth patterns reflect not only differential short-term responses to a suite of biotic and abiotic conditions but also long-term intra- and interspecific evolution. The preceding chapters described the energetics of fish related to growth, as well as patterns of growth among fish. In addition to appreciating the mechanisms leading to differences in growth among individuals and populations, in many cases it is informative to quantitatively summarize and describe growth of individual fish or groups of fish. Quantitative descriptions of growth are necessary to compare growth rates across different natural environments or to evaluate growth responses to various laboratory stimuli.

Growth of a fish is simply change in size over time, and growth can be either positive or negative. When size of an individual can be measured at two known time periods, growth can be readily computed as the change in size over this known time span. Such quantification may be possible when individual fish are tracked in the laboratory or when fish can be individually identified (through a unique tag) and recaptured in a natural environment. In many cases, it is not feasible to repeatedly measure size of a known individual, and mean growth rates of a group of individuals may be estimated by examining size and estimated age of multiple individuals.

Measuring Size

Similar to humans and other animals, the size of a fish is generally quantified in terms of whole body length and mass. One-dimensional size of an individual fish can be indexed along other axes besides length, such as body width, depth, or even girth, but length is by far the most common measure. While length may seem like a simple set measure of fish size, various measures of length have been developed (Figure 7-1). The total length is simply the maximum straight line length that can be quantified for a fish, from the tip of its snout to the very end of its caudal fin. The two lobes of the caudal fin are often slightly squeezed together to determine this maximum linear length. Fork length is the linear length from the tip of the snout to the fork of caudal fin lobes. This method overcomes any potential bias related to measuring where the tip of the caudal fin ends. In addition, this method overcomes fraying of the edges

Biology and Ecology of Fishes, Third Edition. James S. Diana and Tomas O. Höök.
© 2023 John Wiley & Sons Ltd. Published 2023 by John Wiley & Sons Ltd.

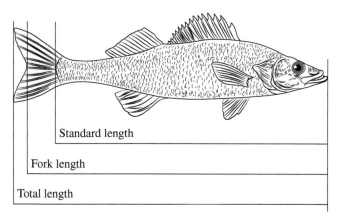

FIGURE 7-1 Walleye (*Sander vitreus*) demonstrating the three most common measures of fish length: standard length, fork length, and total length. *Source: Joseph R. Tomellerri.*

of caudal fin, which may be particularly common in some hatchery environments and can prevent simple measure of total length. Standard length is measured as the length from the tip of the snout to the base of the caudal fin. This measurement is often used for long-preserved fish, such as in a museum collection, as such preserved specimens often experience considerable damage to their caudal fins.

Mass of a fish can be measured as wet or dry mass depending on whether one wants to account for the weight of water contained in a fish. Wet mass has the advantage that it can be measured right after a fish is collected with minimal processing and allows a fish to remain alive. This can allow repeated measures of wet mass of the same individual over time. In contrast, dry mass involves quantifying mass of a fish after the body has been dried for some time. This may provide a more direct measure of the health of an individual, as water content may fluctuate, and water does not directly contribute to the structural size or energy content of a fish. However, measuring dry mass obviously involves sacrificing the fish and does not allow for repeated measurement of the same individual.

Estimating Age

With the exception of stocked or previously tagged individuals, the ages of fish captured from natural environments are generally unknown. For some populations with distinct hatching dates and rapid growth rates, there is very limited size overlap among ages, and age can be estimated as a function of size. Determining age based on length may be appropriate for short-lived species or populations dominated by young age classes. As fish grow older and their growth declines, each age class becomes less distinct in size so it becomes difficult or impossible to separate age classes on the basis of size alone for older fish.

When it is not possible to determine age based on size, it is common to use various hard parts like bones or scales to estimate age in years. Growth of such chronometric structures is seasonal, with contrasting seasons of rapid positive growth and very slow (or negative) growth, leading to distinct annual marks on these chronometric structures (such annual marks are also referred to as annuli). Aging fish using chronometric structures can be thought of as analogous to aging trees using rings visible through trunk cross-sections. Aging of fish was developed for fish from temperate environments. However, any fish that shows some seasonal pattern of growth or reproduction can be aged. Tropical and marine fish, particularly species that spawn once a year, can be aged as well as temperate freshwater fish. Slow growth and depletion of energy reserves due to reproduction may also leave a mark on hard parts, and this may be as important a seasonal event as slow winter growth. In fact, the presence of annual marks in many temperate freshwater fish may be largely due to the presence of limited winter growth and early spring spawning depletion as both provide distinct marks in the same location on the structure. In contrast, tropical fish that breed continually and occupy habitats without seasonal change (either temperature or wet-dry seasons) may be difficult to age by analysis of bony structures.

A variety of structures have been used to age fish, including cross-sections of vertebra, spines, fin rays, cleithra, and opercula (Figure 7-2). The basic approach involves identifying annuli on the chronometric structure and then counting annuli along a transect, from the core of the chronometric structure to the outer edge of the chronometric structure. An annulus is generally expressed as a dark band or ring on the chronometric structure, while the spaces between annuli tend to be lighter and correspond to periods of more rapid growth. Annuli near the edge of the chronometric structure can be particularly difficult to observe. It may be important to consider when a fish was captured and when a chronometric structure was collected in order to determine age. For example, if a fish is captured in early spring the final annulus may

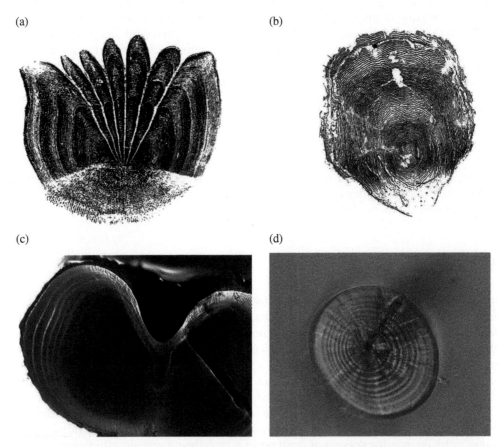

(a)

(b)

(c)

(d)

FIGURE 7-2 Example chronometric structures. (a) Pressed image of a yellow perch scale from Lake St. Clair, estimated to be five years old (image credit: State of Michigan). (b) Pressed image of a steelhead trout scale collected in the Manistee River, a tributary to Lake Michigan, estimated to have spent two years in a stream and four years in Lake Michigan (image credit: David Swank). (c) Cross-sectional image of an Oneida Lake walleye dorsal spine, estimated age of nine years (image credit: Hui-Yu Wang). (d) Image of an otolith from a larval alewife collected in Lake Michigan, showing daily growth increments, estimated age 11 days (image credit: Glenn Carter).

be expected to fall on the outer edge of the chronometric structure and be difficult to identify. However, if a fish is collected in the fall, it is more likely that substantial growth has taken place since the last annulus formation, and hence, the last annulus would be expected to be more visible.

While various structures have been used to age fish, historically the most commonly used chronometric structures are scales. Age estimates using other chronometric structures, such as otoliths, have been shown to be more accurate than scale-determined ages, and scales may be particularly biased when attempting to age older and slower growing fish. Older fish may have few scales that have not been regenerated, or they may have such crowded annuli that they are difficult to age. Nonetheless, scales have some distinct advantages as aging structures. First, unlike many bones and otoliths, removal of scales does not involve killing the individual fish. Second, multiple scales can be removed from the same individual to compare resulting age estimates.

Any aging is somewhat of an art, particularly with scales. The best method to ensure accuracy of aging is to validate the aging process by collecting scales from fish of known ages or scales from the same individual fish collected across different years. While validating chronometric structure aging of known aged fish has been accomplished for many species, such validation is not common across all studies. Other ways to improve accuracy are to compare aging techniques using other experienced scientists (corroboration) or to compare multiple hard parts for their similar age estimates (cross-validation). These latter methods may confirm that standard techniques are used and might produce comparability of aging by different structures, but they do not validate that annuli distinguished on these structures are actually yearly in nature. Since aging is somewhat of an art, and interpretation may change with further study, it is very important to preserve scales for later use. These scales may also become useful as the basis for historic comparisons in future studies.

Given the shortcomings of scales for age determination, various monitoring and research programs now make use of other chronometric structures. In particular, otoliths are commonly used. These calcium carbonate structures are found in the heads of fish and play a role in environmental sensing and balance. Otoliths grow like crystals as fish grow. Unlike scales, otoliths are biologically inert, and once material is deposited onto an otolith it is not sequestered back into the body. Hence, the potential shortcomings of scale regeneration or fraying of scales are not concerns for otoliths.

Another useful feature is the presence of daily growth rings on the otoliths of many species. Daily growth rings are most distinguishable in the portion of otoliths corresponding to growth during early life. These features reflect a daily alternating cycle of deposition of slightly different materials onto otoliths during daily growth and result in alternating patterns of light and dark rings. Daily growth rings have proven particularly valuable for the study of early life stages of fish as they allow for estimation of specific hatch dates and daily growth rates. This is a very valuable technique in estimating the success of young fish from various hatching cohorts and habitats.

Back-Calculating Size at Age

Back calculation uses the growth pattern recorded in scales or bones to estimate growth. It is based on the assumption that the distance between annuli on a hard part is related to the growth of a fish over that time so size can be calculated at each annual mark from those measures. Back calculation requires a large sample of fish that can be aged. Along with aging, the distance from focus of the scale or bone to each annulus, and to the outside edge of the hard part, must be measured. Back calculation has long been used to estimate growth history, proceeding from a nomograph (direct proportionality) to regression techniques.

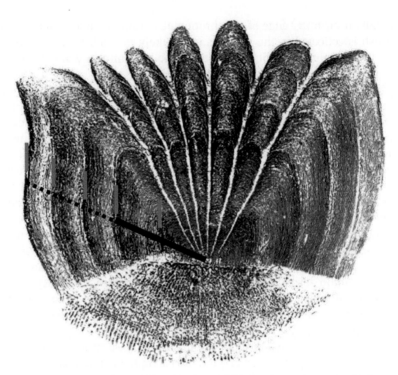

FIGURE 7-3 Pressed image of a yellow perch scale from Lake St. Clair, estimated to be five years old (image credit: State of Michigan). The red lines on the scale image mark the annuli. The solid and dashed lines depict the scale radius at the age-2 annuli and at the time of capture, respectively. The distances of these scale radii along with the length of the fish at the time of capture can be used to estimate (back-calculate) the size of the individual fish at age-2 as described in the text.

The basis of back-calculating size at age is to determine the size of a scale (or other chronometric structure) at the current time (the time when the fish was captured) compared to some earlier time. Generally, this is accomplished by measuring the current radius of a scale and comparing this to the distance from the scale center to an earlier annual mark. For example, Figure 7-3 shows a scale from a five-year-old yellow perch. One can compare the distance from the scale origin to the edge of the scale (the combined length of the solid and dashed lines) versus the distance from the origin to the second annual mark (solid line). It is important that these scale measurements are made along the same axis so they are comparable. These scale measurements can then be combined with current total body size measurements (e.g., total length) to estimate earlier size (e.g., total length at age-2 for the individual yellow perch).

There have been a number of methods developed to back-calculate size at age. Francis (1990) provided a nice critical review of different methods for back-calculating size at age. These methods have included (a) developing a regression of fish length versus total scale radius for a large number of fish and assuming all individuals essentially fall along this regression or (b) assuming direct proportionality between fish scale radius and fish length. A method known as the Fraser–Lee approach has been widely applied. While this method continues to be used in published studies, Francis demonstrates that the Fraser–Lee approach is a misuse of regression by using an estimated intercept independent of a slope. Instead, Francis argues for using body-proportional or scale-proportional methods.

One of the powers of back calculation is that it allows one to collect fish in the present and determine changes in growth that occurred in the past. For example, if a special fishing regulation were placed on a lake in the past, current collections of fish that were alive at the time of that change could be used to assess the success of that regulation in improving growth. The growth prior to the regulation for an age group, which can be back-calculated, should be less than current growth of that same age group if growth improved after the regulation.

One bias involved in back calculation is presence of size-selective mortality, but back calculation actually provides a means to estimate its importance. There is a tendency for back-calculated lengths at an early age, determined from older fish at capture, to be smaller than when they are determined from younger fish. This tendency was first noticed by Rosa Lee, the first woman to be employed by the UK's Marine Biological Association (Lee 1920). This phenomenon has several potential bases, including incorrect method for back calculation, nonrandom sampling (e.g., tendency when sampling young individuals to collect relatively large individuals), or size-selective mortality. For size-selective mortality, larger fish of an age class die more rapidly, so older fish include mainly individuals that grew more slowly when they were young. This may be the case in a size-selective fishery that takes faster-growing fish first. The opposite of Lee's phenomenon has also been described (back-calculated lengths at age-1 become larger for older fish), which may indicate better survival rates for faster-growing fish. If one is careful in back-calculating and sampling techniques, the presence of these phenomena can indicate size-selective mortality, which may influence interpretation of growth curves determined by methods described earlier and may also clarify important aspects of the survival of young fish.

Growth in Length

The most common means to evaluate growth is to examine the length of individuals over time. As described in Chapter 6, growth of most fish is indeterminate, meaning they continue growing throughout their life. However, growth in length of fish is generally not constant throughout life. Instead, fish often increase in length relatively rapidly during early life and then growth in length slows later in life as fish reach maturity and divert energy to gonadal growth and metabolic scaling with body size limits growth potential (Figure 7-4). In quantitatively describing growth of fish, it is common to aim to describe lifetime growth, including the slowing down of growth at older ages (for example, see the von Bertalanffy growth in length model described below). However, for certain periods of a fish's life it may be of interest to quantitatively describe growth over a shorter time period.

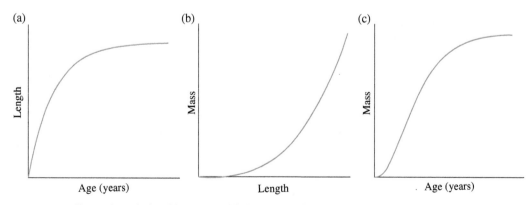

FIGURE 7-4   Illustrative relationships among lifetime age and length and mass for individual fish surviving from early life to late adult age.

Lifetime Growth in Length

For fish that survive early life and continue to grow as relatively old adults, growth dramatically slows down. As fish mature and divert some of their energy toward gonadal growth and reproductive activity, the energy available for somatic growth decreases. In addition, the mass-specific maximum consumption rate of fish tends to decrease with size, limiting maximum growth potential. As a consequence, whole lifetime growth (from larva to old age) can generally not be described by a single linear relationship. Some authors, such as Quince et al. (2008), advocate for using two growth stanzas (or phases) to describe lifetime growth in length. Under this biphasic model, fish increase in length fairly rapidly during early life (high slope relating dependent variable, age in years, to independent variable, length). Then, with maturation fish switch to much slower growth. A potential advantage of the biphasic growth model is that it can also be used to estimate age and size as maturation by statistically estimating the age and size where fish switch from rapid growth to slower growth in length.

In addition to the biphasic model, several other models have been developed to describe lifetime growth in length of fish. However, by far the most common model used to describe lifetime growth in length of fish was developed by von Bertalanffy (1938). While this model (Equation 7-1) was initially developed to describe lifetime growth of various organisms, including humans, it has in particular been widely used to describe fish growth:

$$L_t = L_\infty \left(1 - e^{-K(t-t_0)}\right)$$
(Equation 7-1)

The von Bertalanffy model relates age (t, independent variable) and length at age (L_t, dependent variable) using three parameters (see Figure 7-5). (1) Asymptotic length is often referred to as maximum length or length-infinity (L_∞). This is the model length fish grow toward but never reach if growth strictly follows the model. An asymptotic size term may not seem consistent with the idea that fish growth is indeterminate. In fact, the length of individual fish in a population may at times exceed the L_∞ value estimated for that population. It is important to recognize that this is a model parameter, statistically estimated to best fit observed data. (2) The von Bertalanffy growth coefficient or Brody's growth coefficient, K, describes the rate at which model fish grow toward L_∞ (with typical units of yr^{-1}). High values of K mean that model fish will relatively quickly grow toward L_∞, while low values mean growth toward L_∞ will be relatively slow. Some authors have used estimated K values to compare growth rates among populations. However, one should be careful in doing so, as K and L_∞ interactively determine growth rate in length per time. If L_∞ is low, then a relatively high K value will not necessarily equate to rapid growth in length (e.g., low mm of growth per year); vice versa, if L_∞ is high, then even a relatively low K value may reflect a rapid growth in length. (3) Finally, t_0 reflects the model age corresponding to a length of zero. Again, this parameter is statistically estimated and may take either positive or negative values (Figure 7-5).

Estimating parameters of the von Bertalanffy model is most appropriately accomplished using a nonlinear fitting approach. Under this approach, a computer program iteratively searches for combinations of the three parameter values which lead model predicted lengths at age, L_t, to most closely match observed lengths at age. An advantage of this approach is that it allows for all three parameter values to be estimated. Using a linear regression approach, it is possible to estimate two of the three parameter values, L_∞ and K, while setting t_0 equal to zero. Under this approach – also known as the Ford–Walford approach – mean length at time t, L_t, is plotted against mean length at time $t + 1$, L_{t+1}. As fish become larger, the difference between L_t and L_{t+1} should become smaller and smaller. By model definition, if length reaches L_∞, L_t will equal L_{t+1}, as there would be no additional growth in length beyond L_∞. A fitted linear regression line relating L_t (x-axis) and L_{t+1} (y-axis) should have an intercept and positive slope, with a

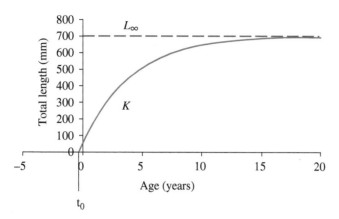

FIGURE 7-5 A hypothetical von Bertalanffy growth in length curve. Note that the model includes three parameters, shown in gray. L_∞ is the asymptotic length (in this case corresponding to a total length of 700 mm). K is the Brody growth coefficient and is expressed per unit time (in this case per year). K determines how quickly size approaches L. Finally, t_0 corresponds to the age of no length.

value less than 1. Projecting this regression line forward to the point where it crosses a 1:1 line allows for estimation of L_∞:

$$L_\infty = \frac{\text{intercept}}{1 - \text{slope}} \qquad \text{(Equation 7-2)}$$

The slope of this line is related to how quickly fish grow to L_∞. In other words, the slope of this line is related to K. Specifically, K can be estimated as

$$K = -\ln(\text{slope}) \qquad \text{(Equation 7-3)}$$

Short-term Growth in Length

Given that growth in length of most fish is not constant throughout life, it is generally not appropriate to describe growth in length as linear over long growth stanzas. When considering growth over very short time periods or during very early life, it may be appropriate to report growth as linear, for example estimating growth in mm per day. If we assume fish hatch out of their egg at a fixed length, then for an individual young fish of known length and estimated age, we can estimate the growth rate of that individual. For example, Höök et al. (2007) compared growth rates of larval alewife across Lake Michigan habitats by measuring their lengths and counting daily growth increments on otoliths. They determined age by assuming that alewife begin depositing daily growth rings two days after hatching, so daily age was estimated by adding two days to the number of daily rings counted. They estimated growth assuming that all larval alewife hatch at a length of 3.5 mm. Hence, an individual's growth rate was estimated as the difference between current length and 3.5 mm divided by daily age. By estimating growth rates of multiple individuals, one can calculate an average estimated growth rate as well as the variability around this average.

For some studies, the assumption of constant hatch size may be inappropriate, and it may be more appropriate to estimate growth rates for cohorts of fish using a regression-based method. In such a case, it is necessary to compile the measured lengths and estimated ages of a number of young fish. Fish may be grouped together based on various criteria, such as habitat of capture, time of hatch, or some other factor depending on the question of interest. Then, growth is simply estimated as the slope of the linear regression line between age (independent variable) and length (dependent variable).

Growth in Mass

As fish grow in length, their mass will also increase. However, length and mass do not necessarily increase proportionally. If the density and relative dimensions of a fish remain constant, then as it increases in size its mass should increase in direct proportion to its volume. A unit increase in length will lead to greater than one unit increase in volume, and hence mass. Consider a $1 \times 1 \times 1$ mm cube with a volume of 1 mm³. If one dimension increases by 1 mm, one would expect the other dimensions to also increase by 1 mm in order for it to remain a cube. The resulting $2 \times 2 \times 2$ mm cube only increased by 1 mm in length, but increased to 8 mm³. Similar to this relationship between volume and length, the relationship between the length (L) and mass (M) of fish should be nonlinear:

$$M = aL^b \tag{Equation 7-4}$$

If fish density remains constant and if fish grow in proportion across all dimensions, the exponent (b) in the above equation should equal 3. Such proportional growth is referred to as an isometric relationship between length and mass. However, for many fish, length and mass do not increase isometrically. The density of fish may change as they grow because their proximate composition changes. Proximate composition refers to tissue makeup of an individual fish. As a fish grows, the proportion of their mass composed of lipids or skeletal tissue, or any constituent, may change. Such changes in composition can affect density. In addition, the shape of most fish is neither cubic nor constant with age. Thus, due to changes in density and shape, the parameter, b, may be expected to be greater than or less than 3. This is referred to as an allometric relationship between length and mass. For almost all fishes, the exponent relating length to mass will be between 2.5 and 3.5.

The nonlinear relationship between length and mass implies that the lifetime relationship between individual age and mass will look somewhat different than the relationship between age and length. Specifically, mass will increase more rapidly with age at young ages, and growth in mass will be slower at older ages (Figure 7-4).

Short-term Growth in Mass

In many laboratory studies of fish – as well as in mark-recapture studies of wild fish – changes in individual mass may be tracked over time, allowing for estimating growth in mass for a distinct time period. Short-term growth in mass of fish can be calculated in a variety of ways. Most simply, absolute growth can be calculated as the change in mass (M) from one time (t) to another, divided by the amount of time that has passed. Absolute growth may be an appropriate measure if all individuals start an experiment at roughly the same size. However, if comparing growth of individuals of different sizes, absolute growth may be misleading. As fish increase in size, their capacity for increasing in size over a certain time period changes. In an attempt to standardize growth among individuals of different sizes, growth can be expressed on a relative basis. Relative growth is often expressed as grams of growth per gram of initial body mass per day or percent body weight per day.

Another approach for standardizing growth involves expressing growth in mass on an instantaneous basis. Instantaneous growth (G) is expressed per unit time (e.g., per day) and allows for direct comparison of growth rates for individuals with different measured masses. Instantaneous growth also allows for direct comparison with other instantaneous rates. For example, for a cohort of fish instantaneous growth can be directly compared with instantaneous mortality to see if the total biomass of the cohort is increasing or decreasing at a point in time.

Relative and instantaneous growth indices allow for standardizing growth by body mass, but as fish increase in size, their relative growth rate will naturally decline. While a large fish may have a higher absolute growth rate compared to a smaller fish, its relative and instantaneous growth rate may be lower. Relative and instantaneous growth indices may allow for standardizing growth by body size, but care should be taken when selecting a growth index to compare fish of vastly different sizes or considering growth over very long time periods:

Absolute growth:

$$\text{Growth}_{\text{Abs}} = \frac{M_t - M_0}{t} \qquad \text{(Equation 7-5)}$$

Relative growth:

$$\text{Growth}_{\text{Rel}} = \frac{M_{t+1} - M_t}{M_t} \qquad \text{(Equation 7-6)}$$

$$\text{Growth}_{\text{Rel}} = -1 + \left(\frac{M_t}{M_0}\right)^{1/t} \qquad \text{(Equation 7-7)}$$

Instantaneous growth (G):

$$M_t = M_0 e^{Gt} \qquad \text{(Equation 7-8)}$$

$$\ln M_t = \ln M_0 + Gt \qquad \text{(Equation 7-9)}$$

$$G = \frac{\ln M_t - \ln M_0}{t} \qquad \text{(Equation 7-10)}$$

Lifetime Growth in Mass

Similar to lifetime growth in length, various models have been developed to describe whole lifetime growth in mass. The most broadly applied model involves a modification of the von Bertalanffy growth in length model:

$$M_t = M_\infty \left(1 - e^{-K(t-t_0)}\right)^b \qquad \text{(Equation 7-11)}$$

Essentially, this model is analogous to the growth in length model with the addition of an exponent term (b from Equation 7-4). Estimating the four parameters of this model is generally accomplished through nonlinear regression.

Condition

Growth rates represent changes in size over time. In general, faster growth rates are considered advantageous for fish and reflective of individuals performing well. However, growth rates are by no means the only way to evaluate individual performance and vigor. There are various other ways to assess the robustness of fish, which do not necessarily require aging fish or capturing individuals over time. For example, it is common to assess the composition of individual fish, e.g., the percent of a fish's mass that is composed of water, the amount of lipid mass a fish contains, or more detailed measures, such as RNA:DNA ratios or the relative composition of a suite of macromolecules, such as fatty acids. A more straightforward and broadly accepted index of body condition of fish involves examining mass-at-length. This measure of condition is analogous to body mass index calculated for humans. Mass-at-length condition indices involve comparing observed mass of a fish of a certain length to expected mass at that length. Specifically, there are two broadly accepted indices of condition, Fulton's K and relative mass (also referred to as relative weight). Various studies have demonstrated that both of these indices of condition are correlated with growth. That is, fish that inhabit favorable environmental conditions and feed well will not only grow relatively fast but will also express relatively high Fulton's K and relative mass condition indices.

Fulton's K is a universal index of condition and can be calculated for any individual fish with no a priori information. Fulton's K is based on the assumption of an isometric relationship between length and mass. Given that many fishes display an allometric relationship between length and mass (most often $b > 3$), Fulton's K values may increase as fish increase in length. In addition, given that different fish species have very different body shapes and dimensions, it is not very insightful to compare Fulton's K values among species. Fulton's K (calculated based on length measured in mm and mass measured in g) is

$$K = \frac{M}{L^3} \times 10{,}000 \qquad \text{(Equation 7-12)}$$

In contrast to Fulton's K, relative mass (M_r) involves comparing measured mass (M) versus a standard mass (M_s). Standard mass is calculated for an individual based on an established species-specific relationship between length and mass. Standard mass equations allow one to calculate the expected mass of a fish in good condition at an observed length. This standard mass can be compared to observed mass to calculate relative mass. If a fish is heavier than the calculated standard mass, it will have a relative mass greater than 1.0, and if it is lighter than the calculated standard mass, it will have a relative mass less than 1.0. Since standard mass equations are species-specific, they allow accounting for allometric relationships between length and mass. In addition, relative mass more readily allows for comparison of different sized fishes or even different species. However, this index is dependent on the existence of an accepted standard mass equation:

$$M_r = \frac{M}{M_s} \qquad \text{(Equation 7-13)}$$

To calculate relative mass (M_r), one must first calculate standard mass (M_s) using a species-specific equation. Various standard mass equations have been developed using somewhat different approaches. In general, standard mass equations are based on a very large number of observations of length and mass of individual fish. Nonetheless, standard mass equations are generally intended for a certain size range of fish (i.e., it is inappropriate to use an M_s equation to quantify standard mass for individuals outside of the length range for which the equation was developed). Some M_s equations are also developed for certain regions, and it may not be appropriate to apply these equations for populations outside of specific geographical regions.

Murphy et al. (1991) provided an overview of the concept of relative mass and presented various standard mass equations, including one developed by Wege and Anderson (1978) for largemouth bass with a minimum total length of 150 mm:

$$\log_{10} M_s = -5.316 + 3.49 \log_{10} \text{TL} \qquad \text{(Equation 7-14)}$$

Here, the standard mass (g) is allometrically related to total length (TL; mm) as the slope is 3.49. Consider two individual largemouth bass, both with total length of 250 mm, but bass A with a mass of 1200 g and bass B with a mass of 1000 g. Both bass have a standard mass of 1129 g. However, since bass A is heavier at the same length it has a greater relative mass of 1.06, while bass B has a relative mass of 0.89. Bass A would also have a greater Fulton's K index (7.7) as compared to bass B (6.4).

Changes in Condition of Native Fishes in the Illinois River

Mass-at-length indices of condition can be used to compare fish among and within systems, including how fish performance has changed over time. In many systems, non-native species have invaded, become established, and increased in abundance. There is often concern that such invasive species may negatively affect native species by preying upon or out-competing native species. In many freshwater systems in North America, there are particular concerns regarding the effects of invasive carp. For example, invasive bighead and silver carp have spread to broad areas of the Mississippi River drainage, including the Illinois River. The Illinois Natural History Survey (INHS), a unit within the University of Illinois, has monitored fish communities in the Illinois River for several years. Such long-term monitoring allowed Kevin Irons and other INHS researchers to compare the condition of native fishes in the Illinois River before and after the invasion of bighead and silver carp (Irons et al. 2007). These carp species are voracious planktivores. Their prey composition seemingly overlaps with some native planktivores, and they may therefore represent strong competitors with native planktivores.

Irons and colleagues focused on the condition of two native planktivorous species, gizzard shad and bigmouth buffalo. Both of these species had been collected in the LaGrange Reach of the Illinois River as part of multi-method monitoring programs conducted before and after invasive carp establishment. In addition, as part of these monitoring programs, the lengths and masses of individual gizzard shad and bigmouth buffalo had been recorded, which facilitated the calculation of mass-at-length indices of condition. Specifically, Irons and colleagues compared the mean condition of these two species before and after 2000 (1983–1999 vs 2000–2006). While invasive carp had been detected in the Illinois River during the 1990s, they used 2000 as a cutoff because they suggested that the carp had become well established and abundant by this time.

Irons and colleagues were not aware of a published standard mass equation for bigmouth buffalo, and the published standard mass equations for gizzard shad were outside of the length range of gizzard shads they measured. Thus, they developed their own standard mass relationships for both species by using the individual length and mass measurements for fishes collected during 1983–1999. The resulting standard mass relationship for bigmouth buffalo was

$$\log_{10} M_s = -5.3256 + 3.2054 \log_{10} \text{TL} \qquad \text{(Equation 7-15)}$$

And the relationship for gizzard shad was

$$\log_{10} M_s = -4.9038 + 2.9516 \log_{10} \text{TL} \qquad \text{(Equation 7-16)}$$

They then calculated relative mass for individual fish by dividing observed mass by an individual's standard mass (calculated based upon its length). They found that for both species, mean relative mass decreased from before carp establishment in 2000 to after carp establishment 2000–2006. For bigmouth buffalo, mean relative mass declined from 1.00 ($n = 817$) to 0.95 ($n = 1,008$), and for gizzard shad M_r declined from 1.00 ($n = 624$) to 0.93 ($n = 2,550$). It is difficult to unequivocally claim that such decreases are entirely due to competitive effects of invasive carp.

However, Irons and colleagues did examine several other environmental conditions that may have contributed to declines in condition, such as water temperature, chlorophyll *a* concentration, and river discharge. These alternative predictors generally did not seem to be strongly related to bigmouth buffalo and gizzard shad condition. Thus, these declines were highly suggestive that carp had indeed negatively affected the condition of native species.

Summary

Previous chapters have largely focused on how fish grow. However, in experiments and surveys of wild populations, we are often most interested in simply being able to quantify rates at which fish grow. Measuring and estimating size, age, growth, and condition of fish may initially seem rather straightforward, but there have been numerous approaches developed to quantify these attributes. Size can be measured as mass or along various single dimensions; growth can be estimated in terms of length or mass and over short or long time frames; and condition can be quantified using different base equations. While chronometric structures are also analyzed to determine ages of other animals, these structures have a particularly long history for estimating ages of fish and facilitate the subsequent estimation of past growth and size at earlier ages. Accurate individual and population-level information on size, age, and growth in turn allows for estimation of processes such as survival and recruitment. Thus, size and age measures help us understand population-level processes and inform various aspects of fish population management.

Literature Cited

Francis, R.I.C.C. 1990. Back-calculation of fish length: A critical review. Journal of Fish Biology 36:883–902.

Höök, T.O., E.S. Rutherford, D.M. Mason, and G.S. Carter. 2007. Hatch dates, growth rates, survival and over-winter mortality of age-0 alewives in Lake Michigan: Implications for habitat-specific recruitment success. Transactions of the American Fisheries Society 136:1298–1312.

Irons, K.S., G.G. Sass, M.A. McClelland, and J.D. Stafford. 2007. Reduced condition factor of two native fish species coincident with invasion of non-native Asian carps in the Illinois River, U.S.A. Is this evidence for competition and reduced fitness? Journal of Fish Biology 71(Suppl. D.):258–273.

Lee, R.M. 1920. A review of the methods of age and growth determination in fishes by means of scales. Fishery Investigations, Series 24(2):32 pp.

Murphy, B.R., D.W. Willis, and T.A. Springer. 1991. The relative weight index in fisheries management: Status and needs. Fisheries 16:30–38.

Quince, C., P.A. Abrams, B.J. Shuter, and N.P. Lester. 2008. Biphasic growth in fish I: Theoretical foundations. Journal of Theoretical Biology 254:197–206.

Von Bertalanffy, L. 1938. A quantitative theory of organic growth (on growth laws. II). Human Biology 10:181–213.

Wege, G.J., and R.O. Anderson. 1978. Relative weight (W_r): A new index of condition for largemouth bass. Pages 79–91 *in* G. Novinger and J. Dillard, editors. New Approaches to the Management of Small Impoundments. American Fisheries Society, North Central Division, Special Publication 5, Bethesda, MD.

CHAPTER 8

Bioenergetics Models

Earlier sections of this book outlined the various aspects of an energy budget. There are many components involved in this budget, and they differ in their difficulty of measure. The simplest characteristics to measure are growth rate and food habits. These can be done with field-based techniques to a fine degree of precision and can be used to extrapolate growth or food types over time with relative precision. In comparison, the most difficult characteristics to measure are ration and metabolic rate. Some of the difficulties in measuring ration are detailed in Chapter 18. Metabolic rate is relatively simple to measure in a laboratory but difficult to extrapolate to the field with all the variables related to standard metabolism, cost of activity, and temperature. This is particularly critical since activity may be a major part of the energy budget.

Bioenergetics models are simply mathematical expressions of the balanced energy equation as in Equation 4-2 and Chapters 4 and 5. Such a model pays special attention to how these parameters would be expressed for a specific situation in the field. A number of such models have been produced, and the most widely used form is attributable to James Kitchell and his colleagues at University of Wisconsin (henceforth referred to as the Wisconsin model).

Many aspects of metabolism are physiological parameters that may be consistent across species. These physiological constants are included in Chapter 5. They can be extrapolated over sizes of fishes, possibly over species of fish, and over temperatures. They may require some species-specific information, but several can also be generalized. Thus, extrapolation of lab-based physiological data to the field may not be that difficult for standard metabolism at different temperatures. Inputs to the model generally include growth rate (possibly in body and gonad tissue) as well as temperature conditions under which the fish exist. An example of a bioenergetics model is demonstrated in Chapter 5 (Box 5-4), where it is used to extrapolate a couple of energetic measures for largemouth bass. In addition, the relative balance of a bioenergetics model is demonstrated for northern pike in Chapter 6 (Figure 6-1). Ration predictions are made based on the model and compared to measured ration data in order to corroborate a bioenergetics model. This can also be done for growth predictions. Such comparisons can be made from field estimates of growth and ration or from controlled lab experiments. These are all commonly used methods to corroborate bioenergetics models. One study can show that a model behaves appropriately in predictions (corroborate the model), but it would take a large number of such studies to ultimately validate the model for all conditions. A considerable amount of field- and lab-based energetic information on the species in question is required to develop and corroborate such a model.

The purpose of this chapter is to review bioenergetics modeling and its use in field energetics of fishes. It proceeds through the forms of basic bioenergetics models and attempts to corroborate the models and practical uses of the models.

Biology and Ecology of Fishes, Third Edition. James S. Diana and Tomas O. Höök.
© 2023 John Wiley & Sons Ltd. Published 2023 by John Wiley & Sons Ltd.

Bioenergetics Models

The bioenergetics model began with some knowledge of energetics for a given species and with extrapolations from energetics of other species (Kitchell et al. 1977). The model can be used to predict consumption from data on growth, temperature, and prey types consumed. It can also be used to predict growth at a given level of consumption.

The basic form of the model is the balanced energy equation with variables to account for temperature, fish weight, and ration. Early modeling attempts suffered from the lack of good bioenergetics data on the species in question, but these values have improved dramatically in later years and the models now are seldom based on data from other species. Similarly, early models had awkward means to estimate the effects of temperature on ration and on metabolism, and the more recent models are more proficient in these areas as well. The model for metabolism as presented in Adams and Breck (1990) is

$$MR = a_2 B^{b2}\, e^{mT}\, e^{gS} \qquad\qquad \text{(Equation 8-1)}$$

where MR is total metabolic rate (kJ/d), B is fish mass, a_2 is a constant defining the level of metabolism for a 1 g fish at 0 °C and swimming at 0 cm/s, b_2 is the coefficient for the standard respiration–mass relationship (Figure 5-9), e^{mT} defines the curvilinear relationship between temperature (T) and standard metabolism (Figure 5-4), and e^{gS} corrects for activity-dependent metabolism at a known swimming speed (S). Often fish behavior in the field – and the energetic costs of swimming at a certain speed – is not known well enough for a species to allow the use swimming speed as an input. In these cases, e^{gS} may be substituted with an activity multiplier that increases metabolic rate above standard to account for activity. An example of this would be the Winberg multiplier of two described in Chapter 4. Similarly, the relationship between temperature and standard metabolism takes on several forms; the most common is a pattern that uses a curve defined between 0 °C and lethal temperature for the fish (Kitchell et al. 1977).

Estimates for the losses of energy in egestion and excretion are also required to compute a ration. Most commonly H, F, and N are considered constant fractions of the food consumed, but F also depends on food quality. If all of these are known, a ration can be calculated for a given growth rate ($B_1 - B_0$ for size at times 1 and 0) by balance, where $C = (B_1 - B_0 + MR)/[1 - (F + N + H)]$. This produces a total ration in kJ/d but cannot be converted to grams of each species of prey eaten without much more information.

If growth rate is calculated from assumed ration, more complications arise. In these cases, values for consumption at different temperatures, fish weights, and prey species, energy values are also needed. These equations follow the same form as respiration, where consumption is a function of maximum consumption, a temperature adjustment, and an adjustment for the fraction of food actually consumed ($C = C_{max} P r_c$). C_{max}, or the maximum weight-specific ration at optimum temperature, equals $a_1 B^{b1}$ where a_1 and b_1 are functions to relate consumption to body mass. P is the proportion of ration actually eaten and varies from 0 to 1. This model parameter is often used to allow the observed and predicted growth to match by adjusting P. r_c is a value to change maximum consumption for various temperatures and again is related to the minimum, optimum, and maximum temperatures for consumption (Kitchell et al. 1977). Of course, r_c could also be estimated mathematically by fitting a curve. For that case, the best fit is a polynomial, which could be substituted for the calculation of r_c.

Calculation of the bioenergetics model requires species-specific details for as many parameters as possible. The values for the yellow perch model are presented in Table 8-1.

One parameter worthy of more detail is the P value. P is necessary to balance growth that occurs at maximum ration to growth that occurs at an estimated ration. In other words, it can be visualized as the fraction of the maximum ration actually eaten. When a bioenergetics model is run without a

TABLE 8-1   Parameter Values for a Perch Bioenergetics Model.

Symbol	Description	Value
a_1	Intercept for C_{max}	0.25
b_1	Slope for C_{max}	−0.27
T_o	Optimum temperature for consumption (°C)	29
T_m	Maximum temperature	32
Q	Slope of temperature dependence of consumption (Q_{10})	2.3
a_2	Intercept for MR	0.035
b_2	Slope for MR	−0.20
T_o	Optimum temperature for standard metabolism	32
T_m	Maximum temperature for standard metabolism	35
Q	Slope of temperature dependence for standard metabolism	2.1
S	Specific dynamic action (H)	0.172
Act	Activity multiplier	1
$F + N$	Fecal and urinary losses (N and F)	$f(T, °C)$

Source: Adapted from Hewett and Johnson (1992).

known ration as an input, various P values are tried until predicted body size at a given time equals measured body mass (the input value). This balancing gives some idea of the relative amount of food eaten over that time and would allow the extrapolation of growth levels under different conditions. Since the P value is forced to balance ration and growth at different times, any inaccuracies in the model are also included in the P value. It cannot be simply thought of as the fraction of ration actually eaten. The P value is not necessary to calculate ration for a given month from known temperature and growth rate. If values for energy density and proportion in the diet are known for all prey species, ration can even be converted into the mass of each species of prey eaten. P values are useful for extrapolation of bioenergetics models into new conditions such as under varied temperature regimes. However, they are only as accurate as the other parameters in the model.

Earlier, we described how temperature, food consumption, and body size are overwhelmingly important factors influencing ecology of fish species. For bioenergetics, we should add activity level as a very important variable to understanding the relationship between consumption and growth. The important variations in the way scientists have fitted bioenergetics models have to do with how they deal with these basic characteristics for the fish population and ecosystem under consideration.

Corroboration of Bioenergetics Models

Bioenergetics models are based on physiological parameters well established in literature. Extrapolating these values to the field requires good data on temperature occupation and growth rates of the fish in question. Both of these are relatively easy to determine in many situations although actual temperature occupation is not often measured and may be in error in situations where there are temperature complexities in an ecosystem. Bioenergetics models can be used to predict total ration, but the estimate of ration is simply the outcome of these physiological processes. To corroborate such a model, independent data on rations are necessary to allow one to test the differences between predicted and measured rations. In addition, data on the activity rate of the fish in question may be needed to adequately test the model. All of these data values are rarely available.

Two general approaches have been used to corroborate bioenergetics models: field comparisons and lab comparisons (Madenjian et al. 2000). Field comparisons involve estimation of ration in the field and using that value to compare with predicted ration from the model. Lab comparisons involve experiments that provide food to fish under controlled conditions, measure growth rate, and then use the measured growth and other values from the experiment as model inputs to predict ration and compare it to the actual ration given. Madenjian et al. (2000) describe a third method to corroborate a bioenergetics model by evaluating retention rates for markers such as PCBs or radioactive isotopes. These materials are taken up from prey items consumed, and the efficiency of marker retention can be estimated from lab experiments. The same efficiency can also be estimated using field measurements of marker concentration coupled with a bioenergetics model and data on concentration of markers for natural prey. The two transfer efficiencies could therefore be used as another corroboration of the model. Madenjian et al. (2000) believe that two or more types of corroboration should be used in model evaluations.

Many attempts have been made to corroborate bioenergetics models in spite of the limitations of data. Some of the corroborations have been relatively successful, and others have been relatively unsuccessful. Some of the evidence on the accuracy of bioenergetics models, as well as questions these studies have provided, is reviewed in the following sections.

Good Fits of Bioenergetics Models to Field Data

Several studies have found relatively good fit between bioenergetics model and field values for growth and ration. The first of these was a study by Rice and Cochran (1984), which evaluated a bioenergetics model produced for largemouth bass. The test was on a field site at Rebecca Lake, Minnesota, where Cochran and Adelman (1982) determined body size and daily ration of age-3 largemouth bass on 10 dates in 1978. Rice and Cochran (1984) defined statistical techniques to test goodness of fit and used these techniques to corroborate the model in several ways. The first of these was to predict the actual mass of fish at various points using ration estimates linearly interpolated between the dates of sampling. Since the mass of fish has variation among individuals, it can be tested statistically against the predicted mass. Of course, variation in mass of the same age individuals is a characteristic established even at the very earliest age of a year class, whereas the bioenergetics model generally utilizes one average-size fish at the start date as input. Rice and Cochran's model appeared to predict body size relatively well over the summer (Figure 8-1). In addition, they tested variations in predicted swimming speeds, as well as predicted temperature occupation, to see how much these values would change the results. In the end, fit was not very sensitive to either change.

A second test was to look at the prediction of ration from growth in the field and compare that to measured ration. Some difficulties with this comparison were previously stated; they include the fact that there are no variation estimates on daily ration measured in the field as well as ration being a point estimate in time rather than a continual function over the modeled time period. The results of those evaluations (Figure 8-2) show that cumulative ration was predicted fairly well by the model, while daily ration was not predicted with much precision. Cumulative ration would be the summed ration over the entire interval. Daily ration is compared for those times when ration was actually measured to point estimates from the

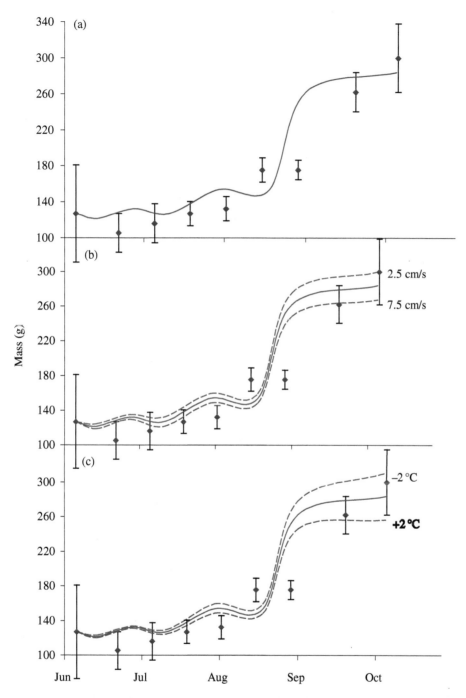

FIGURE 8-1   Body mass predicted by the bass bioenergetics model (lines) as well as observed bass masses (mean ± 2SE) at various times of collection. (a) The overall model fit, (b) the variation caused by changing swim speed from 5 to 2.5 or 7.5 cm/s (dashed lines), and (c) the variation caused changing temperature ± 2 °C (dotted lines) are shown. *Source: Adapted from Rice and Cochran (1984).*

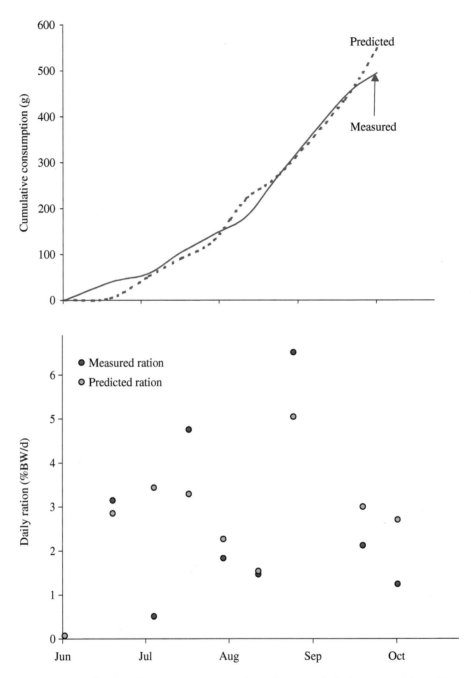

FIGURE 8-2 Measured and predicted rations over various time periods for largemouth bass. The upper graph shows cumulative ration and the lower graph shows point values of daily ration. *Source: Adapted from Rice and Cochran (1984).*

bioenergetics model on that date. There are many reasons why measured ration on a given date may be very different from that on a succeeding date, due to differences in weather, predator, and prey behaviors, and many other factors that can differ over short time intervals. This variability may help explain why comparisons of daily ration may be inaccurate, while cumulative ration expressed over longer time intervals may fit better between field and model data. Rice and Cochran (1984) took their analysis to indicate that the model is fairly accurate predicting

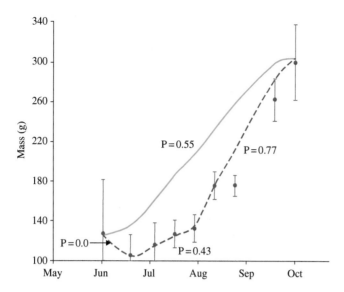

FIGURE 8-3 Observed mass (mean $\pm$ 2SE) and predicted mass from the bass bioenergetics model, using an overall P value (solid line, P = 0.55), or periodic changes in P value (dotted line). *Source: Adapted from Rice and Cochran (1984).*

ration when used conservatively – over cumulative periods of time – but not particularly accurate predicting day-to-day variation in ration.

Finally, Rice and Cochran (1984) conducted an evaluation where they estimated growth using *P* values and compared it to observed body size. In this case, using a *P* value over the entire summer resulted in a poor estimate for most data points in between. However, breaking down the *P* value into monthly or slightly longer time periods resulted in a much better fit of the data to observed growth (Figure 8-3). In other words, *P* value was a reasonable estimator for growth as long as the time periods over which *P* was calculated were not excessive. Of course, since *P* is calculated by adjusting the equation until the observed and predicted mass is equal, it would result in a perfect fit if it were recalculated for every day. In the end, Rice and Cochran accepted this body of information as a reasonable corroboration of the bass bioenergetics model. Whitledge and Hayward (1997) also provided corroboration of the bioenergetics model for largemouth bass in a lab-based study described later. Both of these studies indicated that the model for bass appears to perform well in field and lab settings. There have been a number of other corroborations of bioenergetics models with field evaluations, including Beauchamp et al. (1989) for sockeye salmon, and Madenjian et al. (2000) on lake trout, using PCB uptake.

These evaluations indicate that the bioenergetics model approximates the actual field energetics of some fishes well, when the model was developed on good physiological data and good knowledge of fish growth and behavior. The model was much more accurate predicting cumulative ration than daily ration. It is not as accurate and is subject to much more variability in interpretation when used to predict growth from ration estimates. Due to corroboration using multiple methods, both the lake trout and largemouth bass models have been more thoroughly tested and corroborated than other versions.

Poor Fits of Bioenergetics Models to Field Data

In contrast to previous studies, a number of evaluations have also indicated a mismatch between bioenergetics models and field ration and growth estimates. An example for esocids by David Wahl and Roy Stein from Ohio State University demonstrates this point. Wahl and Stein (1991) quantified daily ration and growth for pike, muskellunge, and tiger muskies in five Ohio reservoirs. These reservoirs were dominated by gizzard shad, which were important items in the diet. Wahl and Stein found the bioenergetics model overestimated observed food consumption and underestimated observed growth. The errors were considerable for all three species (Figure 8-4) with the largest variance in northern pike and muskellunge. These overestimates in ration led Wahl and Stein to believe that metabolic parameters of the esocid

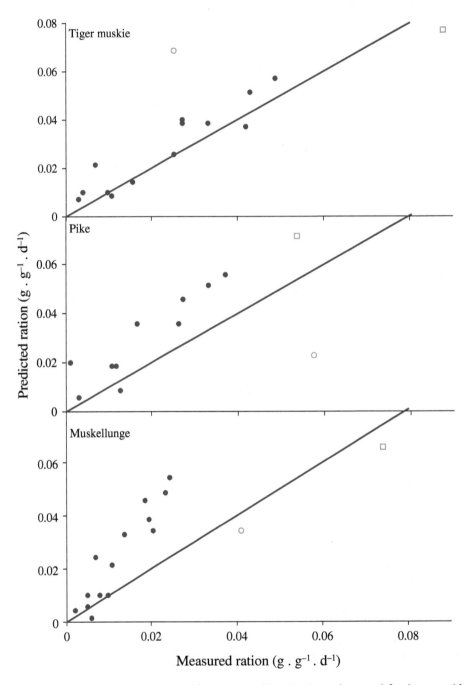

FIGURE 8-4 Comparison between estimated food consumption (circles and squares) for three esocids and predicted ration from the bioenergetics model. Solid line: 1:1 relationship where predicted = estimated ration. Solid circles: North Reservoir; open circles: Madison Reservoir; open squares: Kokosing Reservoir.
Source: Adapted from Wahl and Stein (1991).

bioenergetics model were at fault. In particular, they believed that metabolic rate was not only a function of temperature but a function of season, and the lack of seasonal testing in most studies of metabolic rate may produce poor estimates of actual metabolism at a given temperature. Chipps et al. (2000) also found that the metabolism of muskellunge varied seasonally when tested at the same temperature, and winter metabolic rates were much lower than spring or fall values, again supporting this idea of seasonal changes in metabolism independent of temperature. Other analyses indicating poor fit of the model to field data were done by Diana (1983) for northern pike in an inland lake and by Schaeffer et al. (1999) for yellow perch in Saginaw Bay.

The most controversial corroboration of a bioenergetics model was the work of Daniel Boisclair and William Leggett from McGill University. Boisclair and Leggett (1989) studied ten lakes and determined ration using stomach contents and gastric evacuation rates for yellow perch of each age class in each lake. They also estimated the growth rate in each population as well as temperature and prey consumed.

There was considerable variation in the relationship between growth and food consumption for the 28 subpopulations (lakes and age classes). Boisclair and Leggett believed that this variation to be due in part to differences in activity rates of the fish in each lake. In support of this, they used Kerr's (1982) model that proposed an alternative method that, rather than using a constant multiplier for activity, varied activity as a function of energy intake since Kerr believed that energy expended in foraging was related to energy consumed. To resolve this question, Boisclair and Leggett (1989) predicted activity costs from the Kerr (1982) model and evaluated the imbalance between growth and consumption in the Wisconsin bioenergetics model as an indicator of activity cost. In this case, they assumed that all the imbalance between estimated and predicted ration was due to activity. They calculated a ratio of active to standard metabolism $[(S + A)/S]$ to evaluate the elevation in metabolism required for activity to balance the energy budget.

The result of this analysis was that activity was an extremely variable component of the bioenergetics model using either the Kerr or Wisconsin estimates. On average, Boisclair and Leggett (1989) found that 3.4 times more energy was allocated to activity than to growth, and the activity ratio ranged from 0.14 to 3.88. Both the Kerr and Wisconsin models indicated a strong linear relationship between activity levels and observed feeding levels for all ages of perch. Of course, an activity ratio less than 1 indicates an impossible reduction in standard metabolism or, more likely, a model or ration estimate with a considerable error.

Overall, Boisclair and Leggett proposed that activity was an important and unknown input to most bioenergetics models. They agreed that in cases with piscivores, activity cost was probably negligible, but hypothesized that activity was an unknown and relatively important parameter for pelagic and benthic fishes. They proposed that activity should not be a standard multiplier but might vary with age class as well as characteristics of a given system. They also believed that the accuracy of a bioenergetics model, when lacking all the data to corroborate the model (including activity), would be questionable.

The imbalance in the Boisclair and Leggett corroboration of the perch bioenergetics model created much more controversy than the studies described earlier. Hayward (1990), in a provocatively titled paper ("Can eating really stunt your growth?"), questioned the food consumption estimates of Boisclair and Leggett. In particular, Boisclair and Leggett (1989) had concerns with Hayward and Margraff (1987) and their comparison of rations and growth among locations in Lake Erie. Boisclair and Leggett (1990) defended their methodology for estimating ration for comparisons across sites and provided data to demonstrate that in only one case (for 15–20 g fish) did there exist a significant negative relationship in their data between consumption rate and growth rate. Using a multivariate analysis, they determined that the strongest effect on growth rate was numerical density and consumption rate (explaining 30% of the variance). They proposed that much of the effect of density was on activity and the actual characteristic differing among populations was activity. They stated that their work was not a rejection of the relationship between growth and consumption, but a reminder that many other factors, including activity rate, could strongly interfere with this relationship.

A second debate erupted between Hewett et al. (1991) and Boisclair and Leggett (1991). In this case, Hewett et al. (1991) questioned three areas of the work: (1) that the relationship between growth rate and food consumption was spurious and the food consumption estimates were inaccurate; (2) that the model of gastric evacuation used by Boisclair and Leggett had a size-dependent bias; and (3) that Boisclair and Leggett used the bioenergetics model inappropriately to estimate activity.

These criticisms led to another interesting paper by Boisclair and Leggett (1991), entitled "If computers could swim or fish could be programmed." Their assumption for this article was that comparisons of ration by actual field methods and evaluating energy budgets with field measures were more illustrative than using bioenergetics models to predict values. Boisclair and Leggett defended their methodology comparing growth rate and food consumption. Much of the concerns and responses relate to methods of comparing fish populations from different lakes, where more than one factor differs among lakes and, therefore, controlling for these factors is difficult. This process is easy in a bioenergetics model; one simply holds constant all variables in the model except the one to be tested. In reality, this is not simple because all of these characteristics vary synchronously among lakes and cannot be separated. Boisclair and Leggett attempted to counter the Hewett et al. (1991) arguments in detail.

In the end, Boisclair and Leggett defended the concept that testing of bioenergetics models should be conducted with field data rather than with model-generated data. They believe that, in their case, they have reason to question the lack of activity measurements among populations and that activity is an important component in the fish bioenergetics models. The current method of an activity multiplier appears to work well in some cases, but still does not indicate that activity is not an important parameter overall. Clearly, the result of this argument should be the compilation of more data on activity and metabolic rate in the field, yet these have not become commonly incorporated into bioenergetics models.

Of course, all of these comparisons, and many made using other statistical models when compared to actual data, demonstrate errors as well, and the important point in doing such comparisons is to try to understand why there are errors in the comparisons. A couple of these have been proposed above, including the seasonal difference in metabolism that may be independent of temperature and differences in activity that may exist among populations. Other reasons for errors may be strictly statistical in nature. When using statistics to test relationships, it is important to not extrapolate values beyond the range of conditions under which they were measured. Yet bioenergetics models really have little choice in doing this given the variables that may need to be tested in each habitat analyzed. Extrapolating temperature relationships to winter conditions when these values were only determined in warmer conditions would be one such possible cause of error in model application. Another possible reason for poor corroboration may be genetic differences between the population being evaluated and the fish used to derive metabolic data. In Chapter 5, we spent considerable time evaluating differences in fish populations due to acclimatization to local conditions, which is both physiologically and genetically based. Yet expecting every model application to use data derived only from the target population is unrealistic as well. Overall, the corroboration of bioenergetics models has been useful in better understanding the models themselves, the dynamics of natural fish populations, and the statistical issues in completing corroborations.

Laboratory Fits of Bioenergetics Models

Probably the best method to analyze the validity of a bioenergetics model is to assess growth rate and ration in the lab rather than in the field. This is true because, in the lab, one can measure activity of the fish and feed the fish a known ration at a known temperature and a measured activity level. This gives much better control than estimates of these parameters in the field. Such an approach was used by Whitledge and Hayward (1997) in a study on largemouth bass. They evaluated fish at two temperatures (22 and 27 °C) and two feeding levels (ad libitum and 2% wet body mass) in the laboratory. They used a variety of statistical techniques to test fit from their data. The bioenergetics model did a good job in general predicting actual ration for the fish. In particular, predicted cumulative food consumption over nine weeks was only about 9% less than observed consumption. The model predictions were relatively accurate when fish were not fed to excess but began to underestimate actual rations at ad libitum feedings. Such an error may be the result of the glutton effect mentioned earlier in the consumption and growth relationship determined for carp. This sort of bioenergetics approach seems to be the most valid in testing the physiology of the model and, in this particular case, corroborated the model for largemouth bass.

Examples of Uses

If one accepts that corroborations of bioenergetics models indicate that they are relatively accurate in predicting energetics of a population, then using the models with good growth and population abundance estimates could give meaningful information to managers. Some of these uses might be to evaluate the effects of stocking on forage fish populations, to compare alternative predation by different species of predator, to evaluate uptake of materials such as PCBs and others through the forage-based contamination, to evaluate growth potential under different conditions of prey abundance and temperature, or to evaluate differences among individuals of a species in their energetic performance.

Salmonids in Lake Michigan

One of the first management applications of a bioenergetics model was from Stewart et al. (1981). Don Stewart and colleagues, then from the University of Wisconsin, used a bioenergetics model as a means to evaluate relative predation by different salmonid species in Lake Michigan. The history of the Great Lakes will be covered later, but it is important here to know that salmon in the Great Lakes were introduced in the late 1960s in an attempt to control overpopulation by exotic alewife. Alewife were abundant in Lake Michigan because of predatory effects of sea lamprey, which had removed most top predators. Alewife became the most abundant forage species, competing intensely with adults of many native species, preying upon their eggs and young, and possibly driving some to extinction. They became a pest so predators were introduced in an attempt to reduce overabundance of alewives as well as to produce a sport fishery.

Stewart et al. (1981) modeled the relative forage consumption by different salmonid predators and used this to predict predatory demand of salmon on alewife and other prey species. Stocking numbers, predatory demand, and prey numbers were then used to forecast the status of prey populations under different management scenarios. This has been transformed into a more complex multispecies model by Iyob Tsehaye and a research group from Michigan State University. Their model essentially uses a salmonid bioenergetics model combined with stocking, reproduction, and mortality rates of each salmon species, and then estimates predatory demand on the main prey species (alewife and rainbow smelt) in Lake Michigan. This demand per predator individual is accumulated through a dynamic model of salmon and prey abundance to assess status of each prey species. An additional component of the model since the early studies of Stewart is that recruitment of salmonines has not only been due to stocking levels from various management agencies but also due to natural reproduction.

Tsehaye et al. (2014) corroborated their overall model by comparing outputs for populations from 1970 to 2010 to predicted prey populations as well as predator abundances and harvests for actual and modeled data, and all of these values balanced reasonably well. They then used the model to assess the ecosystem of Lake Michigan and make some forecasts for fishery management. They found that growth and survival of Chinook salmon is density dependent with poor growth and survival in years of high abundance, and these years also relate to years of low prey abundance. Chinook salmon were predicted to cause more than 60% of the total alewife and smelt consumption in Lake Michigan, and overall consumption is pushing the capacity of these prey populations to maintain some stability. Recent reductions in stocking of Chinook salmon have reduced predatory demand, but increasing natural reproduction has increased this demand, so these two changes seem to break even. Probably, the most disturbing part of the model predictions is there appears to be no functional response by Chinook salmon to prey abundance. A functional response is a reduction in consumption of a particular species as its prey declines in abundance and replacement of that demand by increased consumption of other prey. Without such a response, there is likely no natural mechanism to stop Chinook salmon from driving their alewife prey to very low levels unless the population of Chinook is maintained at an appropriate level by the combined recruitment due to stocking and natural reproduction.

Spatially Explicit Energetic Models

One interesting application of bioenergetics models is the evaluation of variation in growth rate spatially or temporally. These models are termed spatially explicit models because they evaluate differences in habitat potential over space (Brandt et al. 1992). The first applications were in Chesapeake Bay, where the temperature and plankton populations varied both seasonally and spatially within a season. These variations were used along with a bioenergetics model to predict the regions of various fish growth potentials, where growth rate would be high or low because of the combined temperature–prey abundance effects. This approach has since been expanded to other locations to evaluate the potential effects of other habitat-related events and its influence on fish growth.

One interesting application of this was by Logerwell et al. (2001) on Pacific sardines. The sardine is a species with major variations in growth rate and population numbers believed to be due to changes in weather or other stochastic events. One hypothesis has been that offshore mesoscale eddies (large areas with cycling currents of water) have a strong effect on the abundance and survival of larval sardines. This is a very difficult issue to study because of the size of the marine system and variability in sampling. It is a perfect one for the use of a spatially explicit energetic model to test such hypotheses.

Logerwell et al. (2001) tested the spatial component of their model over five zones: the inshore area, slope, normal offshore areas, offshore areas with cyclonic eddies, and offshore areas with anticyclonic eddies. The California Cooperative Oceanic Fisheries Investigation (CalCOFI) has sampled this region since 1949, and data on temperature, zooplankton, and fish abundance are available since then. Mesoscale eddies are on the order of 100 km in diameter with current patterns and temperature differences from surrounding waters. They often occur in areas adjacent to major currents. Cyclonic eddies are counterclockwise (in the northern hemisphere) and usually have colder water, while anticyclonic eddies are clockwise and usually contain warmer water. These eddies are ubiquitous features in the oceans and have been identified and mapped often with the advent of satellite observations. Both types occur off the California coast.

There are large differences in these five zones that may affect sardine growth, survival, and abundance. Mesoscale eddies of either type entrain water in their midst, eliminating circulation back to outside areas and accumulating materials in them. The CalCOFI studies revealed that temperatures were maximum in anticyclonic eddies and lowest in cyclonic eddies. Mean copepod densities (prey for the sardine) were greatest in the cyclonic eddy, followed by the inshore, slope, and anticyclonic eddy, and then the offshore region. Therefore, the combination of temperature and prey could cause different growth in each of these habitats. Logerwell et al. (2001) used a slightly different bioenergetics model described by Jobling (1994) to estimate the energy budget. They also modeled consumption geometrically from prey density and reactive distance of the predator (distances over which they can see and react to prey). To make this fit the physiology of larval sardines, the maximum prey consumption was limited to 20% BW/d.

The reactive distance model predicted that consumption was 20% BW/d for all sites except the offshore area. Growth potential, from the consumption model and the bioenergetics model, was between 11.2% and 12% per day in the inshore, slope, and eddy regions, and was lower in the offshore region. Sardine larval biomass was greatest in the cyclonic eddy, followed by the offshore region, anticyclonic eddy, inshore, and slope habitats. The product of the growth potential and the number of larval sardines gave a sardine production potential, which was greater in the cyclonic eddy by one or two orders of magnitude (10–100×). It was greater than the summed totals of all other areas (on a total – not per area – basis) combined by a factor of 4.2. While the overall model, especially with spatial and temporal components, is even less precise than the basic bioenergetics models described earlier, the large difference in production potential between habitats gives support to the importance of eddies as a major factor in the variable recruitment of sardines.

Individual-Based Models

Bioenergetics models have traditionally been based on growth for an average fish although spatially explicit models may compare average details for individuals that occupy different locations with variable conditions. Another type of model emphasized recently is an Individual-Based Model (IBM). IBMs can be used to look at the variability in growth and production of individual fish selecting from variable types of conditions like different prey sizes, locations, and thermal conditions. Since these models use simulations to predict the outcomes in growth for a large number of individuals that experience different conditions randomly or selectively, they can help better understand stochastic conditions in the environment with unknown or random variability.

An example of this is the study by Höök et al. (2008) on alewife in Lake Michigan. They used a model based on simulations for 20,000 "super individuals" (which could represent variable numbers of individuals as well, depending on the relative production of each type in an ecosystem) to estimate the contributions to annual recruitment for alewife occupying three habitat types – offshore waters, nearshore waters, and tributaries. They estimated variance in each habitat type for temperature, prey abundance, and water clarity, and also evaluated variance among those habitats with two different levels of parameters for offshore locations and four for nearshore and tributary locations. They then simulated recruitment from fish occurring in each habitat type over annual periods from 1992 to 2002 based on different individual uses of each habitat type.

These models predicted that the largest contribution of larval alewife to recruitment came from nearshore spawners, but this was based more on the larger amount of inshore habitat than on superior conditions for growth and survival in nearshore compared to tributary waters. When thermal conditions were colder, tributaries became more important to annual recruitment, while in warmer conditions, late emerging individuals survived better due to shorter time intervals that they were exposed to predation. Overall, they found high variability in recruitment depending on both climatic and habitat-related variables and predicted that fish populations at their latitudinal extremes and those in large heterogeneous environments would show the largest variability in recruitment.

Other Uses of Bioenergetics Models

Many other authors have used bioenergetics models for a variety of assessments in fish biology. Some assessments include understanding of basic bioenergetics concepts, accumulation of contaminants, effects of environmental variables, evaluation of life history strategies, effects of predators on their prey, use in fish culture and management, and estimation of energy budgets for larval and young-of-year fishes (Adams and Breck 1990). To these we can add quantifying nutrient regeneration by fishes and spatially explicit comparisons of temperature and growth. Fish culture is a particularly important application as many culture systems are artificially fed with pellets and fish reared to a large size for consumption or stocking. Utilizing different feeding strategies, food types, or temperatures are ways bioenergetics models could predict the actual consumption and growth expected and produce hypotheses regarding the efficiency of growth under different culture systems. These could be combined with cost of food and other manipulations to become economic models of aquaculture production. The closer an aquaculture system is to entirely being intensive with food provided and water quality manipulation done, the more simply a bioenergetics model could be applied.

One example of expanded uses of the model was done by Nislow et al. (2000) from Dartmouth College. They developed a spatially explicit model to assess habitat quality for young-of-year Atlantic salmon. The results indicated limited habitat with positive growth in spring but high occupation of that habitat by the fish. They also used this model, along with published observations, to predict that faster growing fish would have a lower mercury concentration than smaller, slow-growing fish due to somatic growth dilution. Ward et al. (2010) tested this by releasing salmon at 18 sites varying in mercury contamination. They found expected results that fish consuming higher concentrations of mercury in their prey had higher body burdens. They also showed that faster growing fish had lower body burdens with Hg concentration of prey explaining 59% of the variation in salmon body burdens, while somatic growth dilution explained 38% of the variance. This combination of bioenergetics modeling and targeted field studies was powerful in understanding mercury contamination in juvenile Atlantic salmon.

Summary

Bioenergetics models have been developed to accommodate fish physiological characteristics and use them to predict energy budgets for natural fish populations. These models rely on estimates of growth, temperature occupation, consumption, and activity from field fish populations in order to be corroborated. When these values are well known and the physiology of the fish is well understood, a bioenergetics model can be used to accurately estimate various components of an energy budget. Corroboration attempts for bioenergetics models have resulted in cases of good balance for largemouth bass and lake trout and others of poor balance for esocids and yellow perch. Sometimes the imbalance in the model may be an indicator of problems with the model itself, such as the problems of model predictions at high temperature for yellow perch, seasonally adjusted metabolic rates for pike and muskellunge, and uncontrolled activity levels among perch populations. These are important points that deserve further study. At other times, the model imbalance may signify problems with field data collection, such as poor ration estimates for northern pike or yellow perch. Overall, a bioenergetics model can be used to better understand a variety of conditions, including effects of predators on their prey, stocking strategies, and aquaculture production efficiency, all of which are important applications to fishery managers. They represent a significant step in the physiological ecology of fishes and in use of laboratory physiology for field populations.

Literature Cited

Adams, S.M., and J.E. Breck. 1990. Bioenergetics. Pages 389–415 *in* C.B. Schreck and P.B. Moyle, editors. Methods for Fish Biology. American Fisheries Society, Bethesda, Maryland.

Beauchamp, D.A., D.J. Stewart, and G.L. Thomas. 1989. Corroboration of a bioenergetics model for sockeye salmon. Transactions of the American Fisheries Society 118:597–607.

Boisclair, D., and W.C. Leggett. 1989. The importance of activity in bioenergetics models applied to actively foraging fishes. Canadian Journal of Fisheries and Aquatic Sciences 46:1859–1867.

Boisclair, D., and W.C. Leggett. 1990. On the relationship between growth and consumption rates: Response to comments by R.S. Hayward. Canadian Journal of Fisheries and Aquatic Sciences 47:230–233.

Boisclair, D., and W.C. Leggett. 1991. If computers could swim or fish could be programmed: A response to comments by Hewett et al. (1991). Canadian Journal of Fisheries and Aquatic Sciences 48:1337–1344.

Brandt, S.B., D.M. Mason, and E.V. Patrick. 1992. Spatially-explicit models of fish growth rate. Fisheries 17(2):23–31.

Chipps, S., D.F. Clapp, and D.H. Wahl. 2000. Variation in routine metabolism of juvenile muskellunge: Evidence for seasonal metabolic compensation in fishes. Journal of Fish Biology 56:311–318.

Cochran, P.A., and I.R. Adelman. 1982. Seasonal aspects of daily ration and diet of largemouth bass *Micropterus salmoides* with an evaluation of gastric evacuation rates. Environmental Biology of Fishes 7:265–275.

Diana, J.S. 1983. An energy budget for northern pike. Canadian Journal of Zoology 61:1968–1975.

Hayward, R.S. 1990. Comment on Boisclair and Leggett: Can eating really stunt your growth? Canadian Journal of Fisheries and Aquatic Sciences 47:228–230.

Hayward, R.S., and F.J. Margraf. 1987. Eutrophication effects on prey size and food available to yellow perch in Lake Erie. Transactions of the American Fisheries Society 116:210–223.

Hewett, S.W., and B.L. Johnson. 1992. Fish bioenergetics 2: A generalized bioenergetics model of fish growth for microcomputers. University of Wisconsin Sea Grant Technical Report No. WIS-SG-92250.

Hewett, S.W., C.E. Kraft, and B.L. Johnson. 1991. Consumption, growth, and allometry: A comment on Boisclair and Leggett (1989a, 1989b, 1989c, 1989d). Canadian Journal of Fisheries and Aquatic Sciences 48:1334–1337.

Höök, T.O., E.S. Rutherford, T.E. Croley II, D.M. Mason, and C.P. Madenjian. 2008. Annual variation in habitat-specific recruitment success: Implications from an individual-based model of Lake Michigan alewife (*Alosa pseudoharengus*). Canadian Journal of Fisheries and Aquatic Sciences 65:1402–1412.

Jobling, M. 1994. Fish Bioenergetics. Chapman and Hall, London.

Kerr, S.R. 1982. Estimating the energy budgets of actively predatory fishes. Canadian Journal of Fisheries and Aquatic Sciences 39:371–379.

Kitchell, J.F., D.J. Stewart, and D. Weininger. 1977. Applications of a bioenergetics model to yellow perch (*Perca flavescens*) and walleye (*Stizostedion vitreum vitreum*). Journal of the Fisheries Research Board of Canada 34:1922–1935.

Logerwell, E.A., B. Lavaniegos, and P.E. Smith. 2001. Spatially-explicit bioenergetics of Pacific sardine in the Southern California Bight: Are mesoscale eddies areas of exceptional prerecruit production? Progress in Oceanography 49:391–406.

Madenjian, C.P., D.V. O'Connor, and D.A. Northrup. 2000. A new approach toward evaluation of fish bioenergetics models. Canadian Journal of Fisheries and Aquatic Sciences 57:1025–1032.

Nislow, K.H., C.L. Folt, and D.L. Parrish. 2000. Spatially-explicit bioenergetics analysis of habitat quality for age-0 Atlantic salmon. Transactions of the American Fisheries Society 129:1067–1081.

Rice, J.A., and P.A. Cochran. 1984. Independent evaluation of a bioenergetics model for largemouth bass. Ecology 65:732–739.

Schaeffer, J.S., R.C. Haas, J.S. Diana, and J.E. Breck. 1999. Field test of two energetic models for yellow perch. Transactions of the American Fisheries Society 128:414–435.

Stewart, D.J., J.F. Kitchell, and L.B. Crowder. 1981. Forage fishes and their salmonid predators in Lake Michigan. Transactions of the American Fisheries Society 110:751–763.

Tsehaye, I., M.L. Jones, J.R. Bence, et al. 2014. A multispecies statistical age-structured model to assess predator–prey balance: Application to an intensively managed Lake Michigan pelagic fish community. Canadian Journal of Fisheries and Aquatic Sciences 71:627–644.

Wahl, D.A., and R.A. Stein. 1991. Food consumption and growth of three esocids: Field tests of a bioenergetic model. Transactions of the American Fisheries Society 120:230–246.

Ward, D.M., K.H. Nislow, C.Y. Chen, and C.L. Folt. 2010. Rapid efficient growth reduces mercury concentrations in stream-dwelling Atlantic salmon. Transactions of the American Fisheries Society 139:1–10.

Whitledge, G.W., and R.S. Hayward. 1997. Laboratory evaluation of a bioenergetics model for largemouth bass at two temperatures and feeding levels. Transactions of the American Fisheries Society 126:1030–1035.

PART 3

Population Processes

CHAPTER 9

Abundance and Size Structure of Fish Stocks

The study of population ecology of fishes has developed with a large emphasis on fisheries. Fisheries science and management have historically focused around the stock concept of fish populations. In brief, fish stocks can be thought of as fish populations that experience very limited emigration or immigration. Therefore, internal population processes have strong influence on stock dynamics. Preceding chapters described growth of fishes, and subsequent chapters emphasize other internal population processes, such as mortality and recruitment. In this chapter, we provide an overview of fish stocks and then focus on the aspects of fish stocks that define their current biomasses.

At a particular point in time, fish stocks are chiefly defined by their abundances and individual sizes. While it may seem rather straightforward to count fish and measure their sizes, this is often not the case. According to John Shepherd of the University of Southampton, counting and managing fish is analogous to "managing a forest where the trees are invisible and keep moving around" (from an unpublished lecture at Princeton University, ca. 1978).

Fish Stocks

A fish stock can be considered analogous to a population in which primarily intrinsic rates affect the stock's biomass dynamics, and extrinsic processes – such as emigration and immigration – are considered negligible. The primary biomass gains to a fish stock are through growth of individuals and reproduction leading to addition of new individuals, while individual mortalities encapsulate losses to a closed fish stock. In fisheries management, it is common to differentiate between natural mortality rates and fishing mortality rates. Fishing mortality is fish death related to direct harvest or at least capture (e.g., hooking mortality in a catch-and-release fishery). Fishery managers have some ability to affect fishing mortality by controlling the intensity of fishing effort or the type of gear used to harvest fish. In contrast, natural mortality is fish death not related to fishery harvest (see Chapter 10).

The overall biomass of a fish stock may increase if the number of fish in the stock increases or if existing individuals grow and increase in size. The abundance of a fish stock can increase through reproduction and birth of young fish. In general, stock-level reproductive success is termed "recruitment." A new recruit is a young fish that has survived long enough and grown

Biology and Ecology of Fishes, Third Edition. James S. Diana and Tomas O. Höök.
© 2023 John Wiley & Sons Ltd. Published 2023 by John Wiley & Sons Ltd.

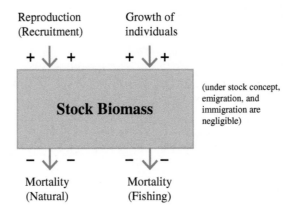

FIGURE 9-1   Russel's (1931) stock dynamics diagram. For a closed stock, growth and recruitment are the primary processes adding biomass to a stock, while mortality (both fishing and natural) may remove stock biomass.

to a sufficient size that it can be considered part of the adult or fishable stock. Recruitment has the potential to be highly variable from year to year, which is often related to physical processes affecting ability of young fish to survive during very early life. As individual fish grow, their lengths and mass increase, and this increase in mass can contribute to an overall increase in biomass of the stock.

The key components defining stock biomass dynamics were encapsulated conceptually by Russel (1931). Russel argued that biomass of a fish stock in a given year could be calculated as stock biomass the previous year, plus recruitment and growth, and minus the sum of losses through natural mortality and fisheries harvest. These dynamics are summarized in Figure 9-1. Again, note that Russel's concept is based on a closed stock, whereby there is no immigration or emigration into or out of the stock. Russel's concept and the diagram in Figure 9-1 are quite simplistic, whereas the processes they include can be quite complex and respond to a variety of abiotic and biotic factors. Nonetheless, they are useful for summarizing the key processes determining stock dynamics.

Stock Discrimination

A key step in the study and management of fish stocks is being able to define a fish stock in space and time. Defining the boundaries of a fish stock is rather simple in some systems. For example, when considering largemouth bass in an isolated lake, boundaries of the lake also define boundaries of the largemouth bass stock. In larger and less isolated aquatic systems, multiple fish stocks of the same species may inhabit the same system. There are multiple stocks of Atlantic cod in the northern Atlantic Ocean, and there are multiple stocks of lake white-fish in Lake Michigan. Stocks may inhabit the same areas and overlap in space during certain seasons. However, through migrations, they may reproduce in different areas and thereby increase genetic differentiation among stocks. Other stocks of the same species may reproduce in the same area but at different times. For example, in some streams draining to the Pacific Ocean, different stocks of anadromous rainbow trout – steelhead – migrate to spawn in the same stream but during different seasons. Stocks can be reproductively isolated based on either spatial or temporal isolation.

In order to study and manage fish stocks, it is important to be able to define the spatial and temporal extent of each stock. When multiple stocks overlap seasonally in space and time, it may be important to discriminate among stocks or at least know when multiple stocks are present in an area. If two stocks mix and one stock is endangered while the other is abundant, it may be risky to allow harvest as one could accidently overharvest the endangered stock. For migratory stocks, it may be important to quantify the areas that a stock occupies during different seasons to appreciate the suite of environmental conditions and potential fisheries harvest the stock experiences throughout the year.

Various methods have been applied to define the spatio-temporal extent of stocks and discriminate among stocks when they mix. A suite of genetic methods has been utilized to differentiate among populations with different levels of reproductive isolation. In addition, various approaches to physically and chemically mark fish have been developed. Fish have been marked by clipping select fins in a recognizable manner, and external, readily visible tags have been applied to fish. Tags inserted internally in fish (e.g., passive integrated transponders) have also been utilized and may include electronic tags that can be read externally using specialized detectors. In addition, one can take advantage of natural chemical marks of fish. For example, several hard parts of fish will take on chemical attributes of the environment they inhabit. Fish may retain the chemical signature of previously occupied environments as they move to another area. Using measurements from thin sections of otoliths (ear stones, crystalline structures used by fish for balance and hearing), scales, or bones may allow one to reconstruct the past history of locations, and even relate them to age of the fish. These types of manual or natural tags allow researchers to ascertain the environments currently and previously occupied by individual fish, and thereby piece together the spatio-temporal extent of fish stocks.

Physically marking fish has often required that researchers catch, mark, and then recapture individual fish in different environments. More recently, novel methods for tagging fish do not require recapturing individual fish. For example, acoustic tags and ultrasonic telemetry are now common tools for studying fish stocks. These involve attaching a tag that emits a unique acoustic signal. This signal can then be detected by acoustic receivers placed in the water at specific locations or by actively towing a receiver through the environment. Radio tags can communicate locations to satellites and are now used to track global locations of individual fish that surface periodically. Pop-up tags attached to fish store data – such as temperature and depth – then release from the fish at some time and float to the surface, where they transmit location and possibly other data to a satellite. Acoustic, pop-up, and various other types of tags now allow for archiving a suite of data to track experiences of individual fish, including depth, internal and external temperature, oxygen, activity, and even location inferred from time of day when sunset and sunrise occur (geolocation). These emerging methods (also addressed in Chapter 16) allow for inferring a variety of information about fishes beyond stock discrimination.

Characterizing Stocks: Relative Abundance

Essentially all methods for collecting fish are biased in terms of the numbers, species, and sizes of individual fish captured. Different methods for collecting fish generally rely on different collection gear, such as nets, hooks, and tools for stunning and collecting fish. Specific gear may target specific habitats, such as distinct vertical zones of the water column, while fish may be distributed across multiple habitats, and habitat usage may vary by size and sex of fishes. Sampling gear may be more effective under certain environmental conditions. For example, the ability of passive nets to capture fish may vary with water clarity and temperature as fish may detect and avoid nets under certain water clarity conditions and temperature may affect their activity level and hence likeliness of encountering nets. The suitability of electrofishing – capturing fish by stunning them with an electric current passed through the water – depends

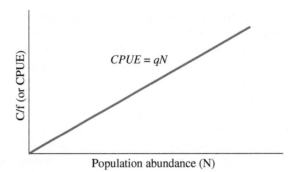

FIGURE 9-2 General assumed relationship between population abundance and catch per unit of effort (C/f or CPUE). The slope of this relationship, q (catchability), is the proportion of a population that will be caught with one unit of effort. While q is almost never known, assuming this linear relationship between abundance and CPUE allows for comparisons of CPUE as indices of relative abundance. Several other fish stock models build from this general relationship as well.

on conductivity of water. Almost all fishing gear are able to collect some species more effectively than others, and within a species, likelihood of capture by fishing gear will vary with size. Gillnets, which depend on fish swimming into a mesh net and becoming entangled, may neither capture very small fish that swim through the mesh nor very large fish that are too large to become entangled in the mesh.

Despite these well-established biases in the ability of fishing gear to collect fish, capture rates are often used to index relative abundances of fish across space or over time. Capture rates are generally expressed as catch (C) per unit effort (f) or CPUE. Catch can be expressed in biomass or numbers, and effort is generally specific to the gear and method used. Typical units of effort include time (e.g., minutes spent electrofishing; hours of trapnets in the water), number of events (e.g., number of overnight lifts of a gillnet), and volume or area (e.g., hectares swept by a bottom trawl). A linear relationship is often assumed between the capture rate of fishes and their abundance (N) in the environment. Specifically, this assumption is based on the following equation:

$$C/f = qN. \qquad \text{(Equation 9-1)}$$

Here, q is the term catchability or the proportion of the fish stock that will be captured with one unit of effort. Based on this equation, the relationship between CPUE and abundance is linear, and catchability is the slope (Figure 9-2). It is not necessary to know the value of q to make use of this relationship. Rather, by assuming q is constant (an assumption that will not always hold), one can infer that a doubling of CPUE represents a doubling of population abundance or density. In this way, CPUE is an index of relative abundance and allows for comparison of abundance from one time or location to another. On its own, CPUE does not allow one to index actual abundance.

Characterizing Stocks: Estimating Actual Abundance

While several assessments rely on CPUE as an index of relative abundance, in many cases it is important to quantify actual abundance of a stock in a system or habitat. There are few cases where it is possible to count all individuals and truly quantify total abundance of a fish stock. In most cases, it is necessary to count a subset of individuals and extrapolate to total abundances. A large variety of techniques has been developed for such extrapolation, and the appropriate approach will vary depending on the characteristics of (a) the habitat surveyed, (b) the fish stock of interest, and (c) the sampling method used.

Whole Counts and Swept-area Method

As mentioned, there are some systems where it is truly possible to count all individuals in a fish stock. It may be possible to drain a small pond or treat it with a poison like rotenone to kill all fish in the pond. This could allow for counting all fish but at the cost of dramatically altering the system or killing all individual fish. Some fish stocks are seasonally concentrated, thereby facilitating counts of the whole stock. For example, many anadromous fishes will swim through the mouth of their natal river system as they return to spawn. This can facilitate counts of the entire spawning stock. Historically, abundances of some salmon stocks have been assessed using count-towers positioned near the mouth of shallow rivers, where an observer can count individuals as they swim by. There are currently more sophisticated approaches for enumerating individual fish as they swim past a location in a river, including resistance-based counting (where a bypassing fish alters the resistive characteristics between electrodes in a river), optical counters (where the interruption of a beam of light by swimming fish is detected), and hydroacoustic methods.

For most fish stocks, all individuals do not swim through a specific constricted location, and it is not practical to count all fish in a system. However, it may be possible to count all individuals in a known proportion of the system and then extrapolate to the entire system. Such a swept-area approach can be used to extrapolate catches from bottom trawls. Bottom trawls generally have a known opening width. As they are pulled along the bottom, the distance they travel (measured with GPS and other methods) can be multiplied by their opening width, therefore allowing for quantification of the bottom area surveyed during a particular trawl event. If one assumes all fish that the bottom trawl passes over are caught, the number of individuals caught divided by the area covered by the trawl provides a measure of two-dimensional density of fish on the bottom (e.g., number of fish per ha). Several trawling events take place during a survey, and these would be integrated to generate an overall measure of density. This overall measure of density can then be simply multiplied by the total bottom area of the system to quantify overall abundance. For example, the swept-area method is used by the USGS Great Lakes Science Center to generate system-wide estimates of prey fish abundances based on an annual bottom trawl survey that takes place in multiple locations around Lake Michigan (Madenjian et al. 2005). Similar to the swept-area method with trawls, it may be possible to block off certain segments of streams, and through very aggressive collections count all individual fish in such segments. Therefore, an estimate of density for each segment can be generated and, these can be combined to quantify abundance of a stock for the entire stream network.

A critical challenge with the swept-area method or the segment sampling of streams is the assumption that one can actually count all individuals in the area. For most species of fish, they can detect a bottom trawl approaching and move horizontally, up in the water column, or even burrow into the sediment to avoid the trawl. Even if fish are not able to detect the trawl, some individuals may be passed over by the trawl, and some small individuals may pass through the mesh of the trawl and escape. Consequently, the swept-area approach for estimating abundance generally leads to underestimates of actual abundance. If the efficacy of fishing gear can be quantified, the probability of actually catching individual fish can be accounted for when estimating abundance. For example, if it is known that a bottom trawl will catch on average 50% of the individual fish it passes over, this information can be incorporated when estimating abundance (by multiplying observed catches by 2). While studies of gear efficiency may not seem exciting, they are critical for being able to estimate true abundances of fish. Such studies can also be very important when attempting to compare abundances across locations or time periods when different gear types are used to collect fish.

Hydroacoustic and Sonar for Estimating Abundances

There is an ever-increasing number of approaches for improving the ability of researchers and managers to quantify the abundances of fish. Hydroacoustic methods for quantifying fish abundances have been in existence for decades, but they continue to be refined. Such methods can not only be used to count fish as they move past a point in a river or stream, but also to quantify fish abundances and overall biomasses in open systems, such as lakes, estuaries, and marine environments. The basic idea behind hydroacoustics is that brief sound waves are emitted from a transducer. These waves then travel through the water and, if they encounter an object such as a fish, some of the sound is reflected back toward the transducer. The time it takes for sound to come back to the transducer can be used to quantify the distance from the transducer, and the characteristics of the returning sound waves can be used to elicit characteristics of individual fish. In using hydroacoustics for quantifying fish abundances, it is key to not only be able to emit and receive sound waves, but to amplify and filter out specific sound waves and record and analyze these waves. Individual fish differ in terms of their acoustic properties that are related not only to the sizes of fish but also their anatomy. In particular, swim bladder characteristics can have strong influence on a fish's acoustic properties, and since species differ in terms of swim bladder morphology, species also differ in terms of acoustic properties. This has allowed for the development of species-specific relationships between total length and acoustic target strength. Even so, potential acoustic target strengths will often overlap, and a challenge with hydroacoustics is being able to differentiate acoustic echoes among species. This is particularly challenging in very species-rich systems. In such cases, hydroacoustic surveys are often accompanied by netting surveys, in which acoustic echoes can be portioned among species based in part on species proportional composition in net catches.

Many methods for estimating abundances of fish – either through whole counts or partial counts (see below) – involve physically capturing individuals. The act of being captured can be quite taxing for fish, contribute to various physiological costs, and even lead to death. Such deleterious effects of capturing fish may be of particular concern when studying fish populations at low abundance or species that may be at risk of extinction. Various researchers have attempted to develop methods to enumerate fish without the need to physically capture them. Hydroacoustic methods as described above are a potential method to allow for enumeration without capture. However, if this method does not allow to differentiate among species, or if the method requires supplemental capture of fish to distribute fish biomass among species, then it may not be suitable.

Atlantic sturgeon (*Acipenser oxyrinchus*) are an endangered, anadromous species that migrates from the Atlantic Ocean and estuaries into tributaries to spawn. It is a species of high conservation concern, and to inform conservation efforts, it is useful to (a) have estimates of abundance for specific tributaries and (b) be able to obtain such estimates without actually capturing and handling sturgeon. Clemson University researcher, Joshua Vine, and colleagues developed a method to count Atlantic sturgeon using side-scan sonar (Vine et al. 2019). They focused on sturgeon spawning areas in the Savannah River (approximately 280–300 km upstream from the Atlantic Ocean) and sampled these locations repeatedly during 2017. The researchers used either a boat-mounted or towed sonar unit to survey 2 km stretches of the river. They counted Atlantic sturgeon in their sonar files by looking for objects greater than 1.2 m (the assumed minimum total length of spawning sturgeon) and had morphological features consistent with Atlantic sturgeon. They used an N-mixture Bayesian statistical model to estimate abundances. In estimating abundances, this modeling approach allowed the researchers to account for false negatives (occasions when sturgeon were present but not detected). Ultimately, they detected sturgeon 803 times with the majority of detections (505) coming from a single 2-km stretch of river.

Estimating Abundances from Partial Counts

Several methods have been developed and refined to estimate abundances of organisms based on enumerating only part of populations. There are too many methods to fully describe here. Many of these approaches rely on rather involved quantitative and statistical models. Overviews of certain common methods for estimating abundances are presented below, including the line transect approach, mark-recapture estimation, depletion, and virtual population analysis.

The line transect approach involves traveling along a defined transect and counting organisms encountered along the way. This approach is commonly used to estimate abundances of terrestrial organisms. A line transect approach would not be appropriate to quantify abundance for many fish stocks and systems because it relies on being able to detect fishes visually. However, in some systems, this approach is entirely appropriate. For example, the line transect approach may be useful in very clear water where an investigator may swim along a transect using SCUBA or snorkel and count individual fish along the way. Line transect or other visual survey methods are frequently used in some coral reefs, not only because the water in these systems is often relatively clear, but because methods to collect fish may be destructive of corals.

The line transect approach allows for the estimation of fish density in the vicinity of the survey transect. A key challenge with this approach is the ability to quantify the actual probability of detecting fish along the transect line. Even in a system with very clear water, all individuals will not be visible as some individuals may be obscured by structures or other fish. Individuals closer to the survey line will be easier to detect than those farther away. The decline in detection with distance from transect can be used to develop a detection probability function based on distance from transect. This detection probability function can be used together with the number of individual fish detected (n) to estimate the actual number of fish within a certain distance of the transect. One can think of a line transect surveying a rectangular area with a length equal to the length of the transect (L) and the width equal to two times the effective width of detection (W_e). The effective width of detection accounts for declining probability of detection with distance from the transect and acknowledges that, while all individuals within this width will be detected, some individuals beyond this width will be detected. Estimating this width is beyond the scope of this chapter, but having estimated the effective width, the density (D) can then be estimated as

$$\hat{D} = \frac{n}{2LW_e} \qquad \text{(Equation 9-2)}$$

The mark-recapture approach is commonly used to estimate abundances of fish. The approach involves marking some number (M) of fish in a stock and releasing these individuals back into the environment. Then, returning to this stock sometime later and collecting a new sample of fish and observing how many of these fish are recaptured marked fish (R) and how many are caught (C). By assuming the number of marked fish in the stock is known (it is equal to the number of fish initially marked and released back into the environment) and by assuming the proportion of marked fish in the stock is equivalent to the proportion of marked fish in the sample collected when the stock was surveyed a second time, one can estimate total abundance (N) as follows:

$$\frac{M}{N} = \frac{R}{C}$$

(Equation 9-3)

$$\hat{N} = \frac{MC}{R}$$

(Equation 9-4)

The above description of the mark-recapture method is a simplification. The equations used to estimate abundance through this approach may be somewhat more complex to account for certain statistical considerations. In many cases, it may not be appropriate to survey a stock only twice. Instead, fishes may be marked on more than one occasion, resurveyed, and potentially recaptured on multiple occasions. While such approaches require modifications to the above equations, the general premise of the mark-recapture methods remains the same. In addition, certain assumptions are required regardless of the exact mark-recapture approach used, including that (a) there is no migration into or out of the system, (b) catchability of all fish is equivalent (so one is not more likely to recapture certain individual fish), (c) there is no tag loss and marks (tags) are recognizable, and (d) survival of marked fish is equivalent to survival of unmarked fish. If there is tag loss or unequal survival, this can be quantified and should be incorporated to generate more accurate estimates of abundance.

The depletion approach involves estimating abundance based upon the rate at which catches of fish decline with repeated removal of fish from a stock. Consider a pond with a population of sunfish fished with four fyke nets. As fish are caught, they are removed and placed in holding tanks (they are not returned to the pond), and the population is depleted through consecutive collections. Assume one collects 60 fish (15 fish per 24 hour fyke net set) during the first 24 hours of fishing, and one collects 30 fish (7.5 fish per 24 hour fyke net set) during the second 24 hours of fishing. This reduction in the catch rate is not surprising, given 60 fish had been removed and were no longer available to be caught. Assume one collects 15 fish (3.75 fish per 24 hour fyke net set) during the third 24 hours of fishing. If one were to continue this collection and removal approach, one would eventually catch no fish. However, rather than actually continuing collections until no more fish are caught, one can use the rate at which catch rates decline to estimate how many fish would need to be removed before the catch rate equals zero. In the above example, one would need to remove 120 fish before the catch rate would equal zero, and the estimate of abundance would be 120 sunfish in the pond. This approach is based upon a series of assumptions including that (a) catchability is constant (and therefore C/f is proportional to abundance), (b) all individuals are equally vulnerable, (c) there is no migration into or out of the population surveyed, and (d) there is sufficient removal of fish to meaningfully deplete the population.

Several studies have evaluated the efficacy of different methods for estimating abundances, and some studies have directly compared the performance of different abundance estimation methods. Rosenberger and Dunham (2005) conducted a study to compare the ability of mark-recapture and depletion methods to assess the abundance of rainbow trout (>60 mm) in streams in Idaho. They blocked off stream segments and collected rainbow trout within a segment through electrofishing. They tagged the fish and returned them to the segment. The next day, they collected fish within the segment again, noted if they were tagged, and removed them from the stream. They repeated this removal sampling four times per stream segment. They considered the recapture of marked fish to generate mark-recapture based estimates of abundance. For the depletion method, they knew how many marked fish were in the segment and were able to assess how rapidly the capture of marked fish declined over consecutive sampling events. They found that the two methods differed in performance. Specifically, the depletion method tended to underestimate abundances relative to mark-recapture methods. They found that capture efficiency declined over consecutive collection events, and captures of rainbow trout therefore declined rather rapidly, leading to a relatively low depletion-based estimate of abundance. In particular, they found that capture efficiency declined sharply in larger streams (greater cross-sectional area) and in streams with more woody structure, leading to particularly biased underestimates of abundance, using the depletion method in such streams.

Virtual Population Analysis

Virtual population analysis is related to a broader class of modeling methods to estimate abundances at age based on fish deaths. If harvest numbers are tracked, deaths due to fishing may be known or at least can be estimated. However, non-fishing deaths may often need to be inferred. The actual abundances necessary to support observed and inferred deaths can be estimated retrospectively. Therefore, the approach can generate historical estimates of year-specific abundances of individual cohorts. Methods related to virtual population analysis include statistical catch-at-age and statistical kill-at-age. These methods differ in terms of whether they assume true catch and kill are known or must be estimated with some uncertainty. These models require information on a variety of factors – such as age-specific fishery selectivity – and can be quite statistically complex. However, they are powerful tools for assessing stock abundances and responses to various fishing pressures and are therefore widely used by assessment agencies.

Genetic Methods to Estimate Abundances

Researchers are consistently searching for novel methods to quantify the abundances of fish populations. Recent advances in the field of genetic analysis have increasingly allowed genetic tools to be used in abundance estimation. Environmental DNA, or eDNA, has now been widely used to detect the presence of species in systems. As fish swim, feed, spawn, and generally inhabit a system, they consistently shed material, including DNA from their body. While such eDNA will degrade over time, it persists in water for some time, and more eDNA may be expected to accumulate if more individuals are present in the system, spawning is ongoing, and environmental conditions are such to slow down degradation. Collection and analysis of eDNA may be particularly useful when attempting to detect a non-native species that may only recently have migrated into a system or rare, difficult to capture native species of conservation concern. eDNA analyses are based on collecting water samples and analyzing such samples for the presence of genetic material from specific species, often relying on high-throughput amplification and sequencing methods.

While eDNA analysis was initially used to simply detect species, more recent analyses are developing this method to quantify abundances. For example, Hideyuki Doi (from University of Hyogo), Ryutei Inui (from Yamaguchi University), and colleagues (Doi et al. 2017) evaluated the ability of eDNA to estimate abundance and biomass of ayu in the Saba River, Japan. These researchers defined seven sampling sites on the Saba River and visited each location three times, in May, July, and October 2015. Upon each visit, they first collected a water sample at the downstream location of each site and then conducted a snorkeling survey to detect and enumerate ayu. They also collected some ayu using electrofishing and measured their lengths and masses to allow an estimation of biomass at each site. To quantify ayu numbers using eDNA, Doi and colleagues filtered their water samples, extracted DNA from filters, and quantified the DNA using polymerase chain reaction to amplify specific ayu mitochondrial gene fragments. They quantitatively compared the DNA concentration collected in their water samples with snorkeling- and electroshocking-based estimates of abundance and biomass. They found that for each month there was a strong positive association between the number and biomass of ayu they quantified at a site and the DNA concentration detected in water samples. They suggested that eDNA should be an appropriate method to not only detect fish in a system but also to quantify their abundance.

The abundance of a fish population affects its genetic variability. Genetic diversity tends to decrease as populations decrease in size. Genetic drift – the random change in gene distributions – generally becomes more influential at low population size. In addition, inbreeding tends to increase in small populations, further compromising genetic diversity. These linkages between population size and genetic diversity imply that the genetic variation of a population may be used to infer its true population size. However, such inferences are not straightforward. For example, a population may experience a period of low abundance, leading to a decrease in genetic variation. If the population subsequently increases in abundance, its genetic variation may remain constricted due to the period of low abundance, leading to a population with low genetic variation but relatively high actual abundance. Similarly, if mating in a population is non-random and a small number of individuals contribute the vast majority of offspring, a relatively large population may be characterized by low genetic diversity. In short, population abundance and genetic variation are strongly related but not analogous. This is acknowledged through the calculation of effective population size. This quantity has been defined and calculated somewhat differently by various authors. Effective population size involves relating observed genetic variation in a population to an idealized population. Here, an idealized population has random mating and genetic drift. The effective population size is the number of breeding individuals in the idealized population that would yield an equivalent level of genetic diversity as observed in the actual population. In other words, a population with high actual abundance but low genetic diversity would be characterized by a low effective population size.

Alexandra Pavlova, from Monash University in Australia, and colleagues explored genetic diversity and effective population sizes of Macquarie perch – an endangered freshwater fish species in Australia (Pavlova et al. 2017). They collected and analyzed genetic samples from 20 populations of Macquarie perch from various rivers in Australia. Given the history of management concern of this species, some of these populations had been previously transplanted or supplemented from other populations. However, most of these populations were relatively small and at further risk of decline due to habitat loss and fragmentation. Pavlova and colleagues found that the populations expressed high levels of inbreeding and had very low effective population sizes. Through simulations, they showed that if these populations were left alone they would likely continue to decline due to inbreeding depression. They argued for active management and translocations among remnant populations in the same catchments. Their analyses suggested that such translocations were the best approach for maintaining within-population genetic variation and sufficient effective population sizes to allow the populations to persist. Pavlova et al.'s study demonstrates the need for abundance assessment and management decisions to not only be based on actual abundances but also on genetic variation and effective population sizes.

Characterizing Stocks: Size Distributions

In addition to abundance, key point-in-time attributes of a fish stock are the sizes of individuals and the size distribution of the entire stock. Measuring size and growth of individual fish is described elsewhere. Here, we focus on the size distributions of fish stocks. A typical size distribution of a fish stock based on collections by fishing gear may look similar to Figure 9-3. One might expect that small fish will be most abundant in a fish population as all individuals begin as small, young fish and few individuals survive to a large size and old age (see Chapter 10 and survivorship curves). Since most fishing methods are size biased, catch of very small fish is not a true reflection of their abundance in the environment. In addition, the size distribution may have multiple modes (peaks), a common feature of fish size distributions. Abundant year

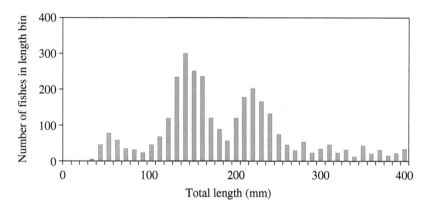

FIGURE 9-3   Hypothetical length distribution of a large sample of individuals from a fish stock. Note that small fishes are often underrepresented in such samples because they are not fully vulnerable to collection gears. Also, note that multiple modes of lengths are often evident from large samples, corresponding to different age classes of fishes.

classes (individuals born the same year) will often be reflected in size distributions, and these may be particularly evident for young ages. While sizes of older ages often overlap, sizes of faster growing fish of younger ages may allow for age discrimination by size.

For management purposes, it may be useful to succinctly summarize size distributions of fish stocks. This is particularly common for assessing recreational freshwater fisheries. More so than commercial fisheries, many recreational fisheries are judged by their size distributions and the prevalence of sought-after, large, trophy fish. As a simple quantification of recreational fishery size distributions, managers often use stock density indices. The most common stock density index is proportional stock density (PSD), which is calculated from a sample of fish lengths as follows:

$$PSD = \frac{NQL}{NSL} \times 100. \qquad \text{(Equation 9-5)}$$

NSL is the number of fish in a sample of stock length or greater. Stock length is defined differently for a variety of species and stocks, but may be considered as the size threshold when fish are adequately sampled by assessment gear, the size above which there is some potential recreational value, or the smallest size when fish may reach sexual maturity. Quality length is the size threshold anglers prefer to catch. This is a length greater than stock length and NQL (number of fish collected quality length or greater) never exceeds NSL. PSD can take values from 0 to 100, and a high PSD would suggest there are a relatively large number of fish anglers like to catch in the population, while a low value suggests the stock is dominated by smaller, less desirable fish. For an example of calculating PSD, see Box 9-1.

The idea of analyzing fish stock size distributions using an approach analogous to the PSD calculation was first proposed by Johnson and Anderson (1974). The general approach has expanded to include the calculation of not only PSD, but also various other relative stock density (RSD) indices based on the proportion of different size cutoffs in the fish stock. In addition to considering the number of fish above stock and quality length, these indices can estimate the proportion of fish above preferred, memorable, and trophy size. Studies have suggested different lengths to use as cutoffs between these length categories in a variety of species, and various PSD and RSD indices have been proposed. These all have in common the idea of collapsing a large number of length measurements into a single value between 0 and

Box 9-1 Example of calculating proportional stock density (PSD)

While the calculation of proportional stock density (PSD) may appear confusing, in reality, it is rather straightforward. Consider the following example for yellow perch. According to Gabelhouse (1984), the length threshold for yellow perch to be considered stock length is 130 mm, and the size threshold for quality length is 200 mm.

Researchers from Grand Valley State University collected 66 yellow perch from Pentwater Lake, Michigan, during April and May 2016. The total length of each yellow perch is provided in Table 9-1. Note that out of the 66 yellow perch, 28 are below the stock length threshold and do not factor into the calculation of PSD, whereas 38 individual fishes are above the stock length threshold (underlined values in Table 9-1). Out of these 38 fishes, 28 are also above the quality length threshold (bolded values in Table 9-1). Thereby, the PSD value for this sample of yellow perch is equal to $100 \times (28/38)$ or 74.

TABLE 9-1 Total Lengths of 66 Individual Yellow Perch Collected in Pentwater Lake During April–May 2016. Individuals' Lengths That are Equal or Above the Stock Length Threshold of 130 mm are Underlined, and Individual Lengths That are Equal or Above the Quality Length Threshold of 200 mm are Bolded.

185	**261**	**225**	**219**	**212**	**246**	**250**	**256**	**287**	**288**	**214**
252	**231**	**203**	**233**	**246**	141	136	132	**210**	**212**	160
261	125	123	135	129	89	135	124	123	125	159
115	128	129	115	124	112	137	111	112	122	142
122	124	121	130	123	127	118	116	118	109	120
116	**213**	**252**	**212**	**244**	**213**	**250**	196	**256**	**232**	**254**

100%. This can be both beneficial and problematic. On the one hand, whenever one calculates an index one loses information (in this case, individual lengths and abundances of fish). On the other hand, the PSD approach is simple and can provide a single value to guide managers. For example, Gabelhouse (1984) provided various targets for largemouth bass stock densities using both a traditional PSD approach and an incremental RSD approach. The targets differ depending upon the management goal. Using the traditional PSD approach, a PSD range of 20–40 may be appropriate if trying to maintain a high density of largemouth bass to feed on small panfish and therefore produce larger panfish. A PSD of 40–70 may be appropriate if largemouth bass are intended to be one of various sport fishes in a balanced community, and a PSD of 50–80 may be appropriate if managing to produce large individual bass with limited consideration for maintaining a large number in the population or for balancing the community among different species.

Management actions in recreational fisheries are often intended to influence the size distribution of fishes and therefore affect PSD. For example, instituting a size limit such that fish can only be harvested if they are above some minimal size is generally expected to increase the size of fish in a population and therefore increase PSD values. However, several factors can influence the actual PSD calculated for a fish stock, including when the fish are sampled during the year, how many individual fish are examined, and the strength of different year classes.

A very strong year class of relatively young fish could lead to low PSD values because the fish stock would be dominated by young, small individual fish. In this context, Michael Allen and Bill Pine of the University of Florida set out to ask the question of whether instituting a minimum size limit in a recreational fishery would lead to detectable changes in PSD values (Allen and Pine 2000). They addressed this issue through a simulation study where they explored the response of simulated populations to different size limits for white crappie (simulated size limit of 254 mm) and largemouth bass (simulated size limits of 305, 356, and 457 mm). They also considered the effect of different levels of recruitment variation and the consequence of evaluating population size structure responses over a three-year versus five-year period. Based on their simulations, they found PSD values were more likely detectably different from populations with no size limit if (a) a larger size limit was instituted (more likely to see an increase in PSD with a 457 mm size limit for largemouth bass than with a 305 mm size limit), (b) recruitment variation was low (high recruitment variation confounded the ability to detect changes in PSD), and (c) a longer evaluation period was used (more likely to detect an increase in PSD with a five-year evaluation period versus a three-year evaluation period). These simulations demonstrated not only how PSD values may respond to management actions, such as institution of a minimum size limit, but also some of the complexities and considerations in being able to actually detect a response in PSD.

Summary

Fish and fisheries have traditionally been studied and managed as discrete stocks. Assuming a closed stock (with no immigration or emigration) and focusing on a single stock (with little consideration for interactions with other stocks and species) may be inappropriate and too simplistic for the management of many systems. However, the single stock paradigm provides a useful construct to conceptualize the processes that affect the dynamics of fish populations. Key attributes of a fish stock include the abundance and sizes of the individuals comprising the stock, and these attributes collectively determine the biomass of a fish stock. As defined by Russel's stock equation, the future biomass of a closed stock increases based upon individual growth rates and recruitment, while biomass decreases due to losses through natural and fishing mortality.

Literature Cited

Allen, M.S., and W.E. Pine III. 2000. Detecting fish population responses to a minimum length limit: Effects of variable recruitment and duration of evaluation. North American Journal of Fisheries Management 20:672–682.

Doi, H., R. Inui, Y. Akamatsu, et al. 2017. Environmental DNA analysis for estimating the abundance of stream fish. Freshwater Biology 62:30–39.

Gabelhouse Jr., D.W. 1984. A length-categorization system to assess fish stocks. North American Journal of Fisheries Management 4:273–285.

Johnson, D.L., and R.O. Anderson. 1974. Evaluation of a 12-inch length limit on largemouth bass in Philips Lake, 1966–1973. Pages 106–116 in J.L. Funk, editor. Symposium on Largemouth Bass Overharvest and Management in Small Impoundments. Special Publication 3, North Central Division, American Fisheries Society, Bethesda, Maryland, USA.

Madenjian, C.P., T.O. Höök, E.S. Rutherford, et al. 2005. Recruitment variability of alewives in Lake Michigan. Transactions of the American Fisheries Society 134:218–230.

Pavlova, A., L.B. Beheregaray, R. Coleman, D. et al. 2017. Severe consequences of habitat fragmentation on genetic diversity of an endangered Australian freshwater fish: A call for assisted gene flow. Evolutionary Applications 10:531–550.

Rosenberger, A.E., and J.B. Dunham. 2005. Validation of abundance estimates from mark-recapture and removal techniques for rainbow trout captured by electrofishing in small streams. North American Journal of Fisheries Management 25:1395–1410.

Russel, E.S. 1931. Some theoretical considerations on the "overfishing" problem. Journal du Conseil International pour l'Exploration de la Mer 6:3–20.

Shepherd, J. 1978. *Thoughts & Sayings.* Unpublished lecture at Princeton University. Available at: http://jgshepherd.com/thoughts/

Vine, J.R., Y. Kanno, S.C. Holbrook, W.C. Post, and B.K. Peoples. 2019. Using side-scan sonar and N-mixture modeling to estimate Atlantic sturgeon spawning migration abundance. North American Journal of Fisheries Management 39:939–950.

CHAPTER 10

Mortality

Survival and mortality rates of fish stocks have direct effects on stock abundance and biomass, and current and future mortality risks have a strong influence on individual fitness, the ability to pass on genetic material to subsequent generations. If an individual does not survive to reproduce, they will not directly pass on their genetic material. At the population level, cohorts and stocks experiencing relatively high mortality will be characterized by low expected lifetime duration and mean age. Consequently, such stocks will often have relatively high abundances of young, small fish and relatively few old, large fish. As larger, older fish are generally more fecund, mortality rates may in turn affect the reproductive potential of a fish stock. Further, since these larger fish are generally more sought after by fisheries, high mortality rates may affect the perceived fishery value of a fish stock.

Most fish populations follow a type III survivorship curve, meaning that most individuals die during early life, and few individuals survive to old age (Figure 10-1). This is in contrast to type II survivorship, whereby the number of deaths is fairly constant across ages (as exemplified by some bird and rodent species), and type I survivorship, whereby there is relatively high survivorship during early life and increased mortality at older ages (as exemplified by many mammal populations, including humans).

Various factors and events can contribute to fish deaths. Fish can die due to processes such as starvation, predation, pathogens, lack of oxygen, stranding out of water, excessively high or low temperatures, physical trauma, osmoregulatory imbalances, and disease. Human activities can contribute to several of the above causes of fish death. For example, human land use can contribute to hypoxia and cause asphyxiation; through stocking of fish, humans can spread pathogens and introduce predators; and human manufactured turbines and propellers can cause physical trauma and death to fish. Nonetheless, fisheries managers often focus on two sources of mortality: mortality through fishery harvest and non-fishing mortality (or natural mortality). Fisheries managers have some control of fishing mortality, through regulations that limit fishing pressure or the size of fish harvested, while natural mortality is generally considered to be outside of the control of fisheries managers.

Survival and mortality rates can be considered proportional rates and are often represented as the proportion of fish that survive from one year to the next. In a closed fish stock, individual fish can either survive from one year to the next or die. Therefore, proportional survival (S) can take values from 0 to 1, and the sum of S and proportional mortality (A) should equal 1 or $A = 1 - S$.

Biology and Ecology of Fishes, Third Edition. James S. Diana and Tomas O. Höök.
© 2023 John Wiley & Sons Ltd. Published 2023 by John Wiley & Sons Ltd.

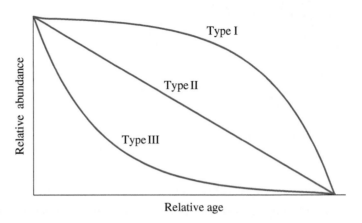

FIGURE 10-1 Typical organism survivorship curves. Most fish populations follow a type III survivorship curve, whereby most individuals die during early life.

Mortality rates can also be expressed on an instantaneous basis based on the following equation:

$$N_t = N_0 \, e^{-Zt} \qquad \text{(Equation 10-1)}$$

Here Z is instantaneous mortality rate, t is time, N_0 is initial abundance, and N_t is abundance at time t. Instantaneous mortality rates are expressed in per time units, in general year^{-1} for adult fish and day^{-1} for eggs and larval fish.

Variability in Mortality Rates

Mortality rates can be highly variable among different species of fishes and different populations of the same species. In addition to inter-specific and inter-population differences in mortality rates, within a population of fish, mortality rates often change dramatically through ontogeny. However, fish ecologists often assume that mortality rates for a population are constant once individuals survive some early life stages. This assumption allows for estimation of mortality rates and subsequently allows for comparison of adult mortality rates among species, populations, and cohorts. For example, adult mortality rates of fish species are highly variable with long-lived species, such as sturgeon experiencing very low mortality rates, while short-lived species, such as herring, experience relatively high rates (Table 10-1).

Even within a species, adult mortality rates can be highly variable, within and among populations. Mortality rates can vary widely within a population over time as sources of mortality change. For example, increased predation pressure or higher risk of starvation could both lead to an increase in mortality rates for an adult fish population over time, while decreased prevalence of a pathogen or parasite could lead to a decrease in mortality rates over time. Several researchers have compared intra-specific mortality rates of different populations occupying different bodies of water. As an example, Heibo et al. (2005) examined variation in mortality rates among different populations of Eurasian perch and documented a broad range of adult instantaneous annual mortality rates, from 0.23 to 1.14 yr^{-1}, which correspond to proportional annual mortality rates ranging from 21 to 68%. If two groups of 100 perch experienced these mortality rates for four consecutive years, the group experiencing $Z = 0.23$ yr^{-1} would still have

TABLE 10-1 Example Estimated Instantaneous Annual Natural Mortality Rates of Different Fish Species and Populations.

Species	Location	Mortality Rate (yr⁻¹)
Acipenser fulvescens	Wisconsin, USA	0.01
Pleuronectes platessa	North Sea	0.12
Esox lucius	Lake Windermere, England	0.26
Perca fluviatilis	Sweden	0.29
Thunnus albacares	Hawaii	0.80
Clupea harengus	North Sea	1.00
Tilapia esculenta	Lake Victoria	1.75
Mallotus villosus	Labrador Sea	2.45
Stolothrissa tanganicae	Lake Tanganyika	5.20

Source: Adapted from Pauly (1980).

40 individuals remaining in the population, while the group experiencing $Z = 1.14 \text{ yr}^{-1}$ would only have 1 individual fish remaining.

In addition to inter-specific and inter-population differences in mortality rates, within a population of fish, mortality rates often change dramatically through ontogeny. Mortality rates for early life stages are generally reported at sub-annual increments (e.g., daily mortality rates), but mortality rates during early life tend to be much greater than during adult life stages when considered in equivalent time units. Ed Houde from the University of Maryland has published several papers comparing growth and mortality rates among larval fishes. In one such paper (Houde 1989), they compiled mortality estimates for 22 species of feeding-stage fish larvae and reported a range of daily instantaneous mortalities from 0.01 to 0.69 d^{-1} (Table 10-2). If these values were converted to annual rates, they would correspond to annual instantaneous mortality rates of 3.7 to 251.9 yr^{-1} or proportional annual mortality rates of 97% to greater than 99.99999999%. Interestingly, Houde also compared these daily

TABLE 10-2 Example Field-Derived Estimated Instantaneous Daily Mortality Rates of Different Marine Larval Fish Species.

Species	Mortality rate (d⁻¹); Ranges from Multiple Studies with Midpoint in Parentheses	Temperature (°C); Ranges from Multiple Studies with Midpoint in Parentheses
Pleuronectes platessa	0.02–0.08 (0.05)	1.0–8.0 (4.5)
Sebastes sp.	0.05–0.07 (0.06)	3.5–12.5 (8.0)
Morone saxitilis	0.13–0.21 (0.17)	12.0–22.5 (17.0)
Engraulis mordax	0.16–0.22 (0.19)	12.0–20.0 (16.0)
Clupea harengus	0.01–0.46 (0.24)	6.0–17.0 (11.5)
Anchoa mitchilli	0.30–0.45 (0.38)	24.0–31.0 (27.5)
Scomber	0.35–0.69 (0.52)	14.0–20.0 (17.0)

Source: Adapted from Houde (1989).

				Stage-Specific Instantaneous Mortality (Z) and Duration for a Hypothetical

TABLE 10-3 Stage-Specific Instantaneous Mortality (Z) and Duration for a Hypothetical Fish Population, Demonstrating how Abundances Would Change Through Different Stages.

Stage	Starting Abundance	Z	Duration	Proportion Dead	Number Dead	Ending Abundance
Eggs	25,000,000	0.05 (day^{-1})	5 days	0.22	5,529,980	19,470,020
Yolk-sac larvae	19,470,020	0.25 (day^{-1})	6 days	0.78	15,125,671	4,344,349
Larvae	4,344,349	0.10 (day^{-1})	40 days	0.98	4,264,779	79,570
Juveniles	79,570	2.0 (year^{-1})	6 months	0.63	50,298	29,272
Adults	29,272	0.5 (year^{-1})	6 years	0.95	27,815	1,457

instantaneous mortality rates for fish larvae to the temperatures they occupied and fitted a relationship between daily mortality rate (Z) and temperature (T): $Z = 0.0256 + 0.0123T$. This relationship suggests that the daily mortality rate would be expected to increase by 0.0123 with every 1 °C increase in temperature, or, as Houde points out, from 0.09 d^{-1} at 5 °C to 0.30 d^{-1} at 30 °C.

Ultimately, the proportion of a cohort that dies during a life stage is a function of not only the mortality rate, but also the duration of that life stage. That is, while larval mortality rates tend to be much greater than adult mortality rates when expressed in the same units, the larval stage is generally much shorter than adult life stages and individuals are only exposed to this high risk of mortality for a relatively short period (Table 10-3).

Estimating Mortality

One can estimate instantaneous mortality rates using information on relative or actual abundance by age. Obtaining data on abundance by age is not straightforward and may involve various assumptions. One approach involves examining the age-specific abundance of a fish stock at a specific point in time. That is, by collecting fish and estimating their ages, it is possible to estimate the relative abundance of different ages in a stock and then analyze the rate at which abundance decreases with increasing age to estimate mortality rate. This point-in-time-based estimate of mortality essentially assumes that recruitment is constant over time. That is, it assumes that the initial abundance at age 0 was the same for all annual cohorts or at least that there is no temporal pattern in recruitment strength. This assumption may not hold for many fish stocks, but the point-in-time method allows for estimation of mortality rate without the need for repeated sampling.

Another approach for estimating mortality involves tracking the relative abundance of cohorts over time. This can involve tracking the abundance of annual cohorts from one year to the next or tracking daily cohorts of larval fish across days. Obviously, this approach requires repeated sampling of the fish stock to estimate how relative abundance changes over time. It is also important to use sampling methods whereby the vulnerability of individual fish to the collection gear is fairly constant over the range of sizes and ages for which one intends to generate a mortality estimate. For example, larval fish are often sampled with ichthyoplankton nets with a certain mouth area and certain mesh. Very small larval fish are generally poor swimmers and may relatively easily enter the net, but due to their small size, they may slip through the mesh and not be captured. Larger larval fish are often much better swimmers and have more developed senses. Therefore, they may be able to sense and actively avoid the net. It is therefore appropriate to limit mortality estimation to sizes with constant vulnerability to capture or sizes for which relative vulnerability is known and can be adjusted for.

Catch Curves

Given estimates of relative abundance by age, one can estimate instantaneous mortality over a range of ages using a transformation of Equation 10-1:

$$\ln\left[N_t\right] = \ln\left[N_0\right] - Zt \qquad \text{(Equation 10-2)}$$

Instantaneous mortality is simply estimated as the slope of the relationship between age (*x*-axis) and the logarithm of age-specific relative abundance (*y*-axis). This approach is termed the catch-curve method for estimating mortality rates. The approach has been widely used to estimate mortality rates of diverse fish populations, including instantaneous annual mortality rates of adult species, such as lake whitefish, and instantaneous daily mortality rates of larval fish, such as Atlantic herring.

In generating catch-curve-based estimates of mortality, it is imperative to only include ages that are fully vulnerable to assessment gear. Unfortunately, the ages vulnerable to assessment gear are often unknown a priori. Instead, by visualizing a catch curve, one can select the ages likely fully vulnerable. If all ages are fully vulnerable, one would expect the youngest ages to be most abundant in catches since all fish start at youngest ages and then numbers decline with increasing age. However, in many catches, the youngest ages of fish are less abundant than some intermediate ages. Since almost all fishing gear is highly size selective, the youngest and smallest individuals are often less vulnerable to capture. An increase in capture abundance from very young ages to intermediate ages generally does not reflect an increase in abundance in the environment but rather an increase in vulnerability. In contrast, a decrease in capture abundance from intermediate ages to some older ages may partially reflect decreases in actual abundance due to mortality. In viewing a catch curve, it is common to only include those ages encompassing the decreasing portion of a catch curve. That is, it is common to include all ages greater than the age of the peak catch.

Other Approaches for Estimating Mortality

In addition to capturing wild fish, aging them, and using the catch-curve method to estimate mortality rate, there are several other approaches for estimating mortality. One approach is to tag a known number of fish and then track the recapture of these fish over time. Regardless of whether fish are recaptured by recreational anglers, commercial fishers, or research and management assessment efforts, because of mortality the number of marked fish in the wild population will decrease over time. One can estimate the mortality rate by tracking the rate of decrease in recapture of marked fish. In addition, such mark-recapture studies can be designed to simultaneously allow for estimation of total mortality rate, natural mortality rate, fishing mortality rate, and population abundance.

Another approach for estimating mortality is based on examining the longevity of a fish population. If a fish population experiences low mortality, individuals will survive to relatively old ages, and the population will be characterized by high longevity. Defining and estimating longevity for a population may not be straightforward. One approach is to define the age to which a very small proportion of the population survives. For example, longevity could be defined as the age at which only 0.001 of the population survives. However, estimating such a proportion would require a very large sample size and involve aging a large number of individuals. John Hoenig, from the Virginia Institute of Marine Sciences, has used a somewhat different approach in relating mortality rate to the maximum age observed in a sample from a population. In order to use this method, it is important to obtain a large enough sample from the population (Hoenig suggested 200). However, it really only requires aging the oldest, largest fish in the population. To develop a relationship between maximum age and total mortality, Hoenig (1983) assembled data from the literature for 134 stocks of mollusks, fishes, and cetaceans, including 84 primarily marine fish stocks. All of these stocks included estimates of both total mortality (Z) and maximum age (t_{max}), which allowed Hoenig to estimate a relationship between the two:

$$\ln Z = 1.44 - 0.982 \ln t_{max} \qquad \text{(Equation 10-3)}$$

Hoenig's fitted relationship suggested a relatively strong negative correlation ($R^2 = 0.82$) between Z and maximum age. That is, stocks with greater mortality rates will experience lower maximum age. This relationship could be used to generate a preliminary estimate of total mortality rate for a population when maximum age is available, but relative abundance by age is lacking.

Fishing and Natural Mortality

As described above, it can be informative to distinguish between different types of mortality, and in particular it may be important to quantify mortality related to fishing. Instantaneous mortality rates are additive, and if one only considers two types of instantaneous mortality rates – fishing, F, and natural mortality, M – then total instantaneous mortality is the sum of instantaneous fishing mortality and instantaneous natural mortality; or

$$Z = F + M \qquad \text{(Equation 10-4)}$$

Given that these rates are additive, if one can estimate Z (e.g., through a catch-curve analysis), one needs to only estimate F or M in order to ultimately estimate all three rates. However, it is generally not straightforward to generate estimates of F or M. Some unique situations may allow for discriminating between these two rates. For example, during World War II, there was limited commercial fishing in the North Sea and North Atlantic. During this time, F was essentially negligible and $Z = M$. However, prior to World War II and immediately after the war, commercial fishing was relatively intense and $Z = F + M$. Assuming M is constant over time (a common assumption), estimates of Z during the war provide estimates of M, and the difference between such estimates and estimates of Z prior and after the war provide estimates of F during these periods.

A potential complicating factor in separating natural and fishing mortality rates relates to density-dependent effects. The concept of density dependence is more fully described in Chapter 11. Briefly, through various compensatory processes, it is possible that if a fish stock is thinned out, remaining individuals will survive better because of factors such as increased access to food and reduced disease transmission. In contrast, a very dense population could experience higher mortality rates, and therefore, the actual natural mortality rate of a fish stock in the absence of fishing could be greater than the natural mortality rate when fishing leads to thinning of the population. Density-dependent effects can also act in the opposite direction through depensatory density-dependence. Through various processes, more dense populations may actually experience lower natural mortality rates.

Some researchers have attempted to develop generalizable estimates of M. Perhaps the most broadly adopted such generalizable model was developed by Daniel Pauly (1980), then at the International Centre for Living Aquatic Resources Management and now at the University of British Columbia, for both freshwater and marine species spanning from the tropics to the poles. Pauly compiled a large number of estimates for M (see examples in Table 10-1) by scanning the scientific literature. In addition, they compiled estimates of species growth patterns (von Bertalanffy growth parameters) and temperatures occupied. They then used all of these data to fit a relationship across species to relate natural mortality to growth parameters and temperature:

$$\ln M = -0.0152 - 0.279 \ln L_\infty + 0.6543 \ln K + 0.462 \ln T \qquad \text{(Equation 10-5)}$$

Here, T is the estimated mean annual temperature experienced by a fish stock, and L_∞ and K are asymptotic length and the Brody growth coefficient from the von Bertalanffy lifetime growth model. In other words, this model suggests that fish species that grow to a relatively large asymptotic length will experience lower natural mortality, while species occupying warmer temperatures and growing rapidly to their asymptotic size will experience relatively high natural mortality.

Pauly's model has been used rather widely. However, it is important to note that it is an estimated model, fitted to assumed temperature experiences, estimated mortality, and growth coefficients with different levels of certainty. As described above, there are several processes contributing to natural variation, and there are many reasons why the actual natural mortality rate of a given fish stock may not closely match predictions from Pauly's model. While the model may be useful as a starting point when no other information is available to estimate natural mortality rates, caution should be taken when using this and similar models.

Separating Mortality Sources

A common goal of various studies is to separate the sources of mortality for a population. In many studies, this has involved separating natural and fishing mortality from total mortality. For example, Sinclair (2001) took advantage of a closure of the Atlantic cod fishery in the Gulf of St. Lawrence to estimate natural mortality rates by comparing estimated total mortality rates between years when there was active fishing versus years when the fishery was closed.

In addition to separating mortality between fishing and natural, in many cases it may be important to separate among other sources of mortality. For example, in several systems, large shipping and power generating turbines or water intake pipes may kill a large number of fish. For cooling purposes, power plants may pump huge volumes of water from aquatic systems and in the process kill many adult and early life stages of fish. Similarly, fish may be killed by turbines operating on hydroelectric dams and may be sucked into huge turbines on barges and large ships. Understanding how these human-related sources of mortality compare to other types of mortality may be important in part because the companies operating the turbines or water intake pipes may be legally and economically responsible for making up for the loss of fish they cause. Similarly, in some systems it may be important to differentiate between mortality caused by predators or parasites. In the Laurentian Great Lakes, invasive sea lamprey act as parasites on a number of large-bodied fishes, including native lake trout. Sea lamprey attach to lake trout using their suction cup-like mouth. Sea lamprey use their tongue and teeth to remove tissue and then directly consume tissue and fluids, including blood from lake trout. This often results in mortality for the lake trout. Since lake trout not only represent an ecologically important species, but also constitute important (and economically valuable) recreational and commercial fisheries, sea lamprey-related mortality of lake trout comes at a potential cost to the fishery. Control programs, primarily relying on lampricides and barriers, target sea lamprey in streams where they reproduce, but these control efforts are quite expensive, and the funds and effort directed at sea lamprey control are most justifiable if they appreciably decrease total mortality rate of lake trout. Therefore, it is of interest to be able to separate mortality rates between not only fishing and natural mortality, but also lake trout mortality related to sea lamprey parasitism.

Other examples of separating different types of mortality involve specifying mortality attributable to fish-eating birds. Wide-spread use of pesticides during the middle of the 20th century led to dramatic declines of several bird species. For example, the previously widely used pesticide, DDT (dichlorodiphenyltrichloroethane), caused eggshell thinning in a variety of bird species in Europe and North America, leading to limited reproductive success and contributed to population decreases of birds such as eagles and pelicans. With subsequent cessation of DDT use, many species of birds have slowly recovered. However, some species have recovered quite rapidly, even seemingly exceeding their pre-affected levels. Some cormorant species, including double-crested cormorant in North America and great cormorant in Europe, have dramatically increased and established large populations on water bodies throughout the continents. These fish-eating birds have the potential to consume a large number of fish and in several systems where fish populations have declined. Anglers and commercial fishers have argued that cormorant predation – and not fisheries harvest – is primarily to blame for fish population declines.

Oneida Lake is a large lake in New York, USA, which has historically supported strong fisheries for species such as yellow perch and walleye. Double-crested cormorant numbers around Oneida Lake increased pronouncedly during the 1980s and 1990s. At the same time, walleye and yellow perch stock sizes decreased, leading to speculation that cormorants could be responsible for fish population declines. To address this possibility, VanDeValk et al. (2002) and Rudstam et al. (2004) conducted studies to estimate the total mortality rates of walleye and yellow perch as well as the proportional mortality attributable to fishing and cormorant predation. They collected and combined data on perch and walleye populations from long-term assessment surveys and angler interviews, with (a) estimates of cormorant consumption of fish from analysis of cormorant stomach and pellet contents, (b) estimates of cormorant population sizes and (c) models of total food consumption by individual birds. They concluded that mortality rates attributable to cormorants were roughly equivalent to mortality rates attributable to fishery harvest. However, while fishing mortality was higher for large adult fish, cormorant-related mortality was greater for smaller, sub-adult fish.

Emerging Methods for Estimating Mortality

The basic concepts related to natural mortality, fishing mortality, and survival of fish involve declines in abundance with age and over time. The general methods for estimating these rates have not changed fundamentally since they were developed. However, advances in statistical approaches now allow for more sophisticated techniques in estimating mortality rates, including considering how mortality rates may change over time, across space, and among different cohorts. In addition, new methods have allowed for the use of novel data types in the estimation of mortality.

An example of statistical approaches allowing for novel methods in the estimation of mortality is incorporation of non-stationarity. Traditionally, most analyses of fish populations, including estimates of mortality, have assumed stationarity. Stationarity implies that a relationship, including the parameters that define the relationship, is fixed. For a linear relationship, this would imply that the slope and intercept are fixed (e.g., these parameters do not vary over time). Most relationships are likely not fixed, and as environmental conditions change, the strength of relationships may be expected to change. However, incorporating non-stationary effects in models is not straightforward; instead, most analyses have assumed that the average stationary relationship is sufficient to capture population dynamics.

Yan Jiao from Virginia Tech University and colleagues developed an analysis to evaluate the consequences of stationary versus non-stationary natural mortality based on Atlantic weakfish (Jiao et al. 2012). These researchers considered weakfish catch data from a 26-year period (1982–2007) and used these data to develop various age-structured models, known as statistical catch-at-age (SCA) models. Such SCA models are intended to estimate age and year-specific abundances and fishery exploitation rates. They generally assume that natural mortality is fixed and does not vary among years or with age. Jiao and colleagues explored this assumption by modeling natural mortality, M, as both stationarity and non-stationary (changing over time). They were better able to match the observed catch data if they allowed M to vary over time, using a statistical approach known as a random walk. In models that allowed M to vary over time, M generally increased during more recent years. Finally, they evaluated if the changes in M over time were associated with any environmental conditions and found that the year-specific estimate of M was positively related to the Atlantic Multidecadal Oscillation index (AMO). This index is a widely recognized measure that encapsulates annual climate and oceanic conditions. It is based on sea surface temperature variability, and warm sea surface temperatures lead to high AMO index values. A relationship between M and AMO could allow for prediction of weakfish natural mortality into the future given observed or predicted AMO values.

In addition to statistical methods and allowing for more sophisticated approaches in the estimation of mortality rates, new types of data are increasingly used to estimate mortality rates. Tag-recapture approaches have been used for many years to estimate mortality rates. Essentially, these methods involve tagging a known number of fish and evaluating how recapture rates change over time. As fish experience mortality, one would expect recapture rates to decrease over time, and the rate at which recapture rates decrease can allow for estimation of mortality rates. By examining marked fish captured by fishers, it is possible to separately estimate fishing mortality versus natural mortality.

Increasingly acoustic tags are used to track the movement of fish in various systems. Acoustic tags repeatedly emit unique signals detected by hydrophone receivers. Receivers can be fixed in specific locations or can be actively moved around to track fish. Several receivers can be deployed in different locations in the same systems, potentially allowing for detection of fish as they move to different areas of a systems. Acoustic tags are often equipped with sensors that can record and transmit environmental experiences of individual fish, such as pressure (to index depth), temperature, oxygen, and acceleration (to index swimming dynamics of fish). There are even tags and sensors that can detect and transmit when a fish is eaten by a predator. Such sensors can assess when tags come into contact with a predator's stomach and transmit this information. Thereby, these acoustic tags allow for documentation of a specific type of mortality events. For example, Weinz et al. (2020) demonstrated the use of such tags to detect predation of yellow perch in a small region of the Detroit River.

Methods to use acoustic tags to estimate mortality have been actively developed over the past few years, and these methods will undoubtedly continue to evolve in the future. While predation-detection tags allow for documenting a specific mortality event, acoustic tags that simply transmit the presence of fish have the advantage of being able to repeatedly document the presence of an individual in a system. Methods based on repeated detections across various acoustic receivers in an area (i.e., an array of receivers) may allow for the detection of not only mortality but also specific types of mortalities. David Villegas-Ríos and colleagues developed such a method to assess mortality of tagged fish in an acoustic array (Villegas-Ríos et al. 2020). Their method is based on analyzing movement of tags in an array and inferring mortality and types of mortality based on these movements. They aimed to detect tagging mortality (i.e., mortality that occurs immediately after tagging as a result of the tagging process), fishing mortality, predation mortality, and other natural mortalities based on patterns of tagged movement. They evaluated their method based on 291 tagged Atlantic cod in a coastal zone of Norway and concluded that they were able to identify the fate of 97% of the cod. Either the cod were determined to have dispersed from the study area, survived to the end of the study, or died through one of various types of mortalities.

Latitudinal Variation in Mortality Rates of Eurasian Perch

As mentioned above, mortality rates can be highly variable among different populations of the same species. Heibo et al. (2005) from the Swedish University of Agricultural Sciences assembled a suite of life history information, including mortality rates, for a diversity of Eurasian perch populations. These authors were particularly interested in examining how mortality rates (and other life history traits) varied with latitude. They collected some data by sampling Eurasian perch populations on their own, but they primarily worked to assemble data from previously published studies. Ultimately, they assembled mortality data for populations from as far north as 66°N (Lake Kiutajärvi, Finland) and as far south as 40°N (Lake Yaskhan, Turkmenistan).

Eurasian perch are quite variable in their feeding habits. In many perch populations, individuals will transition to feed piscivorously when they become sufficiently large, whereas some other populations will feed on invertebrates throughout their life. This can lead to large differences in growth and maximum size between piscivorous populations versus non-piscivorous populations, with potential ramifications on mortality. Thus, Heibo and colleagues separately analyzed mortality rates for piscivorous and non-piscivorous populations.

Even within a population of fish, mortality rates may vary among different groups. For example, male and female fish may experience different mortality rates for a variety of reasons. Activity levels and energetic utilization may differ between sexes, potentially leading one sex to experience greater starvation-related mortality (because of higher energetic expenditures) or causing one sex to be more at risk from predation (because greater activity makes them more conspicuous). The processes of reproduction may also lead to different mortality rates between the sexes. For many fish species, including Eurasian perch, mature ovaries are much larger than mature testes, and the energetic investment to gonads is much greater for females than males. However, males may expend more energy in other aspects of reproductive behavior, such as finding and attracting mates and competing with other males for spawning opportunities. Similar to many other species, Eurasian perch express sexual size dimorphism with females potentially growing to a larger size than males. Size can have a strong effect on various aspects of mortality – both natural mortality and fishing mortality. Larger fish may be less susceptible to certain types of predators, but larger fish may also be more vulnerable to various fishing gears that selectively target large fish. To control for all of these potential differences in size and mortality between male and female Eurasian perch, Heibo and colleagues only used data for female perch whenever such separate data were available.

Another potential difference in mortality rates within a population relates to age and maturation status. Juvenile fish are relatively small and may thus experience greater mortality risk from starvation and predation by size-limited predators. At the same time, larger adult fish may experience reproduction-related mortality and be more susceptible to fishery harvest. Due to these potential effects, Heibo et al. separately considered juvenile and adult mortality rates.

They found that piscivorous Eurasian perch populations grew to a larger maximum size and matured at older ages and larger sizes relative to non-piscivorous, stunted populations. However, there were on average no differences in mortality rates between these two types of perch populations. In contrast, population mortality rates did vary negatively with latitude. More southern populations (lower latitudes) tended to experience higher mortality rates than more northern populations. This pattern was true for both juvenile and adult mortality rates. Further, because northern populations tended to survive at a higher rate, the authors also found a strong positive relationship between latitude and the observed maximum age in a population.

Juvenile mortality rates of Eurasian perch populations were on average much higher than adult mortality rates. Juvenile mortality rates ranged from 4.89 yr^{-1} at a relatively low latitude to 1.61 yr^{-1} at a higher latitude. In contrast, adult mortality rates only ranged from 1.14 yr^{-1} at a low latitude to 0.23 yr^{-1} at a high latitude. Nonetheless, juvenile and adult mortality rates were highly correlated. That is, populations with high juvenile mortality rates also tended to have high adult mortality rates.

Summary

Mortality and survival rates of fishes are highly variable. Most fish populations experience particularly high mortality during early life, while survival increases for older ages. Estimating mortality rates generally involves evaluating how quickly relative abundance decreases with increasing age of fish. It is also possible to discriminate among different sources of mortalities. In particular, in many cases it is useful to differentiate between mortality caused by natural processes and mortality due to fisheries harvest. Fisheries managers generally have the ability to control fishing mortality, while natural mortality is often out of the control of managers.

Literature Cited

Heibo, E., C. Magnhagen, and L.A. Vøllestad. 2005. Latitudinal variation in life-history traits in Eurasian perch. Ecology 86:3377–3386.

Hoenig, JM. 1983. Empirical use of longevity data to estimate mortality rates. Fishery Bulletin 91:898–903

Houde, E. 1989. Comparative growth, mortality, and energetics of marine fish larvae: Temperature and implied latitudinal effects. Fishery Bulletin 87:471–495.

Jiao, Y., E. Smith, R. O'Reilly, and D. Orth. 2012. Modeling nonstationary natural mortality in catch-at-age models: An example using the Atlantic weakfish (*Cynoscion regalis*) fishery. ICES Journal of Marine Science 69:105–118.

Pauly, D. 1980. On the interrelationships between natural mortality, growth parameters, and mean environmental temperature in 175 fish stocks. ICES Journal of Marine Science 39:175–192.

Rudstam, L.G., A.J. VanDeValk, C.M. Adams, et al. 2004. Cormorant predation and the population dynamics of walleye and yellow perch in Oneida Lake. Ecological Applications 14:149–163.

Sinclair, A.F. 2001. Natural mortality of cod (*Gadus morhua*) in the Southern Gulf of St Lawrence. ICES Journal of Marine Science 58:1–10.

VanDeValk, A.J., C.M. Adams, L.G. Rudstam, et al. 2002. Comparison of angler and cormorant harvest of walleye and yellow perch in Oneida Lake, New York. Transactions of the American Fisheries Society 131:27–39.

Villegas-Ríos, D., C. Freitas, E. Moland, S.H. Thorbjornsen, and E.M. Olsen. 2020. Inferring individual fate from aquatic acoustic telemetry data. Methods in Ecology and Evolution 11:1186–1198.

Weinz, A.A., J.K. Matley, N.V. Klinard, A.T. Fisk, and S.F. Colborne. 2020. Identification of predation events in wild fish using novel acoustic transmitters. Animal Biotelemetry 8:28.

CHAPTER 11

Density-Dependence and Independence

Population growth rates and vital rates (such as mortality), reproductive output, and individual growth are controlled by a variety of abiotic and biotic factors. An important distinction between these rates is whether or not they are a function of population density. If oxygen concentrations in a small pond decrease to a low level, a proportion of fish in the pond may die. However, the proportion of fish that die is unlikely to be related to the number of fish in the pond but may simply be a fixed proportion of the population depending on individual susceptibility to low oxygen. Similarly, if the water body inhabited by a fish population increases in mean temperature, individual growth rates may increase (due to increased consumption potential and other bioenergetics considerations). If the population has ample access to food, this temperature-induced increase in growth rates may be independent of the number of fish in the population. In contrast to these density-independent responses, many rates influencing populations are dependent on the density of the population. If a pathogen infects a portion of a population, a larger proportion of the population may ultimately die, if the population is relatively large, because of greater transmission rates in large, dense populations. Similarly, if prey are limited in a system, the amount of food consumed by an individual fish and the mean growth rate of individuals in the population will be lower in relatively abundant populations. More abundant populations will depress the density of prey and lead to reduced individual prey consumption and lower growth.

Whether population processes are density-independent or density-dependent has important implications for population dynamics. Most density-dependent processes serve to regulate the population toward some intermediate population density (but see description of depensatory density-dependence below). If a population becomes very large, density-dependent processes may contribute to decreased individual growth, reproductive rates, and survival, and such effects may push the population to a lower abundance. Similarly, if a population decreases in abundance, density-dependent processes may lead to improved survival, individual growth, and per capita reproductive success, regulating the population by favoring an increase in abundance. This regulation is often called compensation – when the population compensates for reduced abundance by increased growth and survival rates of the remaining individuals. Density-dependent processes encapsulate a feedback component to control population dynamics. Density-independent processes do not encapsulate such feedback and allow key rates to vary independent of density.

Biology and Ecology of Fishes, Third Edition. James S. Diana and Tomas O. Höök.
© 2023 John Wiley & Sons Ltd. Published 2023 by John Wiley & Sons Ltd.

Exponential and Logistic Population Growth Models

Population abundance (N) is often modeled coarsely by tracking how population growth (dN/dt) and per capita population growth ($dN/N\ dt$) vary over time (t). These population growth rates essentially encompass individual vital rates – such as survival and reproductive rate – into a single value. One can think of population growth rate as the difference between birth and death rates. If birth rate exceeds death rate, population growth rate is positive and the population is increasing in abundance. If death rate exceeds birth rate, population growth rate is negative and the population is decreasing in abundance. And, if birth rate equals death rate, population growth rate is 0 and the population is stable.

Under density independence, population growth will follow the exponential population growth model:

$$N_t = N_0\ e^{rt} \qquad\qquad \text{(Equation 11-1)}$$

Population size at time t (N_t) is dependent of population size at time 0 (N_0), based on a single parameter r, the intrinsic rate of increase of the population. If r is positive, the population will grow in perpetuity. This will lead to an exponential increase in population size over time (Figure 11-1a). Under the exponential model, the population growth rate will increase linearly with population size ($dN/dt = rN$), and per capita population growth rate will be constant and independent of population density ($dN/N\ dt = r$; Figure 11-2a&c). As the population grows, its capacity for growth increases. This is analogous to a snowball increasingly gaining size while rolling down a snow-covered hill. A small snowball can only pick up a small amount of additional snow as it rolls down the hill. However, as the snowball becomes larger, it has a greater capacity to accumulate more snow during each rotation.

The above version of the exponential growth model is a continuous time model. N_t can be estimated for any time value, both integers and non-integers. The population also responds instantaneously to changes in population abundance. The exponential population growth

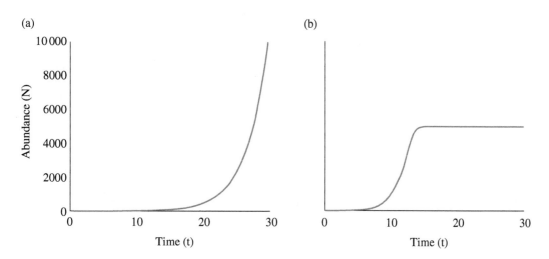

FIGURE 11-1 Example population abundance trajectories over time for (a) exponential population growth model and (b) logistic population growth model. The latter population growth model is constrained at a carrying capacity (K) of 5,000.

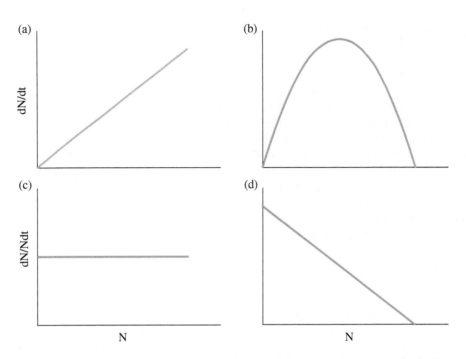

FIGURE 11-2 Relationships between population abundance (N) and population growth, dN/dt, or per capita population growth, dN/N dt, for continuous exponential population growth model (a and c) and logistic population growth model (b and d).

model can also be expressed in a discrete time form. The discrete time exponential growth model is set up so abundance in the next time step is a function of current abundance. The population size next year depends on the population size this year. Discrete time models can be useful if (a) the system has clear annual or seasonal cycles such as reproduction once per year and (b) the data underlying a model are collected at fixed times during the year (which is often the case for fish population assessments). Discrete time models are incremental and not continuous. The discrete time form of the exponential growth model can be depicted as

$$N_{t+1} = N_t + \Delta N_t \qquad \text{(Equation 11-2)}$$

Here, Δ represents the change or increment for N over time t. It should be clear from this equation that population abundance at the next time step is simply equal to abundance at the current time plus the change in abundance ΔN. More formally, we can depict the discrete time form as

$$N_{t+1} = N_t + r_d N_t \qquad \text{(Equation 11-3)}$$

The parameter in this model, r_d, is analogous to the intrinsic rate of increase, r, from the continuous time model, in that it defines the population growth rate. However, they are not equivalent, and one can show that $r = \ln(1 + r_d)$.

One may question if the exponential population growth model is realistic. If a fish population were to continue to increase exponentially, it would eventually run out of space in the body of water it inhabits. Thus, it is realistic to expect that at some population sizes, density-dependent processes must control further growth. However, it is plausible that populations may follow expectations of the exponential population growth model for some time as long as the population does not become sufficiently large to be constrained by limited resources.

For example, invasive fish populations have been suggested to follow such short-term exponential population growth when they initially colonize and expand in a novel environment. During this initial stage, an invasive fish population may be able to exploit resources previously not limited in the system, and the population may be sufficiently low in numbers such that it is not limited by resources. However, as an invasive fish population becomes sufficiently large, resources will eventually constrain its further growth.

The concept of a carrying-capacity limiting population size has been presented in various forms. Perhaps most famously, Thomas Malthus argued in the late 18th century that human population size will inevitably exceed food supply, and such effects will limit population size. In population modeling, a simple carrying-capacity parameter, K, can be added to the exponential population growth model to account for resources limiting population growth. Under the resulting logistic population growth model, populations will increase in abundance until they reach the carrying capacity, when population growth will cease (Figure 11-1b; Figure 11-2b&d). Population growth rate is non-linearly related to population size:

$$\frac{dN}{dt} = rN\left(1 - \frac{N}{K}\right) \qquad \text{(Equation 11-4)}$$

Here, the population growth rate is zero when population size equals carrying capacity and the maximum population growth rate is achieved at 50% of carrying capacity. At this intermediate size, the population is large enough to have increased capacity for growth, but not so large that it is severely limited by resources (as when the population approaches carrying capacity). Unlike the exponential population growth model, per capita population growth under the logistic model is a function of population size and declines linearly with N:

$$\frac{dN}{N\,dt} = r\left(1 - \frac{N}{K}\right) \qquad \text{(Equation 11-5)}$$

Similar to the exponential population growth model, logistic population growth can also be modeled in a discrete time form:

$$N_{t+1} = N_t + r_d N_t\left(1 - \frac{N_t}{K}\right) \qquad \text{(Equation 11-6)}$$

Surplus Production Model

Population growth models are not just theoretical descriptions of how populations may grow under certain idealized conditions. These models can also be incorporated into practical models used for fisheries management. For example, a class of fisheries models, known as surplus production models, builds on concepts of population growth and density dependence to consider how much biomass can be consistently harvested from a fish population. Several variations of surplus production models have been developed. A standard form presented by Schaefer (1954, 1957) builds directly from the discrete time form of the logistic population growth model but focuses on stock biomass, B, instead of abundance, N, with maximum stock biomass, B_∞, analogous to carry capacity, K:

$$B_{t+1} = B_t + r_d B_t\left(1 - \frac{B_t}{B_\infty}\right) \qquad \text{(Equation 11-7)}$$

By adding an additional term for fisheries catch during time t, C_t, we can not only consider the growth of population biomass based on density-dependent effects, but also the loss of biomass in the form of fisheries harvest:

$$B_{t+1} = B_t + r_d B_t \left(1 - \frac{B_t}{B_\infty}\right) - C_t \qquad \text{(Equation 11-8)}$$

Note that in the absence of catch, the change in stock biomass from time step t to $t + 1$ is given as

$$r_d B_t \left(1 - \frac{B_t}{B_\infty}\right) \qquad \text{(Equation 11-9)}$$

This term in Equation 11-9 is analogous to potential production. If this term is positive, population biomass has the potential to increase, while if it is zero, there will be no potential production. When catch is included in this model, the population's biomass would remain constant when production is equal to catch. If the potential production term is equal to fisheries catch during a time step, then $B_{t+1} = B_t$, and there will be no change in population biomass. Thus, one can theoretically remove biomass in the form of fishery catch equivalent to potential production and not cause a change in population biomass. Recall that the population growth rate for the logistic population growth model is greatest at 0.5 of carrying capacity. Similarly, growth in population biomass is greatest at $0.5B_\infty$. If the population biomass is equal to $0.5B_\infty$, one can essentially harvest the maximum amount of *surplus* production without affecting population biomass at this level. Specifically, the surplus production harvested at this biomass level can be calculated by solving the potential production term, when $B_t = 0.5B_\infty$, which is equivalent to $0.25r_d B_\infty$.

The term, surplus, to describe this class of models is based on the idea that fisheries can remove the extra population biomass production that exceeds what is needed to simply maintain population biomass. However, the notion that this population biomass is somehow surplus is not only anthropocentric but may risk the long-term sustainability of a fishery. The maximum surplus production that can theoretically be harvested from a fish stock is often referred to as maximum sustainable yield or MSY. Unfortunately, precisely and accurately estimating MSY is difficult and true MSY may vary from year to year as environmental conditions change (see concept of non-stationarity described in Chapter 10). If harvest levels are set to target a certain MSY level, but the true MSY is lower, harvest levels will contribute to a decline in population biomass. As described in Chapter 32, a huge number of fish stocks have been overharvested during the past century, and many stocks have been greatly depleted and even extirpated. Reliance on concepts such as MSY have seemingly contributed to this repeated pattern of overharvest and is why many fishery scientists have advocated for cessation of the use of MSY to set fishery harvest levels. For example, see "An epitaph for the concept of maximum sustainable yield" by Larkin (1977).

Density-Dependent and Density-Independent Effects

Many studies have explored factors that control fish population growth rates as well as individual vital rates (e.g., survival, growth, and reproduction). Understanding which specific environmental factors control fish populations can facilitate prediction of how fish populations may respond to environmental changes and can inform management actions aimed at affecting population abundances or individual sizes. Some populations and vital rates may respond strongly to density-independent environmental conditions, while others may be largely constrained by density-dependent population controls. For example, annual recruitment (analogous to population-level reproductive success; see Chapter 12) of many marine fish populations seems to be highly responsive to environmental conditions that vary from year to year, while in some smaller, freshwater systems, fish recruitment may be more limited by density-dependent effects.

Compensatory Density-Dependence

Compensatory Growth

Some of the clearest examples of compensatory density-dependent growth come from fish culture ponds. In fact, Fan Li wrote a book in 500 BCE that developed the idea that fishes in dense populations grow slowly. Swingle (1950), who initiated much of Auburn University's work in aquaculture, clearly demonstrated that a balanced population of largemouth bass foraging on bluegills in ponds would limit the bluegill population size and result in rapid individual growth rates of surviving bluegills. If bass did not forage effectively on bluegills in those ponds, the bluegill population increased and their individual growth was limited. Individuals of dense populations were expected to grow less than their counterparts in less dense populations. In this case, the relationship between population density of bluegills and individual growth rate was a compensatory density-dependent effect; the growth rate was dependent on bluegill density.

Diana et al. (1991) examined changes in mass of tilapia stocked in ponds and related those to the number of tilapia stocked. At low population densities, tilapia achieved high individual growth rates, and there was a linear decrease in growth rate with an increase in density (Figure 11-3). This is clear confirmation of density-dependent growth, and this

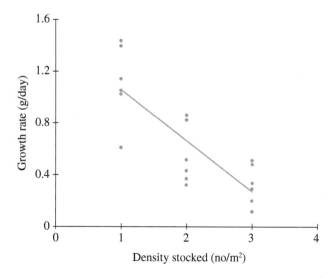

FIGURE 11-3 Density-dependent reductions in daily growth for tilapia in aquaculture ponds. Mean individual growth rate of tilapia is plotted against the number (no) of tilapia stocked per area of pond. *Source: Adapted from Diana et al. (1991).*

phenomenon is common in aquaculture ponds for several reasons. Since the animals are stocked into ponds at an advanced size, they are not likely to die prior to subsequent harvest. Resources are restricted in the pond, and the expected response for animals when stocked at excessive density is to consume less food per capita and therefore grow less. Density-dependent growth commonly observed in small ponds is also observed in larger lakes for similar reasons. Lake environments often provide abundant habitat into which animals can escape. These habitats may not be good feeding sites, but the animals are not necessarily exposed to high rates of predation or strong agonistic behavior in these habitats. The density dependence observed in lakes is therefore often expressed in terms of individual growth.

Compensatory Mortality

Brown trout in a stream can show strong compensatory density-dependent mortality. If the number of animals is artificially increased (by stocking), mortality rate increases (Figure 11-4). At very low density, increased density has little influence on mortality. As the population increases, feeding stations and cover become saturated and mortality increases. The mortality rate increases once numbers exceed a certain threshold. One result is that the population usually has the same number of adult animals regardless of the number of young born. This effect mainly occurs through mortality for trout in streams. Compensatory mortality is the critical process here because small fish born (or stocked) into a population are less likely to survive than larger individuals already present.

A stream is a very space-limited environment, where there is a certain amount of substrate available for colonization. The smaller the stream, the more limited the space. Within a range of densities, increasing numbers by either increased reproduction or stocking might effectively increase a fish population. If the mortality rate does not increase because of the density increase, more individuals will survive. However, at high density, more individuals result in a higher death rate, and the number that survive does not increase. In terms of a management strategy, it may not be effective to stock fishes such as trout into a stream where there already is an abundant, established resident population.

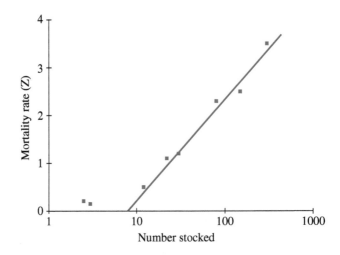

FIGURE 11-4 Density-dependent increases in mortality (Z) for brown trout stocked into a stream. *Source: Adapted from LeCren (1962).*

Compensatory Production

Population production is the increase in biomass related to both growth and survival rates. The production curve for brown trout in a stream shows a density response as well (Figure 11-5). Over low population abundances (between 2 and 12 individuals per square meter for fish of 1, 2, and 3 months in age), an increase in density results in a large increase in production (LeCren 1962). At high population density, production is constant regardless of any increase in density. There are two possible trade-offs involved here. (1) Compensatory density-dependent mortality would likely be expressed in survival of small fish. At the beginning of the year, the number of fry born in an area might be quite high, but the higher the density of young fishes, the higher the mortality rate, so the same number of fry live to the end of the year. These all grow reasonably well in this population, so the same amount of production occurs over a wide range of reproductive outputs. (2) A decline in the growth rate could balance an increase in the numbers of fishes in a pond, resulting in an asymptotic level of production at higher fish densities as seen in the tilapia example above. The density–production curve would essentially be the same as the production curves for brown trout, but the mechanism would be reduced growth rather than reduced survival. Different mechanisms (e.g., via reduced individual growth or reduced survival) can produce similar patterns in population production.

Compensatory density-dependent effects can also affect production potential across generations. Reductions in growth can lead female fish to obtain a smaller size and therefore have a lower gonadal capacity and lower potential fecundity (number of eggs produced). In addition, lower individual energy intake may lead to a lower proportion of consumed energy being directed to reproduction. Tyler and Dunn (1976) studied the growth and reproduction of winter flounder on a variety of rations (daily feeding rates). They examined how much energy was allocated to egg production or body growth under different rations. Their study would parallel density dependence because at higher densities each individual has less food to eat, so its reproductive output might be scaled according to ration. Both fecundity and egg size were very dependent on ration for flounder. Higher densities of fish competing for the same total amount of food resulted in lower fecundity for individuals and smaller eggs, both of which

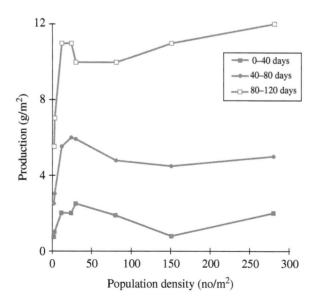

FIGURE 11-5   Density-dependent production of age-0 brown trout populations over three 40-day time periods. *Source: Adapted from LeCren (1962).*

would reduce total survival of young. If those changes occurred in nature, birth rates would decline under high density. Fecundity declined considerably under changes in ration; as few as half the number of eggs were produced at low ration compared to high ration. The actual percentage reduction in fecundity with increasing density will vary among species and populations. The numbers themselves are not important here. As density increases, the number of eggs a female produces declines, which affects the maximum number of offspring that could be produced by the population that year.

Depensatory Density-Dependence

The cases highlighted above are examples of compensatory density-dependence, where individual performance decreases as densities increase. Compensatory density-dependence is relatively easy to conceptualize and is by far the most common kind of density-dependence. As density increases, growth declines, mortality increases, and birth rate declines. However, there are mechanisms in which the reverse relationship between individual performance and population density may be manifest. Such depensatory density-dependence leads individual performance to decrease at relatively low population densities and increase at relatively high population densities. For example, in a large system with low population densities, it may be difficult for reproductively ready males and females to locate each other, leading to relatively low individual reproductive success. In addition, at low population sizes, inbreeding and low genetic variation can negatively affect offspring viability and progeny performance. Many fishes form aggregations (such as schools), which may allow individuals to more effectively avoid potential predators and detect and capture potential prey. At high densities, individuals may be able to take advantage of these benefits related to aggregating, whereas at low densities, individual survival and growth may decrease as fish are not able to effectively aggregate.

It is possible for density-dependent processes to switch from being compensatory to depensatory and vice versa, depending on population density and environmental conditions. For example, individual reproductive success may be influenced by compensatory density-dependent processes over a broad range of densities. However, if densities become sufficiently low and it is difficult for fish to locate other mature individuals or if genetic variation becomes sufficiently low, depensatory density-dependent effects may instead become evident. This notion of depensatory density-dependence becoming evident at low population size is often referred to as an Allee effect. This effect is named after Warder Clyde Allee who observed that goldfish in tanks grew at a faster rate when their densities were increased.

Depensatory Mortality

The dynamics of walleye and yellow perch in Oneida Lake, New York, have been studied by researchers from Cornell University for decades. Consistent annual surveys of these populations using trawls and other gear have allowed researchers to examine long-term relationships between relative abundances (indices of population densities) and vital rates (growth and mortality). Forney (1971, 1974) documented that a high density of perch in Oneida Lake usually resulted in compensatory density-dependent growth. As the population increased, the amount of resources per individual decreased, and the growth rate of individual perch declined. However, they did not observe a similar response in perch survival from age 0 to age 3. Instead, they found evidence of depensatory density-dependence as catch per effort of perch in trawls was inversely related to mortality rate, that is, the lower the abundance of perch, the higher the mortality rate (Figure 11-6).

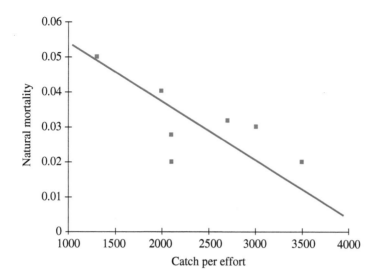

FIGURE 11-6 Depensatory (declining) mortality (per-day mortality coefficient) with increasing density (indexed based on bottom trawl catch rates) for age-0 yellow perch from Oneida Lake. *Source: Adapted from Forney (1971).*

In considering the mechanisms that may have contributed to depensatory density-dependent mortality, Forney evaluated the abundance of fishes in the lake and found that the strongest factor influencing the natural mortality of perch was not the density of perch but the density of walleye. Walleye were abundant in the lake and an important predator. When perch were abundant, walleye fed on perch almost exclusively. When perch were not abundant, walleye could still survive by feeding on alternate prey. Although walleye preferred to consume perch, their numbers did not depend necessarily on the number of perch. Due to the availability of alternate prey, a relatively constant number of walleye remained in Oneida Lake. Therefore, when fewer perch were available, a relatively constant number of predators contributed to perch suffering a higher population-level mortality rate. For example, if 1,000 walleye preyed on a population of 100,000 perch, and the average walleye ate 50 perch a year, walleye would cause a 50% mortality in perch. If the perch population increased to 200,000, walleye might eat a few more perch per individual (maybe 75), but with 200,000 perch existing, and possibly 75,000 eaten by walleye, the mortality rate is 38%. At one million perch, the mortality rate would only be 7.5%. The mortality rate is calculated with a relatively constant number of prey eaten but with a changing total number of prey in the lake. The more individual perch, the lower the fraction is eaten and the lower the mortality rate. Forney suggested that very abundant year classes of yellow perch essentially swamped the walleye's ability to consume this prey and led to greater survival for yellow perch.

Forney's original analysis was primarily based on surveys conducted during the 1960s. Irwin et al. (2009) examined similar long-term survey data of yellow perch in Oneida Lake, covering the period from 1961 to 2003. The inclusion of over 40 years of data led to the emergence of different long-term associations between perch densities and vital rates. While there was still evidence of a compensatory density-dependent growth effect, there was no longer evidence of depensatory density-dependent mortality. In fact, when the longer time series of data was considered, there was no relationship – positive or negative – between mortality rates of young perch and their densities. While difficult to test explicitly, Irwin and co-authors speculated as to why the depensatory effect may have been disrupted. They suggested that environmental changes in Oneida Lake, including increases in alternative prey for walleye, increased water clarity, and changes in the coverage of aquatic plants in Oneida Lake may have altered the predator–prey relationships between walleye and perch and disrupted the depensatory density-dependent effect on perch mortality. This example highlights the potential for density-dependent effects to change in a system as the environment changes, and in particular demonstrates how the rarer form of density-dependence, depensation, may only be evident for a relatively narrow set of environmental conditions and population densities.

Summary

A number of processes influence the vital rates of individual fish and the growth rates of populations. While many factors affecting populations act through density-independent processes, many populations are also strongly influenced by density-dependent effects. Compensatory density-dependent effects are likely common – especially in smaller freshwater systems – and serve to constrain the population. Depensatory density-dependent effects act opposite to compensatory effects, are less common, and may only be evident under subsets of environmental conditions and population abundances. Elucidating density-dependent effects is not only of theoretical interest but can also be important for predicting how fish populations will respond to environmental changes and informing management of fisheries.

Literature Cited

Diana, J.S., D.J. Dettweiler, and C.K. Lin. 1991. Effect of Nile tilapia (*Oreochromis niloticus*) on the ecosystem of aquaculture ponds, and its significance to the trophic cascade hypothesis. Canadian Journal of Fisheries and Aquatic Sciences 48:183–190.

Forney, J.L. 1971. Development of dominant year classes in a yellow perch population. Transactions of the American Fisheries Society 100:739–749.

Forney, J.L. 1974. Interactions between yellow perch abundance, walleye predation, and survival of alternate prey in Oneida Lake, New York. Transactions of the American Fisheries Society 103:15–24.

Irwin, B.J., L.G. Rudstam. J.R. Jackson, et al. 2009. Depensatory mortality, density-dependent growth, and delayed compensation: Disentangling the interplay of mortality, growth and density during early life stages of yellow perch. Transactions of the American Fisheries Society 138:99–110.

Larkin, P.A. 1977. An epitaph for the concept of maximum sustainable yield. Transactions of the American Fisheries Society 106:1–11.

LeCren, E.D. 1962. The efficiency of reproduction and recruitment in freshwater fish. Pages 283–296 *in* E.D. LeCren and M.W. Holgate, editors. The Exploitation of Natural Animal Populations. Blackwell Scientific Publications, Oxford.

Schaefer, M.B. 1954. Some aspects of the dynamics of populations important to the management of the commercial marine fisheries. Bulletin, Inter-American Tropical Tuna Commission 1:25–56.

Schaefer, M.B. 1957. A study of the dynamics of the fishery for yellowfin tuna in the Eastern Tropical Pacific Ocean. Bulletin, Inter-American Tropical Tuna Commission 2:247–285.

Swingle, H.S. 1950. Relationships and Dynamics of Balanced and Unbalanced Fish Populations. Alabama Agriculture Experiment Station, Bulletin 274, Auburn, Alabama.

Tyler, A.V., and R.S. Dunn. 1976. Ration, growth, and measures of somatic and organ condition in relation to meal frequency in winter flounder, *Pseudopleuronectes americanus*, with hypothesis regarding population homeostasis. Journal of the Fisheries Research Board of Canada 33:63–75.

CHAPTER 12

Recruitment

Outside the field of fish ecology, the term recruitment is often used to refer to the hiring of employees, conscription of soldiers, or enrollment of students. In fish ecology and fisheries science, the term generally relates to an index of the number of new individuals. While the specifics of how recruitment is defined can be quite variable, essentially fish recruitment refers to a measure of annual, population-level reproductive success. Inter-annual recruitment variation can also be considered as variation in the strength of year classes (cohorts of a fish population born in a specific year). As demonstrated with the Russell stock diagram (Chapter 9), recruitment is the only way to add new individuals to a closed population, and therefore, recruitment has strong control on population dynamics.

The ability to accurately and precisely predict recruitment could greatly aid fisheries managers by allowing them to anticipate population sizes in the future and adjusting harvest regulations accordingly. However, this has proven incredibly difficult. A common feature for most fish stocks is that they display recruitment that is highly variable and very difficult to predict. In many cases, annual recruitment success – or year-class strength – seems to be dependent on processes occurring during very early life. Most fish species are highly fecund, producing huge numbers of relatively small offspring, the vast majority of which die during early life. Given the large number of offspring, small changes in early-life growth and survival rates can have huge effects on subsequent recruitment success.

The Critical Period and Why Recruitment is so Variable

While density-dependent processes may regulate fish populations, the early-life dynamics of many fish populations are often subject to strong density-independent effects. Individuals that happen to hatch at the appropriate time or location – possibly when availability of suitable prey is high or where the risk of predation is low – may experience relatively high survival. Individuals hatching a day later or a few kilometers away might have poor survival. Density-independent effects (including seemingly stochastic environmental influences) are often important in determining such early-life survival. Most populations controlled by strong density-independent mechanisms will demonstrate dramatic variations in year-class strength, often from 10- to 100-fold or more. In contrast, if density-dependent survival or reproduction are

Biology and Ecology of Fishes, Third Edition. James S. Diana and Tomas O. Höök.
© 2023 John Wiley & Sons Ltd. Published 2023 by John Wiley & Sons Ltd.

strong controlling factors, year-class strength will vary less and the relative age-class structure of fish will remain fairly similar.

For many species, year-class strength is generally set by the time fish have lived one year. Sometime during that first year, between the millions of eggs laid and the thousands of fish that survive, a population bottleneck occurs. That is the time when key density-dependent or density-independent effects occur and cause high mortality. This general span is called the critical period, the point at which most mortality occurs and which largely sets year-class strength.

The critical period was first defined by Hjort (1914), and various other people have since refined the idea. It states that there is a very high mortality at a particular early age. This concept was developed for marine fishes, and most marine fishes are highly fecund animals with very small eggs and larvae. The young are born vulnerable and usually suffer very high mortalities. Part of this vulnerability is due to the ecology of the early lives of fish. For example, many fishes spawn planktonic eggs, which hatch and remain as larvae in the plankton, where they are susceptible to many predators. Small larvae also have limited energy stores, high mass-specific metabolic rates, and cannot survive for an extended period without feeding. Therefore, the likelihood of an egg surviving in plankton is very low. Part of the vulnerability is the life history of the fish itself; a pelagic fish that produces many small eggs dooms most of those eggs because of their small size and vulnerability.

Various potential processes in early life may cause a population bottleneck (a severe reduction in population numbers). While variable fertilization rates may represent population bottlenecks, the most common bottlenecks are believed to occur after fertilization, during early development. There are two stages to examine: from early development prior to hatching (embryos) and from hatching through the larval stage, including the transition from feeding endogenously (feeding on yolk reserves) to exogenously (feeding on environmental prey). This transition period is commonly identified as a bottleneck for early-life survival, but it by no means represents the only ontogenetic stage affecting fish population numbers.

Embryos develop considerably before they hatch. They usually hatch before they are capable of living on their own, but they have yolk reserves to use for energy needs and further development. The level of development before hatching varies among species. Larval fishes are discussed more fully in Chapter 24, but in some situations, larval fishes may hatch, utilize their yolk sacs, and still not have developed functional mouths. With insufficient energy reserves to develop mouths, these individuals will not survive. Early development and hatching could therefore be a major bottleneck, especially in challenging conditions. Salmon that spawn in polluted streams that lack oxygen and have high levels of CO_2 require metabolic regulation from the eggs and embryos. This regulation results in the use of yolk energy for homeostatic purposes rather than for development and hatching, and may reduce the fraction of eggs that hatch successfully. Similar dynamics may be related to temperature. At high temperatures, more energy may go into maintenance and less energy into growth, reducing the yolk available for development.

In addition to rapid development and metabolic costs during the egg stage potentially contributing to mortality during the egg or early larval stage, various other processes could lead to high mortality during the egg stage. For example, many predators target fish eggs and, in some cases, can consume a very large proportion of the eggs produced by a population. Eggs that incubate on the bottom of a body of water can be physically displaced from suitable incubation environments. In systems with high loads of sediment, bottom-incubating eggs can be covered by sediment, leading the eggs to be deprived of oxygen or favoring development of fungal growth, both of which can lead to high egg mortality. In western Lake Erie, many walleye spawn on rocky reefs and eggs incubate on these structures. Roseman et al. (2001) demonstrated that intense storms in the spring – when walleye eggs were incubating – could physically displace a large number of eggs from the reefs, presumably depositing them onto unsuitable soft sediment, where egg survival and hatching success is very low. Such potential negative effects of bottom substrate and sediment cover was demonstrated experimentally by Gatch et al. (2020). The researchers collected and fertilized walleye eggs from a reservoir and brought them to Purdue University for incubation. They incubated eggs on top of sand or silt while covering the eggs with different amounts of sand or silt. They demonstrated that walleye eggs experienced severe reduced hatching success when covered by even a small amount of silt.

Many fish ecologists consider a key population bottleneck to be larval development. In most species of fishes, the degree of metamorphosis during early life is less extreme than that seen in some other taxonomic groups, such as insects. However, there is a gradual ontogenetic change from a fish larva – that does not resemble the adult and is ecologically quite different – to a juvenile. In general, fishes with larger eggs and more energy available in the yolk produce larval fishes that more closely resemble the adult. Fishes with smaller eggs provide fewer resources for development, and larvae are less likely to resemble adults. Rapid ontogenetic changes that follow involve behavioral, morphological, and physiological responses. The ability of a larval fish to successfully navigate these different ontogenetic changes is largely dependent on it experiencing suitable environmental conditions throughout its ontogenetic transitions.

As described in Chapter 10, fish – especially larval fish – are susceptible to various types of mortality. However, starvation and predation are seemingly the most influential for young fishes. These processes are generally size dependent, with small larvae often being at greater risk of mortality than larger individuals (Miller et al. 1988). Small fish have relatively high mass-specific metabolic rates, meaning that per unit mass they use energy very rapidly. Also, given their small size, larval fish have low energy stores. Therefore, larval fish have the potential to quickly deplete all of their energy stores and starve to death, although in many cases, individuals in poor condition after energy depletion may be more vulnerable and ultimately succumb to predation mortality. Larval fish are limited in their ability to capture prey. While swimming ability tends to improve as larval fish increase in size, small larvae are generally poor swimmers, limiting their ability to capture highly mobile prey. While some larval fish do not successfully develop mouth parts early in life, those that do are gape-limited predators, that is, the size of their mouth limits their ability to consume many prey items. Senses, including vision, develop during early ontogeny, and larval fish have limited ability to detect prey, relative to older life stages. Their limited ability to capture prey, together with rapid depletion of low energy stores, render many larval fish highly susceptible to starvation. Their early-life survival is largely dependent on overlapping with sufficient prey of suitable sizes that are readily captured and of high nutritional value.

The small size and limited swimming ability of larval fish also make them vulnerable to many predators. As fish develop and increase in size, they can more readily detect and actively avoid predators. Also, many potential predators on larval fish are gape-limited and can only consume small fish larvae. The faster an animal grows, the quicker it exceeds a vulnerable size for predation, and the better its chances of survival (Werner and Gilliam 1984). Fish that grow quickly and get out of the size range of highest vulnerability should ultimately have lower mortality rates than fish that grow slowly and stay within that size range for a longer time.

Starvation and mortality are both related to the ability of larval fish to obtain sufficient prey and grow rapidly. As fish become larger, they have higher energy stores and lower mass-specific metabolism. Larger larvae can swim faster, consume more prey, and are better able to sense their environment – both potential predators and prey. If a larval fish has limited energy reserves, that larva is not capable of swimming very fast and may be eaten long before it starves. There is also an effect of stage duration. Faster growing larval fish will spend less time as small larvae and more rapidly develop past this vulnerable life stage.

It should be obvious, based on the above description, that the availability of prey during early larval stages can have strong effects on survival. By directly affecting growth and energy storage, the consumption of suitable prey strongly influences risk of both starvation and predation. Variation in the availability of suitable prey can lead to variation in larval survival and, ultimately, recruitment. Cushing (1969) demonstrated that several fish populations spawn at consistent times during the year, while spawning of other populations is more variable. Cushing considered the implications of spawn timing in terms of annual variation in the temporal overlap of prey and first-feeding larval fish. They came up with the idea of the match–mismatch hypothesis. Phytoplankton and zooplankton essentially bloom. That is, they experience periods of rapid and dynamic seasonal expansions in biomass. The timing of these blooms can be variable from year to year and related to processes such as thermal conditions, light availability, and water mass movement. Similarly, for some fish populations, the timing of first feeding by larvae can be variable from year to year. Regardless of whether plankton bloom timing varies from year to year or if larval fish first feeding varies from year to year, the end result is that some years the timing of plankton blooms and larval first feeding will match up in time and some years they may end up being out of synch or mismatch. During match years, the availability of suitable prey should allow larval fish to grow rapidly and experience relatively low starvation and predation mortality. In contrast, during mismatch years, limited prey availability is expected to contribute to slow growth and relatively high starvation and predation mortality. The match–mismatch hypothesis provides a mechanistic model that can help explain the observed high annual variability for recruitment for many fish populations.

The general concepts of the match–mismatch hypothesis can be extended to not only consider temporal matching with suitable prey but also spatial matching. In addition, spatio-temporal matching may not only relate to prey overlap but also overlap with predators and environmental conditions that influence the ability of larval fish to effectively feed and avoid predators. Ultimately, he spatial or temporal mismatch of favorable environmental during early life has the potential to strongly influence whether a particular year class is strong or weak.

Marine Versus Freshwater Larval Dynamics

While concepts like the critical period and match–mismatch hypothesis can help conceptualize and explain recruitment variation for several fish populations, it is important to recognize that fish species and the environments they inhabit are diverse. While the transition from endogenous to exogenous feeding and the spatio-temporal overlap of first-feeding larvae and suitable prey may be key determinants of recruitment success for several populations, other life stages and processes may be more important for other populations.

Ed Houde from the University of Maryland has examined early-life dynamics and recruitment variation of several fish populations. In a particular analysis, they gathered information on various marine and freshwater fish larvae (Houde 1994). Of the larval species Houde considered, they found marine larvae were on average smaller at hatch, displayed higher mass-specific metabolic rates, experienced greater mortality rates, and took longer to develop from hatching until they metamorphosed to juveniles. Houde proposed that taken together, these characteristics of marine fish larvae make this early-life stage particularly important to recruitment variation of marine fishes. Small changes in growth or survival could have huge implications for the number of marine larval fish that survive to juvenile stage. Houde estimated that while only 0.12% of individuals from a marine larval fish cohort survive to metamorphosis, 5.30% of freshwater larvae survive to this point. He suggested that due in part to this relatively high larval survival, processes during the juvenile stage are more likely to be important controllers of recruitment variation for freshwater fish than marine species.

Again, Houde's analysis represents generalized patterns for freshwater and marine fish larvae. There are clearly some freshwater fishes who produce very small larvae that experience very high mortality, and there are marine species that produce very large larvae with high survival. These patterns can vary even within a species. For example, Feiner et al. (2016) examined variation in the size of eggs produced by walleye in the Great Lakes region. Since egg size is strongly related to larval size, individuals that produce larger eggs will yield larger larvae. Feiner and co-authors demonstrated that walleye egg size was related to female size and annual thermal conditions. However, the strongest and most consistent effect on egg size was population identity. Walleye in some populations consistently produced small eggs, while walleye in other populations consistently produced large eggs. Presumably, differential environmental conditions across the habitats experienced by these walleye populations selected for different egg and larval sizes. In terms of influencing recruitment success, the relative importance of processes affecting larval versus juvenile stages might also be expected to differ among these populations.

Recruitment Indices

A first step in considering recruitment for a fish stock is defining how to index annual recruitment. It is important to consider not only the biology of the stock but also available data. Recruitment is often indexed via annual fishery-dependent or fishery-independent assessment programs. An index may quantify the number of new individuals that (a) become available to a fishery or assessment gear, (b) reach a specific age or size, or (c) move to a specific location. A useful recruitment index should balance the goal to detect recruitment as early in life as possible, while ensuring recruitment is set by the time (or age) it is indexed. It is useful to detect recruitment as early in life as possible so managers can have sufficient time to potentially adjust management practices in anticipation of a particularly strong or weak year class. However, if an abundance index of a year class during early life is not a good predictor of how abundant that year class will be in subsequent years, this is likely not a very useful index of recruitment. An index of recruitment should not target an age or size before recruitment is set.

Ivan et al. (2011) examined long-term (1970–2008), age-specific catches of walleye and yellow perch in an annual fishery-independent trawling survey conducted by the Michigan Department of Natural Resources during fall in Saginaw Bay, Lake Huron. They found that the CPUE of age-0 walleye in the fall was very strongly, positively related to the CPUE of age-1 walleye collected the following year and the CPUE of year-2 walleye collected two years later. Recruitment of walleye appeared to be set by the time age-0 walleye abundances was assessed in the fall. Therefore, CPUE of age-0 walleye was a useful index of recruitment because it both indexed year-class strength early in life and was reflective of abundance of subsequent ages. In contrast to walleye, the CPUE of age-0 yellow perch was not strongly positively related to CPUE of age-1 perch the following year or CPUE of age-2 perch 2 years later. However, CPUE of age-1 yellow perch was strongly positively related to CPUE of age-2 perch the following year. Thus, recruitment of yellow perch was seemingly set by fall of age 1, but not fall of age 0. Perhaps age-0 perch CPUE was not a good index of recruitment because these young fish were not always fully vulnerable to the sampling gear at this age. Alternatively, inconsistent annual mortality of young yellow perch – perhaps due to differential predation pressure or climate-related overwinter survival – could serve to disrupt the relationship between age-0 relative abundance and subsequent relative abundances at older ages.

Stock–Recruitment Relationships

Recruitment has traditionally been predicted as a function of adult stock size. Reproductive adults are necessary to produce offspring. Stock–recruitment (S-R) relationships involve relating some annual index of recruitment (e.g., mean number of age-1 yellow perch collected per hour of trawling) to an index of stock size. Some of the considerations for selecting an appropriate index of recruitment are provided above. Similarly, selecting an appropriate index of stock size is important and should be based on the biology of the stock as well as data availability. Stock size is often indexed based on annual CPUE of fish above a certain age. However, as size and fecundity tend to increase with age, more precise indices of reproductive stock size may consider annual differences in population size and age structure to estimate the spawning stock biomass or the overall stock egg production. After defining appropriate indices for recruitment and stock size, it is also critical that time series of these indices are properly matched temporally. For example, if recruitment is indexed as CPUE of age-1 fish in the fall, this should be related to the stock size the previous year. That is, age-1 fish in year t were age-0 fish in year $t - 1$, and these fish were produced by the adult stock in year $t - 1$.

Bi-plots of indices of stock size versus recruitment have often been described as shotgun blasts. There tends to be a high degree of variation in recruitment indices, and that variation is generally not well described by stock size. Some of the inter-annual variation in recruitment indices is likely related to difficulty in accurately quantifying abundances of young fishes (measurement error). However, much of the variation evident in stock–recruitment plots simply reflects that annual recruitment is sensitive to a variety of processes and not fully determined by stock size. However, there must be some relationship between stock size and recruitment, given that if there are no reproducing adults, there can be no recruitment. Myers and Barrowman (1996) considered a multitude of stock–recruitment relationships to ask the question, "Is fish recruitment related to spawner abundance?" They compiled 364 stock–recruitment data sets, encompassing a variety of marine and freshwater fish populations. They analyzed all these data sets to simply evaluate if recruitment tends to be relatively high when stock size is relatively high and, in contrast, if recruitment tends to be relatively low when stocks size is relatively low. They found that as long as there was sufficient contrast in stock size across years, there was on average a detectable effect of stock size on recruitment. Recruitment tended to be relatively high when stock size was high and, vice versa, recruitment tended to be low when stock size was low. Thus, even though there may be a high level of unaccounted for variation, it does make sense to use stock size as a predictor of fish recruitment.

Stock Recruitment Models

Numerous models have been developed to relate recruitment to stock size. These discrete time models assume that recruitment is nonlinearly related to stock size and there is no recruitment when stock size is equal to zero. Also, these models are based on various forms of compensatory density-dependence. Therefore, as stock size increases, there is a decrease in R/S (recruits per spawner or recruiting units per unit stock size). Different stock–recruitment models have been developed based on studies of different fish species with different mechanistic controls on recruitment. Therefore, the biological processes initially hypothesized as the mechanisms underlying stock–recruitment relationships differ among models. As these models have become widely adopted, they have been applied for a variety of species and populations, likely with diverse biological controls. In fact, for most fish populations, there are likely multiple density-dependent controls acting simultaneously, and isolating the specific processes contributing to a specific stock–recruitment relationship may be impossible. Instead, stock–recruitment models are routinely fit with little consideration of the biological relationship underlying such a relationship. The statistical fit of different model forms and the mathematical and statistical appropriateness of the models are often the basis for selecting a specific model. However, as models make different assumptions regarding stock–recruitment relationships, it is useful to at least contemplate the biological suitability of specific models. For example, consider the two most widely adopted stock–recruitment models: the Ricker and Beverton–Holt stock recruitment models.

William Ricker from the Fisheries Research Board of Canada developed a stock–recruitment model based on a variety of species, including observations of Pacific salmon in western Canada (Ricker 1954). In a particular stream or larger river system, there will be some limited amount of suitable habitat for salmon to excavate redds, where they deposit eggs. If a salmon run is very large, all of the suitable habitat may be used for spawning. As additional spawners move into the system, they may excavate redds where other salmon have already spawned. Therefore, they may displace incubating eggs or cover eggs in sediment, depriving them of oxygen and leading to mortality of eggs. High numbers of spawning salmon may also enhance the spread of pathogens and facilitate the development of fungal infections on eggs. When salmon larvae hatch, emerge, and begin to feed, the number of young salmon in the stream will affect the per capita prey availability, and large numbers of young salmon may attract predators or enhance pathogen transmission, collectively affecting growth and survival. When adult population size is low, increasing numbers of spawning salmon may lead to more recruits as the population is able to take advantage of a larger proportion of suitable spawning habitat. However, there are multiple mechanisms where recruitment may reach an asymptote – or even decline – as the number of spawners becomes very high.

Ricker's stock recruitment model nonlinearly relates recruitment (R) to stock size (S) based on two parameters α and β:

$$R = \alpha S e^{-\beta S}$$

(Equation 12-1)

The parameter α influences the magnitude of recruitment, while β determines the nonlinear shape of the relationship. The model assumes compensatory density-dependent controls on recruitment, and recruits per spawner (R/S) declines with increasing stock size, i.e., $R/S = \alpha e^{-\beta S}$ (Figure 12-1).

Estimating the parameters describing a stock–recruitment relationship can be challenging. First, given the high variation in recruitment, no single set of parameter values may be able to describe a large proportion of the variation in recruitment. Second, the error distribution of observed recruitment relative to expected recruitment has not been shown to follow the commonly assumed normal distribution. Instead, a log-normal distribution appears more appropriate to describe recruitment residuals. Under a normal error distribution, observed recruitment would be equally likely to be greater than or less than predicted recruitment, and the distribution of residuals would follow a symmetrical, bell-shaped curve with most observed recruitment values slightly less than or greater than predicted recruitment (and only occasional observed recruitment values deviating greatly from predicted values). In comparison, a log-normal error distribution is skewed to the right, meaning that observed recruitment values will not be symmetrically distributed around predicted values and observed recruitment values much greater than predicted recruitment would be relatively common. It is common to account for a log-normal error structure by predicting the logarithm of recruitment, $\ln(R)$, instead of actual recruitment. The error distribution of observed $\ln(R)$ minus predicted $\ln(R)$ follows a normal distribution, making it much more straightforward to estimate stock–recruitment parameter values.

Given a time series of stock recruitment indices, there are two approaches for estimating parameters for Ricker's S-R model. One option is to use an iterative, nonlinear fitting method to estimate α and β values that minimize the sum of squared residuals (observed $\ln(R)$ minus predicted $\ln(R)$). Another approach involves linearizing the relationship and using linear regression methods to estimate parameter values. Specifically, Ricker's S-R model can be converted to a linear model by dividing both sides of the model equation by S and taking the log of both sides of the equation (see Box 12-1):

$$\ln\left(\frac{R}{S}\right) = \ln(\alpha) - \beta S \qquad \text{(Equation 12-2)}$$

Consistent with compensatory density dependence, the logarithm of R/S (dependent variable) declines with increasing S (independent variable), following a linear relationship with $-\beta$ as slope and $\ln \alpha$ as intercept (Figure 12-1).

Similar to William Ricker, Ray Beverton and Sidney Holt were highly influential fishery scientists from the other side of the Atlantic Ocean in the Lowestoft Lab in England, and their book, *On the Dynamics of Exploited Fish Populations*, published in 1957, has guided a great deal of subsequent quantitative treatment of fisheries and population dynamics of harvested populations (Beverton and Holt 1957). Beverton and Holt presented a stock–recruitment model similar to Ricker's model based upon compensatory density-dependence and including two parameters a and b:

$$R = \frac{aS}{1 + bS} \qquad \text{(Equation 12-3)}$$

Under the Beverton–Holt model, recruits per spawner decrease as the number of spawners (stock size) increases, i.e., $R/S = a/(1 + bS)$. Unlike the Ricker model, the total number of recruits increases toward an asymptote at high stock size and does not decrease (Figure 12-2).

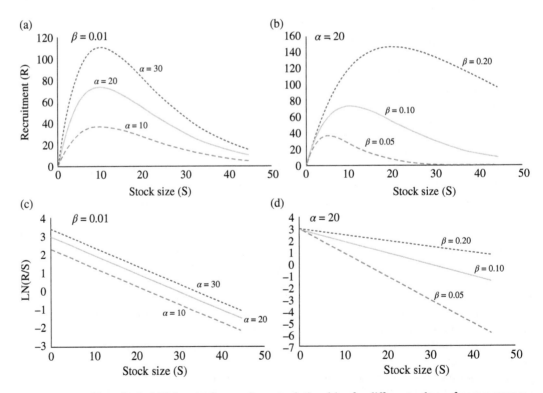

FIGURE 12-1 Hypothetical Ricker stock–recruitment relationships for different values of parameters α (a and c) and β (b and d). Note that changes in α affect the magnitude of recruitment, while changes in β affect the shape of the recruitment relationship. The two bottom plots (c and d) show a linearized compensatory density-dependent relationship between stock size and the natural logarithm of recruitment per stock size.

Box 12-1 Fitting a Ricker Stock–Recruitment Model

To demonstrate how to estimate stock–recruitment relationships, consider yellow perch in central Lake Erie. The Ohio Department of Natural Resources (ODNR) has conducted annual fall bottom trawling of this population for several years. Upon collecting yellow perch, the ODNR also ages a subset of individuals leading to annual age-specific estimates of relative abundance. These data allow for annual estimates of adult relative abundance based on the area swept by bottom trawls (indexed as mean catch per hectare). If recruitment by yellow perch occurs when they reach age-3, the recruits observed during a particular year were actually produced by the adult stock three years earlier. Therefore, an index of stock size during year t should be related to an index of recruitment from catches during year $t + 3$, as shown in plot (a).

Assuming yellow perch recruitment in Lake Erie is described by a Ricker style stock–recruitment model, one can estimate these parameters using a nonlinear fitting approach or by linearizing the relationship. To accomplish the latter, plot S versus $\ln(R/S)$ and fit a regression line through this relationship (see plot (b)). The negative slope is an estimate of β, i.e., $\beta = 0.053$, and the intercept is an estimate of $\ln(\alpha)$, i.e., $\ln(\alpha) = 1.878$, and thus, $\alpha = 6.540$. The estimated model is depicted in plot (c).

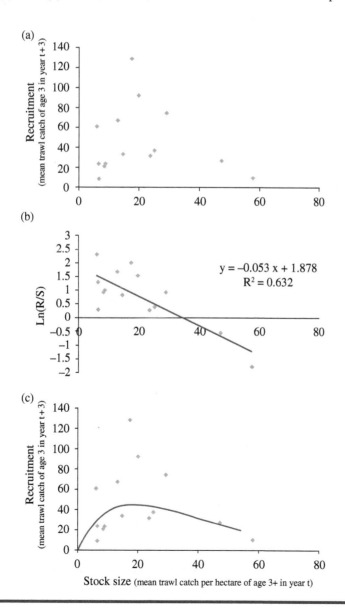

(a)

(b)

$y = -0.053\,x + 1.878$
$R^2 = 0.632$

(c)

Stock size (mean trawl catch per hectare of age 3+ in year t)

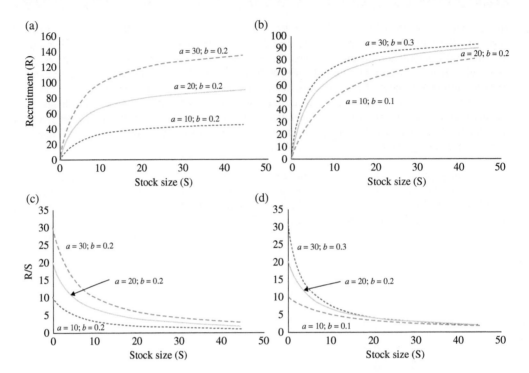

FIGURE 12-2 Hypothetical Beverton–Holt stock–recruitment relationships for different values of parameters a and b. Note that changes in a affect the asymptote of the relationship (a), while simultaneous changes in a and b affect how quickly the asymptote is approached with each unit increase in stock size. The two bottom plots (c and d) demonstrate the compensatory density-dependent relationship between stock size and recruitment per stock size.

Beverton and Holt developed their model when considering marine species in the North Atlantic and North Sea, in particular focusing on plaice and to a lesser extent on haddock. Their model is based upon the notion that the juvenile mortality rate would increase as the number of juveniles in a nursery environment increases. Therefore, if recruitment takes place after the juvenile period, the number of recruits produced will not simply be a fixed proportion of stock egg production. Beverton and Holt suggested various mechanisms where juvenile mortality could decrease with increasing abundance, including prey availability and predator attraction. One could envision several other mechanisms similar to Ricker's model that could contribute to such compensatory density-dependence.

Unlike Ricker's S-R model, there is no suitable approach for linearizing the Beverton–Holt model. Therefore, estimating parameter values for this model generally involves using a nonlinear, iterative approach to select parameter values that minimize the sum of squared residuals. Given the log-normal error distribution of observed recruitment minus predicted recruitment, it is appropriate to estimate parameter values with residuals indexed, as observed $\ln(R)$ minus predicted $\ln(R)$.

Year-Class Strength from Catch Curves

Fishery researchers and managers often define a year-class as a strong year-class or weak year-class, and determine the environmental conditions that contribute to year-class strength. Estimating stock–recruitment relationships can be a step toward defining year-class strength, and environmental variables can be incorporated into S-R models to help improve ability to describe recruitment variability (see below). However, there are other approaches for defining year-class strength.

In Chapter 10, the catch-curve approach is presented as a method to estimate the mortality rate. This approach involves examining the relationship between age (independent variable) and the logarithm of age-specific relative abundance (dependent variable). If one only includes ages fully vulnerable to capture gear, this relationship should be roughly linear with a negative slope reflecting the negative value of Z (total instantaneous mortality). Mortality rates can be estimated through the catch-curve method either by examining the CPUE of a single year class over several consecutive years or by examining the CPUE of multiple ages at a particular point in time. Under the latter approach, changes in CPUE with age will not only reflect decreases in abundance with increasing age but also differences in recruitment across year classes. While such variation in recruitment can lead to increased uncertainty in estimates of mortality, year-class variation around a catch-curve can also be used to index year-class strength.

Michael Maceina from Auburn University demonstrated a simple method of how catch-curve residuals can be used to index year-class strength (Maceina 1997). This approach involves sampling a population on a single occasion, aging individuals, and quantifying age-specific catch rates (C/f, catch per effort). A regression line relating age versus $\ln(C/f)$ not only reflects an estimate of instantaneous mortality, but also provides an expected value of $\ln(C/f)$ at specific ages. Age-specific residuals – difference between observed $\ln(C/f)$ and expected $\ln(C/f)$ – indicate year-class strengths, with positive residuals reflecting relatively strong year classes and negative residuals reflecting relatively weak year classes. An advantage of this approach is that it only requires sampling a population on a single occasion. While combining catch-curve information across a few years could improve estimates of year-class strength, this method has the potential to provide cost and effort savings compared to long-term annual sampling often required to develop a stock–recruitment model.

Maceina used the catch-curve residual method to examine year-class strength of largemouth bass in reservoirs in Alabama. By sampling a population two years apart, he was able to demonstrate that the same year classes identified as strong and weak during an initial sampling event were also identified as strong or weak two years later. Maceina also incorporated annual environmental variation into his analysis and showed that some of the variation in catch-curve residuals was attributable to annual differences in key hydrologic variables.

Incorporating Additional Explanatory Variables to Predict Recruitment

The number of new fish added to a population through recruitment has potential to strongly influence the trajectory of a population. However, the variability and uncertainty of predicting recruitment of many fish populations based on stock size alone not only makes it difficult to predict future population sizes, but also these features of fish recruitment can confound the management of fisheries. Therefore, improving predictions of recruitment has been a goal of a multitude of analyses over the last several decades. A large number of population-specific stock–recruitment relationships have been developed for marine, freshwater, and estuarine populations. These models have only described a small portion of recruitment variation. Many researchers have incorporated additional predictors into stock–recruitment models to improve predictive ability. For example, in any stock–recruitment model, it is possible to not only use stock size to predict recruitment, but also some annual measures of key environmental conditions (e.g., an index of temperature, productivity, or prey availability). Charles Madenjian from the USGS Great Lakes Science Center (GLSC) worked with colleagues to model the recruitment of alewife in Lake Michigan from 1962 to 2002 (Madenjian et al. 2005). The USGS-GLSC has monitored the abundance of alewife in Lake Michigan using an annual daytime fall bottom trawling survey. They used catches of adult alewife to index spawning stock size each year. However, young alewife do not remain near the bottom of the lake during the day and are not fully vulnerable to the bottom trawl gear until they reach age 3. Therefore, while they used catch of adults during a specific year to index spawning stock, recruitment was indexed by the catch of age-3 alewife in the bottom trawl survey three years later. Madenjian and colleagues used a Ricker type stock recruitment model to relate alewife recruitment to adult spawning stock size. They also considered the influence of a variety of other types of annual indices, including cumulative predation pressure by salmon and trout on alewife, an index of winter severity, an index of spring and summer temperatures, and an index of primary production in Lake Michigan. The final model they chose included not only spawning stock size as an index of recruitment, but also salmon and trout predation (PRED) and spring and summer temperatures (TEMP). This model explained 75% of the variation in alewife recruitment in Lake Michigan from 1962 to 2002:

$$R = 2.18S\,e^{-0.0091S - 0.01406\,\text{PRED} + 0.00015\,\text{TEMP}} \qquad \text{(Equation 12-4)}$$

The addition of environmental variables often improves the ability of stock–recruitment models to match observed recruitment time series. However, the development of S-R models and consideration of environmental influences are based on retrospective analyses. Myers (1998) considered several previously identified relationships between recruitment and an environmental variable. They re-examined these relationships using additional data compiled since the recruitment–environment relationship was first identified. These retests were not very successful; only 28 of 75 environment–recruitment relationships were robust when reanalyzed. Myers identified relationships between temperature and recruitment of populations near their latitudinal edges as most robust. In contrast, relationships between recruitment and indices of wind or temperature for populations in the middle of their geographic range were not very strong. The inconsistency of these environment–recruitment relationships may in some cases suggest that the initially identified relationship was in fact spurious; or perhaps conditions in the environment changed so the effect of one environmental variable was altered or trumped by the effect of another variable.

Another approach to deal with inconsistent environment–recruitment relationships is to abandon the assumption of stationarity or temporally consistent relationships. Stock–recruitment relationships are based on estimating constant parameter values (e.g., for the Ricker model, estimating constant α and β values), but this may not always be appropriate. For example, David "Bo" Bunnell and colleagues from the US Geological Survey's Great Lakes Science Center demonstrated an inconsistent relationship between stock size (egg production) and recruitment (subsequent age-3 abundance) for bloater (*Coregonus hoyi*) in Lake Michigan (Bunnell et al. 2006). During the 1980s, the spawning population produced an intermediate number of eggs, which led to consistently very high recruitment. However, during the 1990s when spawner egg production was at very similar levels, the recruitment levels were consistently low. This inconsistent stock–recruitment relationship seemed to be related to sex ratios, and when the population was dominated by females during the 1990s, recruitment levels were consistently lower than expected based on egg production. However, there are many reasons why stock–recruitment parameters may not be consistent over time. For example, if there is a long-term trend in nutrient loading to a water body, the productivity of the system may also trend over time, and this may affect recruitment potential. In these cases, it may be appropriate to allow parameter values to be non-stationary to reflect changing system conditions. For example, when considering stock–recruitment relationships for rainbow smelt in Lake Michigan, Feiner et al. (2015) considered multiple Ricker-based stationary models, including various environmental variables. However, the model that by far fared the best allowed α to vary over time. Feiner and co-authors speculated as to the mechanisms that may have contributed to changes in recruitment potential (variable α), but they were not able to identify a clear driver of this changing relationship.

In addition to considering the influence of various environmental variables, a number of recent analyses have considered the potential effects of changing demographics of fish populations on recruitment patterns (Hixon et al. 2014). Older, larger female fish not only produce more eggs, but for many species, there is evidence that they also produce larger eggs. Larger eggs yield larger larvae with greater yolk reserves. Such larvae may experience greater early-life growth and survival as they would have lower mass-specific metabolic rates, greater swimming abilities, and the ability to survive longer without feeding during early life. Therefore, recruitment potential may be strongly altered if the age structure of a fish population becomes truncated (e.g., through intense size-selective fishing) and shifts from containing adults of diverse ages to primarily containing young, small adults.

While differing in the size of offspring they produce, female adult fishes of different sizes and ages may also differ in timing of spawning. Older females may spawn earlier than younger females in many populations. As a result, larvae from older females may hatch and begin feeding relatively early. Given the high spatial and temporal variability of prey and other suitable environmental conditions for larval fish, a population with a diverse age structure that spawns over a protracted period and across multiple habitats is more likely to produce at least some offspring that encounter favorable conditions during early life. In contrast, a population dominated by small, young females may produce larvae that almost all start feeding within a very narrow time window. This leads to the prediction that populations with more diverse age and size structure will express less variable recruitment from year to year, while populations with limited age and size diversity will experience higher recruitment variations. Hsieh et al. (2006) considered this prediction by examining a long (~50 years) time series of larval fish catches in the Pacific Ocean off California. They compared catches of 29 species and assumed that larval catches were proxies for adult abundances. The species they considered included stocks that were exploited and not exploited by fisheries, and they expected that exploited stocks would be characterized be lower age and size diversity. Even after accounting for effects such as species' life-history traits and abundances, they showed that heavily exploited fish species displayed more variability in larval abundances than unexploited species.

Synchrony in Recruitment

Another approach to identify potential environmental conditions contributing to recruitment success is to evaluate recruitment synchrony and consider if there are any broad scale environmental or climatic factors that lead to high recruitment across multiple populations. There are several mechanisms by which annual environmental conditions in aquatic systems are strongly related to climatic conditions. For example, some years may be characterized by (1) unusual warming, which may not only lead to warmer waters but also earlier blooms of plankton and potentially greater biological production and (2) high precipitation that can lead to greater stream and river discharge, potentially contributing to elevated nutrient loading, but also potentially dislodging incubating eggs from stream beds. These types of climatic effects on aquatic systems do not tend to be highly localized instead they tend to be structured over broad spatial scales. So, different populations inhabiting different lake or river systems may all experience relatively warm conditions or high river discharge during a particular year. In large marine systems, populations of the same species may be separated spatially, but experience similar annual changes in environmental conditions.

If recruitment of fish populations is influenced by broad climatic conditions, it is possible that recruitment success will be synchronized among spatially distinct fish populations. There are multiple mechanisms by which spatially distinct populations may display a level of synchrony in abundance, including a low level of dispersal among populations or some broad scale biological driver (e.g., wide spread disease or highly mobile predator). The potential for environmental conditions to synchronize distinct populations is referred to as the Moran effect, after Pat Moran, an Australian statistician (Moran 1953).

Myers et al. (1997) explored the spatial scale of recruitment synchrony among various marine, anadromous, and freshwater fish stocks. They focused on intra-specific comparisons by fitting Ricker stock–recruitment models to time series for each stock and matching the residuals by year to compare across stocks. A positive residual suggested that year was a strong year-class for a particular stock, while a negative residual indicated a weak year-class. Two stocks displayed positive synchrony in recruitment if they tended to exhibit positive or negative residuals (indicating strong or weak year-classes) during the same years. Myers and co-authors found that several marine stocks displayed positive synchrony in recruitment and calculated that the average scale of within-species recruitment synchrony for marine species was 500 km. In contrast, the average scale of within-species recruitment synchrony for freshwater species was 50 km. The authors suggested these patterns were consistent with the idea that broad scale environmental factors influence recruitment of marine stocks. In contrast, they speculated that localized biological interactions were more important determinants of recruitment success for freshwater systems.

Myers and co-authors suggested that size differences between large marine and smaller freshwater systems were likely one reason for the difference in scales of recruitment synchrony. The Laurentian Great Lakes are interesting to consider in this regard because they are much larger than most freshwater systems, and recruitment of some fish populations in these systems may be influenced by similar processes as in marine systems. Honsey et al. (2016) examined inter-annual variation in year-class strength of different yellow perch populations from throughout the Great Lakes. Rather than fitting stock–recruitment models as Myers et al. (1997), Honsey et al. used Maceina's catch-curve residual approach to identify strong and weak yellow perch year classes. They found evidence of year-class strength synchrony for yellow perch at a scale of 150 km, a greater distance than Myers et al. found for freshwater fish in general. Honsey et al. evaluated what environmental variables might be contributing to the synchrony in recruitment of yellow perch in the Great Lakes and found a relationship with spring and summer temperatures. In particular, during years when spring and summer temperatures were relatively high across the Great Lakes region, yellow perch year-classes tended to be particularly strong and vice versa.

As part of Honsey et al.'s (2016) analysis, 1992 was noteworthy as a particularly cold year during spring and summer with weak year-classes of yellow perch across the study populations. Globally, year-classes of many fish populations of various species were weak during 1992, which has been related to the eruption of Mount Pinatubo, a volcano in the Philippines. During 1991, Pinatubo experienced a huge eruption, by far one of the largest volcanic eruptions during the past century. The massive expulsion of ash and chemicals and the coincident weather patterns led to a sulfuric haze spread throughout the Earth's atmosphere and contributed to a global cooling during 1992 and even into 1993. In turn, some fish populations whose year-class strengths are positively related to temperature experienced weak year-classes during 1992 and 1993, including walleye in Minnesota lakes (Schupp 2011; Honsey et al. 2020).

Summary

Recruitment is of central importance to stock dynamics; in closed fish stocks, recruitment represents the sole process for increasing abundance. However, annual recruitment is highly variable. Most fishes are very fecund and produce larger numbers of rather small offspring that experience high mortality. Due to their numerical abundance, even relatively small changes in survival rates can lead to large increases in the number of young fish that survive through early-life stages and ultimately recruit to the adult population. Due to the importance of recruitment, several models have been developed to describe and predict recruitment variation. In general, these models assume that recruitment is (a) controlled by compensatory density-dependent processes and (b) related to adult stock size (without spawning adults, there can be no recruitment of young fish). Model analyses have identified a variety of factors associated with recruitment variation, including factors influencing a variety of life stages and both biotic and abiotic factors. Physical processes affecting the spatio-temporal overlap of early-life stages with suitable environments for both growth and survival are often influential for recruitment success.

Literature Cited

Beverton, R.J.H., and S.J. Holt. 1957. On the dynamics of exploited fish populations. Fishery Investigations Series II 19:1–533.

Bunnell, D.B., C.P. Madenjian, and T.E. Croley III. 2006. Long-term trends of bloater (*Coregonus hoyi*) recruitment in Lake Michigan: Evidence for the effect of sex ratio. Canadian Journal of Fisheries and Aquatic Sciences 63:832–844.

Cushing, D.H. 1969. The regularity of the spawning season of some fishes. ICES Journal of Marine Science 33:81–92.

Feiner, Z.S., D.B. Bunnell, T.O. Höök, et al. 2015. Non-stationary recruitment dynamics of rainbow smelt: The influence of environmental variables and variation in size structure and length-at-maturation. Journal of Great Lakes Research 41:246–258.

Feiner, Z.S., H.-Y. Wang, D.W. Einhouse, et al. 2016. Thermal environment and maternal effects shape egg size in a freshwater fish. Ecosphere 7(5).

Gatch, A., S. Koenigbauer, E. Roseman, and T.O. Höök. 2020. The effect of sediment cover and female characteristics on the hatching success of walleye. North American Journal of Fisheries Management 40:293–302.

Hixon, M.A., D.W. Johnson, and S.M. Sogard. 2014. BOFFFFs: On the importance of conserving old-growth age structure in fishery populations. ICES Journal of Marine Science 71:2171–2185.

Hjort, J. 1914. Fluctuations in the great fisheries of northern Europe, viewed in the light of biological research. Rapports et Procès-Verbaux des Réunions du Conseil Permanent International Pour L'Exploration de la Mer 20:1–228.

Honsey, A.E., D.B. Bunnell, C.D. Troy, et al. 2016. Recruitment synchrony of yellow perch in the Great Lakes Region. Fisheries Research 181:214–221.

Honsey, A.E., Z.S. Feiner, and G.J.A. Hansen. 2020. Drivers of walleye recruitment in Minnesota's large lakes. Canadian Journal of Fisheries and Aquatic Sciences 77:1921–1933.

Houde, E.D. 1994. Differences between marine and freshwater fish larvae: Implications for recruitment. ICES Journal of Marine Science 51:91–97.

Hsieh, C-h., C.S. Reiss, J.R. Hunter, et al. 2006. Fishing elevates variability in the abundance of exploited species. Nature 443:859–862.

Ivan, L.N., M. Thomas, D. Fielder, and T.O. Höök. 2011. Long-term and interannual dynamics of walleye (*Sander vitreus*) and yellow perch (*Perca flavescens*) in Saginaw Bay, Lake Huron. Transactions of the American Fisheries Society 140:1078–1092.

Maceina, M.J. 1997. Simple application of using residuals from catch-curve regressions to assess year-class strength in fish. Fisheries Research 32:115–121.

Madenjian, C.P., T.O. Höök, E.S. Rutherford, et al. 2005. Recruitment variability of alewives in Lake Michigan. Transactions of the American Fisheries Society 134:218–230.

Miller, T.J., L.B. Crowder, J.A. Rice, and E.A. Marschall. 1988. Larval size and recruitment mechanisms in fishes: Towards a conceptual framework. Canadian Journal of Fisheries and Aquatic Sciences 45:1657–1670.

Moran, P.A.P. 1953. The statistical analysis of the Canadian lynx cycle, II. Synchronization and meteorology. Australian Journal of Zoology 1:291–298.

Myers, R.A., and N.J. Barrowman. 1996. Is fish recruitment related to spawner abundance? Fishery Bulletin 94:707–724.

Myers, R.A., G. Mertz, and J. Bridson. 1997. Spatial scales of interannual recruitment variations of marine, anadromous, and freshwater fish. Canadian Journal of Fisheries and Aquatic Sciences 54:1400–1407.

Myers, R.A. 1998. When do environment-recruitment correlations work? Reviews in Fish Biology and Fisheries 8:285–305.

Ricker, W.E. 1954. Stock and recruitment. Journal of the Fisheries Research Board of Canada 11:559–623.

Roseman, E.F., W.W. Taylor, D.B. Hayes, R.L. Knight, and R.C. Haas. 2001. Removal of walleye eggs from reefs in western Lake Erie by a catastrophic storm. Transactions of the American Fisheries Society 130:341–346.

Schupp, D.H. 2011. What does Mt. Pinatubo have to do with walleyes? North American Journal of Fisheries Management 22:1014–1020.

Werner, E.E., and J.F. Gilliam. 1984. The ontogenetic niche and species interactions in size-structured populations. Annual Review of Ecology and Systematics 15:393–425.

CHAPTER 13

Social Behavior

It is a common observation to anyone who has held fishes in an aquarium that as the number of animals in a group increases, their behavior changes. For example, one tilapia in a tank is not content; while it will live, it will neither feed very voraciously nor will it be very active. Two male tilapia in a tank will fight; one often ends up killing the other. A male and a female will sometimes attempt mating, although sometimes they will fight, depending on whether they are ready to breed or not. If three or four tilapia are in a tank, the amount of agonistic behavior decreases. With 10 in a tank, they seldom fight and are relatively calm and content to be in a group. There are large shifts in tilapia behavior, depending on their density. One of the reasons why tilapia are so favored in aquaculture is because they tolerate high density without overt aggression. Aquaculture would not be as suitable for certain other species that maintain strong territorial defenses at all densities. The shift from agonistic behavior at low density to more docile behavior at high density in tilapia, and in most cultured species, is important for aquaculture.

A different example of social behavior is demonstrated by most minnows. Two minnows together in a tank will attempt to school, but two fish do not make a very good school, so they become disoriented. Three minnows in a tank will school much better, while nine will school very well. The nine minnows swim in tightly arranged patterns, with their bodies a fixed space apart, and orient to each other in a pattern that is considered a school. Each minnow's behavior does not change in terms of agonistic encounters with density, but their ability to school increases dramatically with the number of individuals present.

These social behaviors are important to fish ecology. The basis for competition or predation may differ considerably depending on such social interactions. It is clear from the examples above that a general statement about how density influences social structure cannot be made because it depends on the kind of social system the fish population exhibits. This chapter details the common social behaviors of fish.

Individual Behavior

Social behaviors are usually based upon some type of ritualized display or conduct that functions to reduce aggression and injury yet allow differential access to resources. These rituals permit competition to occur so the fittest individual may get the best resources (territory and position in the social hierarchy), yet the dominant fish does not have to continually fight every fish nearby. Fish dominance is communicated more by displays and ritualized behavior than by aggression.

Biology and Ecology of Fishes, Third Edition. James S. Diana and Tomas O. Höök.
© 2023 John Wiley & Sons Ltd. Published 2023 by John Wiley & Sons Ltd.

A fish does not have many anatomical structures that can be manipulated in such a way to allow other individuals to recognize a signal. Basically, fishes can flare gills, flare fins, or approach from certain directions or with certain types of swimming. The same basic displays occur in most fish species. One type is a frontal display in which the aggressor approaches directly at another fish, flares its opercula, expands its fins, and attempts to look as large as possible. A lateral display is similar, but it exposes the displaying fish's side to the other fish. Most dominance displays are based upon fishes attempting to look larger, which could be correlated with being more aggressive. Chemical communications through pheromones are also common social signals. For example, groups of bullheads recognize dominant fish and subordinate fish based on chemical cues. Bullhead do not tend to flare fins or opercula since they do not see very well and often live in murky water. The dominant fish emits a certain odor into the water that other fish can recognize and respond to appropriately. Coloration can be another behavioral display in addition to body postures. Even in tilapia, the dominant tilapia develops dark black bars down its side, while the subordinates become much more bland in color. Other communication techniques may be important in electric-generating fishes, where social position may be expressed by electrical signals. Basically, any signal that can communicate and identify an individual may be an indicator of dominance.

Ritualized Displays

An important component of social behavior is ritualization. Ritualization means that fishes demonstrate exactly the same behavior each time they use it and do so repeatedly. For these behaviors to effectively reduce aggression, they must become innate. Strongly ritualized behavior is common in communicating social position through aggressive or breeding encounters. A subordinate's posturing to communicate submission does not necessarily mean the dominant fish will recognize and respond to that signal. In ritualized displays, fishes learn to pattern displays appropriately, and some components of the behaviors signal responses to reduce aggression. Use of known behavioral patterns in appropriate ways allows other individuals to recognize the meaning of the displays and behave accordingly. A naive fish that was reared by itself and suddenly placed into a social hierarchy would probably be attacked frequently because it would not know signals to communicate. The naive fish does not realize how to alter its fins, opercula, or swimming pattern correctly, so it becomes the subject of much aggression.

The breeding behavior of stickleback is an excellent example of ritualized behavior (Tinbergen and van Iersel 1947). Sticklebacks are small fish that often live in slow, stagnant streams or lakes. When three-spined sticklebacks are ready to breed, the male is thin and develops a bright red throat. The female becomes swollen with eggs when she is ripe and develops a gray or silver belly. When the male is ready to breed, he builds a nest in his territory that is essentially a tube or cave, so the fish can swim through the middle of the tube but is surrounded by vegetation. If the male sees a fish with a red throat coming near his nest, he attacks and drives it off. When he sees a fish with a gray swollen shape approach, he tries to attract it to the nest. Stickleback behavior has become so ritualized that experiments can induce a male to attack a pencil painted red, or court any large, round gray shape near the nest. The swollen gray shape is important, as is the red color, rather than the overall dimensions of the animal.

The male stickleback attracts the female by going through what is called a "zigzag dance"; he flutters back and forth above the nest and gradually moves toward it. He does this repeatedly, trying to attract the female. If the female starts to follow the male, the male will move into the nest. If the female follows him, he will go through the nest and try to stimulate the female to move inside. The nest is constructed in such a way that the female, who is usually larger than the male because of her swollen belly, can move into the nest but cannot get out. The male then pokes at the female's abdomen to encourage her to release eggs. When she does, she will leave the nest, then the male moves through it and fertilizes the eggs. The male will then drive off the female and protect the fertilized eggs against any male or female that comes along. This whole sequence of behavior is a ritualized pattern that leads to breeding in the stickleback.

If a female stickleback suddenly appeared in a male's nest before he had gone through the zigzag dance and led the female into the nest, he would most likely chase her out. Disruption of the behavior by not going through the appropriate steps confuses the male, and he does not recognize the chance to spawn. In ritualized behaviors, the fish must perform the proper behavior in the correct sequence in order for another individual to respond appropriately. Breaking the chain of ritualized behavior is confusing to a fish because much of the pattern requires appropriate releasers, some of which are behavioral and others physical. Some releasers (for example, the red color for the male stickleback) are physical patterns that release the next step in the behavior sequence. A ritualized behavior is then a pattern of events that are stereotyped, occur in the same way consistently, and they are not recognized if they occur out of sequence.

Nonreproductive Social Behaviors

There are at least five distinct forms of spacing behaviors in groups of fishes: personal space, home range, territoriality, social hierarchy, and schooling. The simplest type of spacing behavior common to a fish is personal space. A good example of personal space is the spacing of birds on a telephone wire. Birds will not generally sit in contact with each other; they will sit a distance apart. If it is cold, they may sit closer together to warm each other, and on a warmer day they may sit farther apart, but in any case they protect the spaces around themselves. Personal space also has been described in schools of cod, which maintain open spaces around themselves. Even though a cod is schooling, if another cod comes too close, it will drive the intruder off. This defense of personal space has been termed "territoriality" in cod, but it does not really fit the definition of a territory. That space around the individual is not a fixed geographic area; it is not the same physical place. It is the area around the animal, wherever that animal may be. Cod, by defending this space, do not sequester resources for breeding, hiding, or feeding. Some fishes will defend personal spaces but are not territorial or aggressive in any other way.

The second type of spacing behavior common to animals is called home range. A home range is the area an animal uses in its regular activities. In the case of fishes, they may have been hatched in one area, lived as juveniles in another, and then finally settled in a third area as adults. It becomes unclear in such a case what would be considered the home range. Obviously, if an animal is observed for a day, it demonstrates a limited home range. The longer an animal is observed, the larger an area it is likely to cover. Researchers sometimes describe two or three home ranges for one animal based on more extensive movements. It is not clear if such home ranges have any functional significance for most fishes. An animal may or may not defend its home range from animals of the same or other species. The home range may just be an area with which that animal is familiar, so it knows where to escape from predators, where to seek cover for the night, and where to find food. That level of knowledge is probably common in a home range. However, for fishes such as largemouth bass, it is less clear if the home range has a certain size and is defended against other largemouth bass, or whether the size of the home range differs with resource richness.

The third kind of spacing behavior is territoriality. Territoriality is observed when an animal acquires a physical territory (space) and defends it against other animals. The owner of a territory attempts to defend it against all individuals of its own species and possibly other species as well. Even if a larger fish intrudes, the territory owner may try to defend its territory. A territory is a fixed geographic space. It is different from personal space because it does not follow the animal around if the animal leaves. As the density of a population increases, territories may become smaller because there are more individuals to defend against. Territory size is correlated with many other ecological characteristics, such as food density. Unlike a home range, there is a functional definition: a territory allows relatively exclusive access to food and space.

The fourth kind of spacing behavior is a social hierarchy. In this social system, the dominant individual does whatever it wants. The next individual in the hierarchy can dominate or take resources from all other members except the dominant. This continues in more or less a straight line from the highest (alpha) to the lowest (omega) ranked individual. In the case of a social hierarchy, social position does not depend upon the animals' geographic location. The dominant animal remains dominant in that group even if the group moves to a new location. It is not like a territory, where dominance will change depending on whose territory is being contested. Hierarchies and territories can become difficult to differentiate if a social hierarchy is found in a situation in which each animal also chooses to stay in one place regularly. It is hard to say whether each animal is actually defending a territory or whether it is a dominant for a bigger region but simply prefers to stay in one spot.

The final spacing behavior is schooling. Schooling is a regular orientation of one individual to the next without any aggressive interaction. It is the most stereotyped spacing behavior of fishes and is not demonstrated by many other animals. Flocks of birds might show a similar kind of spacing behavior when they are migrating. Schooling is a spacing behavior that has been related predominantly to hydrodynamic advantages – swimming better in a school than alone – but there are also other functions of schools.

The importance of size variation presents a dilemma in sorting out hierarchies and territories (Keenleyside 1979). For example, a mixed group of fish, such as rainbow trout in a stream, includes young-of-year fish to large adults. A small trout is not likely to be able to defend a territory against a very large adult. In a mixed group with large size differences, a social hierarchy is most likely to occur where larger fish are dominant and can suppress smaller fish regardless of where they are. In contrast, in a social system of trout fry in a stream, there is little likelihood that one individual will dominate because the fry are very similar in size. Trout fry will most likely become territorial. Size structure makes a big difference. Fishes continue to grow larger as they age, so older fishes have better chances of being dominant. Coral reef fishes, which tend to grow rapidly to a certain size then remain that size (usually fairly small), demonstrate a high degree of territoriality, which makes sense based on their size distributions.

Fishes generally do not have many ways to express aggression, so success in agonistic encounters is often strongly correlated with size. Larger fishes generally win over smaller fishes in aggressive encounters. A second characteristic sometimes linked to aggressive success is prior residence. If a fish lived in an area before, it might achieve a higher position in the social hierarchy or a better territory than a larger animal that came in later but had to fight all the territorial holders to find a place to live. Therefore, prior residence may become important.

These brief introductions to each form of spacing behavior should clarify the main distinctions among categories. The following sections will detail behavioral studies on social hierarchies, territoriality, and schooling. Home ranges are covered in Chapter 16 on movement. Personal space appears fairly easy to understand and will not be covered further. Reproductive behaviors will be covered in Chapter 22.

Social Hierarchies

A social hierarchy is a group of animals geographically associated, that is, they live together in the same area. Those same individuals interact with each other regularly, meaning they remain in that area. Fish form a hierarchy based on agonistic encounters (fights), from an alpha (dominant) individual through an omega (subordinate) individual. In cases with relatively few animals, this hierarchy is most likely to be linear. In cases with many individuals in an area, there are probably some reversals of dominance, so the hierarchy does not follow a straight line. It is less clear what happens in the middle ranks, but the dominant and the subordinate ranks are clearly defined in most hierarchies. As a social hierarchy becomes established, there may be many agonistic encounters. Later, there is a reduction of aggressiveness and a replacement with ritualistic behavior – more threats, more postures – and less actual nips and fights.

Hierarchies in Stream-Dwelling Trout

Visual observations of trout over short periods of time would probably describe their behavior as territorial because each fish tends to remain in a similar location and to have agonistic interactions with other individuals. Closer observation of the behavior of trout in such a pool would indicate that trout actually have a social hierarchy. Trout remain in moving water feeding on invertebrates that drift downstream. The dominant or alpha fish most commonly moves to the head of a pool, right under the major current that enters the pool, and watches for drift. This location provides first access to any drift. The alpha remains there much of the day. In an almost linear fashion downstream, the subordinate fish line up by rank, along the major current (Figure 13-1). When there is one major current moving through the pool, the beta, gamma, delta, and omega fish are lined up downstream. This downstream arrangement reduces agonistic encounters as the alpha fish, which can displace any other individual, does not see subordinates regularly unless they are in front of it. Subordinate fish, by remaining downstream, do not swim in front of dominants and are not attacked. They remain along the major current to get whatever food is transported to them, and they remain downstream of fish dominant over them. They also choose to be upstream of any fish subordinate to them because then they have

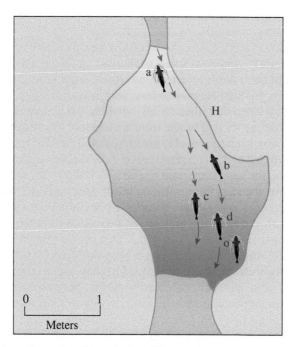

FIGURE 13-1   Social hierarchy and position choice of five cutthroat trout in a small pool (Pool 3 from Table 13-1). Alphabetic order denotes social rank. *Source: Diana (1975) / John Wiley & Sons.*

TABLE 13-1 The Social Hierarchy and Size of Cutthroat Trout in Three Pools.

Social Position	Size (mm)		
	Pool 1	Pool 2	Pool 3
alpha	220	222	246
beta	210	215	205
gamma	170	200 or 178[1]	196
delta			230
omega		158	

[1] Appeared equally dominant.
Source: Diana (1975) / John Wiley & Sons.

earlier access to food in the drift. The omega has last access to food as all other fish must pass up food before it drifts that far. From a resource acquisition point of view, the dominant should get the best resources (food). It probably also has a location behind a rock, so it can remain out of the current yet look up into the current and observe drift.

Hierarchies from different pools are largely based on size, where the dominant fish is the largest (Table 13-1). When a large fish is fairly far down in the hierarchy, it may indicate the fish's late arrival into that pool (for example, see Pool 3 in Table 13-1). That fish may have been displaced (perhaps caught by an angler) and then returned to a new location. Normally, this large fish should have a high social rank, but when this new resident came into the pool, all fish were aggressive toward it, and it could not take the position it normally would have. As a consequence, the new resident ends up down in the hierarchy.

There might also be some differences in social structure depending on currents in a pool. Some pools might have a major current going one way and another current going in a different direction. The hierarchy might be difficult to sort out in such a pool because there is not necessarily a single best location (Figure 13-2). There is still a lining up downstream by rank, but this situation may include several dominant fish – each heading up a current. These fish may not interact often, so it could be difficult to determine which is dominant, but they are definitely dominant over the others. Hierarchies may be less clear in such large or complicated systems.

Social hierarchies have been described for all individuals of the same species of trout or for mixed species groups. When brook and brown trout are together in a stream, brown trout (which tend to grow larger than brook trout) probably are more dominant, but there will be a mixture of species in the hierarchy, based mainly on size of each individual, even though they are different species. Size appears to be the most important characteristic defining hierarchies in trout.

At a certain size, trout abandon this kind of behavior. Jenkins (1969) described large brown trout as nomadic because they seemed to appear in a pool, remain for several days, and then disperse. These individuals did not seem to have any geographical homes but wandered. Clapp et al. (1990) and Diana et al. (2004) have shown that large brown trout demonstrate regular and predictable behavior and movement, but both are very different from those of smaller trout. These large fish have home sites they use regularly, but they might move 2 or 3 km foraging at night, returning to their same home sites the next morning. They probably do not have agonistic encounters with other trout along the way because they are large enough that other trout avoid them. This behavior pattern appears to be related to large trout feeding on different food. When brown trout get very large, they feed on other fishes such as sculpin or other trout. If these other fishes are resident in hierarchies or territories, large brown trout must actively forage to encounter many individuals. These studies have shown that this active movement begins

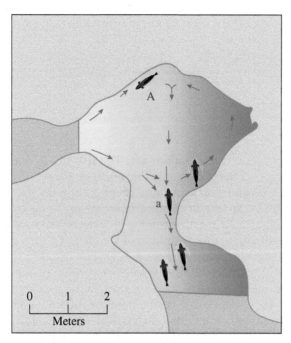

FIGURE 13-2 Social hierarchy and position choice of five cutthroat trout in Pool 1. Trout A and a appeared co-dominant. *Source: Diana (1975) / John Wiley & Sons.*

when brown trout are large enough to feed on other fishes. They have also shown that foraging activities for large brown trout peak around dawn and dusk and remain high at night. Both studies used telemetry to determine trout behavior; obviously, visual observations would not easily detect such activities. The movement and foraging of other predatory fishes is covered more thoroughly in Chapter 16.

If dominance hierarchies are related to fitness, there should be differences in fitness between dominant and subordinate fish. Carl Schreck and his colleagues from Oregon State University have evaluated stress in fish populations. For trout, they evaluated whether the alpha and omega showed differences in these measures (Ejike and Schreck 1980). Tail beat frequency indicates how fast a fish is swimming; an advantageous feeding site should be one in which the fish is required to swim less to maintain station. In a pool, the dominant fish had a low tail beat frequency, while the subordinate had a higher tail beat frequency, indicating that it had to work harder to maintain its location. Liver glycogen is an indicator of how much energy is stored in the liver; liver glycogen was high in the alpha fish and low in the omega fish, indicating that the alpha fish was in better physical condition and had more energy storage. Plasma cortisol is an indicator of stress; higher levels in blood indicate more stress. The alpha had low plasma cortisol levels, while the omega had high levels. This general pattern indicates that there is a fitness correlate to being high in the dominance hierarchy.

Finally, fishes that demonstrate seasonal behavior may re-establish hierarchies regularly. In the winter, trout may change their behavior, do not feed much, and seek cover. Cunjak and Power (1986) from the University of Waterloo in Ontario described non-aggressive aggregations of trout during winter. These aggregations occur in areas of winter refuges from temperature, and fish are more concentrated in harsher winters. In the following spring, a new hierarchy is likely to be established and individuals might move up in that hierarchy, while the large fish might become more nomadic or predatory, and others might die. It is not clear whether hierarchical trout, which migrate to other locations to spawn, necessarily return to their former pools or hierarchies after spawning. If they do not return to the same site, the hierarchies would re-assort regularly.

Territoriality

Territoriality is the most rigid level of space use. The fish that holds a territory is trying to maintain exclusive rights over the resources in that territory. There are behavioral aspects of territorial use – how to defend and hold a territory. There are also resource aspects – how rich the territory is in food, in hiding spaces, or in breeding locations. Some species of fish will defend territories that satisfy all of their needs including feeding, breeding, and hiding. Other fishes will defend different territories at different times of year because they are undergoing different activities. Bluegill build nests and defend them in spring for breeding, but they will not defend those territories later in the year. Some animals will move from breeding territories to other locations where they will feed and hold territories. Some territories may accommodate all the fishes' needs, while some may support only the specific behaviors they are undergoing at that time.

A series of factors influence territory size, all of which could be used in a model to understand the energetics of territorial defense. The most important factor influencing territory size is probably the function of the territory. If it is solely a breeding spot, it could be extremely small. The more functions a territory has, the larger it is likely to be. The size of a territory should obviously be related to resource distribution; the kinds of food utilization may be one example. Carnivores are more likely to have much larger territories than herbivores. Another factor influencing territory size is resource abundance; as the abundance of prey per square meter increases, territory size should decrease. Territories might be expected to increase in size as fishes grow larger. There are several reasons for the increase in territory size with animal size. The first is that size means the animal is physically more capable of defending a territory and can possibly defend a larger area. Second, a larger animal has higher total food needs. Finally, the density of animals should influence territory size. Since a territory is an area that an animal occupies and defends against others, more animals make the defense more difficult.

Territories of Coral Reef Fishes

Coral reefs are perhaps the most studied ecosystems for the behavioral ecology of fishes. Many reef fishes are territorial and their territories are often all-inclusive. Studies conducted in the Virgin Islands tried to define what kind of spaces might be limited on coral reefs: hiding spaces, feeding spaces, or breeding spaces (Smith and Tyler 1972). On this particular reef, 53 species of fish were resident, 13 were transient, and 7 were occasional. Total space seemed to be the biggest limit to the number of fishes that occupied a given coral reef. Smith and Tyler defined a hierarchy of space occupation in which the home ranges were usually fairly large areas that were not defended. The animals that held territories moved off into other areas at times. Usually within a territory, there was a feeding space that was much more limited in size, and even smaller than the feeding space was the area used for shelter. On the Virgin Island reef, shelter appeared to be the major limiting factor for fishes. Shelter included small holes or caves in the coral in which to hide if a predator approached or in which to remain when inactive. Feeding spaces without shelters were quite abundant, but fishes could not persist in these locations over long periods of time because they could not find places in which to hide.

Robertson et al. (1981) researched territoriality in damselfish as related to feeding. Most territorial fishes on coral reefs are small and feed on some part of the reef system. They may feed on periphyton on the coral, the coral itself, or on animals that colonize the coral. These particular damselfish fed on algal mats which are flat reef areas often composed of dead coral, where algae could build up. Damselfish are grazers and they utilize larger mats of algae. Damselfish breed in the same territories in which they feed, and they defend those areas strongly. They expand the sizes of their algal mats by chewing off pieces of coral. They bite off a piece of coral and move it away to expand the size of the open area for mats and may even displace the mats across the coral over time. Since live coral is growing at the same time, there is a conflict between the fish and coral as to whether the flat area becomes larger or smaller. Damselfish do not get nutrition from chewing these corals but use it only to expand territory size.

Most food that damselfish consume (95%) comes from the algal mats on their territories. On some occasions, damselfish may move out of their territories and feed in other areas but not very often. Most of the identifiable mats (96%) had fishes on them, so there appeared to be a high level of competition for feeding space. Robertson et al. (1981) did two large-scale experiments to test territory size limits. One experiment removed one half of the algal mats; the other removed coral and algal mats over one half of the entire reef area. The purpose was to see how fish responded to this manipulation with only half a reef (or half of the algal mats) and the same number of fish. Two extremes could be expected: either fish that formerly resided in a territory on the destroyed half of the reef would die or disperse or those fish would remain and compete with the fish that still had territories. When algal mats alone were removed, resident damselfish re-established the mats on the reef within 100 days, and the same number of fish remained. When the coral was completely removed, fish density on the remaining half of the reef increased, indicating that space was not necessarily limiting. The growth of individuals after the manipulation was about the same as before. Since growth is based on how much food is available, the increased density (twofold) did not appear to decrease individual food consumption. Territories that were being held may have been larger than necessary to survive, or alternate sites for territories were available so space was not limiting.

There are some problems with this kind of experiment. In removing habitat from half a reef but leaving all the adult fish, a very different situation is set up than would occur in nature when spaces are occupied. Usually, a juvenile fish comes to a reef and must either find a territory or leave, so the competition for territories occurs among these juvenile fish. Changing the population by leaving adults is not the way territories are set in nature. In addition, the study was done only over a few months, so cropping effects by the fish might not have been severe enough to drive down food resources over that time. It would be interesting to go back one or two years later and measure the density of fishes and their growth rates on those reefs. The ability of algae to reproduce and maintain consistent populations might be impaired over the long term but not over the short term. While this experiment demonstrated that space did not seem to be limited in this reef, it also left questions as to whether this is generally true.

As mentioned earlier, the density of fish may be important in territorial defense. Will residential fish on a territory defend that territory against other species, and if so, can we predict that defense based on some characteristics of those other species? Ebersole (1977) studied such interspecific relationships. He defined potential competitive index as a measure of competitive potential between different species and resident territorial fish (*Eupomacentrus leucostictus* was the resident species in this study). He studied pomacentrids (butterfly fish) and measured attacks by territorial inhabitants of one species on various intruders. Potential factors causing a high potential competitive index could be the degree of food overlap and physiological requirements for the two species. If the intruding species eats food similar to what the residents consume, there is potential for competition, and the intruders would be potential competitors that might be excluded. For physiological requirements, the more food the intruding fishes are likely to eat, the more they should be excluded from a feeding site. Food consumption is related to metabolism, and Ebersole used Weight$^{0.75}$ as an index of physiological food requirement. A high potential competitive index occurred for an animal that overlapped strongly in food and was large. This is interesting because it predicts the resident would attack a large fish more vigorously than a small fish. From a psychological point of view, just the opposite would be expected since small invaders would be easier to displace. Animals with a low potential competitive index ate different foods or were so small they would not eat much.

Ebersole observed different species intrude on the territories of butterfly fish and estimated how vigorously the resident attacked intruders. He then tried to correlate intensity of agonistic encounters – which he defined as the natural log of the number of attacks per unit of encounter time – to competitive measures. There was a significant correlation between agonistic acts and food overlap with an explained variance (r^2) of approximately 16%. When agonistic acts were correlated to physiological size ($W^{0.75}$) and food overlap, the correlation coefficient (r) increased to 0.613 or about 36% of the variance explained. Since this was significant but still relatively low, he then searched for other explanations that could further strengthen the fit of his model. One such explanation was that a fish might more strongly defend a territory against a species that would consume its eggs than against a species that does not eat eggs, even if the remainder of the food niche overlap with the egg predator was not very high. If Ebersole removed egg predators from the species list, then R increased to 0.74, or about 50% of the variation was explained. Egg predators were more vigorously attacked then would be predicted on food overlap alone, which agreed with his original idea. Territory-holding butterfly fish can apparently recognize other species and the kinds of foods they eat and defend their territories appropriately.

How important is the potential competitive index, and what does it mean in terms of attack rates? Interactions among conspecific butterfly fish (individuals of the same species) have attack rates of 0.2–0.8 attacks per minute encountered. For other species with high PCIs, there might be 2–20 attacks per minute. In other words, butterfly fish often defended their territories more vigorously against other species than they did against members of their own species. There are several reasons why this might occur. Although food overlap is important, intruder species may mainly rely on one kind of food. There may be 20 kinds of food in a butterfly fish diet. One kind of prey might be the most important, but butterfly fish are more or less generalists. Butterfly fish might attack more vigorously against specialists that use their main prey preference. For example, a competitor might consume prey that grow out of crevices in the coral. A butterfly fish may be unable to consume these until they reached the surface, while an intruder with a more pointed snout might be able to remove these prey earlier and reduce the availability for the resident.

A second explanation for these high attack rates is that attacks may not be the most appropriate measure of interactions among individuals of the same species because attacks become displaced by ritualized behavior. Butterfly fish may threaten with postures to exclude conspecifics from their territories. These ritualized behaviors probably work much more effectively between individuals of the same species than between different species. A conspecific may recognize the threat and depart without being attacked, whereas a member of a different species may not recognize that behavior and may remain until it is attacked. This would suggest that the attack rate is partly an artifact of the experimental methods rather than an indicator of only more vigorous territorial defense.

Shelter was shown by Smith and Tyler (1972) to be the major limiting resource on coral reefs. Nocturnally and diurnally active species might use shelter at different times: nocturnal species during the day and diurnal species during the night. Certain territory holders might hold very similar territories and just reverse the pattern of use: the nocturnal species come out of shelters and start foraging and defending their territories at night, and diurnal species go into shelters simultaneously. These two residents may well tolerate each other without any real aggression between them. There appears to be a changeover between the two territory holders, usually synchronized around dawn and dusk.

Schooling

Most classes of vertebrates show some type of grouping behavior. A school is different from these groupings. It is a very particular, definitive group of three or more fish, in which each member constantly adjusts its speed and direction to match the speed and direction of other school members. It is not just a loose organization or a group of fish near to each other but a definitely organized group – organized by each individual swimming in coordination with other individuals. Schooling must have a fairly significant role in the lives of fishes because out of more than 20,000 known fish species, 10,000 of them show schooling at some portion of their life histories (Partridge 1982) and many school throughout their entire lives. While there are schools of large top carnivores, schooling most commonly occurs in smaller individuals. That leads one to suspect that schools may protect these small individuals from predators. Much work on schooling has investigated protection from predation.

Schooling and Predation

Schooling can affect predator–prey encounters in several ways. The first and simplest is that schooling may reduce the chance of discovery (Figure 13-3). If eight individuals are randomly spread throughout an area with one predator, the predator can detect prey only within

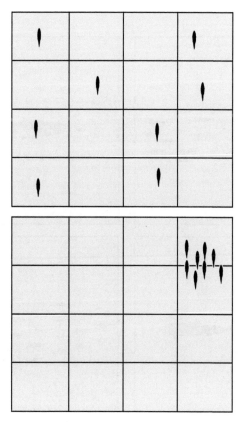

FIGURE 13-3 Schematic of reduced chance of discovery for fishes in a group. Each square indicates a predator's detection distance (in nature this would be a sphere). Upper pattern is random spacing and lower is for a schooling group.

a certain radius. The predator would be in visual contact with at least one prey item half of the time and would be able to attack regularly. On the other hand, if all eight prey remained in a school in one small area, for 15/16 of the time this predator would be out of visual contact with the prey because prey are grouped in a small area and much of the area the predator traversed would have no prey. Fish could reduce their chances of discovery by remaining in a group. This group would not have to be a school to reduce discovery opportunities; it would only have to be a group of fish close together. When encounters occurred, if the predator could only eat one prey item, then the size of a group might still successfully reduce the predation rate. For schools in the open pelagic environment, geography is not one dimensional as in Figure 13-3; depth is also involved. Predator and prey can be swimming at different depths as well as in different areas, and both would avoid detection.

Schools of fish continue and their geometry may even improve in the presence of a predator. If reduced discovery were the only function of schools, then once a predator observed the school, the school might break up and scatter. Why should a fish remain in a school after discovery if the school's only function is to reduce initial detection by a predator? Also, some predators continually follow fish schools. Partridge (1982) described two kinds of behaviors that schools of fish use to confuse predators after an encounter: one is called the fountain effect and the other is flash expansion. If a school of fish observes a predator, it may use the fountain effect where the school splits into two halves, swims around the predator, and then rejoins on the other side (Figure 13-4). With the fountain effect, the school avoids the predator by equally dispersing around it in an organized fashion. If the predator manages to get fairly close and attack the school, flash expansion occurs by a series of rapid starts by school members. When suddenly the predator attacks, fish move rapidly

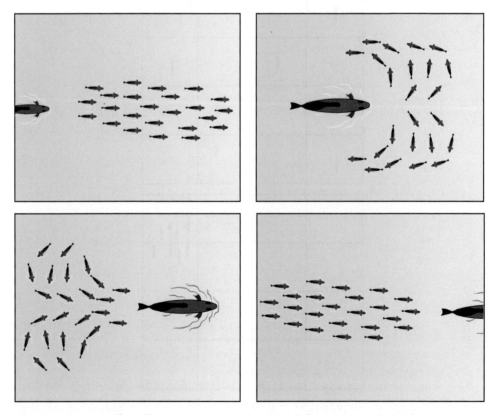

FIGURE 13-4 Fountain effect of herring to a barracuda approach from the rear of the school. *Source: Adapted from Partridge (1982).*

FIGURE 13-5   Flash expansion of herring to a barracuda attack from the side of the school. *Source: Adapted from Partridge (1982).*

away; the predator charges in, but cannot necessarily orient or pinpoint prey because all are accelerating in different directions (Figure 13-5). Since schooling fishes are often silvery colored, this erratic movement may couple with the reflection of sunlight off their scales as they move about, further confusing the predator. One interesting characteristic is that flash expansions are rapid reactions by many fish (possibly using C-starts), yet in a large school of fish, none bumps into another when they accelerate. The fish seem capable of orienting in some way to avoid collision, so there must be some degree of orientation even during flash expansion.

These two behavioral reactions are means schooling fishes use to react to predators: one is an organized movement around the predator and the other is a rapid chaos of movement. Another means for predator evasion, called the confusion effect, takes place because the predator becomes confused by the large number of targets in the school and is unable to fix on one fish to attack. The predator may attack a school believing that with so many prey around it will be able to catch something. Without targeting on one individual, the predator is generally unsuccessful. More individuals reduce the likelihood of predatory targeting on an individual. This explains why unusual behaviors are detrimental to individuals in a school. A fish with fungus on its side, or one that is darker or larger than other individuals, might be more vulnerable because the predator can focus on it.

Combining these ideas indicates that schools should comprise individuals that are similar in size. A mixture of too many sizes causes individuals to become more conspicuous and more likely to be picked out by a predator. There are also hydrodynamic reasons why individuals should be similar in size, which are covered later. Similarly, schools should largely be composed of only one species.

Several experiments confirm the general statement that schools reduce predation. In one, several predators and schooling prey were placed into tanks together with different numbers of prey as a treatment, while the behavior was recorded (Neill and Cullen 1974). If a single prey was put into the tank, it was readily captured. For 6 or 20 prey, the captures per contact declined with school size (Figure 13-6). Each predator's behavior also modified capture success. Eurasian perch (active predators) appeared to display confusion, often switching targets. The other three predators (sit-and-wait predators) spent less time in predation behavior and more time in other activities, thus reducing the encounter rate. Considering successful attacks per school member, not only did predation success decline with school size, but attacks per individual school member declined even more.

Sean Connell from the University of Adelaide questioned the generality of the idea that larger schools always result in reduced prey mortality (Connell 2000). He conducted field experiments exposing schools of small reef fishes to predation by small and large predatory fishes. By caging areas to eliminate large predators from these locations, he could determine

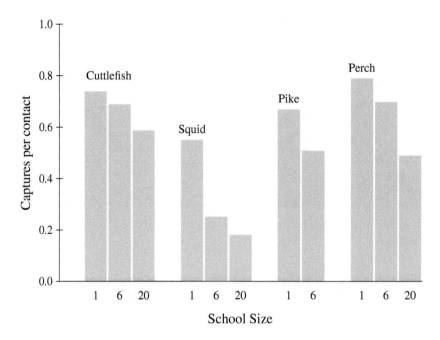

FIGURE 13-6 The effect of school size on the ratio of captures per contact for four predatory species. *Source: Adapted from Neill and Cullen (1974).*

what happened to prey mortality when only small predators were in the area or when large and small predators were common. He found that larger schools resulted in more prey consumed under all treatments, but the effect on prey mortality depended on predator size (Figure 13-7). While small predators consumed more schooling prey at higher prey density, their predation resulted in about the same prey mortality rate in schools of different sizes. However, the presence of large predators not only increased prey consumption in larger schools of prey but also resulted in increasing prey mortality with increasing school size. He proposed that the results of these interactions depended in part on the functional and numerical responses of predators, and that larger predators that gulp numerous prey in an encounter could cause density-dependent mortality in schools, while those that focus on one prey at a time more likely would not be able to cause such a density effect. Overall, this casts some questions on the generality of the earlier ideas related to increased safety of prey in larger schools.

Hydrodynamic Advantages of Schools

Another commonly accepted reason for fishes to school is hydrodynamics; fishes in schools may swim faster or more efficiently than fishes out of schools. Danny Weihs, an Israeli scientist, has observed mackerels in a water tunnel. Weihs (1973) found that distance between individuals on the same plane (that is, individuals parallel to each other, not above or below) was usually between one and two widths of the fish; the distance from one individual's head to the tail of the next forward individual was usually about 3/4 of a body length. He termed this a diamond pattern. One fish in front of the school was not always the leader, but regularly another fish went to the front and the front fish moved back. The individual in front of the school had a higher tail beat frequency than individuals farther back in the school, indicating that it had to swim harder, even though all fish were moving at the same speed. Individuals farther back in the school appear to receive some advantage from the swimming of individuals in front of them. Weihs examined water currents near the fish in a school and found that

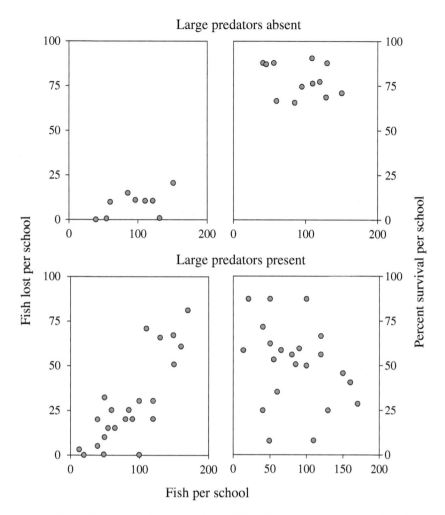

FIGURE 13-7   The effect of presence of large predatory fishes (lower charts) compared to absence (upper charts) on the number of prey lost (left charts) and survival of prey (right charts). Each point represents a school. *Source: Adapted from Connell (2000).*

a fish moving its tail back and forth in swimming caused a current vortex, with water moving forward (Figure 13-8). If two fish were swimming in the diamond pattern at a given speed, the rear fish could achieve acceleration from the forward water movement caused by swimming of the front fish. This makes it easier for the rear fish to swim at the same speed. This difference in efficiency may be 30% between good and poor positions. For this advantage to occur, the swimming speed of a school should affect distance between individuals. The faster the fish swim, the more rapidly they move their tails, and the distance between these waves changes. The pattern of distribution of fish Weihs measured was consistent with the concept that this current forward was being used to reduce drag.

Schooling of Predators

Not only do forage fishes school, but large predators may also school. Bluefish and tuna, for example, clearly move in groups. The schooling pattern of bluefin tuna has been studied from the air, and it is different from the pattern one would expect because the school moves somewhat in reverse.

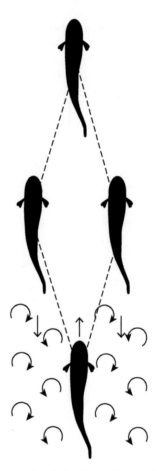

FIGURE 13-8 Vortices created by swimming fish (dots and curved arrows) and water movement (straight lines) for schooling fishes. The diamond pattern is also shown. *Source: Adapted from Weihs (1973).*

Tuna tend to move like a big U through the water (Partridge 1982). Tuna follow schools of prey that are surrounded by the U-shaped school and driven to the middle where individuals can be attacked more easily. Flash expansion of the prey may not work well in this interaction because the prey are crowded between tuna. The interaction has not been well studied, but the regular pattern of tuna schools leads to such an interpretation. In addition to the U-shaped schools, predators also drive prey to the water surface and may use a school or pack to push their prey near the surface, where prey cannot use all three dimensions to escape. Predators may school, or at least forage, in a concentrated fashion to take advantage of the defense mechanisms inherent in these schools. Attacks on prey at the surface have led to the evolution of flying fishes, which can burst out of the water and escape predation, so this pattern must have had long and important significance.

Summary

The social behavior of fishes is often the mechanism for population regulation because spacing behaviors and space use often define the population reactions to perturbation. Many aspects of this social behavior may be tightly fixed by ritualization, which means these responses will be relatively consistent over time or among populations. For example, density dependence is commonly observed in territorial species, while density independence may be more common in schooling species that are less antagonistic. These behaviors also affect the patterns of fish distribution in space and time, and help explain the randomness or patchiness of space utilization by fishes. The social structure of a given fish species may be a useful predictor for the outcome of various intra- and interspecific interactions.

Literature Cited

Clapp, D.F., R.D. Clark Jr., and J.S. Diana. 1990. Range, activity, and habitat of large, free-ranging brown trout in a Michigan stream. Transactions of the American Fisheries Society 119:1022–1034.

Connell, S.D. 2000. Is there safety-in-numbers for prey. Oikos 88:527–532.

Cunjak, R.A., and G. Power. 1986. Winter habitat utilization by stream resident brook trout (*Salvelinus fontinalis*) and brown trout (*Salmo trutta*). Canadian Journal of Fisheries and Aquatic Sciences 43:1970–1981.

Diana, J.S. 1975. The Movement and Distribution of Paiute Cutthroat Trout, *Salmo Clarki Seleniris* Snyder, in the North Fork of Cottonwood Creek, Mono County, White Mountains, California. Master's Thesis, California State University, Long Beach.

Diana, J.S., J.P. Hudson, and R.D. Clark, Jr. 2004. Movement patterns of large brown trout in the Mainstream Au Sable River, Michigan. Transactions of the American Fisheries Society 133:34–44.

Ebersole, J.P. 1977. The adaptive significance of interspecific territoriality in the reef fish *Eupomacentrus leucostictus*. Ecology 58:914–920.

Ejike, C., and C.B. Schreck. 1980. Stress and social hierarchy rank in coho salmon. Transactions of the American Fisheries Society 109:423–426.

Jenkins, T.M., Jr. 1969. Social structure, position choice, and microdistribution of two trout species (*Salmo trutta* and *Salmo gairdneri*) resident in mountain streams. Animal Behavior Monographs 2:57–123.

Keenleyside, M.H.A. 1979. Diversity and Adaptation in Fish Behaviour. Springer-Verlag, Berlin.

Neill, S.R.J., and J.M. Cullen. 1974. Experiments on whether schooling by their prey affects the hunting behavior of cephalopods and fish predators. Journal of Zoology, London 172:549–569.

Partridge, B.L. 1982. The structure and function of fish schools. Scientific American 246:114–123.

Robertson, D.R., S.G. Hoffman, and J.M. Sheldon. 1981. Availability of space for the territorial Caribbean damselfish *Eupomacentrus planifrons*. Ecology 62:1162–1169.

Smith, C.L., and J.C. Tyler. 1972. Space resource sharing in a coral reef fish community. Pages 125–170 in B.B. Collette and S.A. Earle, editors. Results of the Tectype Program: Ecology of Coral Reef Fishes. Natural History Museum, Los Angeles County, Science Bulletin 14.

Tinbergen, N., and J.J.A. van Iersel. 1947. "Displacement reactions" in the three-spined stickleback. Behaviour 1:56–63.

Weihs, D. 1973. Hydromechanics of fish schooling. Nature 241:290–291.

CHAPTER 14

Competition

Competition is the struggle between organisms for a resource in limited supply. The resources of importance to fishes usually include food, space, and mates. The main components in the definition of competition are the means of inhibition and the limited supply of resources. Inhibition can occur in two ways: interference and exploitation, which may occur together or separately. Interference competition takes place when activity of one organism directly or indirectly limits access to the resource. It is most commonly expressed in terms of behavior: one fish physically excludes the other from resources. Exploitative competition takes place when an animal – by using a resource in limited supply – reduces its abundance and removes it from use by another animal. This sort of competition may involve disparate animal species, such as stonefly and trout. One would not observe a stonefly chasing a trout out of its territory, but by exploiting the same food a trout eats, the stonefly may limit availability of food for trout and, therefore, compete with it by exploitation.

The other important part of the definition of competition is the necessity for resources to be limited. Two fish species that eat similar food do not necessarily compete. If that food is abundant enough so neither species suffers from lack of food, competition does not occur. Resource managers and the public often consider competition for food in terms of food overlap. For example, white sucker and trout may compete in a lake since both eat benthic invertebrates. This can lead managers to reduce the sucker population. For competition to occur, the sucker population has to either eat enough benthic invertebrates to limit the amount the trout can eat or it has to exclude the trout from use of that food. Only then will the removal of sucker cause competitive release for trout.

Competition is probably one of the most studied, controversial, and important areas in ecology. Competition arguments often lead to concepts of fitness and animal evolution to avoid competition. Animals of various taxonomic levels may compete. Competition can occur between two very similar species, very different species, or even different genotypes of one species. For example, competition for mates between a fish that spawns when it is one year old and a fish that spawns when three years old may make a difference in terms of which genotype is most successful in reproduction.

Competition studies begin with the concept of a niche, the *n*-dimensional hypervolume of environmental factors influencing survival of an animal (Hutchinson 1957). The Hutchinsonian Niche is one of the fundamental paradigms of ecology. A paradigm is a series of "rules" or expectations used to guide studies and generate hypotheses. G. Evelyn Hutchinson was a very influential limnologist at Yale University, who not only formalized the idea of a niche but wrote a pioneering book on limnology. The Hutchinsonian Niche is based on the concept that an animal or plant has characteristics under which it can survive, and its persistence in a location

Biology and Ecology of Fishes, Third Edition. James S. Diana and Tomas O. Höök.
© 2023 John Wiley & Sons Ltd. Published 2023 by John Wiley & Sons Ltd.

or within a habitat is based on these survival characteristics, as well as the influence of other plants and animals on its survival (Hutchinson 1957).

When an animal competes with another for resources, the result can be evaluated based on the overlap in their two niches. If the realized niche (volume of niche in the presence of competitor) is different than the fundamental niche (volume of niche possible in isolation), this niche compression may be evidence of competition between the two species. An example of this will be detailed later for sunfishes, where bluegill in isolation will forage in the vegetation and plankton, but when green sunfish are present and vegetation prey become limiting, bluegill switch to forage on plankton. One concept related to the Hutchinsonian Niche is that two animals cannot utilize the exact same niche: either one – or both – of the species will evolve to utilize a somewhat different niche or become extinct. The more niches overlap, the stronger the potential for competition, and the stronger the force for either change or extinction. The Darwinian concept of evolution indicates that survival of the fittest – and changes in gene pools – should cause evolution of previous phenotypes to animals that exist today. Scientists believe competition was one of the main factors influencing evolution of species. A similar argument is sometimes used to explain why we do not expect competition to be as important today as it was when closely related species first began to separate. This idea is the "ghost of competition past." A more scientific statement of the same idea is that animals may vary based on interactive segregation (current behavioral interactions that cause them to occupy different niches when they are sympatric with a similar species) or selective segregation (where animals have evolved differences in niches based on previous interactions in evolutionary times, and their niches do not change when sympatric). In this chapter, we proceed from some niche studies in fishes that include interactive and selective segregation to competition studies analyzing interactive segregation between species.

Niches of Intertidal Fishes

A good example of the changes in niche breadth of interacting species can be found in the distribution of tidepool fishes. Animals living in tidepools have a variety of factors influencing their abundance and survival in any particular location. The intertidal zone is periodically flooded or exposed to air depending on tidal heights. In most coastal areas, the tides are regular; both high tides and low tides occur twice a day. The magnitude of the tide changes with the periods of the moon. Tidepools are depressions in the rock shelf that may contain water even after the tide recedes but do not maintain direct contact with oceanic waters during low tide events. Stranding during low tide can result in large changes in the habitat conditions in a tide pool. Temperature may increase as much as 15 °C during exposure, and at night, pools may become anoxic due to lack of photosynthesis and high levels of respiration. Shallow tidepools may expose resident animals to predation by birds or terrestrial mammals. Deeper areas in the intertidal zone expose animals to predation by fishes. Among all sites, competition for space may influence which pools a particular individual may occupy, and whether that pool has good cover habitat or exposure to predators. So, the distribution and abundance of animals in these tidepools are a combined effect of physical conditions, predator abundance, and competitor abundance in the intertidal area. All of these are characteristics of an animal's niche and subject to interactions with other animals.

Karen Martin at Pepperdine University in California has conducted research to understand mechanisms behind the distribution of intertidal fishes, focusing on factors causing these distributions, such as temperature tolerance and survival during times of air exposure. These factors are important in understanding the fundamental niches of tidepool fishes. In one study, Martin (1996) evaluated the gradient in ability of marine cottids to survive air exposure. The

TABLE 14-1 Differences Among Five Sculpin Species in Their Ability to Tolerate Aerial Conditions. Data From Martin (1996).

Species	Habitat	Behavior Under Hypoxia	Aerial Metabolism	Survival Out of Water
Chinotus pugetensis	Mainly subtidal	Succumbed	Declined over time	0% after 2 h
Jordania zonope	Mainly subtidal	Respired at surface	Declined over time, became anaerobic	10% after 2 h
Icelinus borealis	Only subtidal	No response	Declined over time, became anaerobic	50% at 2 h
Oligocottus maculosus	Throughout intertidal	Willingly emerged	Constant over time and equal to aquatic	100% for 2 h
Ascelichthys rhodorus	High intertidal	Willingly emerged	Constant over time and equal to aquatic	100% for 2 h

family Cottidae (or sculpins) is common in the Puget Sound area, with over 30 species. Martin chose seven species with varying habitat preferences, ranging from strictly subtidal to widely intertidal. For all of these species, she tested respiration in air and water, ability to survive exposed to air, and desiccation losses when exposed.

The seven species showed predictable changes based on their habitat selection. Species commonly intertidal had better ability to respire in air, showed minimal anaerobic metabolism in aerial conditions, voluntarily emerged from hypoxic water to utilize air breathing, and survived easily with two hours of air exposure (Table 14-1). Subtidal species had much poorer ability to tolerate aerial conditions. All of the species demonstrated desiccation rates similar in the air, indicating that there was no adaptation to reduce desiccation rates. These results indicate that intertidal species were much better adapted to the challenges of life in upper tidepools than subtidal species. This adaptation, it could be argued, is an example of the ghost of competition past or selective segregation, that is, evolution that occurred as a result of the competitive interactions resulted in intertidal fishes residing higher in the tidal zone and subtidal ones lower. In either case, the result would be that the niches of these species are fundamentally different, apparently because of their evolved ability to withstand exposure to tidepool conditions.

Another study on tidepool fishes was conducted by Jana Davis at Scripps Institute of Oceanography, University of California, San Diego. She was interested in choice of habitat by tidepool fishes in the San Diego area. She determined habitat characteristics for a number of natural tidepools, particularly the surface area, depth, intertidal height, rugosity (amount of rocky substrate in the pool), and algal cover. The intertidal height was the location in tidal elevation for each pool and was estimated by determining the time when each pool became isolated from coastal water as a tide fell. She evaluated the species of fish that occupied these tidepools by completely draining each and removing all fishes several times over a two-year period. After each determination, water was replaced and the fishes were returned. Finally, she evaluated tidepool selection by manipulating the conditions in experimental pools and evaluating the change in species composition.

The five species of fish showed differential distributions and selection of certain habitat characteristics (Table 14-2). Overall, all species favored deeper pools, with rock or algal cover, in the lower intertidal zone (Davis 2000). Pools higher in the intertidal zone were generally shallower, less structurally complex, and had fewer and smaller individuals than lower pools. Only two species occupied upper pools. One of these, *Clinocottus analis*, is a well-known species for

			Correlations of Tidepool Conditions with Fish Abundance				
Species	**Intertidal Distribution**	**Relative Abundance (%)**	**Surface Area**	**Depth**	**Intertidal Height**	**Rugosity**	**Algal Cover**
Clinocottus analis	Wide, especially lower	48–72	Positive	None	Negative	Positive	Positive
Girella nigricans	Wide	8–22	Positive	Positive	None	Positive	None
Gobioesox rhessodon	Only lower	11–30	Positive	None	Negative	None	Negative
Hypsoblennius gilberti	Only lower	4–8	Not consistent	None	Negative	Positive	Negative
Gibbonsia elegans	Only lower	2–4	Positive	None	Negative	Negative	None

TABLE 14-2 Distribution in the Intertidal Zone, Relative Abundance, and Correlations Between Habitat Characteristics and Abundance of Five Fish Species in Tidepools Near San Diego, California. Data From Davis (2000).

its air-breathing characteristics. The other was juvenile opaleye *Girella nigricans*, which settle in the intertidal zone for one to two years and then move out to complete adult life in the subtidal environment.

When pools were modified experimentally, abundance and size of occupants changed as was predicted from the results of natural pool selection. For rugose pools that had rocks removed, the number of species declined, and the abundance of the two most abundant species – *Clinocottus* and *Gobiesox* – also declined. However, *Gobiesox* showed a much more dramatic decline than *Clinocottus*. When rocks were added to smooth pools, density increased and *C. analis* abundance also increased. These results provide more support to the cause of the observed distributions being pool condition and demonstrate cause and effect better than correlations between pool types, species, and number present.

Davis attributed the different occupation of pools by each species to reflections of physiological tolerance, habitat preference, and avoidance of predators. She believed predation pressure from terrestrial animals excluded many fishes from the upper, shallower, and less complex pools. Inability to tolerate aerial conditions also played a part in this absence. Davis also hypothesized that territoriality among resident fishes excluded smaller individuals of each species from preferred pools, so they were forced to reside in pools more exposed to predation. In this example, the fundamental niche for many tidepool species would be the entire tidal range although some of them might be less capable of tolerating conditions in the upper intertidal zone. The realized niche was changed, however, by two processes. Predation from terrestrial animals in the upper zone would restrict larger and more exposed species from pools in the upper zone, where they would be more vulnerable to predation. Predation by aquatic species in the lower zone could exclude some sizes or species of fishes from the lower intertidal zone. Competition with residents in each pool could also exclude some species or sizes of fish from preferred habitats of deeper pools. As a result, the fundamental niche of these intertidal fishes would include locations from the subtidal to the intertidal zone, with the upper limit affected by their ability to tolerate aerial conditions. The realized niche would be the reduced space accounting for avoidance of predation by subtidal fishes and intertidal animals as well as avoidance of competition by other intertidal fishes. The final distribution of species in the intertidal zone is the result of both physiological capabilities and biotic interactions with predators and competitors. Similar results were found by Elliott and Mariscal (2001) for anemonefishes and their utilization of hosts.

Tests of Competition

There are several ways to evaluate if competition may be occurring in an ecosystem. The most common method is to examine competitive displacement. This may also involve a "natural experiment." If an animal is introduced – by natural or human influences – into a location inhabited by a potential competitor, any changes in resource use by the original resident could indicate competition. A scientist may observe changes in the breadth of food use, so in the presence of this new competitor, the original species eats fewer types of food. If the original resident is excluded by the new animal from eating as many different types of food as before, the two species may compete. Change in niche breadth is strong support for the idea of competition, but other information on the resources and their exploitation is needed to prove competition. Also, this observation does not indicate whether exploitation or interference is occurring. The change in the food niche could be the result of the new animal foraging more efficiently or the new animal aggressively displacing the original one. Such a reduction in resource use in the presence of a competitor is called competitive displacement. The opposite can also occur – an expansion of resource use in the absence of a competitor. This is often termed ecological release.

A second test of competition is character displacement, which is a change in the morphological characteristics of one animal species in the presence of a competitor. One example would be change in the number of gill rakers on a certain species of fish, when that species is sympatric with a competing species. Gill rakers are projections from the gill arch that aid in filtering plankton from water passing over the gill. The two competing species may eat plankton of a slightly different size, and gill raker spacing affects the size of prey they can remove from the water. Character displacement may involve physical changes due to the nature of food eaten, or they may be evolutionary changes due to natural selection for individuals in each species that perform best under competition. This level of natural selection is not yet sufficient to evolve a different species, but change has occurred. Obviously, character displacement takes longer than competitive displacement to develop.

Character Displacement in Stickleback

Animals often evolutionarily adapt their anatomical and physiological functions to better exploit resources. As resources in different ecosystems can vary dramatically, local adaptations could be the result of different resources being available as well as interactions with other animals that may exploit those resources. Character displacement is the change in morphology of an organism in the presence of a competitor that allows the organism to better exploit some portion of the limiting resources. These changes are believed to be controlled genetically and at least represent natural selection of different phenotypes based on their fitness to the environment. Dolph Schluter from the University of British Columbia evaluated the trade-offs in feeding performance and growth for different populations of stickleback in an attempt to evaluate character displacement. In the lakes near the coastal area of British Columbia, there exist two forms of stickleback (*Gasterosteous*) as well as hybrids. The two forms have not yet been defined as species but include a benthic form and a limnetic form. The differences between these forms are genetically based and correlate well with differences in the habitat and diet of each form. Limnetic forms of stickleback are slender, narrow-mouthed, and have longer and more numerous gill rakers that assist in filtering plankton from the water. Benthic forms have more robust bodies with larger mouths and utilize larger benthic prey by selective foraging. Schluter (1995) conducted experiments to determine the linkage between these morphological traits and fitness of the two stickleback forms.

In order to clearly demonstrate character displacement, evidence must be provided for a link among adaptive changes in morphology, changes in feeding efficiency due to morphological adaptation, and improved fitness due to better feeding performance. In most studies of this sort, growth rate is believed to be directly linked to fitness. This was the assumption of Schluter as well. He attempted to demonstrate improvements in growth rate based on morphology and feeding efficiency, but not that improved growth rate resulted in enhanced fitness by production of more young in the next generation. This is a common difficulty in a variety of studies evaluating fitness – they cannot evaluate actual contribution to future generations of the genotypes with differing growth rates.

Schluter used cages – either on the bottom in the littoral zone or suspended in the limnetic zone – as habitats to test for different foraging and growth efficiencies of the two genotypes. Littoral cages were on the bottom and allowed fish to forage on benthos found in the littoral zone, while limnetic cages were suspended in the water and allowed fish to forage on plankton. Schluter used 36 cages – 18 in the littoral zone and 18 in the limnetic zone – in which he placed one individual of either genotype into a cage. He then measured growth and performance over 21 days in spring 1991 and 1992.

As predicted, the two genotypes had considerably different feeding efficiencies and ultimate growth rates in the two environments (Figure 14-1). The littoral zone provided more forage and better growth rates for the benthic form but also provided comparable growth to open water for the limnetic form in 1992. While the open water zone provided better growth for the limnetic form in 1991, it caused much worse growth for the benthic form. This shift in growth rate is exactly what one would predict based on morphological characteristics of the two and their foraging efficiency. The trade-off in growth rate between habitats was that benthic forms had about double the growth rate of limnetic forms in the littoral zone, and limnetic forms had about double the growth rate in open water (Figure 14-1c). One interesting component of this is that Schluter also evaluated hybrids, which are intermediate between the two forms and would be expected to be worse than either form in terms of foraging efficiency in the preferred habitat. The growth rate of hybrids was poorer than either form, indicating that their intermediate characteristics did not allow them to utilize either the benthic or limnetic areas as well as either of the parental types.

Schluter also exposed each species to laboratory feeding conditions and evaluated feeding efficiency on open water and benthic prey. Growth rate in the different habitats is the result of differential feeding efficiencies of the two forms, but in nature, it is simpler to demonstrate growth differences than feeding efficiency differences. Feeding efficiency for the benthic form in the laboratory was highest on benthic prey, and growth rate was highest at highest feeding efficiency. Similarly, feeding efficiency for the limnetic form was highest utilizing limnetic prey and resulted in the highest growth rate for that form. Using this link, Schluter completed the cycle of morphological differences leading to foraging differences and then to differences in growth rate necessary to prove character displacement.

These studies were done in isolation, so differences were completely based on the ability of individuals to forage. When the two species are sympatric, not only would differences in foraging be a consideration, but also interactions between the two. One would expect the benthic form to forage more effectively on benthos and reduce its abundance, further reducing the growth rate of the limnetic form on benthic prey. Also, it may be more aggressive and displace the limnetic form. In other words, this test should demonstrate the minimal amount of the difference in growth that would appear for the two species in natural conditions, where agonistic behavior, food availability, and foraging efficiency all influence growth rate and fitness.

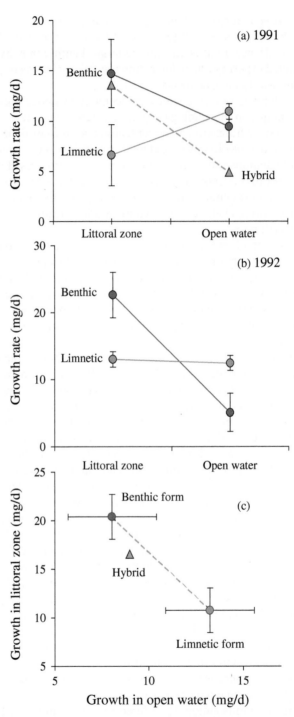

FIGURE 14-1 Comparative growth rates for each species of stickleback in two habitats of Paxton Lake: (a) each habitat and species in 1991, (b) each habitat and species in 1992, and (c) mean + SE of growth for each species in open or littoral zone. *Source: Adapted from Schluter (1995).*

Competitive Displacement by Sunfishes

Behavioral experiments using fish in seminatural environments can be used to determine the outcomes of interactions between species and support or refute the idea of competition between the species. Experiments conducted by Werner and Hall (1979) provide excellent examples of behavioral shifts or competitive displacements. Werner and Hall studied species interactions of sunfishes in ponds at the Kellogg Biological Station at Michigan State University. The ponds were small, approximately 30 m in diameter. Each pond could be considered a replicate ecosystem into which different combinations of animals could be placed. Another replication technique was to divide a pond in half with a net and use one half for a treatment and the other half for control. Nearly natural experiments can be done in such systems, manipulating fish density or species combinations, determining feeding efficiency and growth of species under experimental conditions, and then evaluating these outcomes to better understand species interactions and competition.

One of the first questions they asked was related to competition among sunfishes (Werner and Hall 1979). This was addressed by stocking different species of sunfishes – bluegill, green sunfish, and pumpkinseed – into ponds and evaluating behavior in the presence or absence of congeners. Evidence for competition would require that food was limited, which could be assessed by growth rates. Competition for food would occur if fishes affected the abundance of food or the ability of other species to gain access to food.

The habitat preferences of the three species were similar in the ponds. The three available habitats – vegetation, sediment, and open water – contained different kinds of invertebrates; certain insects predominated the vegetation, different insects in sediment, and zooplankton in open water. Werner and Hall could categorize a large fraction of fish diet as being exclusively from one of those three zones, so by examining stomach contents of a sunfish, they could determine where it had foraged. This is extremely useful because visual observations of foraging would be difficult. Not every prey species segregates by this technique, but enough do that a reasonable evaluation of past behavior can be based on stomach contents.

Werner and Hall evaluated species interactions in four ponds by stocking 900 bluegill in one pond, 900 pumpkinseed in another, 900 green sunfish in a third, and 2,700 from all three species in the last pond. There is an obvious difference in overall density among these treatments. When the fish were stocked by themselves, all three species remained in vegetation. These ponds were shallow – less than 2 m – and did not have much open water or plankton. The weedy shoreline was the richest habitat. Therefore, all three species by themselves stayed in emergent vegetation near shore, which had the most food.

When all three species were included together in the same pond, there were shifts in habitat use. The large increase in fish density in this treatment was done in an attempt to hold intraspecific competition constant by maintaining 900 of each species and manipulate interspecific competition to determine these competitive outcomes. For this experiment, green sunfish stayed in vegetation, bluegill moved more into open water, and pumpkinseed were found with benthic prey. One might predict that green sunfish won in competition over the other two and drove them out of their preferred habitats. Another alternative would be that, as prey declined due to intense predation, bluegill and pumpkinseed moved to habitats where they could forage more efficiently at the new prey densities, while green sunfish remained where they were most efficient at feeding.

In order to better understand foraging efficiencies of each species, Werner and Hall did a caging study in which the fish could not move out of the vegetation, similar to the caging studies by Schluter. There is a need for this kind of growth control treatment in such an experiment. Without such a control to evaluate the capability of each species alone, it is difficult to make a definitive argument about cause and effect. Green sunfish alone, restricted to vegetation, grew 24% faster than bluegill treated similarly. Innately, green sunfish were apparently more effective at eating vegetation prey than bluegill. When included together in vegetation, green sunfish grew faster than bluegill by about 44%. This additional 20% increase in growth was due to the competitive superiority of the green sunfish.

Werner and Hall then conducted a series of experiments to examine these interactions more carefully and evaluate whether or not predictions could be made on changes in habitat use due to foraging efficiencies of each species. These foraging differences are often termed optimal foraging and will be covered more thoroughly in Chapter 18. Vegetation was the richest habitat in terms of food, and all species were predicted to remain there if the opportunity existed. At some point, remaining there would become a poor strategy due to increased competition for declining food resources. At this time, fishes should shift to alternate habitats where they could forage more efficiently than in depleted vegetation.

They conducted another experiment with 2,700 fishes (three species) in a pond where all vegetation had been manually removed and 2,700 fishes in a normal pond. Two different predictions on behavior may be made for foraging in the altered pond: (1) green sunfish may be more aggressive and utilize any habitat they choose or (2) green sunfish may not be as effective in using other habitats and could lose in competition for benthic or pelagic prey. The latter results would be a condition-specific outcome, where the winner in competition would depend on conditions during the interaction rather than an innate superiority under all conditions. Bluegill most often consumed prey from open water in both ponds, but they shifted farther away from vegetation prey in ponds without vegetation (Table 14-3). Pumpkinseed rarely ate prey from vegetation in either pond class. Green sunfish mainly used vegetation prey, and when vegetation was not present, they continued to forage on similar organisms (even though these organisms were less abundant), although they also increased consumption of benthic and pelagic prey.

Growth rates of each species also indicated competitive shifts (Table 14-4). Bluegill did not grow as well in ponds without vegetation. Pumpkinseed did not show any change in growth rate in either type of pond. Green sunfish declined in growth when vegetation was absent. In terms of the original predictions, green sunfish won over bluegill and pumpkinseed in vegetation, but pumpkinseed dominated green sunfish and bluegill for growth in nonvegetated ponds.

These results led to three outcomes: green sunfish do best in vegetation, bluegill and pumpkinseed are equal in vegetation, and both grow better in vegetation than in open water. Pumpkinseed do best in ponds without vegetation, while bluegill and green sunfish are about equal in sediment habitat. Werner and Hall used these generalizations to predict switches in habitat use by each species in a vegetated pond with poor plankton populations.

TABLE 14-3 Percent Contribution of Prey Categories Consumed by Each Sunfish Species in a Pond With or Without Vegetation.

		Prey Type			
Species	**Pond**	**Vegetation (%)**	**Benthic (%)**	**Open Water (%)**	**Other (%)**
Bluegill	Vegetated	15	15	33	37
	Nonvegetated	3	23	33	42
Pumpkinseed	Vegetated	5	34	6	55
	Nonvegetated	4	37	3	56
Green sunfish	Vegetated	40	12	4	44
	Nonvegetated	14	24	13	49

Source: Adapted From Werner and Hall (1979).

| **TABLE 14-4** | Average Dry Weight (SE) at the End of Each Experiment for Each Species Stocked in Ponds with or without Vegetation. | |

	Pond Type	
	Vegetated	**Nonvegetated**
Bluegill	1.29 (0.02)	0.89 (0.03)
Pumpkinseed	1.21 (0.03)	1.17 (0.03)
Green sunfish	1.34 (0.03)	1.01 (0.04)

Source: adapted from Werner and Hall (1979).

They hypothesized that all fishes should remain in vegetation in early spring, when food was abundant. As vegetation prey became limited because so many fishes consumed it, species more effective at foraging elsewhere should move out of the vegetation. One prediction they made was that pumpkinseed should move out into sediment first because they grow better there than either bluegill or green sunfish. They believed bluegill should shift later because they grow poorly on benthic prey. Finally, green sunfish were predicted to never shift out of vegetation.

They then measured habitat use in these ponds and found their predictions borne out (Figure 14-2). All three species utilized vegetation prey early in the year. As the abundance of food declined, pumpkinseed moved out first (in early July) and consumed sediment types of prey. As vegetation continued to decline, bluegill moved out to forage on benthos in late July. Green sunfish never made a shift. There may have been some improvement in food resources for green sunfish when bluegill and pumpkinseed moved out of the vegetation. In general, predictions made based on foraging capabilities held up in behavioral switches of these fishes.

One assumption is that all of these changes occurred because abundance of food declined. The dry weight of food in stomachs, plotted against months of the year for the three species, did show such declines in food consumed over time (Figure 14-3). Food consumption early in the year was very high, and as summer progressed, consumption declined dramatically. This decline correlated with the time (July) when species started to move away from the vegetation to forage. While cause and effect could not be shown, a reasonable correlation between abundance of food in the stomach and shifting of habitat was demonstrated.

One interesting comparison is between bluegill and pumpkinseed. Green sunfish liked vegetation and remained there. The bluegill and pumpkinseed did not do particularly well in vegetation when green sunfish were present, and they moved offshore to feed in the sediments or open water. That movement should have had some effect on their growth. The two species grew at about the same rate until the pumpkinseed moved offshore. At that point, pumpkinseed grew better because they were better at foraging on benthic prey than bluegill (Figure 14-4). This again correlated well with the relative foraging capability and competitive ability of each species.

The results of these experiments provide documentation of mechanisms responsible for sunfish competitive interactions and can be used to hypothesize similar mechanisms for other species. They may provide good explanations for competitive outcomes in natural habitats and, as such, provide information useful to managers who manipulate ecosystems to provide better conditions for one or all of these species. However, extrapolations to natural environments also depend on the nature of the experiments, the experimental system, and the species. Experiments based on one age class during one season may not reflect what happens in natural systems with multiple age groups and annual seasonality. Such experimental limitations are typical of all controlled experiments and indicate that experimental design is extremely important to make more generalizations from the lab to the field.

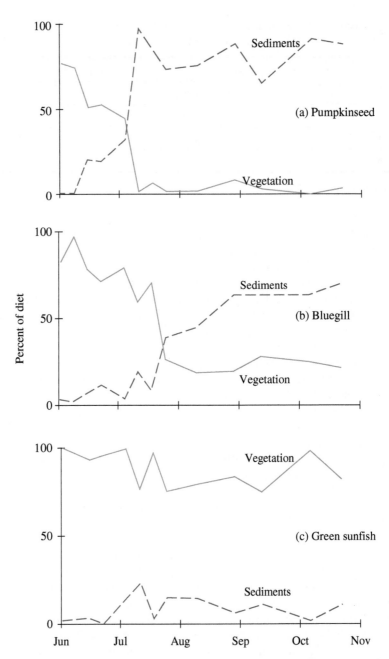

FIGURE 14-2 Percentage of vegetation or sediment prey eaten by (a) pumpkinseed, (b) bluegill, or (c) green sunfishes during the summer in ponds at Kellogg Biological Station. *Source: Adapted from Werner and Hall (1979).*

Competition in Coral Reef Fishes

An alternative to controlled experiments in settings like ponds is to manipulate species in natural habitats. This proves difficult in many cases because environmental damage may result from such manipulations. Such studies tend to be less controlled than ones in managed ecosystems

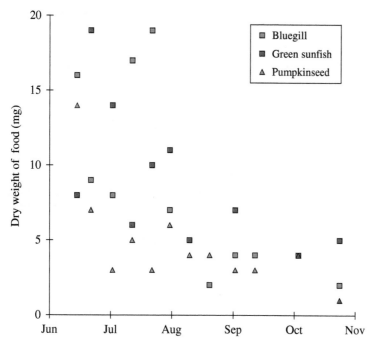

FIGURE 14-3   Seasonal changes in dry weight of food eaten by three centrarchid species in a pond experiment. *Source: Adapted from Werner and Hall (1979).*

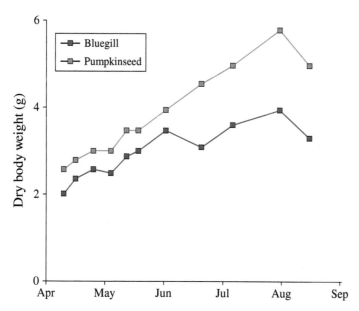

FIGURE 14-4   Gains in body weight (dry g) by bluegill and pumpkinseed in a pond, indicating the growth advantage for pumpkinseed upon leaving the vegetation (in July). *Source: Adapted from Werner and Hall (1979).*

but can provide good information on species interactions. Competitive interactions among coral reef fishes have been studied extensively in natural settings. Many fishes that reside on coral reefs are territorial and occupy very limited space throughout their lives. They produce pelagic larvae that settle on territories when larvae move from the pelagic lifestyle to a normal juvenile existence. Philip Munday from James Cook University in Australia studied the interactions between two gobies – *Gobiodon histrio* and *G. brochus* – that inhabit two species of coral – *Acropora nasuta* and *A. loripes*. Both species of *Gobiodon* maintain exclusive territories in a coral area (Munday 2001). *G. histrio* is most commonly found associated with *A. nasuta*, while *G. brochus* is most commonly found associated with *A. loripes*. However, larger individuals of *G. brochus* may be found in association with *A. nasuta* as well. Preliminary studies have shown *G. histrio* competitively excluded *G. brochus* from preferred corals except in the case of the larger individuals of *G. brochus* or *G. brochus* which were prior residents in an area. Therefore, Munday hypothesized that the distribution of *G. histrio* was in its selected habitat, while the distribution of *G. brochus* was limited by competition with *G. histrio*. In order to demonstrate this, he conducted transplant and exclusion experiments to evaluate growth and survival among different conditions.

Munday removed all fishes from coral areas and placed one or the other species in association with different species of coral. He removed fishes, as well as crabs and shrimps, from the coral manually after anaesthetization in order to reduce interactions between transplanted fishes and other potential residents. He then tagged both the coral and the individual fish to allow identification of residents and measured them at zero days, four months, and ten months.

These experiments demonstrated that both species of *Gobiodon* grew faster on *A. nasuta* than *A. loripes* (Figure 14-5). In fact, the elevated growth rate was approximately threefold for *G. histrio* and 2.5-fold for *G. brochus*. In addition, survival of *G. histrio* was much better

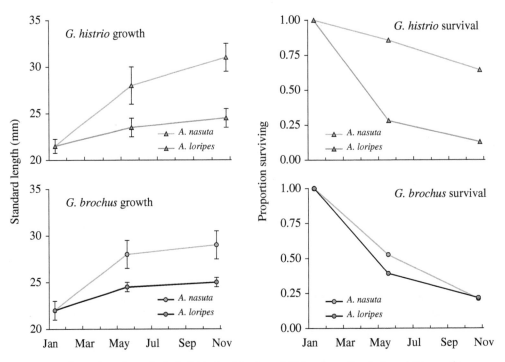

FIGURE 14-5   Growth and survival of *Gobiodon histrio* and *G. brochus* after 0, 4, and 10 months inhabiting *Acropora nasuta* and *A. loripes* in the Great Barrier Reef, Australia. Upper graphs, *G. histrio*; lower, *G. brochus*; left, growth; right, survival. *Source: Adapted from Munday (2001).*

on *A. nasuta*, while survival of *G. brochus* was poor on both coral species (Figure 14-5). His hypotheses were supported in that *G. histrio* grew better and survived better on *A. nasuta* than *A. loripes*, and *G. brochus* grew better than *G. histrio* on *A. nasuta*. Apparently, *G. brochus* was displaced from its favorite coral by *G. histrio* and, after that, suffered poor growth. Munday correlated the habitat produced by the two species of coral with the growth of fishes. *A. nasuta* had a larger branching structure with more space between branches, and he believed this resulted in the ability of *Gobiodon* to grow larger and occupy that space. In comparison, *A. loripes* had much smaller space between branches, which constrained resident fishes there. By comparing the results of transplanted fishes in the two habitats, this experiment demonstrated an effect on the behavioral competition in fitness characteristics of growth and survival, at least for one of the species. In comparison to the studies of Werner and Hall, no direct measures of foraging efficiency or other mechanisms for the competitive outcome could be determined. Similar field experiments allow for the natural interactions with competitors and predators to be included in the outcome along with competitive interactions but also rely on experimental protocols. For example, in this study the recruitment of fishes to their coral locations occurred for adult fishes, while in nature the occupation of reef habitats by fishes with pelagic larvae largely occurs through young fish settling in new habitat.

Summary

Competition can play a large role in structuring aquatic communities. Both exploitation and interference can influence foraging behavior and habitat selection. Competition studies conducted in natural systems or under controlled ecosystems all have a role in evaluating competitive outcomes. Character displacement of sticklebacks provides evidence on the functional nature on natural selection and its influence on foraging characteristics of fish species interacting with other phenotypes of their species. It also demonstrates how evolution can allow coexistence of competing species in a system. The experiments with sunfishes show strong ecological differences among three very similar species not only related to prey consumption and growth but also to habitat selection under varying environmental conditions. The study on coral reef fishes demonstrates that conditions can be manipulated to better understand competitive outcomes in natural settings, but mechanisms for these outcomes remain unresolved. Overall, extrapolating competition studies to field conditions relies on the nature of the experiments, the species, and how well experimental protocols relate to natural populations. Many managers strive to reduce competition between target sport and commercial species by removing competing species or manipulating environmental conditions to favor the target species. Experiments on competition can help determine what manipulations may be successful as well as better understand what biological mechanisms actually influence the outcome.

Literature Cited

Davis, J.L. 2000. Spatial and seasonal patterns of habitat partitioning in a guild of southern California tidepool fishes. Marine Ecology Progress Series 196:253–268.

Elliott, J.K., and R.N. Mariscal. 2001. Coexistence of nine anemonefish species: Differential host and habitat utilization, size and recruitment. Marine Biology 138:23–36.

Hutchinson, G.E. 1957. Concluding remarks. Cold Spring Harbor Symposia on Quantitative Biology 22:415–427.

Martin, K.L.M. 1996. An ecological gradient in air-breathing ability among marine cottid fishes. Physiological Zoology 69:1096–1113.

Munday, P.L. 2001. Fitness consequences of habitat use and competition among coral-dwelling fishes. Oecologia 128:585–593.

Schluter, D. 1995. Adaptive radiation in sticklebacks: Trade-offs in feeding performance and growth. Ecology 76:82–90.

Werner E.E., and D.J. Hall. 1979. Foraging efficiency and habitat switching in competing sunfishes. Ecology 60:256–264.

CHAPTER 15

Positive Interactions

Positive interactions are ubiquitous in nature but have lacked a quantitative and systematic approach to their study and integration into ecological theories compared to negative interactions such as predation, competition, and parasitism. The significance of positive intraspecific interactions among fishes – especially shoaling and schooling – is relatively well understood, but positive interspecific interactions were previously considered as curiosities, and their importance in fish population and community dynamics has been in doubt. Positive interactions involving fishes were reported earlier from marine ecosystems. These include cleaning interactions with wrasses and shrimps, multispecies shoals, and feeding or transportation interactions involving smaller fish species such as remoras on large and long-distance travelers such as rays and sharks. Positive reproductive interactions among freshwater fishes were discovered much later and are continually being recognized as a more widespread and distinct kind of positive interaction (Silknetter et al. 2020). Fish dispersing seeds of riparian plants, bioturbation by benthic feeders and lithophilic nest builders (that build nests with or on cleaned rocky surfaces), bioluminescence in deep sea fishes, and fish-gut microbiome relationships are positive interactions involving fishes. Common terms used to describe positive interspecific interactions include facilitation, mutualism, and commensalism (defined below). In this chapter, we explore the typology and specific examples of positive interactions involving fishes in the context of fitness consequences, evolving theoretical frameworks to study and explain why these interactions occur, and demonstrated effects of interactions on population and community dynamics, and conservation and management implications.

Stachowicz (2001) defined positive interactions as "encounters between organisms that benefit at least one of the participants and cause harm to neither", therefore clearly excluding parasitism. It is important here to highlight that positive interactions involve a participant acting in its own benefit that may result in benefit of another organism. Benefits and costs for both participants are determined in terms of net fitness consequences, measured in effects on survival, growth, or reproduction. The terminology of positive interactions warrant special treatment because many terms can be confusing, given the historical lack of research attention to this subject and absence of unified research frameworks. Definitions of the various terms used in positive interactions are provided in Table 15-1. Definitions are based on Bronstein (2009, 2015) and Silknetter et al. (2020) with slight modifications, unless otherwise indicated.

Biology and Ecology of Fishes, Third Edition. James S. Diana and Tomas O. Höök.
© 2023 John Wiley & Sons Ltd. Published 2023 by John Wiley & Sons Ltd.

TABLE 15-1	Definitions of Terms Used in Positive Interactions of Fishes.	
Term	**Definition**	**Notes and Examples**
Mutualism	Interaction in which both species involved receive a measurable net benefit	
Commensalism	Interaction in which one species benefits, while the other has no net cost or benefit	
Facilitation	Interaction in which the presence of one species alters the environment or reduces interaction with enemies to enhance fitness for a neighboring individual	Often discussed in relation to ecosystem engineering or habitat amelioration, but can apply to other fitness benefits
Symbiosis	Intimate (and not exclusively positive) interspecific relationship with prolonged physical contact	Terms are largely synonymous, with the only distinction here is the presence or absence of prolonged contact
Partnership	Interspecific association with beneficial fitness consequences for at least one organism, but which is not biologically obligatory or lacks prolonged physical contact	
Phoresis	Commensal interaction in which an animal (phoront) superficially attaches itself onto a host animal for the purpose of dispersal	Considered commensalism, but context may affect outcomes
Host	The interacting species that provides the physical space or resource initiating the interaction; the host always provides a measurable benefit to its partner	Not all positive interactions require a host; particularly for interactions within a single trophic level
Partner beneficiary phoront epibiont	Interacting species that benefit from the physical space or resource provided by the host (or other interacting species when no host exists)	Terms are synonymous, but apply to one of three different interaction types; beneficiary can be used to generally apply to all three

Source: Silknetter et al. 2020/John Wiley & Sons.

Common Types of Positive Interactions

Most positive interactions involve the provision or exchange of nutrients, energy, transportation, or protection, referred to as key resources or services (Bronstein 2015). It is better to consider this categorization as a heuristic framework.

Commensalism

Commensalism is a two-species interaction in which one species receives net fitness benefit, while the other receives no net benefit or cost. Perhaps the best example of marine commensalism involves the remora – also called suckerfish (family Echeneidae) – and their phoretic interaction with long-distance traveling animals. Phoresy is a two-species interaction in which a phoretic animal (or phoront) superficially latches itself onto a host for the purpose of dispersal (White et al. 2017). Remoras engage in such interactions with sharks, manta rays, whales, and turtles, and have even been reported to attach to boats and divers. The apparent objective of the remora is transport for dispersal although enhanced feeding opportunities and protection from predation risk could also be inferred. Hosts do not appear to benefit from phoront in this relationship. It is speculated that remora may move around on the host and perform some cleaning services; in which case, the relationship might tend toward mutualism. Conversely, a large number of phoronts on a host could be energetically costly to transport, and parasitism may be the result. We have found no experimental study that establishes a net benefit or loss for the host in this relationship.

Another example of commensalism – feeding-associative interactions – is found in streams. Certain benthic feeding fishes are well adapted to rummaging through gravel and debris or flipping pebbles and small boulders. In flowing water, such substrate disturbance dislodges insects and detritus in which other small fishes show great interest. The northern hog sucker in North America with its robust head and benthic mouth feeds by flipping pebbles and boulders in riffles, and is often followed by pods of smaller benthic fishes that catch dislodged insects (Matthews 1998).

Mutualism

Mutualism is a two-species interaction that confers net fitness benefit to both partners. A common example of mutualism reported from both marine and freshwater systems is multi-species shoals or schools. Heterospecific shoals are usually composed of individuals of similar size, and participants can include more than a pair of species. This behavior is hypothesized to afford the same protection gained by shoaling conspecifics, such as enhanced hydrodynamic efficiency and ability to access and exploit food patches, protection from predators, and overwhelming territorial defense of competitors.

Another well-studied mutualism is for fish that clean the skin surfaces, wounds, gills, and even between the teeth of client (host) fish. The cleaners feed off dead tissue, scales, mucous, and ectoparasites, therefore offering protection or grooming and receiving nutrition in return. Fish cleaner–client examples are widespread, especially in marine ecosystems. Fishes known to perform cleaning include wrasses, cichlids, catfishes, pipefishes, and gobies. Some tropical shrimp species are also known to be fish cleaners. Cleaning clients are not limited to fish species. In addition to groupers, ocean sunfish, sharks, rays, and eels, fish cleaners are also known to clean sea turtles, marine iguana, manatees, whales, octopuses, and hippopotami. Cleaning is such a widespread phenomenon that there are often dedicated cleaning stations where cleaners wait and large animals go to be cleaned. The extent of cleaning interactions can be enormous. In an experimental study, a single Indo-Pacific blue-streaked cleaner wrasse consumed about 1,200 parasites (mostly isopods) from upward of 2,300 fish per day. In the absence of the wrasse, isopods on caged fish increased fourfold in 12 hours (Grutter 2010).

Another interaction widespread in freshwater ecosystems and considered to be mutualistic is nest association. The resource or service involved here primarily is protection – from a hostile abiotic environment and brood predators. Nest association involves the use of a host's nest for spawning by one or more associate species, while the host continues to spawn or provide some form of parental care. Common species involved as hosts or associates in North America are lithophilic nest-building minnows in the genera *Campostoma*, *Nocomis*, and *Semotilus*. The minnows sometimes associate with saucer-nest builders in the centrarchid family, and still others associate with gars and other fishes. Associates may achieve better brood survival from the enhanced abiotic environment and parental care provided by the host (Johnston 1994; Peoples and Frimpong 2013). In exchange, the nest-building host can achieve a dilution effect on brood predation if eggs consumed from a disturbed nest include both host and associate species. Associate eggs accounted for up to 97% of all eggs in a nest in one study (Wallin 1992), and the average was about 85%. As a measure of prevalence of nest association, approximately 1/3 of minnows in the United States with confirmed modes of reproduction were nest associates (Johnston and Page 1992). In order for this association to be adaptive, the dilution effect on predation must exceed other negative effects of association, such as potential disease, lower dissolved oxygen, or increased fungal growth in the associated nests.

Facilitation

Facilitation is an interaction where one or both species alter their habitat or modify interactions between other organisms (e.g., predators or competitors) that provide a benefit to another organism. While this can include both commensalism and mutualism, facilitation is exemplified by less direct interactions between participants. A good example of facilitation studied in freshwater systems is habitat modification by a host species (Moore 2006). Fishes may be facilitators or beneficiaries of facilitation, including by other taxa. A common example of facilitation in North America is the effect of beavers, which create dams from trees and provide habitat for numerous aquatic organisms including fishes. Similarly, migrating salmon construct redds for spawning in headwaters, and since they are semelparous, the adults die after spawning, resulting in an annual pulse of major nutrient releases. This nutrient subsidy increases production in nutrient poor streams, enhancing production of young salmon, as well as other species present. Bioturbation – the process by which organisms disturb and release benthic materials – is another example of facilitation through habitat modification. A benthic foraging fish may remove insects and disturb sediments, which releases nutrients and sediments to the water while cleaning the substrate. All of these interactions involve resources or services that do not fit neatly in a dual or multispecies interaction.

Nest construction by lithophilic fishes is another example of facilitation. For example, adult male chubs (*Nocomis* spp.) construct large gravel and pebble nests for spawning. These nests often occur in high densities and usually differ starkly from the surrounding substrate as the only sources of concentrated, unsilted gravel in heavily embedded or sediment-starved reaches. Besides the spawning habitat provided for other lithophilic fishes, a large number of other organisms begin colonizing *Nocomis* nests immediately after they are constructed (Swartwout et al. 2016).

Costs of Positive Interactions

Positive interactions involve costs for one or both participating species although the net effect of the interaction is positive. This is consistent with the modern view that many symbiotic interactions lie on a mutualism–parasitism continuum, and these interactions are reciprocally exploitative. Historic reexaminations of presumably negative interactions – especially parasitism – have revealed conditions under which positive outcomes occur. Costs of positive interactions can be broken down into at least three main categories. The first are costs associated with a behavior or action by hosts or facilitators that reduce their own performance, such as energy expended during habitat modification (e.g., nest building in minnows or damming behavior in beavers) or energy expended in defenses.

The second type of cost is incurred when one partner in the interaction – through excessive numerical response or aberrant behaviors – causes a cost to the other who has to take repeated corrective counter action. An example from nest association occurs when participants dig spawning pits in locations where eggs are already buried. Such actions expose and dislodge eggs causing the host to ward off individuals building nests nearby and rebury exposed eggs before they are eaten or float away. Excessive numbers of partner species can create chaos and overwhelm an orderly interaction, thus necessitating antagonistic regulation of partner numbers by the host. Mating disruptions and unintended hybridization between hosts and associates or among different associate species are additional risks or costs in nest association. An aberrant behavior considered an example of cheating is common in cleaning interactions when there is an excess of cleaners and not enough hosts. The cleaners may resort to feeding on the host's live tissues, resulting in a parasitic interaction. Hosts or partners may also exhibit intentional selfish behaviors to maximize their gain from interactions, therefore costing the other more than they perceive of the relationship.

The third way mutualism may incur cost is through the exploitative action of species external to the mutually beneficial relationship, such as mimics of the true partners that contribute no benefits to the partners. For example, in *Nocomis* nest association, other fish species and crayfish frequently visit nests; their only interest being to prey on eggs. Species assumed or confirmed to be nest associates but also confirmed to feed on eggs on nests is the reason nest association has occasionally been described as nest parasitism.

Context Dependency in Positive Interactions

Although previously conceptualized as discrete, static categories, the strength and outcomes of positive interactions are dynamic, changing predictably with abiotic (the environment in which the interaction occurs, varying across space and time) and biotic (traits, ontogeny, or abundance of participants) context (Figure 15-1). Outcomes of interspecific interactions lie along a continuum that ranges from antagonistic to mutually beneficial. Recent work has shown that even in interactions generally considered mutualistic, outcomes can be highly

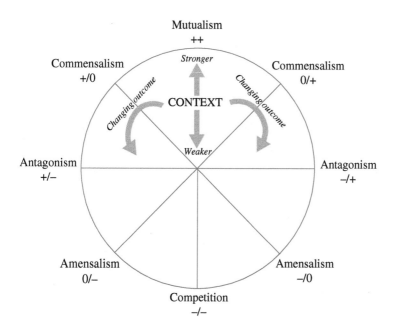

FIGURE 15-1 All possible pairwise species interactions can be depicted using an interaction wheel modified from the "coaction compass." By adding context to the compass (black arrows), we can better understand how sign (changing outcomes are represented by movements around the wheel) or strength (movement toward the edge of the circle) of an interaction may vary. *Source: Reproduced with permission from Silknetter et al. (2020).*

variable. Environmental conditions establish the abiotic context within which interactions occur. For example, Peoples and Frimpong (2016) found that in the presence of egg-eating suckers, decreased substrate quality caused the mutualism between two nest associative fishes to switch to commensalism.

Biotic factors may also set the context for the outcome of positive interactions. For example, larger individuals of fruit-eating *Pacu* provide better survival to the seeds they ingest because large *Pacu* pass seeds through their digestive tract intact, while smaller fishes damage the seeds and significantly decrease germination rates (Galetti et al. 2008). In other interactions, a critical mass of participants is needed for an interaction to benefit all species. For example, Silknetter et al. (2018) found that increased partner density caused nest association to switch from parasitism to mutualism; low densities of associates attracted egg predators without providing a sufficient dilution effect, and chubs were better off spawning without associates than with them at low densities.

Probably, the most unusual example demonstrating evolution of symbiotic relationships is the one between freshwater fishes and mussels. Mussels produce young, called glochidia, which attach to the gills of a host fish. These glochidia are then transported upstream by the host fish and settle out to become benthic juveniles. Since drift of the mussels results in downstream movement, this interaction allows mussels to recolonize upstream areas. Glochidia are generally considered parasites of the fish, although this is more likely a commensal relationship since the fish are not substantially harmed. The relationship can be even more convoluted as in the bitterling of Europe that spawn their eggs into mussel shells, which provide shelter for the eggs, while glochidia infect the bitterling, making the relationship mutualism (Reichard et al. 2012). In some cases, it can even become parasitic for the fish on the mussel, when glochidia do not attach to the fish, yet eggs are held and protected by the mussels. Since many mussels must have a fish host to complete their life cycle, it is clear this is a long-term and evolved relationship.

Positive Interactions and Community Organization

Doubts previously held by scientists on the importance of positive interactions have been erased by evidence from both marine and freshwater fish assemblages. Grutter et al. (2003) showed experimental evidence of the disproportionate importance of cleaner fish on the abundance and diversity of reef communities. For this study, they removed cleaner fish from nine reefs and resampled the reefs several times to remove any recruiting cleaner fish to these reefs. They also had nine control reefs with cleaner fish. They sampled the reefs after 18 months, using video cameras, snorkelers, and scuba divers to assess presence and abundance of 78 species of visitor fish, as well as the resident fish species. There were significantly more species on reefs with cleaner fish – the elevation in visitor species abundance was two- to fourfold, depending on survey method. Of the 78 visitor species, 38% were only found on reefs with cleaner fish, 52% were found on both types of reefs, and 9% were found on reefs without cleaner fish. The most important cleaner was the cleaning wrasse *Labroides dimidiatus*, and communities without them retained only half the species diversity of fish and one-fourth the abundance of individuals. Resident fish species were not affected by the removal of cleaner fish and averaged around 50 species per reef. The lost diversity and abundance of species that move among reefs were predominantly large species that selected reefs for inhabitation, based on presence of cleaner fish, and can regulate the population of other reef species by their predation and herbivory.

Evidence is also accumulating on the importance of nest-building fishes and nest associa-tion in the dynamics of stream communities. In the mix of forested, historically agricultural, and currently urbanized segments of streams in the middle New River basin, Virginia, more than 50% of individual fish sampled were central stoneroller *Campostoma anomalum*, mountain red-belly dace *Chrosomus oreas*, or bluehead chub *Nocomis leptocephalus*. Stoneroller and bluehead chub are nest builders, while the stoneroller is also a nest associate of bluehead chub. Redbelly dace associates with the other two species, and all three species combined make up the most abundant species in the system (see nest in Figure 15-2). Other abundant fishes are nest building centrarchids and species, such as darters and suckers, that are egg predators (Figure 15-3). In that system, fish communities shift from being dominated by a diverse set of reproductive guilds in more forested streams to being dominated by chubs and associate species in more degraded streams. This is presumably due to associates taking advantage of facilitation by chubs as habitat conditions become more harsh (Peoples et al. 2015). Analyses of historical changes in composi-tion of fish assemblages in the New River basin over the last 60–70 years have found that blue-head chub expanded its local range in watersheds undergoing development, while other species experienced range contraction. As the chub expanded its range, so did its nest associates (Hitt and Roberts 2012; Buckwalter et al. 2018). Mutualistic nest association with bluehead chub has been implicated to facilitate a rapid population and range expansion of the rough shiner *Notro-pis baileyi*, which is an introduced species in the Chattahoochee River system of Alabama and Georgia (Walser et al. 2000; Herrington and Popp 2004). As the introduced species spread, the proportional abundance of bluehead chub also increased in assemblage samples.

FIGURE 15-2   Nest of bluehead chub with redbelly dace associates. Image taken by Tod Pusser.

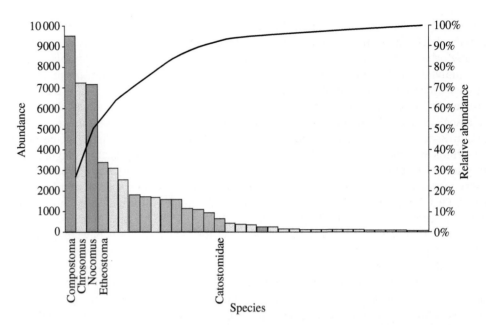

FIGURE 15-3 Abundance distribution of 49 fish species collected in surveys of the middle New River basin, Virginia, in 2008. Left axis represents the number of fishes collected and right axis represents cumulative abundance. Orange, nest-builder–*Compostoma*, *Nocomis*; green, nest-builder–*Centrarchidae*; yellow, nest associate – *Chrosomus*; blue, egg predator-*Etheostoma*, *Catostomidae*. *Source: Unpublished data from Frimpong (2022).*

Processes that are strong regulators of population and community dynamics must be the focus of designing conservation and management strategies for fish populations. Due to the historical lack of research attention and under appreciation of positive interactions, many management and conservation strategies inadequately account for them (Frimpong 2018). For example, *Nocomis* nest association likely (1) regulates persistence and spread of native species in the face of environmental changes, such as siltation in streams; (2) facilitates persistence and spread of native associates or interaction partners; (3) explains rare-common patterns in species-abundance distributions; and (4) predicts establishment and success of some invasive species in novel habitats. Understanding these relationships and their outcomes is vital in developing successful management strategies for fish assemblages in many locations.

Summary

There are a large number of positive interactions among fish species that have received far less research attention than negative interactions, such as competition and predation. However, positive interactions are not only important to the species involved in the interaction but may also have strong implications for the assemblage of fishes existing in their locations. Species involved in mutualism are potentially fundamental to population and community dynamics (such as bluehead chub in streams and cleaner wrasse in reefs) and fit the definition of "keystone" species. A deep understanding of positive interactions would allow us to effectively identify keystone species, better predict the effects of natural perturbations and human interferences, and manage for naturally functioning populations and communities.

Literature Cited

Bronstein, J.L. 2009. Mutualism and symbiosis. Pages 233–238 *in* S.A. Levin, editor. The Princeton Guide to Ecology. Princeton University Press, Princeton, NJ.

Bronstein, J.L. 2015. The study of mutualism. Pages 3–19 *in* J.L. Bronstein, editor. Mutualism. Oxford University Press, Oxford, UK.

Buckwalter, J.D., E.A. Frimpong, P.L. Angermeier, and J.N. Barney. 2018. Seventy years of stream-fish collections reveal invasions and native range contractions in an Appalachian (USA) watershed. Diversity and Distributions 24:219–232.

Frimpong, E.A. 2018. A case for conserving common species. PLOS Biology 16:e2004261.

Frimpong, E.A. 2022. Fish Surveys in the New River Basin. Unpublished data, Department of Fish and Conservation, Virginia Tech University, Blacksburg.

Galetti, M., C.I. Donatti, M.A. Pizo, and H C. Giacomini. 2008. Big fish are the best: Seed dispersal of *Bactris glaucescens* by the pacu fish (*Piaractus mesopotamicus*) in the Pantanal, Brazil. Biotropica 40:386–389.

Grutter, A.S. 2010. Cleaner fish. Current Biology 20:547–549.

Grutter, A.S., J.M. Murphy, and J.H. Choat. 2003. Cleaner fish drive local fish diversity on coral reefs. Current Biology 13:64–67.

Herrington, S.J., and K.J. Popp. 2004. Observations on the reproductive behavior of nonindigenous rough shiner, *Notropis baileyi*, in the Chattahoochee River system. Southeastern Naturalist 3:267–276.

Hitt, N.P., and J.H. Roberts. 2012. Hierarchical spatial structure of stream fish colonization and extinction. Oikos 121:127–137.

Johnston, C.E. 1994. The benefit to some minnows of spawning in the nests of other species. Environmental Biology of Fishes 40:213–218.

Johnston C.E., and L.M. Page. 1992. The evolution of complex reproductive strategies in North American minnows (Cyprinidae). Pages 600–621 *in* R.L. Mayden, editor. Systematics, Historical Ecology, and North American Freshwater Fishes. Stanford University Press, Stanford, CA.

Matthews, W.J. 1998. Patterns in Freshwater Fish Ecology. Chapman and Hall, New York.

Moore, J.C. 2006. Animal ecosystem engineers in streams. BioScience 56:237–246.

Peoples, B.K., and E.A. Frimpong. 2013. Evidence of mutual benefits of nest association among freshwater cyprinids and implications for conservation. Aquatic Conservation: Marine and Freshwater Ecosystems 23:911–923.

Peoples, B.K., and E.A. Frimpong. 2016. Context-dependent outcomes in a reproductive mutualism between two freshwater fish species. Ecology and Evolution 6:1214–1223.

Peoples, B.K., L.A. Blanc, and E.A. Frimpong. 2015. Lotic cyprinid communities can be structured as nest webs and predicted by the stress-gradient hypothesis. Journal of Animal Ecology 84:1666–1677.

Reichard, M., M. Vrtílek, K. Douda, and C. Smith. 2012. An invasive species reverses the roles in a host – parasite relationship between bitterling fish and unionid mussels. Biology Letters 8:601–604.

Silknetter, S., Y. Kanno, K.L. Kanapeckas Métris, et al. 2018. Mutualism or parasitism: Partner abundance affects host fitness in a fish reproductive interaction. Freshwater Biology 64:175–182.

Silknetter, S., R.P. Creed, B.L. Brown, et al. 2020. Positive biotic interactions in freshwaters: A review and research directive. Freshwater Biology 65:811–832.

Stachowicz, J.J. 2001. Mutualism, facilitation, and the structure of ecological communities. BioScience 51:235–246.

Swartwout, M.C., F. Keating, and E.A. Frimpong. 2016. A survey of macroinvertebrates colonizing bluehead chub nests in a Virginia stream. Journal of Freshwater Ecology 31:147152.

Wallin, J.E. 1992. The symbiotic nest association of yellowfin shiners, *Notropis lutipinnis*, and bluehead chubs, *Nocomis leptocephalus*. Environmental Biology of Fishes 33:287–292.

Walser, C.A., B. Falterman, and H.L. Bart, Jr. 2000. Impact of introduced rough shiner (*Notropis baileyi*) on the native fish community in the Chattahoochee River system. The American Midland Naturalist 144:393–405.

White, P.S., L. Morran, and J. de Roode. 2017. Phoresy. Current Biology 27:578–580.

CHAPTER 16

Movement and Habitat Use

Fish exhibit a wide range of movements and displacements ranging a few meters from the "home" or geographic area where they were spawned to hundreds or even thousands of kilometers during their life time depending on species and their life history. Therefore, the fish species observed in a defined habitat at any given time constitute only a representation of what could be there. In addition to spatial extent, types of fish movement are distinguished in terms of temporal scale ranging from seconds to years and probability of returning to original location after displacement. Some movements are unpredictable, short, or limited to small geographic areas, whereas other movements are predictable daily, seasonal, or generational patterns. Movement entails exposure to predators and energetic cost and therefore would rarely be embarked on without a fitness benefit to the individual in the form of enhanced growth, survival, or reproduction. Some movement may be a social behavioral routine, such as shoaling, schooling, or territoriality. To track and explain fish movements require techniques well beyond short-term visual observations, and an ever-increasing toolkit of methods for fish movement studies indicates just how important ecologists view an understanding of fish movements. In this chapter, we use some key conceptual frameworks and case studies to explore patterns of fish movements and behaviors characteristic of various habitats and species, the reasons fish move, methods for studying movements, and the ecological and evolutionary implications of movement disruptions mainly caused by humans through habitat alterations.

Patterns of Movement

Most movement and social behaviors are driven by environmental cues, habitat quality, and configuration, as well as the presence, density, and behavior of other animals, especially conspecifics. Differences in movement patterns among populations of the same species may reflect size of their home system and connectedness to other systems as fish can move farther in larger lakes and rivers compared to smaller streams or lakes. While an individual may move to find food, shelter, or mates or to avoid predators or unfavorable habitat conditions, its movement affects gene flow and the spatiotemporal structure of populations, as well as interactions and species composition observed in its home and receiving communities. To quantify how movement affects individuals, populations, and communities, studies have focused on questions such as who moves, how far, and why? The questions of who and how far fish

Biology and Ecology of Fishes, Third Edition. James S. Diana and Tomas O. Höök.
© 2023 John Wiley & Sons Ltd. Published 2023 by John Wiley & Sons Ltd.

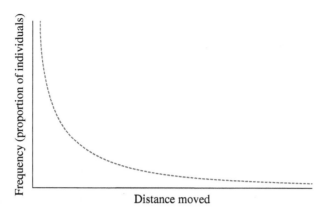

FIGURE 16-1 Leptokurtic pattern of fish movements frequently observed in radio-tag and mark-recapture studies.

move have frequently led to the conclusion that many individuals of a population do not move permanently away from a home location while some individuals undertake long-distance movements – the so-called leptokurtic movement distributions common from telemetry and mark-recapture studies (Figure 16-1).

Many arguments have been advanced to explain leptokurtosis of observed movements, including a purely statistical artifact of a positive correlation between the spatial and temporal scales of a study and the observed distances moved (Radinger and Wolter 2014). There are plausible ecological explanations as well. The long tail of the distribution likely represents movement outside of a home range due to competitive displacement, predation risk, or other unfavorable conditions. In an experimental study of Trinidad killifish, Fraser et al. (2001) explained leptokurtosis as a consequence of heterogeneity among individuals in a population. They tested boldness in a tank experiment that later predicted the amount of movement when fish were released in a 19-month mark-recapture field study. They found that movement was positively correlated with individual growth after accounting for size and sex in a stream, where movement was threatened with a high risk of predation and the killifish population was highly fragmented. This correlation was not found in an area of the stream where predators were absent and the population was not fragmented. Their conclusion was a conditional fitness advantage exists for "dispersers" over "stayers" in each population. These results indicate that some individuals have a specific inherited propensity to disperse long distances (Saastamoinen et al. 2018).

However precisely a fish's movement is documented, it may still be difficult to explain what was accomplished and therefore the best descriptor for that movement, which would then help to infer its ecological or evolutionary function. All movements fit under at least one of three categories: resource tracking, migration, and dispersal. Paul Angermeier of US Geological Survey and Virginia Tech has proposed an unpublished framework for classifying movements into these categories based on relative distance moved and probability of returning to original location after displacement. Resource-tracking movements occur within the home range to access food as a patch becomes congested or depleted, safer locations to avoid predators, optimal temperature, better water quality, mates or spawning sites. This occurs continually over short periods during a fish's life and short relative distances, individually or in groups, with little or no net displacement. Diel vertical migration of reef and lentic freshwater planktivores, movements of marine littoral species with the tidal cycle, and pool-to-pool movement of stream-dwelling minnows on subhourly time scales are a few common examples of resource tracking.

According to Angermeier's framework, migratory movements are relatively long distance from home but involve a return to home at some point in the life cycle, thus entail no lifetime displacement. Migrations are often undertaken in large schools at a specific life stage to spawning, feeding, or refuge habitat, which differs from the origin habitat. Since migrations are tied to life stages, they are predictably periodic in nature. Migrations are well studied among salmonids, especially spawning migrations (see Chapter 21) and poorly studied in small fishes. Five major types of migratory movements are anadromy, catadromy, amphidromy, oceanodromy, and potamodromy. *Anadromous* fish are spawned in freshwater, migrate to the ocean as juveniles where they grow to sexual maturity and then migrate back to their natal freshwater stream to spawn. Examples of anadromous species are Pacific salmon, American shad, striped bass, and gulf sturgeon. *Catadromous* fish are spawned in the ocean, migrate into freshwater streams as juveniles where they grow to sexual maturity, and then migrate back into the ocean to spawn. American eel, European eel, inanga, and striped mullet are examples of catadromous species. *Amphidromous* fish are spawned in either freshwater or estuaries, drift into the ocean as larvae, and then migrate back into freshwater where they grow to sexual maturity and spawn. Examples of amphidromous species are Dolly Varden, mountain mullet, and torrentfish. Anadromous, catadromous, and amphidromous fishes are collectively known as *diadromous* fishes due to their movement between freshwater and marine systems to complete their life cycle. *Potamodromous* fish are spawned in tributaries of freshwater rivers or lakes, then drift (or migrate) downstream to grow in freshwater systems, and migrate back upstream to spawn. Examples of potamodromous fish include redhorses, lake sturgeon, walleye, and freshwater lampreys. *Oceanodromous* fish are born on spawning grounds, then drift or migrate on ocean currents as larvae, and settle into new locations to grow into sexual maturity before migrating back to spawning grounds. Examples of oceanodromous species include black grouper, mutton snapper, and goliath grouper,

Many fish movements are casually – and often incorrectly – described as dispersal in the literature. Paul Angermeier's model considers dispersal as a relatively long-distance movement with low probability of returning. Dispersal functions to colonize open habitats, maintain gene flow among connected subpopulations, and rescue populations. It is an infrequent or sporadic occurrence in an individual fish's life. Dispersal may occur singly or in groups by lost migrants, drift on current by eggs and larvae, and adults forced downstream by high stream flow. Dispersal thus can occur at any life stage and may or may not be predictable. True dispersal is accomplished when an individual contributes to the gene pool of the next generation of the recipient population. Rapid recolonization after fish kills, spread of introduced species, and the lack of fine-grained spatial structure in population genetics (Roberts et al. 2016) all suggest significant and effective dispersal movements, which can occur over long distances.

In his challenge to the concept of home range, Linfield (1985) presented length distribution data for roach and European dace from River Witham showing that large fish are found mostly upstream and small fish mostly downstream. From this and other observations, several inferences were made. These fish exhibited a variable combination of active movements in response to flow and temperature with a tendency to aggregate in winter and disperse in summer. Passive downstream displacement was negatively correlated with size such that early life stages drifted downstream to more sluggish water, while adults actively swam upstream to populate and breed in these areas. With larval and even egg drift being a common phenomenon among stream fishes, it is likely that net upstream migration as a fish goes through ontogeny is a regular feature of life in streams but not well appreciated because it happens sporadically in time.

Movement of fish is significantly influenced by the configuration of habitats and temporal changes experienced in habitats, such as daily and seasonal cycles of light and temperature, tides, food availability, and predator activity. There are also endogenous controls on movement shaped by these environmental variations, for example, circadian rhythms and ontogenetic shifts in habitat selection and use. Schlosser (1995) presented a conceptual framework of habitat use and movement of stream fishes. In a dynamic landscape, feeding,

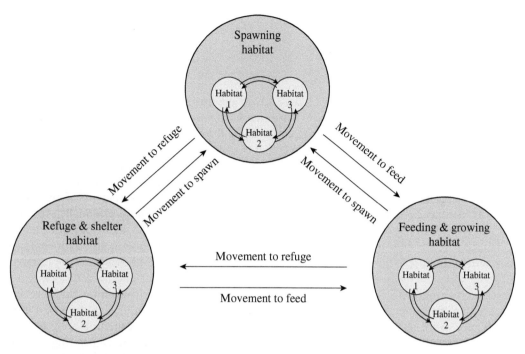

FIGURE 16-2   A generalized model of fish life cycle, habitat use, and movements in a dynamic landscape.
Source: Adapted from Schlosser (1991) and Thurow (2016).

spawning, and refuge habitats change over time and may also be spatially separated, thus requiring movements between habitat patches through a matrix of unsuitable to suboptimal habitat to complete their life cycle. Schlosser's model can be expanded to include a mosaic of feeding habitats (Schlosser 1991), multiple refuge or sheltering habitats (Thurow 2016), a mosaic of spawning habitats, and possible movements between and within any habitat mosaic (Figure 16-2). The three main bubbles in Figure 16-2 represent survival or refuge habitat, feeding and growth habitat, and spawning or reproductive habitat. For many species, these habitats are spatially separated and non-substitutable, and therefore migration is necessary to complete their life cycle.

Movements among habitat types are often strongly tied to ontogenetic changes, where a size-specific shift in food types and growth needs necessitate a shift in feeding habitat. Movement to spawning habitat occurs for sexually mature individuals, whereas optimal habitat for growth and survival may change with life stage, thus requiring movement into a different optimal habitat through life stages. For the simple case of juveniles, Werner and Gilliam (1984) conceptualized ontogenetic changes in habitat use as a trade-off between mortality (μ) and growth (g) to determine the size at which movement into a new optimal habitat should occur that minimizes μ/g.

Movement among patches in a mosaic of refuge habitat, growth habitat, and reproductive habitat is widespread behavior and can be illustrated with a few examples: high summer temperatures may drive many lentic species deeper into cooler water during the day and closer to the surface at night. Many stream fishes rest in calmer backwaters as temporary refuge and feed in open water at night or during the day. Feeding habitat is usually patchy, and the optimal location to feed can change quickly with change in biotic conditions such as an increase in competitors or predators in a patch. Even suitable reproductive sites can exist as a network of patches.

Methods to Study Movement and Habitat Use

Netting or Trapping

A common method used to evaluate activity and movement of fishes is to set a passive net, such as a gillnet, and then examine catch. Fish are caught in passive nets because they are actively swimming, and catch correlates with activity to some degree. This method is commonly used for fish in lakes or the ocean. The problem is that catch is also altered if fish can detect the net. For example, a fish may be more active during the day, but a gillnet may also be much more visible then. Even though the fish is less active at night, it may not be able to see the net at night and may collide with it more often. This presents a bias that affects measurements of diurnal or nocturnal activity, or seasonal activity if visibility and vulnerability change seasonally. Another problem with netting is that certain nets collect particular species well and do not work equally well for others, and this avoidance or attraction may make netting a poor method for comparative activity analyses. In any case, the activity measure for netting studies is based on number of fish caught per some unit of effort, which usually accounts for time and net size (Box 16-1). Such data can be taken from experimentally set nets or from those used in commercial fisheries.

Another method to study activity pattern is to evaluate movements using traps or weirs. This is very common in streams and can be used to determine displacement among habitat types or especially reproductive movements like spawning migrations. Depending on the length of a stream, a series of traps could be used. Fishes moving upstream or downstream into a trap can be collected, counted, and marked, and then passed upstream or downstream. Traps may cover the entire stream width or a subsection. For large river systems, trap nets might be replaced by fixed weirs. Data from trapping studies can include relative activity (time of day or year) and direction of movement. Once again, if fish avoid the trap, activity could be underestimated. Some fish may avoid traps; some may be attracted to traps.

The third method often used is to mark fishes at capture and then release them for later recapture (see Box 16-2). Distances and direction moved can be measured between the two points of capture. Marks vary in style, and the most common type is a fin clip, which actually involves removing all or a portion of a fin (or several fins). When marking is done to recognize fish individually, or to recapture and release fishes several times, details of movement patterns may be gained beyond simply distance moved and time at large.

For larger mark-and-recapture studies, many fish must be marked, usually with group marks. For these fish, the only possible measure of activity is a determination that the fish were marked at one location and time, and then recaptured at another. This presents some idea of general movement and dispersal patterns, but nothing about what happened between mark and recapture. One other bias common to mark-and-recapture studies is that of sampling effort. If a group of fish is marked in one location and later attempts are made to recapture fish in the same location, some of the original fish are usually found there. However, unless all

---

Box 16-1 Measuring Fish Activity by Netting

Passive nets, such as gill nets, trap nets, or fyke nets, collect fishes when they are actively swimming.

Similar levels of net effort (i.e., one net night per month) in similar habitats can be used to demonstrate changes in activity.

Data collected are simply the number of fish taken per standard unit of net effort. Biases include changes in vulnerability to nets due to water clarity or sensory modes.

For further details on passive capture techniques, see Hubert (1983).

Box 16-2 Mark-and-Recapture Studies

Mark-and-recapture studies can give details on fish behavior as well as estimate population abundance.

Fishes are originally captured and marked with some visible sign to identify each individual or groups of individuals with similar characteristics.

Individually recognizable marks include numbered tags, such as spaghetti tags, disk tags, jaw tags, PIT tags, or coded-wire tags. The date of capture and other relevant details are recorded for each individual.

Group tags include removal of a fin (commonly the adipose, pelvic, or pectoral), removal of part of a fin (upper caudal), or some other permanent marking (branding and fluorescent dye injection). Each group may include a location or general time of capture.

Recaptures of marked fish can be made by scientists, the angling public, or commercial fishers. Locations of recapture and time-at-large can be determined for behavioral analyses.

Population estimates can be made based on proportion of marked fishes. One common technique is the Petersen estimate:

$$N = MC/R$$

where N is population number, M is number of marked fish, C is total number of fish captured during recapture operations, and R is the number of marked fish recaptured.

A variety of population estimators exist, depending on the number of mark-and-recapture events and other details. For further information, see Everhart and Youngs (1981). For more details on marking techniques, see Wydoski and Emery (1983) or Voegeli et al. (2001).

---

alternate locations are sampled equally well, the locations of dispersing fish cannot be determined as precisely as the location of resident fish. The percentage of locally resident fish compared to that of dispersers is uncertain, so the description of overall movement pattern may be biased. Similarly, if only a few fish are recaptured, the representativeness of that sample may be questioned.

Obviously, increasing the number of fish marked may help to achieve better recapture rates. However, it may take more automated types of marking systems for large groups of fish, and these may provide no individual identification. For example, Seelbach (1987) marked steelhead smolts by spraying them with florescent dye. When the fish returned as adults, they could be identified by examining their scales under black light where the dye glows. This system can be very effective in marking large numbers of fish, but the effect of the dye and of handling must be estimated, as well as the retention of the mark. Coded-wire and PIT tags can be injected into the fish with a syringe, often in the nose or dorsal musculature. Passive integrated transponder (PIT) tags can be detected remotely with a sensor transmitting radio waves that are responded to passively by the tag. These signals can be decoded for individual fish with transmission distance up to 1 m. Both coded-wire and PIT tags are very small and can be easily implanted in juvenile fish. Coded-wire tags must be removed from the fish for reading and require a fin clip or other visible mark to help identify tagged individuals. However, both tags provide means to mark very small individuals for later recapture and individual identification, and both types of tags are seldom shed from the fish. They are very important devices to aid stock-assessment studies.

Mark-and-recapture studies generally evaluate movement patterns by marking fish in specific areas, using different marks for individuals or for fish from different groups, and then returning later to sample those areas. The number of fish recaptured in each area can then be estimated. For example, Diana and Lane (1978) marked 26 cutthroat trout in area 1 of a small stream and recaptured 14 of them three months later in the same location. However, some fish were recaptured upstream or downstream of the original site. In this case, movement involved overcoming a 1-m waterfall that was considered impassable. As previously discussed, the pattern of some fish moving and others staying is very common in mark-and-recapture studies. In any such study, the number of fish recaptured greatly depends on the environment sampled, sampling effort, and sampling locations. In large lakes or the ocean, marking thousands of fish may only provide a few recaptures, and equally sampling all habitats for recaptures may be impossible. In a small stream, nearly every fish may be recaptured and the entire stream may be sampled. In extensive watersheds, similar challenges of only a few recaptures may be encountered in mark-recapture studies. For example, Roberts et al. (2008) marked and released 485 endangered Roanoke logperch and recaptured only 22 individuals over a nine-year study period. They were able to determine that one Roanoke logperch moved 3.2 km over two years and another moved 2.5 km over five years, distances previously not considered within the range of movement of this relatively small benthic fish. Results from these studies provide some idea about whether fish moved between two time periods but not about their behavior between marking and recapture. The 14 fish in the Diana and Lane study that remained in the marking location might have dispersed widely and then returned to area 1 before recapture; those fine details are unavailable from this sampling method.

Telemetry Observations

A better way of doing long-term movement and activity studies – particularly for fish showing large movements – is telemetry. Transmitters are attached to fishes, which can then be followed from a distance with either mobile receivers or stations with set receivers (Box 16-3). In the case of mobile telemetry studies, relatively few fish may be studied – usually because of the intensity of effort required to follow fish and at any given time even fewer may be at large because it is

---

Box 16-3 Telemetry Studies

Telemetry is the use of a transmitter in a fish to signal its location to a remote observer.

Transmitters are generally radio or ultrasonic. Radio transmitters send their signals through the air, are received by aerial antennas, and are limited by the depth and salinity of water. Ultrasonic transmitters send their signals through water, are received by a hydrophone (underwater microphone), and are limited by physical obstructions or discontinuities in the water.

Transmitters can be attached externally, physically implanted in the stomach, or surgically implanted in the coelom.

Transmitter signals can be received by mobile methods to find fish throughout their habitat or by fixed stations that can be programmed to record when a particular fish is near that location. A combination of fixed stations can also be used to pinpoint locations of fish continually although these systems may only be applied in small areas.

The distance and time duration over which a transmitter functions are related to the battery power of the transmitter, which affects transmitter size. Transmitters should be less than 3% of the fish weight as larger transmitters may be expelled or alter fish behavior.

Recent advances in telemetry systems allow for archiving data by a tag. In these cases, details are logged in the transmitter for later retrieval. Such details may include depth, temperature, and light level. Archival tags may require recapture to allow data acquisition or, in the case of pop-up tags, may detach from the fish after set intervals, float to the surface, and transmit a radio signal to be received by satellite with the encoded and logged data.

For further details on telemetry, see Voegeli et al. (2001) or Adams et al. (2014).

difficult to keep in contact with many fish on a daily or hourly basis. A small number of fish are analyzed in most of these studies, but behavior can be determined continually for each individual. Very detailed information, in terms of diurnal or nocturnal activity or other measures, can be collected using this technique.

There are obviously some concerns using telemetry. The first concern is the opposite of mark-and-recapture studies, which may use thousands of fish and get an average estimate of behavior without much detail. In telemetry, few fish are often used, but many details of behavior can be measured. One question is whether fish are stressed by attachment of transmitters. Another concern, due to low sample size, is whether the fish sampled are normal representatives of the population. With a large fish, such as a lake sturgeon or sharks, there is little concern about transmitter size affecting swimming behavior. In fact, studies of sharks and tunas have used very large, externally attached transmitters that can measure characteristics of the habitats that the fish select and archive these data. Similar measures cannot be done with smaller fishes, due to size limits of the transmitters. Size of transmitter and size of fish both affect telemetry studies – the larger the fish, or the smaller the transmitter, the more typical behavior will likely occur.

Data collected by telemetry also are dependent on sampling design. For example, a typical pattern is to sample by locating fish once each day. This will provide data on long-term patterns of behavior. On the other hand, continually following an animal may provide very short-term measures of activity and swimming speed. This is often done in oceanic studies because the ocean is so large; once an animal is lost, it is not likely to be found again by random search, even with telemetry systems that may transmit signals over a mile. In smaller freshwater environments, especially lakes or streams, one might sample daily, or even less frequently, and still not lose the fish.

Archival tags have revolutionized the study of large pelagic species. These tags can store and transmit details on hourly or even shorter intervals describing light level, depth, and external and internal temperatures. Using light level to estimate day length, the fish's longitude can be estimated by comparing the time of noon to Greenwich Mean Time. Day length can also be used to estimate latitude, given a known date. This process is termed geolocation and is similar to the systems used in nautical navigation. These systems can work fairly precisely for terrestrial animals but are more difficult in aquatic animals that may move deep into the water where light is attenuated. This makes the estimation of day length difficult or impossible.

Transmitters must be attached to the fish in some way, and methods for attachment depend on the goal of the study and its duration. External attachment is commonly used for large fish and for pop-up tags, but since the transmitter is exposed to currents or vegetation, it may become dislodged, entangled, or alter the fish's ability to swim. A transmitter can be implanted in the stomach, which is not very stressful. The transmitter is not exposed to external factors and, over short periods of time, may function very well. A fish might regurgitate the transmitter, or have difficulty feeding with a transmitter in its stomach, so such an attachment might not give unbiased estimates of fish behavior over long time periods. As battery size has declined, surgical implants into the body cavity are most commonly used. Fish implanted in such a way can survive for years, so implants are advantageous for longer studies. However, fish may encapsulate transmitters with body tissue and then expel them through body walls (Summerfelt and Mosier 1984). Such fish may survive, but observations of behavior are truncated. Suturing and surgery also have some effect on a fish, so a week or more of data might need to be discarded while the fish recovers from surgery. For a transmitter that lasts two years, this is a small price to pay.

Habitat Selection

Additional methods are used to evaluate habitat selection. The theory of ideal free distributions (IFD) is often used to construct expected habitat selection or foraging sites for animals.

Gary Mittelbach from Kellogg Biological Station, Michigan State University, wrote an excellent review of the concept and support for IFD. If food is patchy and IFD applies, fish should select a patch based on food input rate to that patch, often called input matching (Mittelbach 2002). An additional rule of IFD is that feeding rates of all individual foragers should be equal, but that rule has rarely been supported empirically. Differential competitive abilities, interference by predators, and other food supply factors are probably related to the imbalance in feeding rates among foragers.

IFD theory, by definition, assumes no cost of habitat choice or free distribution. Of course, predator risk, current velocity, temperature, and other factors can produce differential costs for choosing certain habitat types. In some cases, as previously discussed, the trade-off between gain and risk can be quantified and included in habitat selection models (i.e., μ/g; Werner and Gilliam), but this adds much complexity especially for sexually mature individuals. Experimental studies have tested various predictions of IFD related to predatory risk or competitive interactions. However, extrapolation of such models to the real world is much more complicated, given the large number of risk or benefit factors. At present, IFD studies are contributing more to the knowledge of optimal habitat choice but are not often used in empirical studies to evaluate habitat selection in nature.

Common methodology for the study of habitat preference involves evaluating the frequency of use for different types of habitat and comparing that to frequency of habitat availability in the range. This method is very similar to the food electivity index of Ivlev (1961) and can be used to determine if animals select a particular habitat type more or less frequently than they would by random choice. For example, the frequency of locations in cover for large brown trout could be evaluated and compared to the frequency of cover in a stream, to determine if brown trout are selecting cover as a preferred habitat type. Such analysis cannot determine *why* an animal is using certain habitat types, only that it *is* using that type. In contrast, IFD studies hypothesize distributions based on food abundance, predators, and competitor distributions, which means causal mechanisms of food selection, predator avoidance, or competitive displacement can be evaluated.

Case Studies of Movements of Predators

Sharks

Don Nelson and his students from California State University, Long Beach, applied telemetry to studying shark behavior for many years. Sciarrotta and Nelson (1977) studied blue sharks off Catalina Island in southern California. They used ultrasonic telemetry to describe an interesting pattern of shark behavior. During the day in spring, sharks tended to remain offshore and relatively immobile. As evening approached, they moved rapidly to shore and foraged along the shoreline during the night, moving back offshore again the next day (Figure 16-3). Apparently at night, these sharks were foraging on squid that tended to be nearshore and were vulnerable to capture then. This pattern of onshore/offshore movement was predictable only at certain times of year. During the summer or fall, sharks simply milled around offshore (Figure 16-3d). At this time, squid schools were not common nearshore. In a mark-and-recapture study, diel patterns of behavior would not be easily detected. An individual would probably be marked and recaptured offshore because most researchers would sample during the day. Unless sampling was done at night, the sharks would appear to remain offshore. This demonstrates the differences in data that might be obtained from these two types of studies.

The details on shark behavior indicated regular onshore–offshore movements in spring. However, these sharks did not appear to use any limited home range typical of resource-tracking movements. They did not return to the same offshore site in the day and, observed over several

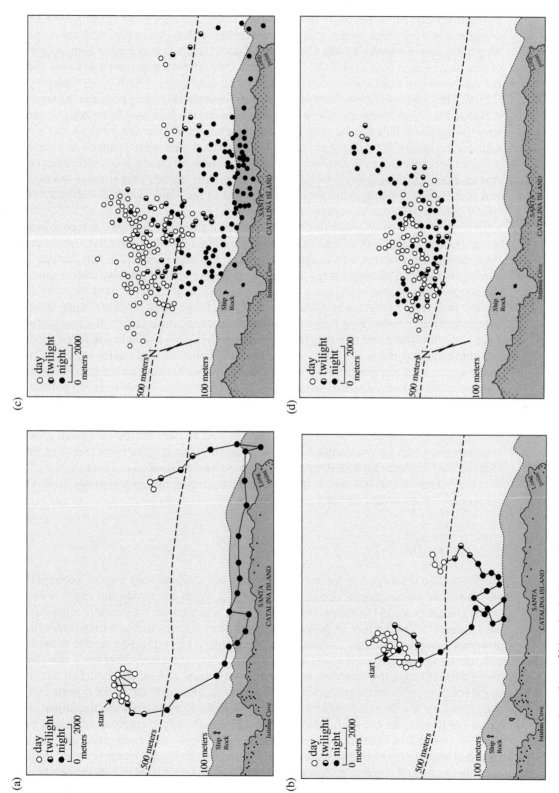

FIGURE 16-3 Locations of blue sharks during the day, twilight, or night: (a and b) one fish tracked during late March to June, (c) seven fish tracked in late March to June, and (d) three fish tracked from late June to October. *Source: Reproduced with permission from Sciarrotta and Nelson (1977)/National Oceanic and Atmospheric Administration.*

days, dispersed to different locations. This pattern was even less limited by space at other times of year when the fish appeared to remain more pelagic and even wider ranging. Blue sharks do not utilize limited home ranges and become familiar with habitat characteristics within those ranges but rather disperse widely. Carey and Scharold (1990) found similar results for behavior of blue sharks from New York to North Carolina. One interesting pattern that Carey and Scharold measured was regular excursions by sharks into deep water (100–400 m), especially during the day. They attributed these dives to foraging on vertically migrating prey and the return to the surface on the need to rewarm after a dive below the thermocline. Minimum body temperatures they measured for blue sharks were 14 °C, while minimum water temperature was 4 °C, indicating some inertia in the cooling of diving blue sharks, which are not capable of endothermy.

Contacts with sharks by telemetry required continual following. Since ultrasonic telemetry was used and ultrasonic transmitters can transmit their signals only through water, the fish had to be followed using hydrophones submerged in the water. These transmitters are commonly used in the ocean or in deep lakes because the alternative telemetry system (using radio signals) does not transmit well through salt water or from much depth. A recent innovation using ultrasonic telemetry in the Great Lakes was done by setting up receiver arrays at various important points within the lakes and then implanting large numbers of fish that are detected when they pass the receiver arrays. This Great Lakes Acoustic Telemetry Observation System (GLATOS) was initiated in 2012 and has been used to follow movements of thousands of individuals of at least 39 species over long time intervals (Krueger et al. 2018). Since the intent of this project was to follow long range movements over long time intervals, the transmitters used have a very low pulse rate and long life. This allows the arrays to determine when a fish passes but eliminates any chance to do on site measures of behavior while following individual fish. These arrays were not positioned to detect fine scale locations but rather passage by an array. However, use of buoy-based receivers in an array can replace labor-intensive pattern of regularly detecting location manually to evaluate activity pattern and swimming speed (Voegeli et al. 2001). These systems using three receivers can locate fish with accuracy up to 1 m and at intervals of one second, so swimming velocity and activity can be calculated much more precisely from such arrays. Coupled with use of depth transmitters or more receivers, the three-dimensional location can be estimated to obtain even better data on distance displaced for fish that may change depth and exist in much deeper water. Of course, use of arrays requires fish to remain within the array field for accurate detection.

Northern Pike

Northern pike are large top carnivores in inland lakes, where they may be ecologically similar to sharks. Northern pike do not tend to localize much day to day, but move rather extensively (Diana et al. 1977; Figure 16-4). They were followed with ultrasonic telemetry in Lac Ste. Anne (a 57,000-ha lake in Alberta). Maps of the locations indicate they prefer to remain nearshore, apparently in shallow water. For example, one fish tracked over two winter months (Figure 16-4d) was originally implanted in a small bay, remained there for a few days, then moved 16 km along the shoreline to another cove, turned around and returned to almost the same location again, and then swam up the shoreline another 8 km or so. It remained in that new area for a few days and finally ended up back in the same place it was implanted, at which point the transmitter ceased functioning. This behavior pattern emphasizes several potential problems related to mark-and-recapture studies. If the pike had been physically marked and nets had been set in the marking area for recapture, that fish would have returned to this same area three times. If it were recaptured during one of those times, it would appear the fish had not moved at all, while in reality, it had moved extensively. This is a bias of mark-and-recapture studies, especially if all other habitats were not sampled as intensively as the marking location.

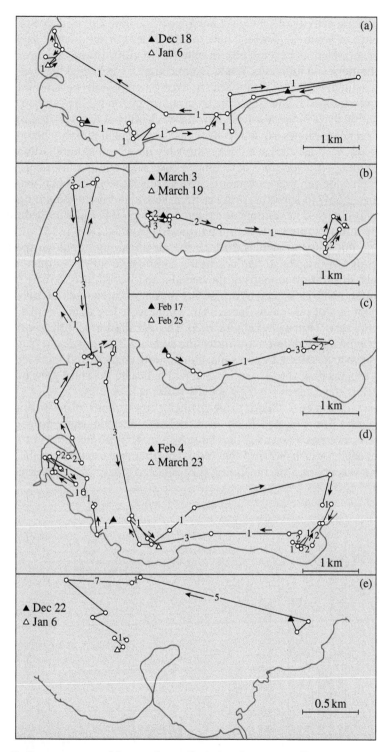

FIGURE 16-4 Daily movements of five northern pike in Lac Ste. Anne, Alberta. Categories a to e represent individual fish tracked over differing time periods. Numbers indicate number of nights between locations. *Source: Reproduced with permission from Diana et al. (1977)/American Fisheries Society.*

The wide-ranging activity of pike does not appear to show any evidence of a territory or limited home range. A home range would have to include tens to hundreds of square kilometers to encompass the movements of some pike in only two months. Home ranges of animals have questionable definitions and functions. While a home range is considered the area an animal traverses in its daily activities, that home range can be fixed geographically or can continue to expand as an animal moves. Limited home ranges are often shown by fish (such as stream dwelling trout in a pool), and within these ranges the fish know the availability of feeding sites, cover, and other needed habitat. As home ranges expand in size, this geographic familiarity becomes less likely. Wide-ranging fishes, such as sharks and pike, probably have limited familiarity with the areas they traverse and the specific locations of feeding or cover sites within those areas.

In spite of their wide-ranging movements, pike do select specific habitats to use. In Lac Ste. Anne, they occurred only in shoreline areas of the lake and did not often use open water. The same behavior was observed in summer and winter although it became more difficult to track fish for very long during summer. This limited tracking resulted from use of ultrasonic transmitters, which do not transmit signals through the air or through photosynthesizing plants. Pike often select densely vegetated habitats in the summer, so they are difficult to follow for long time periods. In addition to maps of movements and habitat use, activity of pike during summer and winter can be estimated by evaluating the distribution of different lengths of daily movements. About 30% of pike displacements in winter were less than 200 m a day, while 5% of them were more than 2 km a day (Figure 16-5). The distribution of different movement categories was not significantly different in winter and summer. The movement patterns of northern pike do not seem to be affected by seasonal temperature changes and show much variability in activity level. Studies described in earlier chapters indicated that pike grew during summer and winter, which is consistent with their activity patterns.

Habitat utilization can be evaluated using telemetry observations by measuring habitat characteristics each time a fish is located. Such measures might include depth, distance from shore, and bottom characteristics. Pike used depths less than 4 m; in fact, most of their summer locations were at depths less than 2 m (Table 16-1). This occurred even though the average depth of Lac Ste. Anne was about 5 m, indicating pike selected shallower water than the average

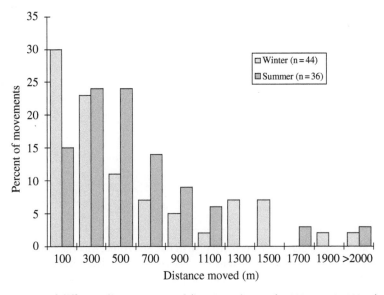

FIGURE 16-5   Percent of different distances moved (in 200-m intervals; 100 m = 0–199 m) on a daily basis for northern pike in Lac Ste. Anne, Alberta. *Source: Adapted from Diana et al. (1977).*

TABLE 16-1	The Frequency and Percent Occurrence of Various Habitat Characteristics for Northern Pike Locations in Lac Ste. Anne, Alberta. Data from Diana et al. (1977).

Characteristic		Summer	Winter
Depth	0–1.9 m	23 (52%)	4 (21%)
	2–3.9	19 (43%)	12 (63%)
	4+	2 (5%)	3 (16%)
Vegetation	Emergent	31 (49%)	2 (12%)
	Submergent	29 (46%)	0 (0%)
	None	3 (5%)	15 (88%)
Distance from shore	0–99 m	27 (40%)	21 (21%)
	100–299	26 (38%)	54 (53%)
	300–599	13 (19%)	23 (23%)
	600+	2 (3%)	3 (3%)

depths available. Similarly, pike selected emergent or submergent vegetation as preferred locations. Ninety-five percent of the locations in summer were in vegetation, and 78% of these locations were within 300 m of shore. It is difficult to compare summer and winter habitat selection for those characteristics. One difficulty is in the selection of shallow water; after a lake is covered with ice, fish cannot move into water shallower than average ice depth (1 m, in Lac Ste. Anne). Similarly, much emergent and submergent vegetation dies and is scoured from the lake during winter, so fish could be selecting exactly the same spot in summer and winter, but in summer the location could be considered submergent vegetation and in winter it might be considered open water because the vegetation would have died off.

Since data on pike were all collected by telemetry, there is some concern how comparable they are to data for average fish in the population. Gillnet catches are another means to measure activity of pike, and data from gillnet catches collaborate the pattern described by telemetry. For pike collected in three-hour gillnet sets, most were collected nearshore or offshore in areas with vegetation, while very few (only 6 out of 152) were collected in open water. These collections, using equal sampling efforts by net, gave parallel results to telemetry, which may indicate that the locations used were typical for pike.

A different sampling schedule is required to evaluate diel activity patterns and swimming speeds of fish. Locating a fish once every day does not indicate details about when or how much it was active. By sampling fish locations simultaneously from multiple stations at close range to the fish, and by locating the fish every few minutes, one could accurately measure displacement of the fish over short time periods. This could be done during the day and at night to estimate diel activity patterns. In order to locate a fish by roving telemetry, one must triangulate on it, which means measure the direction to the fish from several locations and take compass bearings on that location. The fish can then be located on a map at the point where those bearings cross. To do close range monitoring regularly requires multiple stations triangulating at the same time and pinpointing locations simultaneously. This design was used to evaluate diel patterns of activity (Diana 1980).

TABLE 16-2   Number and Percent Occurrence of Active or Inactive Periods During Five-Minute Observations of Pike Behavior at Various Times of Day. Data from Diana (1980).

Time	Active	Inactive
Summer		
Sunrise	19 (22%)	68 (78%)
Day	71 (17%)	351 (83%)
Sunset	16 (21%)	62 (79%)
Night	1 (2%)	63 (98%)
Total	107 (16%)	544 (84%)
Winter		
Sunrise	9 (23%)	30 (77%)
Day	34 (19%)	142 (81%)
Sunset	26 (35%)	49 (65%)
Night	1 (2%)	43 (98%)
Total	70 (21%)	264 (79%)

The concept of crepuscular foraging proposes that predators may use periods near dawn and dusk to selectively forage on their prey during intervals when prey are changing behavior from activity to cover seeking or are less capable to see predators during changing light levels. Northern pike were sometimes considered crepuscular predators in the literature, foraging predominantly at those times. To test that hypothesis, their activity patterns were determined over a 24-hour period. However, it is not clear swimming activity in pike correlates to foraging. Since pike are sit-and-wait predators, stationary behavior may indicate the time they are foraging, and when they are moving they may be searching for new foraging areas. Activity patterns can be evaluated to determine if there are any diel changes in activity that might indicate changes in foraging behavior, but the direct link to foraging cannot be made.

The most obvious expectation for pike would be they are mainly inactive at all times since they are sit-and-wait predators. Categorizing how far they swim is probably not as important as measuring whether they move at all over short time intervals. Their diel patterns of movement indicated that about 80% of the time they were inactive (Table 16-2). Pike were inactive most of the time and almost completely inactive at night. There was no evidence for crepuscular changes in activity and no real increase or reduction of activity at sunrise or sunset. That pattern was true summer or winter. Northern pike were diurnally active, nocturnally inactive, with no crepuscular changes in activity. There were no significant differences in frequency of activity in summer or winter.

Another analysis from close-range telemetry data can estimate swimming speeds and durations. Identifying the locations of a fish over short time periods can help a researcher estimate the distance moved over that time period. A pike may swim continuously in a straight line for the period, in which case this method gives a reasonable measure of swimming speed. The pike could also move erratically, neither in a straight line nor continuously, in which case the distance displaced would be a poor measure of swimming speed. The maximum distances moved may provide an indication of swimming speed when pike are active. For four different fish, the maximum velocity was 0.4–0.9 body lengths per second (BL/s), and the average velocity was 0.45 BL/s (Diana 1980). These measures correlate well with the swimming performance of pike in the lab; on a sustained basis, maximum velocity is about 1 BL/s, and average cruising velocity is believed to be 50% of the critical velocity. Both of these numbers agree with the behavior of pike in the field and could be used for activity measures in an energy budget.

Telemetry advances have automated many of the methods described for pike. Baktoft et al. (2012) from the Technical University of Denmark conducted a similar study of pike using automated telemetry to determine movement patterns and activity. They utilized an Acoustic Positional Telemetry (APT) system which enabled continuous automatic monitoring of fish position. They conducted the study in a 1-ha inland lake and utilized eight acoustic receivers to follow movements of 29 northern pike tracked over two years. In order for tracking to continue that long, the pulse interval for transmitters was 45 seconds. Over each 45-second period they could measure the distance displaced and sum them to get distances displaced per hour or day. Overall, they found that typical movements were between 600 and 1,200 m per day, summed up across the individual 45-second movements. They also found that pike were most active in diurnal periods and inactive at night. Overall, they were able to accumulate a large volume of data on pike movements but only in a very small system. The difficulty with automated systems like ATPs is that they require a number of receivers that can function only over a short-distance range, so the logistics of using them to continually monitor fish in all parts of a large system are prohibitive.

Bluefin Tuna

Bluefin tuna are large predatory fish that inhabit both the Atlantic and Pacific oceans. The fish grow greater than 300 cm in length and achieve masses more than 600 kg. Large bluefin tuna school or travel in groups rather than exhibiting solitary behavior as found in sharks or pike. They have been targeted for commercial fishing for a long time, and there is a large record of their catch and estimated population status. They are also interesting because they have been shown to retain heat from muscle activity and produce warm body temperatures above the ambient temperature of water. Advances in archival tags have dramatically improved the knowledge of bluefin tuna behavior.

Bluefin tuna were historically managed based on the belief that there were two distinct populations of tuna in the Atlantic: one that spawned in the Gulf of Mexico and resided west of 45° longitude and another that spawned in the Mediterranean Sea and resided east of 45° longitude. Both of these groups had distinct ages at maturation and isolated breeding grounds, indicating they were somewhat genetically isolated. However, there was believed to be population mixing at a low level, and this was supported by conventional tagging and mark-and-recapture experiments.

A massive telemetry study was begun on bluefin tuna in 1996 utilizing either implantable or pop-up archival tags. A total of 377 fish were implanted (Block et al. 2001). For fishes whose tags were recovered, geolocation could be used to estimate their location of sighting each day in the Atlantic Ocean. While geolocation has been shown to have an accuracy of ±40 km, given the size of the Atlantic Ocean, this is not a huge issue. In addition, recapture locations of either archival tags or the location of first encounter with pop-up tags could be used as a second indication of long-distance displacement. These locations can be pinpointed better by Global Positioning Systems (GPS). Altogether, data were recovered from 49 of 279 (18%) bluefin tuna with archival tags and 90% of the 98 tuna with pop-up tags. These data not only include site (by geolocation), but also depth, body temperature, ambient water temperature, and light conditions.

The behavior of tuna appeared to be much more variable than originally thought. In particular, tuna were shown to regularly make dives over 1,000 m in depth, although they generally occurred at depths less than 300 m during most tracking intervals. Such dramatic dives also altered their temperature occupation. Bluefin tuna that spawned in the Gulf of Mexico were regularly found in warm and shallow water, and the signature of this high water temperature could be used as an indicator of their residence in this spawning habitat. Tuna in the open Atlantic often moved within the Gulf Stream. Fish originally tagged on the eastern coast of North America spent much of their time around the 75° longitude and commonly did not transgress the 45° longitude boundary. However, out of 19 recovered bluefin tuna marked with archival tags, 3 moved across the boundary, all the way to the Mediterranean Sea. Recapture locations for fish with implanted archival transmitters largely occurred in areas with major commercial fisheries and reflected a similar response as the measurements from traditionally tagged fish. In comparison, the behavior of individual fish with transmitters demonstrated much more large-ranging behavior than would have been exposed by just recapture data. Clearly, the degree of intermixing between the two stocks of tuna was fairly large.

In a second study of Atlantic bluefin tuna, Lutcavage et al. (1999) tagged fish in the Gulf of Maine and looked at their behavior and potential spawning. All of these fish were of spawning size and would be expected to spawn every year. A total of 20 fish were marked with pop-up satellite tags, and 85% of these tags were recaptured. Based on temperature conditions, two individuals remained within the continental shelf area of New England at temperatures less than 16 °C, while 14 individuals traveled to the Gulf Stream waters and resided in temperatures of over 20 °C. The fish were originally tagged in September or October and followed for up to eight months, with five groups of tags released in time periods from late February through July. These bluefin tuna were believed to spawn in the Gulf of Mexico from late April to June. This had been confirmed not only by locations of adult tuna but also by the collection of larvae in these areas. Similarly, the eastern stock of tuna spawning in the Mediterranean was confirmed by adult and larval collections. There are historical accounts that bluefin tuna may spawn in the Azores and Canary Islands but larvae have not been collected from those sites. However, the data from Lutcavage et al. indicated that most of the tuna tracked did not return to the Gulf of Mexico to spawn as expected. All of the recapture sites for pop-up tags occurred in the mid-Atlantic region (Figure 16-6). This location was quite distant from the release site and also distant from the Gulf of Mexico. However, it had temperature conditions comparable to the Gulf of Mexico due to the

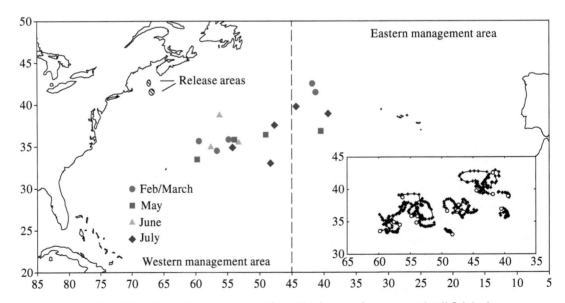

FIGURE 16-6 Map of locations where pop-up tags from bluefin tuna first appeared. All fish had tags attached from 24 September to 6 October 1997. *Source: Reproduced with permission from Lutcavage et al. (1999) / Canadian Science Publishing.*

presence of the Gulf Stream there. Temperature conditions in June, when spawning might have occurred, were 24–27 °C. While spawning could not be confirmed, Lutcavage et al. hypothesized that bluefin tuna may also show mid-Atlantic spawning in the region shown in Figure 16-6.

Use of modern telemetry equipment has dramatically improved our knowledge of bluefin tuna behavior. While physiologists have been interested in the endothermic capacity of tuna for some time, details on their behavior in the wild, their depth distributions, temperature distributions, and roving movements have not been easily collected. As geolocation techniques improve, it is likely that even more consistent data will be collected on the migrations of bluefin tuna. At this point, however, it is clear that some traditional beliefs used in management may be in error and are at least worthy of re-examination.

Summary

It is critical to understand fish movements and the classification or purpose, especially of long-distance movements and the role of movements in life histories, in order to maintain connectivity among critical habitats. Diadromous species make up just about 1% of all known fishes but have some of the highest rates of imperilment due to dams and impoundments disrupting their life cycles. Even within freshwaters, large potamodromous species have suffered visible consequences of flow reduction and impoundments. For example, razorback sucker and Colorado pikeminnow are some of the large potamodromous fishes of the southwestern United States facing risks of extinction from habitat fragmentation as are the giant Mekong catfish from Southeast Asia. Historically, movement needs of smaller stream fishes such as darters and minnows have not been factored into the design of culverts at road-stream crossings. More and more studies are revealing that upstream movements of these species and others are hindered by relatively small culverts during regular flow. The challenge for managers now is to understand stock structure and population genetic implications of such pervasive habitat fragmentation and inform engineers in the design of road and utility infrastructure to balance the needs of humans and fishes.

Literature Cited

Adams, N.S., J.W. Beeman, and J.H. Eiler, editors. 2014. Telemetry Techniques: A User Guide for Fisheries Research. American Fisheries Society, Bethesda, Washington.

Baktoft, H., K. Aarestrup, S. Berg, et al. 2012. Seasonal and diel effects on the activity of northern pike studied by high-resolution positional telemetry. Ecology of Freshwater Fish 21:386–394.

Block, B.A., H. Dewar, S.B. Blackwell, et al. 2001. Migratory movements, depth preferences, and thermal biology of Atlantic bluefin tuna. Science 293:1310–1314.

Carey, F.G, and J.V. Scharold. 1990. Movements of blue sharks (*Prionace glauca*) in depth and course. Marine Biology 106:329–342.

Diana, J.S. 1980. Diel activity pattern and swimming speed of northern pike (*Esox lucius*) in Lac Ste. Anne, Alberta. Canadian Journal of Fisheries and Aquatic Sciences 37:1454–1458.

Diana, J.S., and E.D. Lane. 1978. The movement and distribution of Paiute cutthroat trout, *Salmo clarki seleniris*, in Cottonwood Creek, California. Transactions of the American Fisheries Society 107:444–448.

Diana, J.S., W.C. Mackay, and M. Ehrman. 1977. Movements and habitat preference of northern pike (*Esox lucius*) in Lac Ste. Anne, Alberta. Transactions of the American Fisheries Society 106:560–565.

Everhart, W.H., and W.D. Youngs. 1981. Principles of Fishery Science. Cornell University Press, Ithaca, New York.

Fraser, D.F., J.F. Gilliam, M.J. Daley, A.N. Le, and G.T. Skalski. 2001. Explaining leptokurtic movement distributions: Intrapopulation variation in boldness and exploration. The American Naturalist 158:124–135.

Hubert, W.A. 1983. Passive capture techniques. Pages 95–122 *in* L.A. Neilsen and D.L. Johnson, editors. Fisheries Techniques. American Fisheries Society, Bethesda, Maryland.

Ivlev, V.S. 1961. Experimental Ecology of the Feeding of Fishes. Yale University Press, New Haven, Connecticut.

Krueger, C.C., C.M. Holbrook, T.R. Binder, et al. 2018. Acoustic telemetry observation systems: Challenges encountered and overcome in the Laurentian Great Lakes. Canadian Journal of Fisheries and Aquatic Sciences 75:1755–1763.

Linfield, R.S.J. 1985. An alternative concept to home range theory with respect to populations of cyprinids in major river systems. Journal of Fish Biology 27(Suppl A):187–196.

Lutcavage, M.E., R.W. Brill, G.B. Skomai, B.C. Chase, and P.W. Howey. 1999. Results of pop-up satellite tagging of spawning size class fish in the Gulf of Maine: Do North Atlantic bluefin tuna spawn in the mid-Atlantic? Canadian Journal of Fisheries and Aquatic Sciences 56:173–177.

Mittelbach, G.G. 2002. Fish foraging and habitat choice: A theoretical perspective. Pages 251–266 *in* J.B. Hart and J.D. Reynolds, editors. Handbook of Fish Biology and Fisheries, Vol. 1. Fish Biology. Blackwell Science Ltd., Oxford, UK.

Radinger, J., and C. Wolter. 2014. Patterns and predictors of fish dispersal in rivers. Fish and Fisheries 15:456–473.

Roberts, J.H., P.L. Angermeier, and E.M. Hallerman. 2016. Expensive dispersal of Roanoke logperch (*Percina rex*) inferred from genetic marker data. Ecology of Freshwater Fish 25:1–16.

Roberts, J.H., A.E. Rosenberger, B.W. Albanese, and P.L. Angermeier. 2008. Movement patterns of endangered Roanoke logperch (*Percina rex*). Ecology of Freshwater Fish 17:374–381.

Saastamoinen, M., G. Bocedi, J. Cote, et al. 2018. Genetics of dispersal. Biological Reviews 93:574–599.

Schlosser, I.J. 1991. Stream fish ecology: A landscape perspective. BioScience 41:704–712.

Schlosser, I.J. 1995. Critical landscape attributes that influence fish population dynamics in headwater streams. Hydrobiologia 303:71–81.

Sciarrotta, T.C., and D.R. Nelson. 1977. Diel behavior of the blue shark, *Prionace glauca*, near Santa Catalina Island, California. Fishery Bulletin 75:519–528.

Seelbach, P.W. 1987. Smolting success of hatchery-raised steelhead planted in a Michigan tributary of Lake Michigan. North American Journal of Fisheries Management 7:223–231.

Summerfelt, R.C., and E. Mosier. 1984. Transintestinal expulsion of surgically implanted transmitters by channel catfish. Transactions of the American Fisheries Society 113:760–766.

Thurow, R.F. 2016. Life histories of potamodromous fishes. Pages 29–54 *in* P. Morais and F. Daverat, editors. Introduction to Fish Migration. CRC Press, Boca Raton.

Voegeli, F.A., M.J. Smale, D.M. Webber, Y. Andrade, and R.K. O'Dor. 2001. Ultrasonic telemetry, tracking and automated monitoring technology for sharks. Environmental Biology 60:267–281.

Werner, E.E., and J.F. Gilliam. 1984. The ontogenetic niche and species interactions in size-structured populations. Annual Review of Ecology and Systematics 15:393–425.

Wydoski, R., and L. Emery. 1983. Tagging and marking. Pages 215–238 *in* L.A. Neilsen and D.L. Johnson, editors. Fisheries Techniques. American Fisheries Society, Bethesda, Maryland.

PART 4

Feeding and Predation

CHAPTER 17

Predation and Foraging Behavior

Some population processes, such as competition, critical period, and density dependence, were addressed in earlier chapters. The major factor influencing mortality of most fishes is predation. Since predation is such an important process structuring fish communities, both direct (related to the predator) and indirect (related to the prey) effects are important to a predator's success. The purpose of this chapter is to review the overall process of predation and to examine examples of different predator tactics to capture prey, as well as some prey responses to avoid being captured. Predation plays a very important role in population regulation of prey species.

Predation includes a suite of behavioral, individual, and population effects. Individuals may respond to changes in prey density by eating more of a particular type of prey. The predator population may respond directly by migrating to an area to feed or staying in an area longer, resulting in higher predator density. Since individuals in that area have more food to eat, they should have greater success in rearing young, and there should be more young in the area. These behavioral responses can be translated into population responses. Responses may occur over several years or very quickly as predators continue to move to other areas where there is more food. This chapter develops ideas on predatory interactions and mainly focuses on the foraging behavior of individual predators.

Components of Predation

C.S. Holling, an ecologist from the University of British Columbia, evaluated the complexity of behaviors involved in predation. Several behaviors are involved in predatory decisions, and these behaviors are often sequential. The sequence may include at least five components:

1. Patterns of search – whether animals actively search or wait motionless for prey.
2. Encounter of prey – being able to recognize a prey item. This may mean overcoming cryptic coloration, or being able to detect prey in an environment that might not be very clear visually.
3. Pursuit of prey – how a predator goes about attacking prey, whether it moves slowly to within a close range and bursts out, or it runs down prey.

Biology and Ecology of Fishes, Third Edition. James S. Diana and Tomas O. Höök.
© 2023 John Wiley & Sons Ltd. Published 2023 by John Wiley & Sons Ltd.

4. Capture – how frequently predators capture the prey they pursue.

5. Handling of prey – the time spent swallowing prey and possibly digesting it.

All of these are sequential events; the next behavior requires successful completion of the previous one. Handling time may limit the next search because, although the predator may encounter another prey item while handling a consumed item, it may not pursue it because of appetite or handling pause.

One of the more difficult problems in studying predation is that predator and prey might be in close proximity, but the predator may not be hungry or motivated to feed, and so the prey might not be attacked. Casual observations indicate there are many potential encounters (a predator being close to a prey item) that do not result in predation. While this may be an encounter because predator and prey were in close proximity, the predator did not make it a successful encounter. Spatial models alone may not predict predation rate from encounter proximity because prey detection, satiation, and digestion all influence the predator's decision. Such behavior is more common in piscivorous fish than in planktivores and is probably associated with elusiveness of the prey.

The components may occur in a fixed sequence in the predation act. For example, a barracuda may remain motionless and wait for prey to swim by it. Once a prey item is encountered and the barracuda decides to pursue, it assumes an S-shape, and then darts after the prey at close range. If it misses, it does not chase the prey, but will most likely resume a sit-and-wait posture until the same or another prey item approaches. This sequence of events is important in many behaviors, and if one event occurs out of sequence, the entire pattern may fail. The barracuda will not generally start in active pursuit again after an unsuccessful capture sequence. More active foragers may chase down their prey, and predation sequences may be less fixed for such predators.

Functional and Numerical Responses

Holling attempted to model how predators vary in their encounters of prey. He defined three important predation responses: functional, numerical, and total (Holling 1959a,b). A functional response is a response of an individual predator to increased prey density. As prey density increases, the predator may eat more of that particular species. This is not an evolved response but rather behavioral; it is a function of the predator's foraging decision. The numerical response is the change in number of predators in response to a change in density of prey. Predator populations may change in number by moving into an area, bearing more young, or dying less often in that area. The total response is a summation of these two parameters and is the effect of predation on the prey population. In terms of prey mortality, the number of individual predators and their relative consumption of prey create a certain mortality rate due to predation.

A number of the previous behavioral steps are involved in his analysis. To simulate predation behavior, he first did a series of experiments (called the disc experiments), in which he spread sandpaper discs on a table, and then blindfolded his secretary and asked her to pick up the discs using finger contact with the disc as her cue. When he timed this event, he observed that when more discs were on the table, the secretary encountered discs more rapidly, and her rate of picking up discs increased. At very high density, every time she placed her finger on the table, she contacted a disc, and the time involved in capturing discs was not spent finding them but rather picking them up and setting them aside (Figure 17-1). Therefore, disc removal rate became asymptotic at high density. In this experiment, the blindfold eliminated vision, which would dramatically increase removal rate. Since feeling with a finger was the detection mode, once a disc was felt, it was easy to recognize it and set it aside. Feeling with the finger was

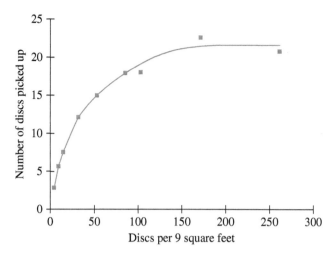

FIGURE 17-1   Functional response of a human picking up sandpaper discs by touch. *Source: Adapted from Holling (1959a).*

search, contacting the table or the disc with the finger was the encounter, and setting the disc aside was the handling. These experiments only involved three of Holling's processes: search, encounter, and handling.

The next step in the experiment was to alter the encounter. Instead of using a finger, the secretary was asked to use a pencil. Now, instead of using the direct sense of touch in her finger to recognize discs, she had to use the pencil, contact the table top, and determine whether a disc was contacted by feeling it through the pencil. This added the component of recognition. As one would expect, increased recognition time resulted in a considerably lower functional response curve. Recognition was not simply a matter of feeling with a finger, which has tactile receptors, but determining through the pencil by use of hearing or indirect touch that the pencil had contacted a disc or the table top. The functional response changed as complexity was increased in terms of these components although the shape of the curve remained largely the same.

Holling was interested in forest insects and how predators might control forest pests. He studied pine sawflies in nature as a test of his model. Sawfly cocoons develop on the forest floor, underneath leaf litter. Predators dig through the leaves and, upon finding a cocoon, break it open and eat it. To estimate predation rate, the number of cocoons present and the number that had been broken open could be counted each day, and then opened cocoons could be removed. Thus, one could compare the density of prey (cocoons present) to the number consumed (cocoons opened). Holling did this with experimental plots, where he sometimes enclosed areas and controlled the species of predator present.

There were three common species of predators in this forest system: *Peromyscus*, a deer mouse; *Blarina*, a shrew; and *Sorex*, another shrew species. Different predators may use different senses to encounter prey and, therefore, have different functional and numerical responses. Holling found that the functional response curves were generally similar to his disc curves. The curves differed by species of predator, particularly which predators were good at finding cocoons at low cocoon density. All three species of predator had functional responses so, as cocoon densities increased, the rates at which they ate them increased (Figure 17-2). For the numerical responses of the same three predators, the number of mammals per acre changed in relation to cocoon density. In other words, predators came into an area – or stayed there longer – when there were more prey. This was measured over a short time interval, so breeding rate probably did not increase density of predators. The predator that had the lowest functional response had the highest numerical response (Figure 17-2), while the animal that had the highest functional response had the lowest numerical response. This is not a general rule for all animals, but in this case, the two responses differed independently. One might expect the two responses to differ because a predator good at finding cocoons may not necessarily immigrate more rapidly to an area with high prey density.

The combined or total response is the percent of cocoons eaten by predators in an area, which was related to the number of predators present and their individual predation rate. This can be estimated by multiplying the number of predators, times the number of cocoons eaten per predator, and then dividing that by the number of cocoons present to estimate a percent of the prey population consumed (Figure 17-2). *Blarina* had an early functional response and no numerical response, so the total response was to consume a high proportion of cocoons

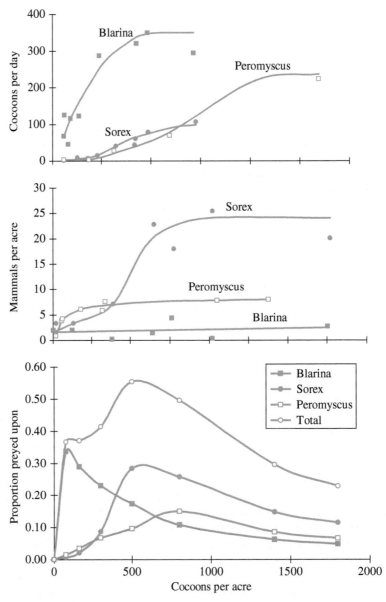

FIGURE 17-2 Functional (upper), numerical (middle), and total (lower) responses of three predators to sawfly cocoon density. *Source: Adapted from Holling (1959a).*

at low cocoon density. There was a large predation mortality (percent of prey consumed) at low cocoon densities, and the prey saturated the predator's ability to increase its functional or numerical response at high cocoon density, so even though more prey were consumed, the percent mortality of the prey population declined at high prey density. The three different species caused higher percent prey mortalities at different densities of cocoons, while the total effect for all three predators combined caused the highest mortality rate at an intermediate density of prey (Figure 17-2). These relationships were not fixed but depended on the functional and numerical responses. Interpreted in relation to forest pests, however, this indicated that the predators may control their prey at lower prey densities, while the prey may swamp the predators and be little affected by predation at high prey density.

Search Behavior

Holling's disc model, using search mechanisms as a means to alter functional responses, indicated that search behavior is an important predation component. Fish foraging behavior varies widely. A good example of this variance is a study by Janssen (1982) on bluegill and herring. Blueback herring are pelagic fish, similar to alewife, which forage in an open environment on zooplankton. Competition studies described earlier indicated that bluegill can forage successfully in vegetation, open water, or sediment. Janssen conducted experiments to evaluate the importance of prey movement on predator encounters. He used *Daphnia* (small zooplankters) and glued them to clear acetate slides. He either moved the slides in the water to mimic prey movement or held the slides stationary to determine whether the predators recognized the prey and whether prey movement had an influence on this recognition.

Types of Search Behavior

Janssen found that herring and bluegill foraged very differently. Herring swam through the water column, continuously searching for prey. Most pelagic fish are schooling, actively moving fish. They move continuously and use the relative movement to detect their prey. (For example, as you are driving, trees near the road may appear to be moving faster than the background. You might notice the trees because of their apparent motion and recognize closer trees because their positions changed in relation to the background.) When herring swim through the water column, they look for something that appears to move, but the prey appear to move because of the predator's actual movement. Janssen termed this behavior swim search. The movement of the prey is relative; it is actually the movement of the predator that causes the prey to be detected. Janssen found movement of prey did not matter at all in prey detection for herring; what mattered was that the predator moved. If the prey were being moved, they were captured equally well because herring still used their own movement to recognize these prey.

Bluegill, on the other hand, did not behave similarly. The bluegill swam to a location, and then remained motionless and looked around. They held a station for some time and, if prey were not detected, moved to the next location, remained there, and searched again. They used what Janssen termed hover search. When the bluegill were motionless and looking around, they searched for moving prey. They would not recognize a *Daphnia* in the water if it remained motionless, but if the *Daphnia* moved, bluegill would recognize the prey and capture it. In the case of bluegill, stationary prey were not detected effectively, whereas prey that were moved were easily detected. The choices for foraging by bluegill include how long to remain on station, how far to move between stations, and what to eat when something is detected.

One question still remains on the reasons bluegill use hover searches instead of swim searches. Hover searches may be much more successful due to the nature of the vegetated habitat in which they forage. A bluegill swimming through vegetation has little likelihood of detecting *Daphnia* by relative movement because all the vegetation interrupts its visual cues and distracts it. In a heterogeneous environment (vegetation or other structure), the movement of a prey item due to swimming of the predator is probably unimportant compared to background distractions. Bluegill use the tactic common of many hunters (fishes and humans alike) – to remain stationary and watch for prey to move. The herring strategy is probably best in the pelagic environment because there are few distractions and it is easier to move. In the littoral environment, where there are sediments, vegetation, and other distractions, swim searching probably would not function very effectively.

Foraging patterns of animals may also function to reduce predator contact as well as increase prey discovery. The foraging pattern of bluegill (hover search) may reduce their chances of discovery by vegetation-dwelling predators – such as pike and bass – which may use sit-and-wait tactics. Such predators rely on prey movement for detection, and the hovering by bluegill reduces their movement rate. Similar arguments could be made for the adaptive value of swim search and schooling by herring in avoiding detection by pelagic predators.

O'Brien and Evans (1991) also evaluated search patterns for planktivorous fishes and found many similarities to the results of Jansen. They described three search patterns – cruise, saltatory, and ambush search – like described above. In the case of five planktivorous fishes they evaluated, all used saltatory search (called hover search by Jansen) and even the patterns of movement, time spent still while watching for prey, and time spent swimming to new locations were fairly similar among all of these fishes. They believed that ambush search is mainly utilized by fishes consuming large prey, and while they believed cruise search could be used by planktivorous fishes, they did not evaluate any species using cruise search.

Many fishes, especially in the pelagic zone, use a variety of sensory cues for locating prey. Many long-range search patterns of pelagic fishes do not involve vision. For example, sharks can recognize the vibrations (sound) in the water caused by struggling fish from a long distance. Similarly, sharks can follow the odor of blood from a dead or dying animal that moves down current for miles in the water. The lateral line sense provides another series of cues sharks use to locate prey. The lateral line senses water movement, and many fishes can sense the presence of another fish by the movement of surrounding water caused by the other fish's swimming activity. Sharks have been shown to close their eyes prior to striking and accomplish the last bit of location through the lateral line sense or electrical sense rather than visually. Fishes transmit electrical signals into the water by muscle contractions. Many fishes can sense this electricity. Some can actually send out signals to communicate with each other on the basis of electrical stimuli. Electric eels can stun and kill fishes by sending out intense shocks through the water.

With all of these sensory systems at play, what is the relative importance of each system in ultimate capture of prey? Jayne Gardiner and colleagues from University of South Florida and Mote Marine Lab conducted an interesting experiment to test this out. They used a 17.5 m swimming flume and placed a tethered prey item at one end of the flume, and then used young sharks about 1 m in length with all senses intact or with one or more senses impaired and determined capture success and attack distance for each predator. They used blacktip shark, bonnethead, and nurse shark as predators, and pinfish or pink shrimp as prey. Senses tested included vision, electrical sense, and lateral line.

Gardiner et al. (2014) found that search and capture behavior varied with senses available and among the three shark species. Ram-feeders, like bonnethead and blacktip shark, detect their prey at long distances then swim rapidly to attack with their mouth opening while swimming. Suction feeders like nurse shark also detect prey at a distance but move close before attacking, when they open their mouth rapidly and suck in prey. Distant detection is commonly through olfaction, and if the sense is blocked, nurse sharks could not detect prey, while

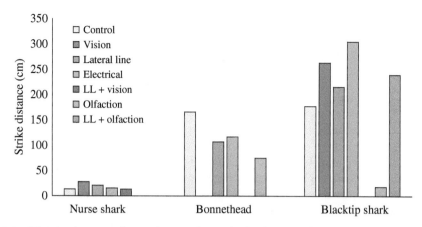

FIGURE 17-3   Distance between the predator and prey (cm) when a strike was initiated, for blacktip shark, bonnethead, and nurse shark. Treatments are for control (all senses intact) and single or multiple senses blocked. Treatments in which striking did not occur were omitted. LL, lateral line. *Source: Adapted from Gardiner et al. (2014).*

the other two species continued to swim through the arena until they detected prey visually at a much closer distance (1–2 m) and then attacked (Figure 17-3). Once prey were smelled, movement toward the prey was coordinated by lateral line sense to swim upstream into the weak current (called positive rheotaxis). This was accomplished by a series of turns, sensing conditions as the sharks turned to help pinpoint direction to the prey. Bonnethead and blacktip shark needed either vision or lateral line sense to complete prey tracking, while nurse shark could track prey without both of those senses. After tracking using olfactory or visual sense, bonnethead and black tip shark attacked using vision or lateral line; electrical sense alone was not sufficient to prompt a strike. These differences among species were linked by the authors to typical feeding behavior of each species, such as the lack of visual importance in nocturnal nurse shark attacking prey.

Capture Success

The fourth component of predation, which involves capture success, can be distinguished from search behavior mainly because it involves close-range interactions (the predator and the prey are within visual distance of each other). Capture success is dependent upon the tactics predators use to capture prey and the behaviors prey use to escape from predators. Planktonic prey have few options to escape capture after detection, so capture by planktivores depends on the feeding mechanisms of the predator, such as suction feeding or swimming capture. Piscivores have a very different problem in capturing prey as the prey are elusive and can avoid capture, so they develop swimming patterns in attack to better capture their elusive prey, given the constraints of their body forms.

Paul Webb and Janice Skadsen from the University of Michigan evaluated predation behavior in a 50 × 50 cm arena designed to film the interaction. There are problems with predator–prey encounters in small arenas. The behaviors observed may be constrained by arena size, so a fish that normally would actively pursue prey may be unable to do so. If a prey gets trapped in a corner and a predator attacks it, the prey cannot show natural escape behavior. Any time a wall or obstacle is part of the interaction, those results must be discarded. Webb and Skadsen (1980) studied tiger musky as predators. Tiger musky are artificially produced in a hatchery, hybridized between northern pike males and muskellunge females. Esocids in general (tiger musky, pike, and muskellunge) show a sit-and-wait (ambush) strategy of predation in which they remain immobile until prey approach closely. Fathead minnow were used as prey.

Predator Attacks

Tiger musky demonstrated two distinct behaviors in attacking prey, exemplified by the tracings in Figure 17-4 of the midline of fish over time intervals (in milliseconds). In a Pattern B strike, the fish coiled in place and then burst out from an S-start to attack the prey. For description, this will be called an S-start. The tracings in Figure 17-4 indicate the musky starts stretched out at about −200 ms and then coils into an S-shape without moving its head forward. It suddenly bursts out from the S-shape and attacks the prey. There was no forward movement during pre-strike behavior. This attack pattern is driven by white muscles and is anaerobic, and thus the number of ambush attacks may be limited by oxygen debt. The other interaction (Pattern A) was somewhat similar in that the fish coiled into an S but did so while moving forward (Figure 17-4). In this interaction, the predator started much farther from the prey than for a Pattern B attack. The distance traveled over time clearly separates these patterns (Figure 17-5). From a shorter distance, musky tended to use Pattern B strike, but there was some overlap. Higher distance traveled per unit time (Figure 17-5) means more rapid acceleration, and Pattern B attacks show more rapid acceleration than Pattern A attacks. The tiger musky initiated an encounter in almost every case, and the fathead minnow did not respond. The tiger musky either caught or failed to catch the prey based on whether the musky attacked properly. The minnow seldom responded, or if it did, the escape was initiated after the attack was over.

All combined, 57% of the attacks were Pattern A and 43% were Pattern B. The Pattern B attack had faster acceleration and faster velocity, at least over shorter distances; Pattern A, with

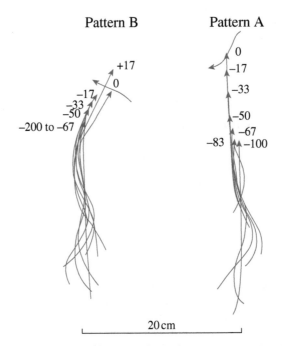

FIGURE 17-4 Tracings of the center line of a musky body during a Pattern A or Pattern B strike.
Source: Reproduced with permission from Webb and Skadsen (1980)/Canadian Science Publishing.

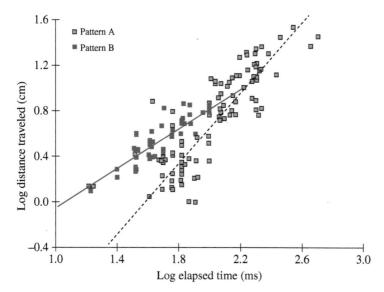

FIGURE 17-5   Relationship between distance traveled and time for Pattern A (open squares) or Pattern B (closed squares) attacks of musky on fathead minnow. *Source: Adapted from Webb and Skadsen (1980).*

lunging attack over longer distances, produced slower velocity. A fish using the S-start assumes a body configuration in which forces in the two lateral directions balance each other out; the force for accelerating during the strike causes fish to go in a straight line. Once the fish formed that S-shape and accelerated, it had no control or adjustment on where it went. If it accelerated correctly and struck the minnow, it was successful; if it accelerated toward the wrong location, it failed.

Only 15% of the minnows showed any kind of an escape response, and they used another stereotyped behavior for escape. In a typical interaction, a minnow would be attacked from the side (at about 90°). The minnow used a response termed a C-start, which is a reflex reaction. This reflex has been studied extensively and initiates in the spinal cord (Mauthner cell). Reflex reactions occur independent of any conscious thought and are responses to some stimulus detected through the central nervous system. With the C-start, the fish bends its body into a C-shape rather than an S. Its movement upon acceleration can be in almost any direction. A predator obviously needs to direct movement toward the prey. For successful escape, the prey's movement should be unpredictable. If prey always moved forward, the predator could anticipate this and aim at the head rather than the center of mass and have a much better chance of capturing prey. However, if the prey can move in any direction, the predator cannot anticipate the escape direction. A C-start allows this unpredictable escape method.

Predators most commonly attack from the side of their prey. A strike angle of 90° indicates the predator attacking prey directly from the side, 0° directly at the head, and 180° at the tail. Attacks from the side expose the largest length of the body to the attack. While 90° attacks were most common (Figure 17-6), predators attacked from almost any angle except from the tail (180°). For Pattern B attacks, there was a wide distribution of angles of attack, and predators were successful from virtually all attack angles (Figure 17-6). Pattern A attacks, on the other hand, were often unsuccessful. Pattern A attacks were most often made at about a 90° angle.

Strike distance affected strike pattern and capture success (Figure 17-7). Distances were fairly short for the Pattern B attacks (2–6 cm) and were almost all successful. There was a decline in capture success at farther distances, which might be expected because the predator

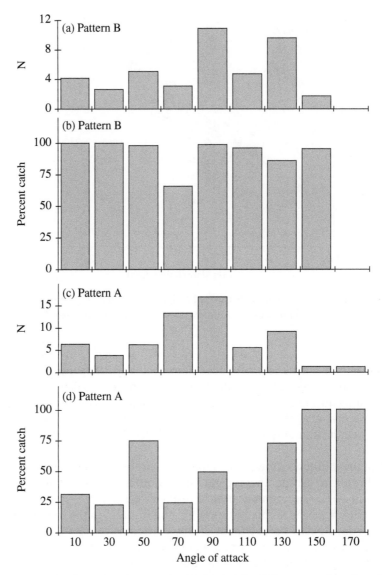

FIGURE 17-6 The number (upper) and percent catch (lower) for different attack angles of Pattern B (a, b) or Pattern A (c, d) attacks of musky on fathead minnow. *Source: Adapted from Webb and Skadsen (1980).*

could not adjust its angle; the farther the distance to the prey, the more likely it would be to miss. Pattern A attacks were all at farther distances and much less successful. There were very few Pattern A attacks at short distances.

Experience influenced the pattern of strike and the success of the predator. Both the percent of prey caught and the percent of Pattern B attacks increased with experience. By the time a predator had five or six attacks, 100% of the attacks were Pattern B, and they were 100% successful. By experience, the fish learned Pattern B attacks were much more successful. This was fairly rapid learning since there were only six or seven encounters. Catch success overall was about 73%, and the only misses involved were due to errors of the predator. The prey never responded rapidly enough to escape predation. Most of the time, they never responded at all, but if they did respond, it was after they were caught or missed.

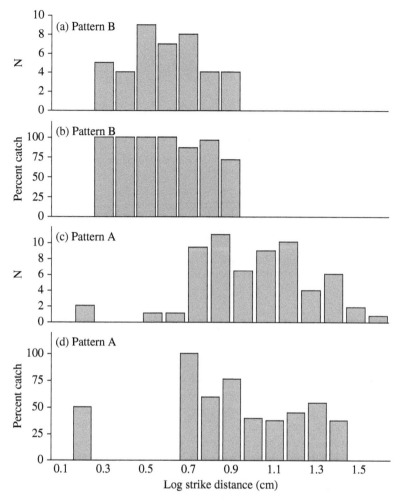

FIGURE 17-7   The number (upper) and percent catch (lower) of different strike distances of Pattern B (a, b) or Pattern A (c, d) attacks of musky on fathead minnow. *Source: Adapted from Webb and Skadsen (1980).*

Prey Escape Response

Webb (1982) further analyzed predator–prey encounters by examining a variety of predators and their attacks on fathead minnows in the same arena. He used four predators: tiger musky, smallmouth bass, rainbow trout, and rock bass. One purpose of this study was to evaluate which characteristics of the predator were detected by the prey for an escape response. Predator body form can be analyzed (or detected) by a variety of measurements. For example, stationary prey would observe a predator attacking from 90° in cross-section, not head to tail. The cross section may include the dorsal fin, pelvic fins, and pectoral fins protruding. Webb measured size of predator as the average of the height plus width to account for this detection. Since these four species have different body forms, there is little variation among individuals of the same species compared to variation among species.

Large predator size resulted in most of the prey responding to the attack, whereas less than 30% responded at small predator size (Figure 17-8). Very few prey responded to tiger musky, while quite a few responded to attacks by the other species. There appeared to be differences

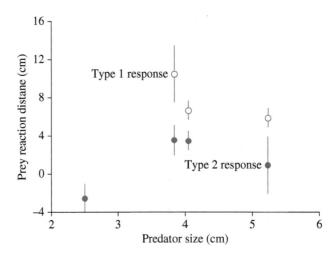

FIGURE 17-8 Prey reaction distance (mean ± 2 SE) for Type 1 (open circles) or Type 2 (closed circles) responses for predators of different cross-sectional sizes. *Source: Adapted from Webb and Skadsen (1980).*

among species in terms of the prey's ability to detect and respond to them. The crucial question is: What causes those differences, and what are the prey reacting to.

Prey demonstrated two kinds of escape maneuvers. If they detected a predator some distance away, they slowly swam away (Type 1 escape). If they detected a predator attacking at close range, they used C-starts (Type 2). Success of the predator and the elicitation of those two types of responses varied among species of predators (Table 17-1). Tiger musky exemplify one extreme for prey responses: 46 attacks were recorded, none resulted in a Type 1 response, and 13 resulted in a Type 2 response. Rainbow trout were the opposite extreme: out of 48 attacks, 18 elicited a Type 1 response, and 25 a Type 2 response. The reactions of prey, particularly Type 2 escapes by reflex action, should be initiated in response to some component of the predator's morphology during a strike. The velocities of the predators differed among species. The speed of a tiger musky in an attack at close range was 94 cm/s, while the velocity for the others was 40–50 cm/s. The tiger musky was more capable of generating

TABLE 17-1 The Number of Responses and Swimming Speeds (Cm/S) for Fathead Minnow Attacked by Four Predator Species in an Arena. Data from Webb (1982).

Species	Number of Strikes	Type 1 Response			Type 2 Response		
		Prey Speed	Predator Speed	Number of Responses	Prey Speed	Predator Speed	Number of Responses
Tiger musky	46	–	–	0	85.3	93.5	13
Rainbow trout	48	35.0	37.7	18	80.2	51.2	25
Smallmouth bass	67	35.8	32.3	22	82.1	52.7	32
Rock bass	51	26.1	12.6	21	61.0	47.5	17

high velocity in attacks. The prey response was about the same in reacting to most species of predator (about 85 cm/s), except the rock bass, which was the slowest of all predators and elicited the slowest response (61 cm/s). Prey responded similarly to the other three predators in speed when using a C-start (Table 17-1). Speed of the predator was considerably less for Type 1 responses (12–38 cm/s). Prey speeds in Type 1 escapes were also considerably less (26–36 cm/s).

Can data from such interactions indicate a detection threshold for prey in terms of the morphological or behavioral characteristics of the strike for each predator? If so, the escape response should initiate at the threshold value for that detection parameter regardless of predator species or size. Predator cross-sectional size at the time of prey reaction (see Figure 17-8) was not the cue since there were significant differences in reaction distance for each predator size. The size of the predator is not the only signal to which prey responded by reflex action because of differences among predator species.

The relationship between predator visual angle and prey reaction distance was also problematic. The visual angle is the angle of the maximum aspect of the predator's size (in most cases it is dorso-ventral) the predator projects on the head of the prey at the instant when the prey responds. A compressed fish with its fins erect (a rock bass, for example) would have a large angle at a given distance compared to a more rounded predator such as the tiger musky. Visual angle is dependent on the predator's dorso-ventral height and the distance from prey to predator. Obviously, visual angle at the time of response applies to prey that respond to an attack. Visual angle did not appear to be the threshold cue, and there was a linear relationship between visual angle and reaction time among the three predators; again, tiger musky fell outside the line.

One other possible cue would be the combination of characteristics, including how fast a predator is approaching prey and the visual angle or size. A moving predator changes its visual angle, and the closer it gets, the more rapidly the angle changes. This change in visual angle (for the predator in this case, but it equally occurs for detection of any moving object) is called the looming threshold. The looming threshold appeared to account for the reaction distance of minnow (Type 2 response, with no significant differences) to three predators, but again not to tiger musky. Looming threshold appears to have the best predictability for reaction distance cues, at least among the three less capable predators. There were still significant differences between the looming thresholds of musky and the other three predators detected and reacted to by fathead minnow.

One reason for the different prey reaction to tiger musky may be its body form. The morphology of the tiger musky is a very common one for top predators, such as gar and barracuda. Medial fins in an anterior area would make a predator appear much larger to prey near the predator's head than fins in posterior locations. An elongated form – rounded rather than compressed body – and posterior medial fins are fairly common characteristics of many top carnivores. Porter and Motta (2004) evaluated strike tactics for three fishes with similar body forms to the tiger musky: Florida gar, redfin needlefish, and great barracuda. They found that all three used S-start kinematics similar to musky but had slight differences in orientation and capture. The gar positioned its head to the side of the prey and then moved its head sideways with a burst to catch the prey in its beak. The barracuda and needlefish used a typical linear attack, with barracuda swallowing its prey on contact, while the needlefish manipulated the prey in its mouth until they were headfirst in the jaw and then swallowed them. This latter behavior is common among predators; musky and gar also show such behavior. This common fin placement and body form appear adaptive for capture of prey and for acceleration performance but not for sustained locomotor efficiency. The predator does not only beat its tail on acceleration from an S-start, but also its dorsal and anal fins. This predatory body form, then, may alter prey reaction distances compared to the size and fin placement for the other three predators.

The Importance of Fast Starts in Prey Escape

Studies by Webb and others base their analysis on the idea that prey that react faster in the formation of a C-start and in acceleration will be more successful in escape than those that react slower. While this is logical, it also omits other possibilities for escapes, such as increased prey vigilance and reaction earlier in the predatory encounter. Walker et al. (2005) evaluated such ideas in a study of pike cichlid foraging on guppy. Using a design and arena similar to Webb's, they found many similar results for predatory encounters. Predators used an S-start when attacking prey, prey responded with a C-start, predators generally could not adjust their attack after initiation of the S-start, and prey in this case often escaped the interaction. They found that three main factors affected the frequency of successful prey escapes: distance from the predator at initiation of the attack, velocity of the predator during the attack, and prey escape velocity. In this case, distance was the most important factor, and an increase in 1 standard deviation of distance increased the odds of survival for the guppy by 26-fold. Increased predator speed of 1 SD decreased odds of survival 11.5 times, while an increase in prey escape velocity of 1 SD increased odds of survival by 2.3 times. All of these factors were related; as distance increases, the opportunity for the predator to reach a faster velocity increases, and the opportunity for the prey to initiate an escape also increases. While their results were not able to support the apparent looming threshold as the cue responded to by the prey, both predator speed and predator distance are components of this threshold, indicating at least some support for that concept. Regardless of this cue, the study showed support for faster escape velocity improving the probability of survival although other factors were found to be even more important.

Evolution of predator escape mechanisms should include increased escape velocity for prey fish exposed to severe predatory stress. Such a response was demonstrated by Langerhans et al. (2004) for western mosquitofish. They studied populations of these fish from habitats with high densities of predators and ones from predator-free environs. In the former populations, they found fish with a larger caudal region, smaller head, more elongate body, and a posterior, ventral position of the eye relative to fish from predator-free populations. They also found that fish exposed to predators had 20% higher escape velocities when startled than fish from predator-free populations. Finally, they completed common garden experiments (fish grown under identical controlled conditions) to demonstrate that the differences in morphology were heritable in the different populations. This study provided clear evidence of the evolutionary response of mosquitofish to reduced survival in predator-rich locations.

Predator Avoidance

As a final point, prey escape behavior so far has been considered only after prey have been encountered. Obviously, with efficient predators, how to escape is a moot point to the prey since it may be too late to react when an encounter occurs. The other components of prey escape are to escape detection or make capture more difficult. Cryptic coloration to blend into the background is one means of escaping detection. The typical coloration of pelagic fishes is dark dorsally and silvery or white ventrally, to match natural light patterns when seen from above or below. Body coloration in fishes that dwell in vegetation may match the vegetation colors. A second kind of defense is warning coloration. Fishes with spines or other toxic components might not be cryptically colored but rather brightly colored. A predator that tries to swallow a brightly colored fish the first time and has difficulty may avoid similarly colored fishes later. Warning coloration, particularly for toxic or unpalatable prey, is common. This has even resulted in the evolution of mimicry, in which a palatable species mimics the color pattern of an unpalatable species. Coloration patterns, as well as the development of spines or unpalatable taste, are all defenses prey might evolve against predation. These defenses do not influence escape mechanisms but are intended to prevent detection or attack.

Summary

Foraging behavior includes searching patterns, prey encounter, pursuit of prey, capture tactics, and handling of captured prey. Predators seek to optimize such behaviors, while prey seek to detect and avoid them. Predators may respond to increasing prey density by selecting that species more strongly (functional response) or by increasing in abundance (numerical response). These combine to cause a predation mortality on prey that also varies with prey density. Foraging can vary from active swim search to more passive hover search or even to sit-and-wait tactics. Such tactics vary in response to habitat foraged in and to other predator characteristics. Close-range success of predators appears to be rapidly learned and involves making a predictable attack on prey while minimizing detection cues. Such cues detected by prey include predator size, visual angle, or looming threshold. Prey responses to attacks involve reflex actions (C-starts) or more gradual avoidance of predators. Additionally, characteristics to minimize detection of prey or to alter capture success of predators also influence these encounters.

Literature Cited

Holling, C.S. 1959a. The components of predation as revealed by a study of small-mammal predation of the European pine sawfly. The Canadian Entomologist 91:293–320.

Holling, C.S. 1959b. Some characteristics of simple types of predation and parasitism. The Canadian Entomologist 91:385–398.

Gardiner, J.M., J. Atema, R.E. Hueter, and P.J. Motta. 2014. Multisensory integration and behavioral plasticity in sharks from different ecological niches. PLoS One 9(4):e93036.

Janssen, J. 1982. Comparison of searching behavior for zooplankton in an obligate planktivore, blueback herring (*Alosa aestivalis*) and a facultative planktivore, bluegill (*Lepomis macrochirus*). Canadian Journal of Fisheries and Aquatic Sciences 39:1649–1654.

Langerhans, R.B., C.A. Layman, A.M. Shokrollahi, and T.J. Dewitt. 2004. Predator-driven phenotypic diversification in *Gambusia affinis*. Evolution 58:2305–2318.

O'Brien, W.J., and B.I. Evans. 1991. Saltatory search behavior in five species of planktivorous fish. Internationale Vereinigung für theoretische und angewandte Limnologie: Verhandlungen 24:2371–2376.

Porter, H.T., and P.J. Motta. 2004. A comparison of strike and prey capture kinematics of three species of piscivorous fishes: Florida gar (*Lepisosteus platyrhincus*), redfin needlefish (*Strongylura notata*), and great barracuda (*Sphyraena barracuda*). Marine Biology 145:989–1000.

Walker, J.A., C.K. Ghalambor, O.L. Griset, D.M. Kenney, and D.N. Reznick. 2005. Do faster starts increase the probability of evading predators? Functional Ecology 19:808–815.

Webb, P.W. 1982. Avoidance responses of fathead minnow to strikes by four teleost predators. Journal of Comparative Physiology 147A:371–378.

Webb, P.W., and J.M. Skadsen. 1980. Strike tactics of *Esox*. Canadian Journal of Zoology 58:1462–1469.

CHAPTER 18

Optimal Foraging and Patch Use

Predation has been evaluated from a behavioral point of view in previous chapters, paying particular attention to the physical encounter and its consequences. The role of prey choice has been largely ignored thus far, but in reality predators can be very choosy about the size and species they consume, as well as the means they use to encounter these prey. This chapter deals with the prey selection and search components of foraging. Foraging may be viewed in an anthropocentric way as the means by which an animal has to make a living, and there are many behavioral decisions to be made. Most ecologists are uncertain about the degree to which these decisions are based on learning or instinct, but it is obvious that fishes do make decisions on where to live and what kind of habitats to use. These decisions may vary with conditions, which indicates learning, as when bluegill move to vegetation while vegetative prey are abundant and then move out when prey are scarce. Foraging decisions must therefore be based on what prey are available to eat.

Another decision besides foraging for animals is how to avoid predators, which was mentioned in the previous chapter and is examined more in this chapter. Some prey behavior is instinctive, such as the reflex C-start escape response, while some is learned.

Returning to decisions on foraging, the question remains why animals eat what they do out of the wide variety of prey available. Holling (1959) evaluated this to some degree based on what was apparently taste, but prey also differ tremendously in energy content, proximate composition, ease to swallow and handle, elusiveness, and density. For years, ecologists have been interested in measuring, predicting, and understanding why animals choose certain types of foods.

Selection of Prey

Ivlev, a Russian scholar, wrote one of the first books on food selection and foraging. Ivlev conducted a variety of experiments in the lab by presenting to fishes different kinds of foods in different ways and measuring which prey they consumed. He controlled density of benthic invertebrates, patchiness, or prey size in the lab and looked at prey selection. Much of his work formed the basis for optimal foraging.

Biology and Ecology of Fishes, Third Edition. James S. Diana and Tomas O. Höök.
© 2023 John Wiley & Sons Ltd. Published 2023 by John Wiley & Sons Ltd.

Ivlev (1961) developed a widely used index of predator foraging selectivity. His electivity index (E) was calculated as $E = (R_i - P_i)/(R_i + P_i)$, where i is the type of prey, R_i is the relative abundance of that prey type in the diet, and P_i is its relative abundance in nature. An item very abundant in nature might have a P_i of 0.8 (80% of what is available in nature; for example, chironomid larvae in the benthos of a lake); R_i would indicate how frequently it appears in the diet. Assuming predators forage randomly and consume prey on the basis of abundance, 80% of the diet would be expected to be chironomids. Even though chironomids may comprise 80% of prey abundance, only about 20% of the diet of perch may be composed of chironomids, for an electivity index of −0.6. Therefore, the predators in this example are not selecting chironomids as preferred prey, but eating them less than they would by random chance alone. Electivity indices vary from −1 (completely avoided) to +1 (strongly favored). Ecologists have often changed calculation of this index somewhat, but the basic idea of selection still applies.

Electivity indices indicate what kinds of foods certain fishes favor; some of these decisions are very obvious. For example, although many invertebrates are available in a lake, a predator, such as an adult northern pike, might never eat one because pike are strongly piscivorous. On the other hand, the question of selection becomes less clear among species of similar organisms, such as northern pike eating bluegill, pumpkinseed, or green sunfish. It is easy to estimate the relative proportion of prey in the diet by sampling fishes and determining their stomach contents. It is much more difficult to define the relative abundance of prey in nature. While prey abundance can be estimated by sampling, species composition will change depending on the location or method of sampling. Even if an entire lake could be perfectly sampled, the abundance of organisms in the lake might be clear, but not necessarily the abundance of organisms encountered by a predator while it was feeding. A foraging fish can only be aware of prey abundance in its immediate vicinity. To define food abundance for a predator in a lake or stream, local abundance may be more important than global abundance. A northern pike remaining motionless in a macrophyte bed would not encounter or avoid preying on schooling minnows in the middle of the lake. It might not have a negative preference for minnows, but it simply may not encounter them. It is difficult to define abundance in nature because it is dependent on behavior of the predator as well as sampling. In reality, the important measure in terms of abundance in nature is the abundance of encounters.

Optimal Foraging

Tom Schoener attempted to solve the question of predicting prey consumption and selection by predators. He tried to relate selectivity to food availability and believed an animal's selectiveness might be largely dependent on abundance of food. A general observation is that feeding niches tend to broaden when food is less available, and they tend to narrow when food is readily available. The earlier work cited on competition demonstrated that food selectivity changed in relationship to food abundance in predictable ways.

Optimal Foraging Theory

Schoener (1971) described the means animals use to maximize their long-term energy gains under normal life constraints like predator risk. He defined two extreme patterns of foraging by animals: time minimizing and energy maximizing. The time-minimizing strategy is used by an animal that may be very vulnerable to predation. That animal would choose a certain time of day when predation risk was low, forage intensively during that time, and return to cover for the rest of the day. Another time-minimizing strategy might be used by an animal that lives in areas where temperatures are reasonable for foraging at certain times but not at others. Desert lizards, for example, may not forage in the middle of the day because of high temperatures. Their foraging is constrained to short time periods. For time minimizers, feeding should be adapted to accumulate food as rapidly as possible per unit time.

The second strategy, energy maximizing, is demonstrated by animals that do not face similar time constraints but forage based more on prey selection. These animals may forage to maximize energy intake. An obvious pattern would be to choose a large item instead of a small item – at least if the cost of handling is about the same – to yield a much higher energy return. Energy maximizers comprise the group in which optimal foraging is most commonly studied.

Several variables differ among prey items for a predator using the energy maximizer strategy. With experience, a predator may learn to expect certain energy returns or handling problems when eating a given type of prey. Each species or type of prey might have a different expected handling time and a different expected energy content. The expected energy content of a prey type can be predicted on the basis of size for most prey species. Both handling time and energy content are variables that can be learned by a predator for its common prey. If each is known, the benefit per unit cost or energy return per unit handling time might be what the predator can maximize by its choice of prey. Prey that provide much energy and take little time to process have a high net energy gain, while prey that yield little energy and take much time to process have a low net energy gain.

Schoener mathematically formalized these relationships into optimal foraging. Optimal foraging theorizes that animals can rank prey species (or sizes) based on expected energy returns and handling times and use these to determine the most favorable species (Figure 18-1). The highest ranked species would have a high energy return per unit handling time, while the lowest ranked species would have low energy return per unit handling time. This seems fairly clear, at least at the extremes.

A predator must know not only how much energy return it will get per unit handling time, but how many prey of that type exist. If a predator decided to eat only the species with the highest energy return per handling time, it might exclude 99% of the prey available. Its total energy return per unit time could be very low. If the predator increased its diet breadth to accept the second-ranked prey species its total energy intake per unit time could increase. This could proceed to lower and lower ranked prey and might still cause an increase in the total energy per unit time (Figure 18-2). However, at some point the low energy return per unit handling

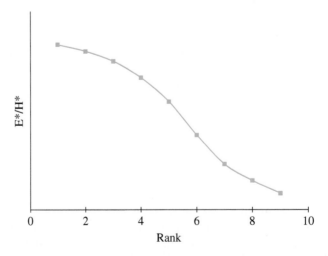

FIGURE 18-1 Hypothetical relationship between expected energy return (E*) per expected handling time (H*) for prey of various ranks. *Source: Adapted from Schoener (1971).*

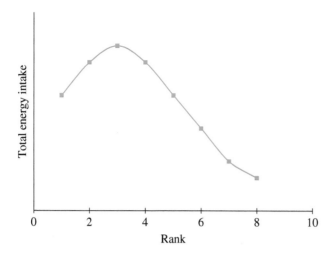

FIGURE 18-2 Hypothetical changes in total energy intake by consuming prey of different ranks (from 1 to n based on highest energy return per unit handling time) from Figure 18-1. *Source: Adapted from Schoener (1971).*

time might not balance off what could be gained if that animal ignored eating such low ranked items and waited for a more highly ranked item. The time this predator spends foraging on low ranked items may actually reduce its ability to consume higher ranked items, and the total energy per unit time would decline.

This optimal diet model can be extrapolated to some rules for optimal foragers. The first is that animals should select – within their optimal sets – foods that result in a higher total energy intake per unit time and ignore those that cause reductions in total energy intake per unit time. This is exemplified in Figure 18-2, where prey up to rank 3 should be consumed. It may be difficult for an animal to sort out precisely which sizes or species to include in the optimal set (for example, the difference between prey ranks 2 and 4), particularly depending on how steep the curve is, but it becomes fairly clear that particular ones should be selected or avoided (prey rank 8, for example).

There are many assumptions in Schoener's optimal diet model. One assumption is that a predator can instantaneously recognize individual species of prey. In other words, a predator can observe prey and decide what species and rank it is without expending energy or time because the only time being accounted for in this foraging model is search and handling time. The second assumption is that handling time is a good indicator of the energy costs of handling. Another assumption is that the predator has good prior knowledge of its environment, meaning it has tried different prey and determined handling difficulties. The predator must also have some idea of the density of prey in nature because these decisions are based not only on individual returns but on abundance of prey species. Many assumptions are involved in trying to simplify the feeding act for optimal foraging.

Schoener further used his model to predict several outcomes of foraging that could be tested to corroborate the theory of optimal foraging. One prediction is if a prey item is in the optimal set (it increases energy return per unit time of a predator), the predator should always try to eat it when it is encountered. Second, a predator should never pursue an item outside of the optimal set. Obviously, this trade-off is difficult because the predator would have no understanding of the expected energy return per unit handling time unless it had pursued novel items. So, in fact, novel items might be treated differently than common items. This may be important to systems where non-native species become common, and predators are not experienced with these prey. For example, lake trout in the Great Lakes did not start to consume round goby for a number of years after the goby became abundant due to their novel status. Once prey have been consumed and can be ranked, predators should not pursue prey items that fall outside the optimal set. These decisions should vary over time as prey size distribution and abundance change seasonally.

The first of three testable extrapolations of this model is that an increase in the density of a highly ranked prey item should shift the entire curve to the left, and the optimal set would narrow. The second prediction is that an increase in density of an item outside the optimal set should have no effect on diet breadth because these prey are not being consumed. The final prediction is that a decrease in abundance of an item within the optimal set should shift the curve to the right and result in a wider optimal set. These predictions are made for changes in prey density. The relative abundance of each species may not have to change. In other words, all food may become more abundant – food within or outside of the optimal set. In this case, the food niche should narrow, even though relative abundance did not change.

Optimal Foraging in Bluegill

Earl Werner and Don Hall became interested in one question regarding foraging: can optimal foraging be used to predict the sizes of prey eaten by bluegill? To test this, they put bluegill into tanks and fed them *Daphnia* (a small zooplankter) of different sizes and abundances reasonably similar to natural *Daphnia* populations (Werner and Hall 1974). Their test of foraging did not evaluate different species of prey, but how bluegill selected sizes of prey of the same species. This goes back to the question from Jansen's work on foraging tactics: once a fish has encountered prey, how does the predator decide whether or not to eat it? Jansen proposed that predators eat the first thing they see, and O'Brien predicted the decision is based on apparent size, while Werner and Hall predicted it is based on maximizing energy return, which is the subject of their study. The experimental design was to use three size classes of *Daphnia* to change the overall density of prey and look at the relative consumption of each size class by bluegill.

The three sizes classes of *Daphnia* were about equally abundant in bluegill diets at low density of prey; in other words, bluegill ate everything encountered (Figure 18-3). The diet shifted as prey abundance was increased, so none of the smallest size of prey class were eaten; only the middle and large size class were eaten at intermediate prey density. The largest size group of prey was mainly eaten as prey abundance was increased further. The breadth of diet declined for bluegill as prey abundance increased, eventually to where bluegill exclusively fed on the largest prey items. It takes little time for a bluegill to swallow any of these size classes of *Daphnia*, so handling time is about the same for all prey sizes. Energy return differs depending on size of prey eaten, and therefore, energy return per unit handling time is directly correlated to prey size. Bluegill diet selection at high prey density was for the highest ranked prey only. Bluegill appeared to behave as predicted by the optimal diet model, that is, when density of food within their optimal diet increased, their diet breadth narrowed.

An interesting component of Werner and Hall's study was to evaluate how these decisions on prey selection might be made. They compared the time bluegill spent in search at different prey densities to the kind of selection pattern bluegill used as a means to elucidate these foraging decisions (Figure 18-4). It took an average of 400 seconds to encounter 10 large *Daphnia* when food was very low in abundance. The bluegill would find a large *Daphnia*, and then they would find another 40 seconds later (on average). With this much time between individual encounters, bluegill ate everything they saw. Search time declined as prey density increased. Selection of only the two largest size groups occurred when search times were between 20 and about 50 seconds to encounter 10 large *Daphnia* (Figure 18-4). As bluegill encountered prey more regularly, they became more selective and chose items of higher caloric content. There was essentially no search time at high density; when a bluegill captured one item, it could

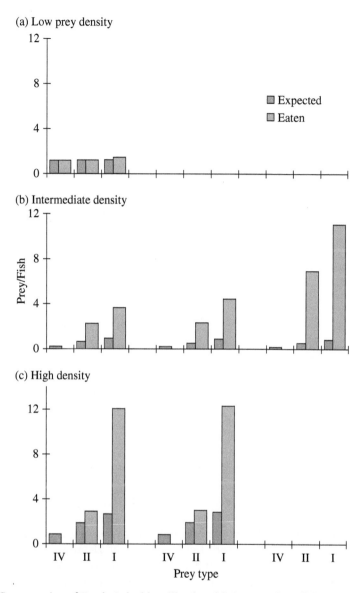

FIGURE 18-3 Consumption of *Daphnia* by bluegill at low (a), intermediate ((b) 50, 75, and 250 prey per class, respectively), or high ((c) 300 or 350 prey per class) density of prey. Prey types are different sizes of *Daphnia*: small (class IV), mid (class II), and large (class I). Dark bars represent expected diet based on abundance; light bars represent actual diet. Each set of three pairs of bars represents experiments at one density of prey. *Source: Adapted from Werner and Hall (1974).*

probably see several other prey simultaneously. Under that abundance, the bluegill became extreme specialists. The prediction of Werner and Hall is that time between prey encounters might be a reasonable measure of prey density and might be the variable that bluegill use to decide how selective to be. The more frequently bluegill encounter food, the more selective they are. This study supports the concept of optimal foraging and confirms that the selection pattern of predators might differ depending on encounter rates correlated to prey density. Therefore, in a patchy environment, where there is high density in some areas and low density in others, each individual might show different patterns of selection depending

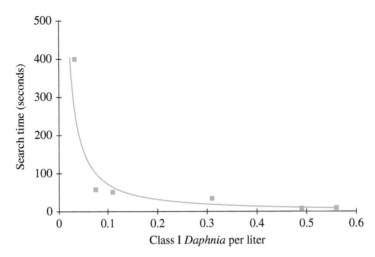

FIGURE 18-4   Average search time to encounter 10 large prey at each *Daphnia* density. *Source: Adapted from Werner and Hall (1974).*

on its history of encounters over recent time. This would depend on whether the individual was in or out of a patch. Finally, this study appears to be evidence for maximizing energy intake as the means of prey selection. Apparent size differences, based on proximity to the bluegill, would result in a less distinct diet at high prey density. However, since the prey were all the same species, differences in visibility (which affect apparent size) were correlated only to size, not species. Finally, this study negates Jansen's idea of first-seen, first-eaten, although that pattern did apply at low prey density.

Is Handling Time the Most Important Foraging Criterion?

The optimal diet model and empirical evidence suggest that fish should optimize their energy returns by maximizing energy consumed per unit handling time and utilize prey that provide the highest ratios. However, handling time is not the only process involved in consuming and digesting fish. Lankford and Targett from the University of Delaware evaluated foraging of weakfish and related their foraging to energy maximization. Weakfish are a temperate sciaenid that inhabit estuarine nursery areas in the US Atlantic coast in spring and summer. Many marine fishes have juvenile stages that reside in estuarine nursery areas and then move as adults to offshore areas to live and spawn. Lankford and Targett (1997) focused on juvenile foraging because they believed juveniles should have the highest growth rate and the highest survival advantage to being a selective forager. Weakfish commonly consume a variety of invertebrate prey, and in Delaware Bay, they focus on *Neomysis americana* and *Crangon septemspinosa*. *Neomysis*, a mysid shrimp, is a pelagic organism, while sand shrimp (*Crangon*) is more benthic in behavior. Both species differ dramatically in behavior and in a number of characteristics that could influence their advantage to a foraging predator. Therefore, Lankford and Targett attempted to evaluate the selection of prey by weakfish and relate that to energy maximization through a series of controlled experiments.

The first experiment involved enclosures to keep the predators and prey in proximity in tank systems. The enclosures were net with mesh sizes to allow water to pass through, as well as a thin layer of sand on the enclosure bottom, which retained the prey items within the enclosure but kept them exposed to any predators there. In the first experiment, they varied relative abundance of the two species of prey and looked at preference for prey by weakfish.

| | **TABLE 18-1** | Effect of Relative Abundance of Prey (Ratio of *Neomysis*: *Crangon*) on Prey Consumption (Mean ± SE) and Prey Selectivity by Weakfish in Laboratory Trials. Selectivity Index = Strauss Index (Varies From −1 To 1). |

| | **Number Consumed** | | |
Relative Abundance	**Neomysis**	**Crangon**	**Selectivity Index**
6:1	21.8 (2.0)	0.1 (0.1)	+0.138
4.5:1	20.4 (3.6)	0.4 (0.3)	+0.164
3:1	11.5 (1.6)	0.4 (0.3)	+0.216
1.5:1	9.1 (1.0)	0.9 (0.3)	+0.307
0.5:1	1.8 (0.6)	0.6 (0.3)	+0.375

Source: From Lankford and Targett (1997)/John Wiley & Sons.

Prey abundance varied from 6 *Neomysis* to 1 *Crangon* in the high treatment, to 0.5 *Neomysis* to 1 *Crangon* in the low treatment. The result of this experiment (Table 18-1) indicated that weakfish preferred *Neomysis* to *Crangon* under all conditions. *Neomysis* were always eaten at a higher rate than *Crangon*, even when they were much lower in abundance, and the relative preference for *Neomysis* was comparable under all treatments.

In order to compare this result with preference behavior of weakfish, the authors decided to present individual weakfish with both *Neomysis* and *Crangon* in an arena and determine which of the two was selected by the predator. Ten fish were used, with 8–10 trials per fish. This resulted in a frequency of 90 completed trials, of which 75 resulted in consumption of *Neomysis* and only 15 in consumption of *Crangon*. Therefore, the preference shown by changes in relative density and consumption of the two prey species were also paralleled by selection by the predator, when given the choice of one or the other prey type in opposition. These two experiments indicate that juvenile weakfish have a strong preference for *Neomysis*, and the authors set out to evaluate what characteristics of *Neomysis* cause this positive selection.

Their first analysis was to determine energy content and handling times for both species, in an attempt to see which should be preferred under optimal foraging theory. They measured these parameters in a variety of ways, including energy, protein, and ash content. They then estimated the profitability for the two prey species and the benefit–cost ratio based on energy return per unit handling time. *Neomysis* was significantly superior over *Crangon* in energy content per gram dry mass. However, *Crangon* was generally much larger than *Neomysis*, and so the energy content per individual was considerably higher for *Crangon*. *Crangon* had a much larger and stouter exoskeleton, while *Neomysis* had a much less rigid exoskeleton. This, in part, caused the higher energy density of *Neomysis*. An individual weakfish consuming 1 *Crangon* would receive 7.5 times more energy than the same individual consuming 1 *Neomysis*. Handling time for *Crangon* was 10.7 seconds, compared to 2.1 seconds for *Neomysis*, and optimal foraging should strongly favor consumption of *Crangon* over *Neomysis* with an energy return per handling time of 21 joules per second for *Crangon* and 14 joules per second for *Neomysis*. These results predicted that *Crangon* should be the highest ranked item in the optimal diet. However, *Neomysis* was the preferred diet item, indicating that either other characteristics of the prey were important or weakfish were not foraging optimally.

The authors conducted a series of sequential experiments, evaluating other characteristics of the two prey species. In particular, they compared short-term feeding rate, gastric evacuation rate, total energy intake rate, growth rate, and growth efficiency, as well as assimilation efficiency for weakfish foraging on either prey species exclusively. When allowed to forage

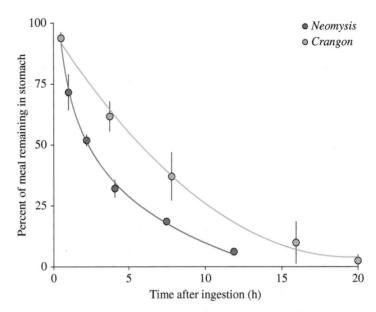

FIGURE 18-5 Gastric evacuation rates of weakfish after consuming 5% BW of *Neomysis* or *Crangon*. Mean + SE for 4–6 individuals sampled at each time interval. *Source: Adapted from Lankford and Targett (1997).*

exclusively on one species, weakfish consumed considerably more biomass feeding on *Neomysis* (855 joules per gram body mass) than on *Crangon* (468 joules per gram body mass). These higher short-term feeding rates reflected differences in the prey morphology with *Neomysis* easier to swallow and pack into the gut than *Crangon*. They could also reflect differences in gastric evacuation rate of one prey species compared to the other.

The gastric evacuation rate and pattern of evacuation for *Neomysis* was significantly different than for *Crangon* (Figure 18-5). When fed *Neomysis*, weakfish had an exponential evacuation pattern, and 50% gastric evacuation occurred in about three hours. When fed *Crangon*, evacuation pattern was more linear, and 50% evacuation did not occur until six hours after consumption. Thus, the rate of food passage for *Crangon* was about one-half the rate of passage for *Neomysis*. These post consumptive differences in digestion time predicted the pattern of prey consumption by weakfish better than optimal foraging and showed that gastric evacuation differences may be important in energy maximization. In this regard, weakfish consuming *Neomysis* could evacuate about 220 joules into the intestine during five hours after ingestion, while they could only evacuate 121 joules during the same time period when fed *Crangon*.

As a final comparative technique, Lankford and Targett evaluated the consumption and growth rate of weakfish fed either of the two prey species. Weakfish fed *Neomysis* had significantly higher growth rates, feeding rates, and daily energy intakes than those preying on *Crangon*. For growth rate, the mean was 1.6 g·d⁻¹ for fish feeding on *Neomysis* compared to only 0.37 g·d⁻¹ for fish feeding on *Crangon*. Assimilation efficiency was not largely different between the two types of prey and averaged 60%.

In the end, the authors used the latter evidence as reasons to consider other characteristics of prey in optimal foraging studies. They believed that post-consumptive handling was an important criterion when comparing prey of different species. This probably does not cause a difference when evaluating at prey of different sizes, but the same species, as was done in the earlier experiment by Werner and Hall (1974). However, in the case of weakfish, the difficulty in digesting made a large difference in predictions for energy maximizing. Of course, this pattern was only true if consumption of *Neomysis* was high enough that consumption of *Crangon* would produce a lower total consumption rate than feeding on *Neomysis* alone.

The authors used evidence from the field to indicate that weakfish often fed exclusively on *Neomysis* and their feeding rates were very high, indicating that *Neomysis* was readily available. However, they also admitted that fish fed upon *Crangon* in the field at times, indicating that density and consumption of *Neomysis* might have been limited under those conditions. One other characteristic that might make a big difference in these consumption patterns is the difficulties in capture success. *Neomysis* is much easier to capture than *Crangon*, and therefore, search-and-capture times would also differ significantly between the two.

Overall, this study cast some doubt on the use of handling time alone as a measure of the energetic cost of prey consumption. In particular, this study shows that extending handling time to include handling after consumption is also important. However, the authors were able to demonstrate that weakfish foraged optimally, in the sense that they selected prey that resulted in the highest energy return per day and the highest growth rates. Expanding optimal foraging to include other characteristics of prey is an important way to better understand the pattern of food consumption in fish with more varied diets.

Does Capture Success Influence Prey Consumption?

While handling time and digestion rate may help explain the prey selection of some species, for predators consuming elusive prey their capture efficiency may also be important. Scharf et al. (2009) assessed the ontogeny of prey consumption for bluefish and striped bass in an attempt to determine the role of predator behavior in the differential consumption of fish prey. They used both field collected data and lab experiments to evaluate feeding performance. Instead of energy return per handling time, they defined profitability as

$$\left[ED_{prey} \times \left(M_{prey}/M_{pred}\right) \times C\right]\Big/H \qquad \text{(Equation 18-1)}$$

where ED_{prey} is the energy density of prey, M is the mass of the predator or prey, C is capture success (%), and H is handling time. A number of studies have shown that fish growth rates increase dramatically when they make the ontogenetic shift from invertebrate to fish prey, so the authors attempted to explain why bluefish shifted more completely and earlier than striped bass in a similar environment.

Their study included predator attack behavior, and they found some differences from the behaviors described in Chapter 17. Both species foraged more actively, while swimming in schools and cruised at about 30–40 cm/s, and then assumed an S-shape for attack of prey. This S-shape generated much higher attack velocities, with bluefish reaching 150–225 cm/s and striped bass 120–160 cm/s. Bluefish were often 2–3 times more likely to capture prey fish than striped bass. The two species had similar handling times when prey were small, but striped bass were much slower at handling larger prey than bluefish. Overall, the profitability of prey for bluefish was considerably higher than for striped bass, and the highest value occurred at a larger prey size than striped bass (Figure 18-6).

The comparative behavior of the predators was quite different. Both had gape sizes that would allow them to consume much larger prey than they actually ate. But capture success was reduced at higher prey sizes, so they mainly consumed smaller prey. The higher capture success of bluefish at all prey sizes resulted in their shifting to fish prey early in their first year, while striped bass did not completely shift for more than a year. Both species grew faster when held in tanks with just fish prey, and bluefish grew much faster than striped bass under this condition. Overall, capture success for these predators was the most important criterion for the prey species they selected, and while this was not a test of the optimal diet model as such, the profitability of different sizes and species of prey varied dramatically based mainly on capture success.

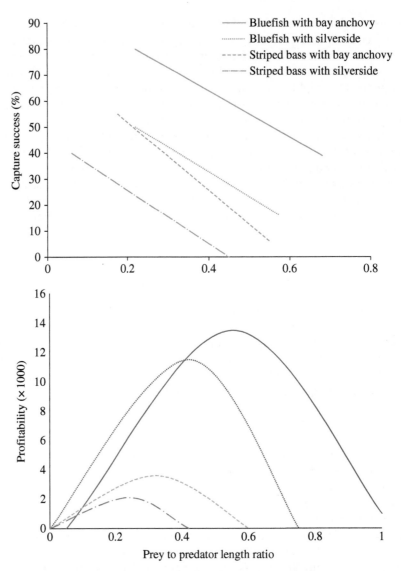

FIGURE 18-6 (Upper graph) Capture success and (lower graph) profitability of bluefish and striped bass when feeding on bay anchovy and Atlantic silversides. *Source: Adapted from Scharf et al. (2009).*

Importance of Prey Defenses and Refuges

Another study by Roy Stein from Ohio State University evaluated prey mechanisms for escape. Stein (1977) examined smallmouth bass foraging on crayfish. This is a very different interaction than between bluegill and *Daphnia* or between weakfish and shrimp. No matter how large *Daphnia* are, they can be swallowed easily by bluegill, so *Daphnia* cannot escape bluegill predation by growing to a size that bluegill cannot consume. Crayfish, on the other hand, have big claws and reach a very large size. Individuals can be difficult to swallow with their hard exoskeleton, claws, and large size.

Stein put smallmouth bass into different types of habitat with different size classes of crayfish and then evaluated the disappearance of crayfish due to predation. Sand and pebbles were

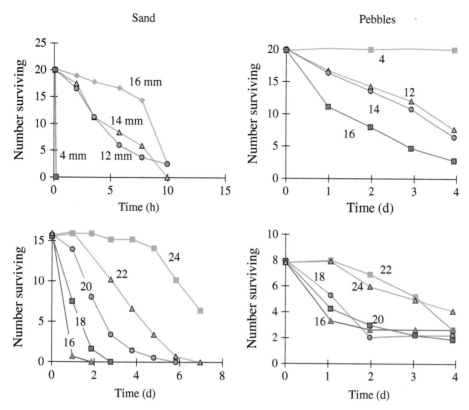

FIGURE 18-7 Disappearance of different size classes of crayfish due to predation by smallmouth bass in sand (left side) or pebble (right side) substrates. *Source: Adapted from Stein (1977).*

the two types of substrate he used. In the sand habitat, the smallest crayfish were eaten first, while the larger ones were eaten later; there was a clear progression to eat small ones then move up in size (Figure 18-7). This is the opposite of what one might predict from optimal foraging because net energy return would be highest when bass ate the largest prey. Of course, in this experiment, abundance declined over time, because once bass ate prey, they were not replaced. In the pebble habitat, pebbles were used as a refuge by the smallest crayfish, which could crawl underneath pebbles to hide. The 4-mm crayfish could not be eaten because they could hide in that refuge; 12- and 14-mm crayfish could hide somewhat using pebble refuge less effectively, while 16-mm prey could not hide there. The foraging pattern in that habitat was to eat the smallest visible prey and then work up the scale.

This pattern of selecting the smallest individuals first was contrary to optimal foraging predictions. Reasons for this selection are obvious when comparing predator behavior to the size of crayfish encountered. Capture behavior varied with prey size. Bass would approach small items and eat them 100% of the time. This behavior became less common for larger prey. An approach, pick up, and reject pattern increased in frequency with larger prey, because bass could not swallow the crayfish easily, as the crayfish would resist. Bass soon recognized that capture success would be very low for the largest crayfish and never even tried to swallow them. Predation was skewed by changes in handling difficulty with crayfish size.

This experiment indicates that it is not just handling time and energy cost that affect foraging decisions but also defense mechanisms. The difficulty of swallowing is a component of handling time, but the success of completing an encounter is not. Optimal foraging predictions

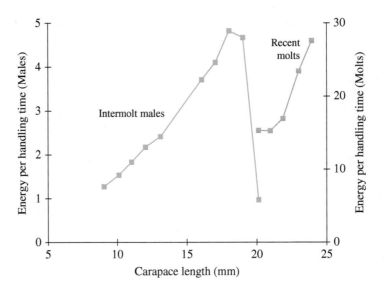

FIGURE 18-8   Energy return per handling time (mg food/s) for smallmouth bass consuming different sizes of male intermolt (solid squares) and recent molt (open squares) crayfish. *Source: Adapted from Stein (1977).*

of return in calories per unit handling time (Figure 18-8) indicated that the highest ranking size of crayfish was 20 mm because this size gave the highest energy return per unit handling time. Yet, in the open environment, bass selected a much smaller size first. Apparently, the problems of capture success reversed optimal foraging decisions in this example. Another untested possibility would be that post-consumptive handling could also be much higher in larger crayfish. An interesting sidelight is that when crayfish molt, they lose their exoskeletons (including claws). For molting crayfish, size selection was purely on the basis of size as would be predicted by optimal foraging. Since molted crayfish had soft exoskeletons, there was no defense mechanism available for molting animals, and crayfish gave much higher returns per unit handling time with increased size when they were molting.

Based on the previous examples – at least in simple environments – fishes do appear to select sizes based on energy maximizing. Prey defenses and differences in digestion time and handling success completely changed the pattern of feeding. Some of the predictions made from laboratory experiments to the field must be made with caution because of the effects of these assumptions.

Optimal Patch Use

Optimal patch use evaluates search techniques and predicts how long to continue foraging in a patch of food. This latter variable is often called give-up density (GUD) – the prey density at which a predator should stop foraging in a patch due to consumption rate declines and move to another patch. Optimal patch use relates well to herbivores because herbivores can move and make choices among relatively immobile patches (unlike animal patches). A good example of patch foraging is a bluegill moving to a spot, hover searching, and then moving to the next location. A planktivore foraging in a patchy plankton population also has the difficulty of plankton mobility and elusiveness.

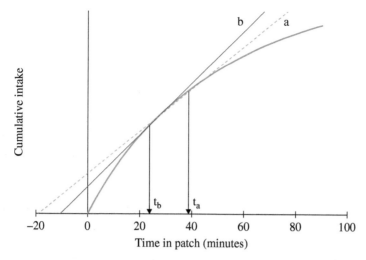

FIGURE 18-9   Cumulative food intake for a consumer feeding in a patch of food. The two tangent lines (a and b) are for longer (a) or shorter (b) travel times between patches. Give-up times for each are t_a or t_b. *Source: Adapted from Charnov (1976).*

Charnov (1976) developed a theoretical model to evaluate time spent in a patch related to gain of foraging in that patch and the overall rate of gain in the habitat (Figure 18-9). His assumption was as an animal spends a longer time in a patch, its rate of energy accumulation from feeding on that patch will continuously decline. As it eats prey, those prey would no longer be available for consumption, and the density of food in the patch would be reduced. The longer the predator stays, the more the density of food in the patch is reduced. The time sequence of depleting food in a large patch with a small predator might be days or weeks rather than minutes. At least for larger animals using smaller patches, depletion of patches is easy to accept.

Charnov's prediction was that the amount of time spent in a patch would be determined by measuring the rate of decline in consumption. The decision of when the decline is unacceptable should be related to the abundance of other patches in nature. In other words, as soon as consumption by the predator causes the consumption rate in the patch to decline below the average consumption rate in the environment the animal should move to another patch. This would be based on knowledge of patch richness in nature. It is not much different from the ideas behind the optimal diet model – that animals can understand and rank their consumption rate of prey in patches and choose to stay or leave depending on that rate. These patch expectations can change seasonally or with location, so animals might have different give-up times even in the same patch depending on their view of the world. Charnov's model gives a prediction of the time at which food consumption rate declines below a global average and the forager should move out.

Other factors influence patch choice and foraging rates. The travel time (and difficulty) to move from one patch to another should also affect patch decisions because the animal consumes no food and incurs moving costs during that time. For Charnov's model, increased travel time is demonstrated by the dashed line (a) in Figure 18-9, and longer travel time results in longer residence time in a patch. This also varies with foraging cues and locomotor techniques. A bird could fly from one place to another and might see the next food patch from the depleted patch, so its time transiting patches might be very short and the cost low. A sculpin in a stream with a patch on one rock might have to move some distance to find another patch; this may take a longer period of time because sculpin may not see the new site. This movement may expose the sculpin to added predation risk as it moves downstream or upstream. In this example, give-up time might be considerably longer because the global pattern of food is not only dependent on how good the patches are, but also on how costly it is to get to those patches. This must be incorporated into the global average. Riskier environments or environments that involve longer or more costly movements should change the expected rates of return.

Patch Use in Sculpin

Patch has been well studied since Charnov's time and has had a number of tests in aquatic and terrestrial habitats. Many animals have been shown to select patches that are better than average patches in their habitat and move from patches when patch quality declines. GUD has been shown to occur for many species, and the allocation of resources varies in two ways (Petty and Grossman 2010). Some organisms fall into an ideal free distribution (IFD), which predicts that animals will distribute themselves among patches so resource utilization by the animals is equal. In other words, fish density will increase in a higher quality patch and decrease in a lower quality patch, so the resources per individual are about the same in each type of patch. An alternative to this is ideal pre-emptive distribution (IPD), which is the result of competition and in which dominant individuals get more resources than subordinate individuals, therefore obtaining higher patch quality than their subordinates. A key to understanding patch use is determining the alternative between IFD and IPD.

Petty and Grossman (2010) evaluated patch use in stream-dwelling mottled sculpin. They used field techniques to determine how sculpin might choose patches and the densities of sculpin in relation to the densities of food in an individual patch. GUD estimates how organisms choose a patch, while IFD and IPD define how an individual chooses, but also how the density of individuals affect patch use. Sculpin have territories on the order of 2 m^2 in size with patches approximating 1 m^2 or less. They are benthic in orientation and tend to remain sedentary, which enables several methods for their detection. Petty and Grossman used visual observations while snorkeling along streams, in addition to small dip nets to collect, measure, and mark individuals. They conducted mark-and-recapture experiments to estimate fish density, estimated the density and colonization rates of invertebrates to define quality of different patches over four seasons in the summer, and estimated patch abandonment by visual observation of sculpin marked with unique pigment brands. They also established a model to estimate GUD under both IFD and IPD predictions.

Petty and Grossman had four tested predictions from patch theory: (1) sculpin would be found in higher quality patches than average; (2) they would have a GUD and depart a patch when it dropped below GUD quality; (3) they would be competitive, showing an IPD, so GUD would be proportional to fish size (which correlates with better competitiveness); and (4) with reduced competition, juvenile sculpin would shift to higher quality patches. They conducted field work on Bull Creek and Shope Fork, North Carolina, to test these ideas.

Petty and Grossman's work resulted in a rigorous test of patch use theory. Their first prediction was supported for adults because the average patch occupied by both large and small adults was higher in prey density than random patches throughout the area. This is particularly supportive because random patches included some patches of high quality as the authors did not evaluate whether or not the random patches were occupied. However, juvenile sculpin had patches approximately the same quality as average random patches. They also used principal component analysis of patch characteristics and found that patch choice by sculpin was more strongly affected by prey density and volume than it was by physical characteristics of the habitat (Figure 18-10). For their second hypothesis, they determined that abandoned habitats were no different in physical characteristics from new habitats chosen after abandonment, but prey density and prey biomass were higher in new patches occupied by adults (Figure 18-11). For the third prediction, sculpin were found to exhibit IPDs indicating competition between individuals, and there was a significant relationship between GUD and sculpin size ($r^2 = 0.73$).

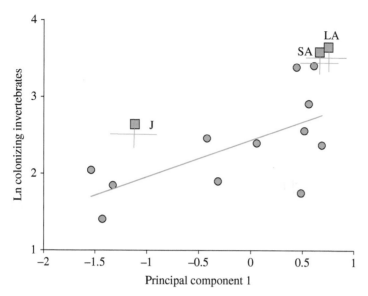

FIGURE 18-10   Macroinvertebrate colonization rates and physical habitat characteristics in Ball Creek during summer 1997. Open circles are random patches, and squares indicate patches chosen by juveniles (J), small adults (SA), and large adults (LA). High PC1 scores indicate locations with higher current velocity, depth, and presence of cobble. *Source: Adapted from Petty and Grossman (2010).*

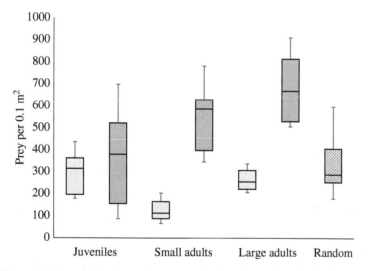

FIGURE 18-11   Box and whisper plot of prey density at abandoned, newly colonized, and random patches for each size class of sculpin in Shope Fork. *Source: Adapted from Petty and Grossman (2010).*

Finally, for their fourth prediction, they found juveniles did switch to high-quality patches in areas where adults had been removed.

This study, in combination with those cited in earlier editions of this book on mayflies (Kohler 1984) and bluegill (DeVries et al. 1989), indicates that search patterns and habitat selection appear to follow the predictions of patch use theory, and the theory can be combined with other interspecific interactions to produce distributions consistent with competitive predictions (IPD) as well as with net positive interactions (IFD).

Summary

This chapter presents the concepts of optimal foraging and patch use and discusses evidence that animals adjust their foraging to maximize energy gains. Mittelbach (2002) reviewed optimal foraging studies in fishes and concluded there is sufficient evidence that fish do prefer to forage on prey that result in higher energy returns. However, he also felt other factors besides handling time were important to maximizing energy return, and his optimal diet model also included probability of capture for elusive prey, energy cost of handling, digestion time, and hunger level. Various foraging theories simplify the act of prey selection by making assumptions regarding cost and benefit of foraging behavior. These assumptions may not always apply, but the important point is that fish appear to forage to maximize long-term energy intake given various constraints. Predators seem to make appropriate behavioral choices to maximize overall energy intake in both constant and patchy environments.

Literature Cited

Charnov, E.L. 1976. Optimal foraging: The marginal value theorem. Theoretical Population Biology 9:129–136.

DeVries, R.D., R.A. Stein, and P.L. Chesson. 1989. Sunfish foraging among patches: The patch departure decision. Animal Behaviour 37:455–464.

Holling, C.S. 1959. Some characteristics of simple types of predation and parasitism. Canadian Entomologist 91:385–398.

Ivlev, V.S. 1961. Experimental Ecology of the Feeding of Fishes. Yale University Press, New Haven, Connecticut.

Kohler, S.L. 1984. Search mechanism of a stream grazer in patchy environments: The role of food abundance. Oecologia 62:209–218.

Lankford, T.E., and T.E. Targett. 1997. Selective predation by juvenile weakfish: Post-consumptive constraints of energy maximization and growth. Ecology 78:1049–1061.

Mittelbach, G.G. 2002. Fish foraging and habitat choice: A theoretical perspective. Pages 251–266 *in* J.B. Hart and J.D. Reynolds, editors. Handbook of Fish Biology and Fisheries. Vol. 1. Fish Biology. Blackwell Science Ltd., Oxford, UK.

Petty, J.T., and G.D. Grossman. 2010. Giving-up densities and ideal pre-emptive patch use in a predatory benthic stream fish. Freshwater Biology 55:780–793.

Scharf, F.S., J.A. Buckel, and F. Juanes. 2009. Contrasting patterns of resource utilization between juvenile estuarine predators: The influence of relative prey size and foraging ability on the ontogeny of piscivory. Canadian Journal of Fisheries and Aquatic Sciences 66:790–801.

Schoener, T.W. 1971. Theory of feeding strategies. Annual Review of Ecology and Systematics 2:369–404.

Stein, R.A. 1977. Selective predation, optimal foraging, and the predator-prey interaction between fish and crayfish. Ecology 58:1237–1253.

Werner, E.E., and D.J. Hall. 1974. Optimal foraging and the size selection of prey by the bluegill sunfish (*Lepomis macrochirus*). Ecology 55:1042–1052.

CHAPTER 19

Diet Composition and Ration in Nature

Energy is the most important resource required for a fish to survive, grow, and reproduce, and thereby achieve fitness. A fish acquires its energy and nutrients through feeding. A fish must feed not only to satisfy its current metabolic needs, but also to grow and store energy for unfavorable times, such as extreme winters and periods of reproduction when feeding is minimal. Different species like piscivores possess adaptations that enable them to exploit certain prey fish resources but not invertebrates, and similar adaptations to diet occur in other consumers. Coupled with environmental variability in availability of certain food items over time and space, fish diets may not just vary among species but also among individuals of the same species over space and time – including diel, seasonal, and ontogenetic changes. Regardless of what a fish eats and how it acquires its food, the allocation of its energy is accounted for through the energy budget. Various aspects of foraging behavior, morphological and anatomical feeding adaptations, and energy budgets have been presented in previous chapters. The purpose of this chapter is to review trophic classifications and the methods used to determine, quantify, and compare diet composition and ration of fishes in the field.

Trophic Categories in Fishes

The range of diet items fish collectively consume is only paralleled by their wide range of adaptations to aquatic habitats as described in Chapter 2. Fish diets range from unicellular algae to aquatic macrophytes, terrestrial vascular plants, zooplankton, macroinvertebrates, fishes, amphibians, birds, and mammals. Some fishes occupy the lowest consumer trophic levels – for example, herring feed on pelagic algae and stoneroller on epilithic algae – while other species are apex predators. Some fishes feed on detritus, others scavenge carcasses, and others are parasites of fishes or other taxa. With this array of feeding habits and known variation in food habits over space and time, it is not surprising that trophic categorization of fishes is daunting and varies by objective. Early trophic level classifications recognized broad ecological types, including detritivores, scavengers, herbivores, and major subcategories of carnivores – benthivores, zooplanktivores, piscivores, and aerial feeders. These classifications identified trophic groups that could be described as guilds. A guild is a group of species that exploit a common resource in a similar way. Such classifications lead to the creation of functional groups – defined as a group

Biology and Ecology of Fishes, Third Edition. James S. Diana and Tomas O. Höök.
© 2023 John Wiley & Sons Ltd. Published 2023 by John Wiley & Sons Ltd.

TABLE 19-1   Trophic Classification Scheme for North American Freshwater Fishes.

Trophic Class	Trophic Subclass	Trophic Mode	Common Example(s)
Herbivores	Particulate feeder	Grazer	*Campostoma anomalum*
		Browser	*Ctenopharyngodon idella*
Detritivores	Filter feeder	Suction feeder	*Cyprinus carpio*
		Filterer	*Hypopthalmichthys molitrix*
	Particulate feeder	Biters	*Notropis texanus*
		Scoopers	*Chrosomus oreas*
Planktivores	Filter feeders	Mechanical sieve	*Hypopthalmichthys nobilis*
		Mucus entrapment	*Lampetra aryesi*
		Ram filtration	*Polyodon spathula*
		Pump filtration	*Dorosoma petenense*
		Gulping	*Brevoortia tyrannus*
	Particulate feeder	Size-selective pickers	*Alosa pseudoharengus*
Invertivores	Benthic predators	Grazers	*Cyprinella lutrensis*
		Crushers	*Moxostoma carinatum*
		Hunters of mobile benthos	*Carassius auratus*
		Lie-in-wait predators	*Noturus insignis*
		Tearers	sponge feeders (marine)
		Diggers	*Percina caprodes*
	Drift predators	Surface feeders	*Hiodon tergisus*
		Water column feeders	*Luxilus cornutus*
Carnivores	Whole body (Piscivores)	Stalking	*Lota lota*
		Chasing	*Morone saxatilis*
		Ambush	*Esox lucius, Amia calva*
		Protective resemblance	*Sander vitreum,*
	Parasites	Blood suckers	*S. canadense*
			Petromyzon marinus

Source: Adapted From Goldstein and Simon (1999).

of co-occurring species that perform a common ecosystem function by their mode of resource utilization (Frimpong and Angermeier 2010). Trophic guild and functional group classification schemes may have overlapping classes, but these are not defined by the same objective.

Goldstein and Simon (1999) attempted one of the most comprehensive – although not exhaustive – expansions of the earlier trophic guild classifications, including 5 guilds and 26 modes of feeding for fishes found in North American freshwaters (Table 19-1). This classification was mainly based on the types of food eaten rather than the method of consuming food. Considering global fish diversity, the Goldstein and Simon table could be expanded greatly – for example, by including tropical fishes. Just as many diets and modes of feeding have been described among the cichlids of Lake Malawi, driven by variation in the shapes of palatine and pharyngeal teeth among these species (see Chapter 29).

William J. Matthews of Oklahoma State University proposed a functional classification of fishes based on interaction of their feeding habits with the ecosystem. He illustrated his

TABLE 19-2	Proposed Fish Functional Groups with Emphasis on Interactions of Fish Feeding with Ecosystem Processes, Examples from North American Fauna.

I. Physically Disturb Substrates

1. Grazers (scrape, pick, or "shovel" materials from hard substrates – stony or woody debris) – *Campostoma*

2. Benthic detritivores – *Prochilodus*

3. Deep burrowers in soft substrates – *Cyprinus*

4. Gravel disturbers – *Hypentelium*, some *Moxostoma*

5. Mud or sand eaters (consume soft sediments, biofilms, and sand) – *Hybognathus*, adult *Dorosoma cepedianum*, *Pimephales*, *Hybognathus*, *Luxilus pilsbryi*

6. Stone turners – some *Percina* (log perch)

7. Scavengers of animal material – *Amieurus*

8. Egg eaters – many small cyprinids or small *Lepomis*

II. Do not disturb substrates

On invertebrate prey

9. Surface feeders (*Lepomis*, many minnows)

10. Drift feeders (*Notropis boops*)

11. Benthic pickers (*Cyprinella venusta*), many riffle-dwelling darters (*Percina*, *Etheostoma*)

12. Water-column particulate feeders (suck – *Menidia*; bite or snap jaws – *Labidesthes*)

13. Snail crushers (*Lepomis microlophus*)

14. Filter feeders (*Dorosoma*)

On fish as prey

15. Suction piscivores (*Micropterus*)

16. Overrun or biting piscivores (*Morone saxatilis*)

On vascular materials

17. Seed crushers (none in North America, and *Colosoma* and others in South America)

Source: From Matthews (1998)/Springer Nature.

classification scheme with examples of stream fishes based on extensive studies of stream fishes of the southeastern United States (Table 19-2). Matthews (1998) identified two main categories of fishes – those that mechanically disturb substrate with their feeding and those that do not. He justified the distinction by the fact that substrate disturbance can potentially resuspend or "activate" otherwise locked or inaccessible material – biofilm, attached algae, detritus, nutrients, and particulate organic matter – therefore contributing to nutrient cycling. This process described here is essentially bioturbation, discussed in Chapter 15 as an indirect positive interaction. Illustrating the potential utility of this type of function-based classification, Matthews compared the feeding modes of suction and overrun or biting piscivores, concluding that the biting predator is likely to generate leftover pieces of meat for other organisms to feed on and release of body fluids to enhance the microbial loop, while the suction predator would not. He also acknowledged what is daunting about this kind of functional classification – requiring an intimate study through direct observation and gut content analysis of every species in the ecosystem. Functional classifications transcend feeding activities and imply deeper and more complex ways of looking at species effects on ecosystems. For example – as indicated by Matthews – gravel nest building by lithophilic species would confer some of the same ecosystem effects as feeding by rummaging through substrate.

Both of these feeding guild methods – as well as a variety of others that have been proposed – attempt to link the wide variety of energy acquisition patterns fishes demonstrate into narrower groups of feeding behaviors and prey consumed, in the hope of generalizing feeding patterns and linking common traits of those patterns. Such guilds are useful in considering trophic relationships in different ecosystems as well as energy flow through various aquatic systems. However, such classifications are never perfect, and behaviors of many fish species make them difficult to fit into any of the groups listed in Tables 19-1 and 19-2.

Determining the Trophic Group of Species

Defining trophic groups and classifying fishes into these categories are two different undertakings. Some species, by virtue of their food habits and plasticity of their food choices, do not fit neatly into any single trophic guild or functional group. Yet practical management concerns, such as calculating indices of biotic integrity for biomonitoring, necessitate associating all species in a community with at least one trophic guild or functional group. One of the major practical challenges is trophic ontogeny. Many species go through at least one major trophic niche shift as they transition from juvenile to adult. Matthews (1998) described the case of creek chubsucker that as adults feed by extracting insects from gravel, while juveniles feed by grazing attached algae. Therefore, adult creek chubsucker is invertivorous, and the juvenile is herbivorous. Similarly, species that are piscivores as adults may go through up to two trophic niche shifts – planktivores as smaller juveniles and invertivores in later juvenile stages – before becoming primarily piscivores as adults. Goldstein and Simon (1999) evaluated the niche shift for a number of species, from invertebrate feeding to piscivory (Table 19-3), and found a wide variety of lengths and weights when this shift occurred. In the development of indices of biotic integrity, trophic ontogeny is accounted for by either focusing explicitly on adult diet only or using size-specific trophic classes. Biological differences among species explain some of this variation, as can the abundance of prey items already being consumed and those available for diet shifts during ontogeny. One can argue that point estimates of size-at-niche-shift conceal significant statistical variation among individuals of a species, and perhaps size-at-niche-shift should be considered more of a range than a point.

In addition to species that undergo ontogenetic niche shift, many other species pose challenges to trophic guild and functional group classification in several ways. For example, bowfin is known to prey on a wide variety of insects, fishes, frogs, and virtually any creature that moves in or around water. Such a species is described as "generalist" or "opportunistic" and does not fit well in a single trophic group. A contrast to this is the example of damselfish – described in Chapter 13 – that feed on algal mats in coral reefs and consume mainly one or two species. These damselfish are considered specialists because of this narrow food niche. Many species of catfishes – with their barbels – are adapted to scavenging through turbid water conditions as well as preying actively on live animals. They may also consume particular vegetative matter or intentionally browse on macrophytes. A species with this food habit is clearly an opportunistic feeder but is also not entirely a carnivore or herbivore. The term omnivore has been used generally to describe species that feed on both plant and animal material. While omnivore may seem a straightforward label, one frequently runs into situations of accidental consumption of both plant and animal matter. For example, a predator on fishes near aquatic plants is likely to frequently

TABLE 19-3   Pivotal Size and Weight Categories that Determine Ontogenetic Niche Shifts for Select Carnivores.

Species	Size (mm)	Weight (g)
Brown trout	300	550
Grass pickerel	100	7
Chain pickerel	100	7
Northern pike	50	5
Muskellunge	50	5
Burbot	500	1,000
White bass	120	20
Striped bass	130	25
Sauger	150	50
Rock bass	120	50
Spotted bass	150	40
Smallmouth bass	50	10
Largemouth bass	80	10
Yellow perch	150	50

Source: From Goldstein and Simon (1999)/Taylor & Francis.
Note: Lengths and weights indicated are classification cutoffs for determining minimum sizes when species are classified as carnivores for calculation of Index of Biotic Integrity. As classified by the Ohio Environmental Protection Agency.

take prey along with some vegetation. Therefore, researchers still grapple with where to draw the line between an herbivore, carnivore, and omnivore. Karr et al. (1986) proposed defining omnivores as species with diets composed of ≥25% plant material and ≥25% animal material.

A significant number of species are known to consume eggs (oovores). Oovory is not strictly associated with herbivorous or carnivorous species and, in fact, may be a trait of all fishes when the opportunity exists. However, certain benthic species – but also some pelagic – are known to actively search and feed on eggs of other fishes and aquatic animals. North American oovores include sculpin, bullhead catfishes, darters, suckers, juvenile and pygmy sunfishes, killifishes, shads, lake and white sturgeons, central stoneroller, and numerous shiners (Frimpong and Angermeier 2009). Since most of these species are opportunistically consuming eggs but mainly consume other prey items, they are usually categorized by their other food consumed.

Determining the Food Habits of a Fish

Accurate classification of fishes into trophic groups requires identification of their food habits, which begins with field studies. The most obvious methods involve examination of stomach contents. This generally requires sacrificing the fish (especially smaller fish) or performing a gastric lavage by inserting a tube in the esophagus and flushing out the gut contents – particularly useful for larger fishes and predators. Observation methods using diving, snorkeling, or video are also common although they may present problems with species identities of prey consumed. Fish stomachs are frequently empty, which is a function of feeding and evacuation rates, as well as sampling technique. If the intent of a study is to simply identify dominant prey in the diet, one can reduce the number of empty stomachs – especially for studies that require sacrificing fish – by sampling at the appropriate time of day for diel, nocturnal, and crepuscular feeders. In many cases, prey consumption may differ throughout the day or in different habitats as a result of prey behavior, so focusing on one time period or location would result in not capturing the full range of prey consumed for fishes that show such variability. It is least productive to conduct diet studies on fishes during the reproductive season as most do not feed actively during that time. The best time to study diet is when temperatures are warm and food is most abundant, unless seasonal diet composition and ration is the objective. A number of species continue to feed through the winter, so seasonal variation in diet and ration remain important in these species, especially considering how human-induced changes in climate may affect fish consumption and growth in the future. Care must be given in handling gut contents and fish sampled for analysis as digestion may continue even after killing and preserving fish, making some diet items unidentifiable. Further complicating accurate identification of diet components is that food items taken may have different digestion rates, so some food types may disappear from the gut content faster than others. To reduce digestion, the gut of dead fish should be removed, fixed, and preserved separately and quickly after collection.

Diet may also be studied with chemical diet tracer techniques and DNA analysis of gut contents. DNA barcoding is a technique where individual items – particularly those that cannot be identified visually – are analyzed for their DNA barcode and classified to species. Metabarcoding is a similar technique, where the entire stomach contents are evaluated using a general barcode marker and high throughput DNA sequencing. These methods can be very useful in identifying rare items in the diet and tend to give more precise identification than visual methods, but do not easily provide information on the number of items eaten or their mass unless physical diet analysis is also used (Jo et al. 2015). Chemical tracer methods include use of bulk stable isotope ratios, fatty acid profiles, and stable isotope analysis of fatty acids. In contrast, stable isotope analysis of bulk tissue in fishes – most typically, white muscle from the dorsal region, but also fin clips, scales, or even mucus – determines the isotopic ratios of H, N, and C to evaluate trophic level of consumers, as well as longer term consumption trends of the fish (Cresson et al. 2014). Field diet analysis only provides the consumption information of a fish on the day of capture, while stable isotopes evaluate the trophic level of that fish over longer time frames. Stable isotopes have been widely used to detect ontogenetic niche shifts of fish. Stable isotope and chemical tracer analyses can get expensive very quickly, both in terms of the large sample sizes needed to account for a high incidence of empty stomachs and the cost of laboratory analyses. Garvey and Chipps (2012) provide extended treatment of field techniques for diet studies, trade-offs in choosing techniques, and quantification of diet information obtained from field studies (Figure 19-1).

Once stomach contents have been estimated, they can be presented in a variety of ways to summarize the food habits of a fish species. Stomach contents may be measured in mass, either wet or dry, to predict the importance of each prey species to the overall diet consumed. These measures give an important representation of overall consumption but can be heavily biased by a few predators that consume large prey items, while the majority may consume small items. Diets can also be estimated numerically for each species, giving a better idea of the predation process. Numeric or mass analyses can be summarized by prey type to determine the trophic level of the predator (piscivore, herbivore, etc.). The number of prey species in stomachs can be used to predict whether the predator is a specialist (small number of prey types) or generalist. Similarly, omnivores can be identified by the great variability of prey

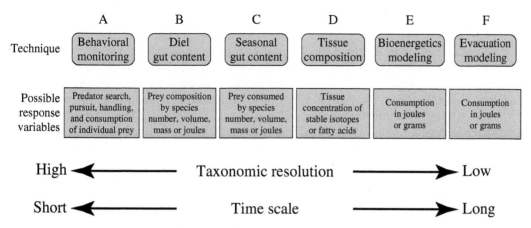

	A	B	C	D	E	F
Technique	Behavioral monitoring	Diel gut content	Seasonal gut content	Tissue composition	Bioenergetics modeling	Evacuation modeling
Possible response variables	Predator search, pursuit, handling, and consumption of individual prey	Prey composition by species number, volume, mass or joules	Prey consumed by species number, volume, mass or joules	Tissue concentration of stable isotopes or fatty acids	Consumption in joules or grams	Consumption in joules or grams

High ⟵——————— Taxonomic resolution ———————⟶ Low

Short ⟵——————— Time scale ———————⟶ Long

FIGURE 19-1   A graphic depiction of the trade-offs associated with different techniques for quantifying diets and trophic status of fishes. *Source: Reproduced with permission from Garvey and Chipps (2012)/American Fisheries Society.*

types and prey trophic levels. As described in Chapter 17, selectivity of the predator can also be estimated by the relative proportion of various prey species in the diet compared to their proportion in nature.

Estimating Ration of Fishes in the Field

To quantify ration of fishes in the field, it is necessary to understand the pattern of feeding by the fish population over diel and seasonal time scales. Ration refers to the quantity of food eaten (kJ/d or %BW/day), while diet is the species composition – or nutritional quality – of prey consumed. Values based on mass of prey consumed can be converted into energy consumed based on energy density and mass consumed of each prey species. Patterns of animal feeding vary dramatically with food type and abundance, and models to estimate ration consider this variance. Animals may feed continuously – always consuming particles of food and digesting them simultaneously. Animals may also feed discontinuously, or eat a large meal, and then digest it later. Also, the patterns in a population may be synchronous, where all animals eat about the same time of day, or asynchronous, where animals catch prey opportunistically, and the timing is not the same for all animals. Marshall Adams and Jim Breck, then both from the Oak Ridge National Laboratory in Tennessee, provided an excellent review of the methodology to estimate consumption rates for fishes in the field (Adams and Breck 1990). A summary of these methods follows.

Continuous Feeders

The earliest and simplest model for estimating ration of continuous feeders was proposed by Bajkov (1935) and modified later into the form

$$C = 24 \cdot S \cdot k \qquad \text{(Equation 19-1)}$$

where C is daily ration (g/d or %BW/d), S is average amount of food in the stomach over 24 hours, and k is the instantaneous rate of gastric evacuation (per hour). This assumes consistent feeding – fish have the same amount of food in their stomachs at the start and end of

24 hours – and gastric evacuation is exponential. A similar model has been developed by Eggers (1977).

The usual method to estimate the amount of food in the stomach using any model is to sample predators at regular intervals over the diel cycle and weigh stomach contents for each prey species. Many fishes have periodic feeding patterns, so the average amount of food varies over the diel cycle, and they therefore cannot be analyzed using Equation 19-1. It is possible to estimate the amount of food consumed in any time interval by determining the difference between the contents at the beginning and end of that time period, and also by the amount of food evacuated from the stomach. Most continuous feeding models use this approach.

Elliot and Persson (1978) produced the most widely used model for continuous feeding fishes. Their formula is

$$C_t = \frac{\left(S_t - S_0 e^{-kt}\right) \cdot kt}{1 - e^{-kt}}$$ (Equation 19-2)

where S_0 and S_t are the stomach contents at time 0 and t, t is the length of the sampling interval (in hours), and C and k are as in Equation 19-1. This equation can be estimated over a number of sampling intervals to include the diel period and thereby measure the daily ration. The model works well when feeding is more or less continuous, but the amount of food in the gut varies over time and the digestion rate is exponential (Adams and Breck 1990). Both Equations 19-1 and 19-2 apply to particulate feeders that feed more or less continuously. They do not apply well to animals that might eat one or two large particles in a meal. An example calculation for this method is given in Box 19-1.

One additional estimate from field stomach samples could be the rate of digestion or k. This can be estimated if the fish do not eat for some period of time, so the change in stomach contents over time can be used to calculate k. For the example of a fish that feeds only during the day, the loss of stomach contents over two time periods at night would estimate k using the formula

$$k = \frac{1}{t} \cdot \ln \frac{S_0}{S_1}$$ (Equation 19-3)

where ln is the natural log base. An example of such a calculation is also included in Box 19-1. This could be estimated either through the natural feeding periodicity of the fish or by taking fish from the field and holding them under conditions with no food, and then evaluating the change in their stomach contents over that time.

Chronology of Feeding

Adams and Breck (1990) presented methods developed by Keast and Welsh (1968) and Nakashima and Leggett (1978) to estimate ration for continuous feeding fishes based on stomach samples over the diel period. These methods use observations of peak stomach contents to estimate daily ration, assuming no gastric evacuation over the time period. They may sum the differences between minimum and maximum stomach contents (Keast and Welsh 1968) or the various peaks of stomach contents (Nakashima and Leggett 1978). They assume noncontinuous feeding, consuming material over one time period and digesting it over another. From this pattern, the peaks in feeding indicate the total food consumed over that time period, and summing them approximates the total eaten over the day.

Box 19-1 Data for Stomach Sampling and the Calculation of Daily Ration for a Continuous Feeding Fish

Samples of fish were collected every six hours over a 24-hour period. Average stomach contents were 1% BW at 06:00, 1.5% at 12:00, 2% at 18:00, 1.5% at 24:00, and 1.2% at 06:00. An estimate for k from lab experiments was 0.20 per hour. The ration for each time increment from Equation 7-2 would be

$$C_{12} = \left[1.5 - 1 \cdot e^{-(0.2 \cdot 6)}\right] 0.2 \cdot 6 / \left(1 - e^{-0.2 \cdot 6}\right) = \left[(0.5 \cdot 0.301)1.2\right] / (1 - 0.301) = 1.138 / 0.699 = 1.629$$

$$C_{18} = \left[2 - 1.5 \cdot e^{-(0.2 \cdot 6)}\right] 1.2 / 0.699 = 2.08$$

$$C_{24} = \left[1.5 - 2 \cdot e^{-(0.2 \cdot 6)}\right] 1.2 / 0.699 = 1.11$$

$$C_{6} = \left[1.2 - 1.5 \cdot e^{-(0.2 \cdot 6)}\right] 1.2 / .0699 = 0.94$$

The total daily ration would be

$$C_{12} + C_{18} + C_{24} + C_{6} = 1.629 + 2.08 + 1.11 + 0.94 = 5.75\% \text{BW per day.}$$

One additional calculation that could be made is the value of k, or instantaneous digestion rate, from field samples. In this case, the fish must not feed over a certain period. For our example above, assume a second group of fish were collected at 18:00 and held without food until 24:00, then measured for stomach contents. If S_{18} was 2% and S_{24} was 0.6, then:

$$K = (1/6) \cdot \ln(2/0.6) = 0.167 \cdot \ln(3.33) = 0.167 \cdot 1.204 = 0.20$$

Non-Continuous Feeders

The previous methods do not apply very well for fishes that eat large (relative to body weight), distinct meals. Instead, methods to determine meal size and meal frequency are preferred. These methods may estimate each separately or all at once, mainly depending on whether the fish has a rapid digestion rate (less than 24 hours) or a slow one. Estimating meal size in these animals – usually large carnivores – is relatively simple because they eat few large items. Even if digestion has occurred to an advanced stage, it is often possible to estimate the size of the prey item when it was ingested based on measurements of length, vertebral diameter or size of other bones or hard parts, and regressing this against total mass for the prey species. So, in these species, the estimation of meal size can be made not only on current weight of prey in the stomach but also on the weight of prey when consumed. Similar extrapolation of weight of prey consumed from partial prey remains can be done for all of the ration methods described here, depending on the type of prey consumed, but is most common with predators consuming large prey items.

Diana (1979) proposed a simple model of consumption for northern pike:

$$C = \frac{M \cdot n}{B' \cdot N} \qquad \text{(Equation 19-4)}$$

where M is the average size of an ingested meal, n is the number of fish with food in the stomach, B' is the number of days required for gastric evacuation, and N is the total number of fish in the

Box 19-2 Example Calculations for Daily Ration of Non-Continuous Feeding Fish

Assume that collections were made on a northern pike population and the following parameters were estimated.

$M = 75$ g/kg. This is the average total mass of food eaten, calculated for time of ingestion for all fish with food in their stomachs.

$n = 50$ fish collected with food in their stomachs.

$B' = 2$ d. This is gastric evacuation rate determined either from a separate lab experiment or by collecting fish and withholding them from food as done in Box 7-1.

$N = 100$ fish collected.

For these data:

$$C = (M \cdot n)/(B' \cdot N) \ (75 \cdot 50)/(2 \cdot 100) = 3750/200 = 18.75 \, \text{g} \cdot \text{kg}^{-1} \cdot \text{d}^{-1}$$

In another experiment, six largemouth bass were collected from a warmwater lake and they contained stomach contents (calculated for weight at ingestion) as indicated below:

Fish	Stomach Contents (g)	Fish Weight (g)
1	50	1,250
2	40	1,000
3	0	1,500
4	100	1,500
5	25	500
6	50	750

The formula for ration calculation is

$$C = 100 \sum \left(P_{Wi}/B_{Wi} \right)/N, \text{ so}$$

$$C = 100 \left[(50/1250) + (40/1000) + (0/1500) + (100/1500) + (25/500) + (50/750) \right]/6$$
$$= 100 \cdot (0.04 + 0.04 + 0 + 0.067 + 0.05 + 0.067)/6$$
$$= (100 \cdot 0.264)/6 = 26.4/6 = 4.4\% \text{BW/d}$$

sample. This model was developed for northern pike – which do not feed synchronously – so a sample of pike taken any time of day has individuals with all combinations of stomach conditions – new meals, partly digested meals, and empty stomachs. Individual pike appear to feed over a short time interval, that is, they have stomach contents with all items in approximately the same stage of digestion. A similar model was produced by Swenson and Smith (1973) with the exception that their model calculates daily ration over several time periods in a day, somewhat more similar to the continuous feeding model. If fish in a population feed synchronously – at the same time of day – or feed over a long time interval, sampling must be done throughout the day, and the model of Swenson and Smith applies better than Equation 19-4.

For fishes with faster rates of digestion, meal frequency can be much less than 24 hours, and therefore the meal size and frequency must be estimated several times throughout the day, whether or not fish feed synchronously. These ration estimators are much more sensitive to rate of digestion than Diana's model. One version of this (Adams et al. 1982) is

$$C = 100 \sum_{i=1}^{N} \frac{P_{Wi} / B_{Wi}}{N} \qquad \text{(Equation 19-5)}$$

where P_{Wi} is the estimated weight at ingestion for prey of predator i over a 24-hour period, B_{Wi} is the weight of predator i, and N is the total number of predators in the sample, including fish with empty stomachs. In this model, time to digestion is less than 24 hours. Example calculations from both of these models are found in Box 19-2.

Difficulties with Ration Estimation

Field values of daily ration are inherently difficult to estimate accurately. Some of the difficulties are methodological, and some are simply the variation in ration that occurs over time and among fishes. The most problematic issue for most of these models is that they estimate a daily ration but do not give variance values for the estimation and, therefore, cannot be compared statistically. Variance may be approximated by considering the variation in mass of stomach contents, but variation in stomach contents is not the only – or even the major – source of variance. For example, Equation 19-4 uses an estimate of meal size and frequency. Variation in meal size could be estimated by boot strapping or some other statistical technique, but the variation in meal frequency could not be estimated. One could possibly measure evacuation rates with variance, for each meal size, particle size, temperature, and predator weight combination to get a better idea of this variance, but the amount of work involved would be enormous. Therefore, variance estimates are not often calculated for field ration data.

In addition to this problem, there are sampling issues that add biases to field estimates of daily ration. Many collection techniques – such as gill netting or fyke netting – collect fish that are active, but they do not collect inactive fish. If the fish are actively seeking food, they may be collected more frequently, and these fish may have less food in their stomachs than others because they are still hungry. If fish are caught in a fyke net, they may eat other fish in the net. They will also digest food from their stomachs during the time they are held in the nets, or they may regurgitate their food if the net causes stress. All of these behaviors would bias the field estimate of ration. For most fishes, use of trawling or some similar technique with more or less instantaneous collection – and with collection not depending on fish activity – would be the least biased method.

One final problem is the daily variation in feeding patterns. Prey consumption is affected both by appetite and prey availability. Availability is not just related to abundance of prey but also behavior. Capture success of predators may vary dramatically from day to day based on weather, prey behavior, or other conditions. Therefore, the daily ration for one day may not be similar to the ration eaten on other days. Most biologists would estimate ration at one point in time and then use linear extrapolation between sampling times to calculate ration each day, although any events that would cause daily ration to change in a nonlinear fashion would make this extrapolation biased.

Uses of Ration Estimates

While rations are difficult to estimate, they are still important in a variety of fish ecology studies. For example, rations are needed to corroborate a bioenergetics model as described in Chapter 8. They can be used in combination with prey abundance estimates to evaluate the impact of predator populations on prey resources and possibly adjust stocking strategies. They may also provide information on the importance of various prey species to predator survival and growth. Diana (1979) found major differences in ration between male and female pike, which correlated well with sex-related differences in reproductive energetics of pike. One interesting component of that diet study was the determination that consumption by pike of rare but large prey like suckers or burbot resulted in a very major contribution to the annual energy budget even though this consumption was infrequent. Much of the growth of pike was driven by consumption of these large items.

Summary

Fishes consume a wide variety of food items and can therefore be categorized into a variety of guilds based on the dominant prey in their diets and foraging behavior. Estimation of diets can be done directly using serial sacrifice of fish and collection of stomach contents or gastric lavage to remove stomach contents from live fish. DNA barcoding or metabarcoding can aid in identification of highly digested prey items, and prey mass can be estimated by determining relative size of remaining body parts from the stomachs. Longer term consumption trends can be estimated using chemical diet tracers. Finally, data on the kinds and sizes of prey consumed can be used to calculate ration for the consumers using models that relate diets and feeding behaviors. When combined, estimates of diet and ration are important in corroborating bioenergetic models as well as evaluating predator and prey dynamics in managed ecosystems.

Literature Cited

Adams, S.M., and J.E. Breck. 1990. Bioenergetics. Pages 389–415 *in* C.B. Schreck and P.B. Moyle, editors. Methods for Fish Biology. American Fisheries Society, Bethesda, Maryland.

Adams, S.M., R.B. McLean, and M.M. Huffman. 1982. Structuring of a predator population through temperature-mediated effects on prey availability. Canadian Journal of Fisheries and Aquatic Sciences 39:1175–1184.

Bajkov, A.D. 1935. How to estimate the daily food consumption of fish under natural conditions. Transactions of the American Fisheries Society 65:288–289.

Cresson, P., S. Ruitton, M. Ourgaud, and M. Harmelin-Vivien. 2014. Contrasting perception of fish trophic level from stomach content and stable isotope analyses: A Mediterranean artificial reef experience. Journal of Experimental Marine Biology and Ecology 452:54–62.

Diana, J.S. 1979. The feeding pattern and daily ration of a top carnivore, the northern pike (*Esox lucius*). Canadian Journal of Zoology 57:2121–2127.

Eggers, D.M. 1977. Factors in interpreting data obtained by diel sampling of fish stomachs. Journal of the Fisheries Board of Canada 34(2):290–294.

Elliott, J.M., and L. Persson. 1978. The estimation of daily rates of food consumption by fish. Journal of Animal Ecology 47:977–991.

Frimpong, E.A., and P.L. Angermeier. 2010. Trait-based approaches in the analysis of stream fish communities. American Fisheries Society Symposium 73:109–136.

Frimpong, E.A., and P.L. Angermeier. 2009. FishTraits: A database of ecological and life-history traits of freshwater fishes of the United States. Fisheries 34:487–495.

Garvey, J.E., and S.R. Chipps. 2012. Diets and energy flow. Pages 733–779 *in* A.V. Zale, D.L. Parrish, and T.M. Sutton, editors. Fisheries Techniques, 3rd Edition. American Fisheries Society, Bethesda.

Goldstein, R.M., and T.P. Simon. 1999. Toward a united definition of guild structure for feeding ecology of North American freshwater fishes. Pages 123–137 *in* T.P. Simon, Editor. Assessing the Sustainability and Biological Integrity of Water Resources Using Fish Communities. CRC Press, Boca Raton.

Jo, H., M. Ventura, N. Vidal, et al. 2015. Discovering hidden biodiversity: The use of complementary monitoring of fish diet based on DNA barcoding in freshwater ecosystems. Ecology and Evolution 6:219–232.

Karr, J.R., K.D. Fausch, P.L. Angermeier, P.R. Yant, and I.J. Schlosser. 1986. Assessing Biological Integrity in Running Waters: A Method and its Rationale. Illinois Natural History Survey, Special Publication 5, Champaign.

Keast, A., and L. Welsh. 1968. Daily feeding periodicities, food uptake rates, and dietary changes with hour of day in some lake fishes. Journal of the Fisheries Research Board of Canada 25:1133–1144.

Matthews, W.J. 1998. Patterns in Freshwater Fish Ecology. Chapman and Hall, New York.

Nakashima, B.S., and W.C. Leggett. 1978. Daily ration of yellow perch (*Perca flavescens*) from Lake Memphremagog, Quebec-Vermont, with a comparison of methods for in situ determinations. Journal of the Fisheries Research Board of Canada 35:1597–1603.

Swenson, W.A., and L.L. Smith. 1973. Gastric digestion, food conversion, feeding periodicity, and food conversion efficiency in walleye (*Stizostedion vitreum vitreum*). Journal of the Fisheries Research Board of Canada 30:1327–1336.

CHAPTER 20

Predation Risk and Refuges

Tactics fish use at close range to capture prey, particularly the interaction between tiger musky and fathead minnow, were examined in Chapter 17. Search mechanisms of blueback herring and bluegill were also examined as well as foraging of bluegill and mottled sculpin in patches of prey. The comparison between blueback herring and bluegill indicated that search behaviors differed between predators depending on the kind of habitats in which they foraged. Herring in open water were more active foragers; they used swim searches and their own movements to detect prey and were swimming most of the time. Bluegill, which commonly foraged in vegetation, used hover searches as a means of detecting prey.

The final component of foraging to be analyzed is foraging in situations more complex than simply open water, such as environments that might include structure and predation risk. Structure adds other elements to the predator–prey interaction: (1) prey may have some refuge; (2) predators may have cover to improve capture success; (3) structure, particularly vegetation, may be an area where food is more abundant for prey species; and (4) structure could alter prey behavior. Predators such as northern pike often hover in vegetation and wait for prey to approach, and vegetation may make the pike less visible to their prey. As a consequence, prey should be more vulnerable to pike in vegetation. For the bluegill competition example, foraging in vegetation was preferred to foraging in open water because food resources in vegetation were much richer. Vegetation may therefore attract prey because of the abundance of food resources in that habitat. Many prey species of fish may school in open water, but they might change to more solitary behavior in vegetation because vegetation could interfere with regular spacing. Vegetation, or any kind of complex habitat structure, can affect many components of predator and prey behavior. The purpose of this chapter is to review several experiments that have evaluated the effects of structure and of other fishes on predator–prey interactions, especially focused on the role of structures such as vegetation on prey refuges and predator behavior.

Biology and Ecology of Fishes, Third Edition. James S. Diana and Tomas O. Höök.
© 2023 John Wiley & Sons Ltd. Published 2023 by John Wiley & Sons Ltd.

Laboratory Experiments on Foraging in Structure

Largemouth Bass and Vegetative Structure

A study of changes in behavior of largemouth bass and predation success with different patterns of structure was done by Jaci Savino and Roy Stein from Ohio State University (Savino and Stein 1982). Savino and Stein used strands of polypropylene rope for cover, attached to the bottom and floated to the water surface. They were not rigid, so if a fish charged into a strand it was not injured. The ropes did not harbor high densities of bluegill food, so the bluegill's only attraction to cover was for refuge. Savino and Stein used stem densities of 0, 50, 250, and 1,000 per square meter and visually observed different kinds of behaviors and recorded both the amount of time animals spent in those behaviors and number of behaviors that occurred.

Most behaviors of largemouth bass changed with stem density (Figure 20-1). There was a slight decline in the number of search events between a density of 50 and 250 stems per square meter although time spent searching was the same at all stem densities. This may imply less active searching behavior at high stem densities. The number of follows, which occurred once a bass recognized a bluegill and began to follow it, demonstrated a general pattern with a decline at intermediate stem density (250). There was also a decline in time spent following at higher stem densities. Pursuit, which is typically a close-range interaction, was independent of stem density, both in number and amount of time spent pursuing bluegill. Number of attacks declined dramatically between 50 and 250 stems per meter. Number of captures also declined between densities of 50 and 250 due to reduced frequency of other predatory behaviors at higher stem density. The presence of structure, at least at low densities (50 or less), did not influence predator behavior in this study, while stem densities from 50 to 250 gave bluegill an effective refuge and altered bass behavior and predation success.

Success rate of bass in various behaviors was also influenced by stem density (Figure 20-2). For a successful follow, an attack must occur, while a successful attack must result in a capture. There was no effect on attacks per follow at low stem density but a continual decline as stem density increased. Once an attack was made, captures per attack did not vary much with stem density. The influence of increased stem density was to decrease the success of bass in following prey and completing attacks, probably because bluegill could find structures in which bass lost visual contact or could not attack them successfully.

Savino and Stein did similar experiments without bass to determine the influence of bass on the behavior of bluegill. Bluegill behavior also changed with stem density (Figure 20-3). As stem density increased, the percent of bluegill schooling declined drastically. There was no difference between densities of 0 and 50 stems per square meter but a large difference between 50 and 250. As the environment became more complex, bluegill shifted from predominantly schooling to more solitary occupation of vegetation. In absence of the predator, bluegill did not school as frequently at low structure density, and there was no significant effect of structure density on schooling frequency (Figure 20-3). Apparently, bluegill responded to predator risk by schooling more in absence of structure and hiding more in presence of structure. Other components of bluegill behavior also changed in the presence of bass, particularly the bluegill's location in the water column (top or bottom of aquarium and in middle or near edge). However, this behavior was so constrained by the experimental apparatus that it may bear little resemblance to similar choices in natural settings. In natural ecosystems, vulnerable prey species respond to a predator threat by retreating to safer areas. An example of such prey response was shown by Magoulick (2004) with experimental addition of smallmouth bass to pools containing various potential prey species. Fish prey response depended on size and species' usual

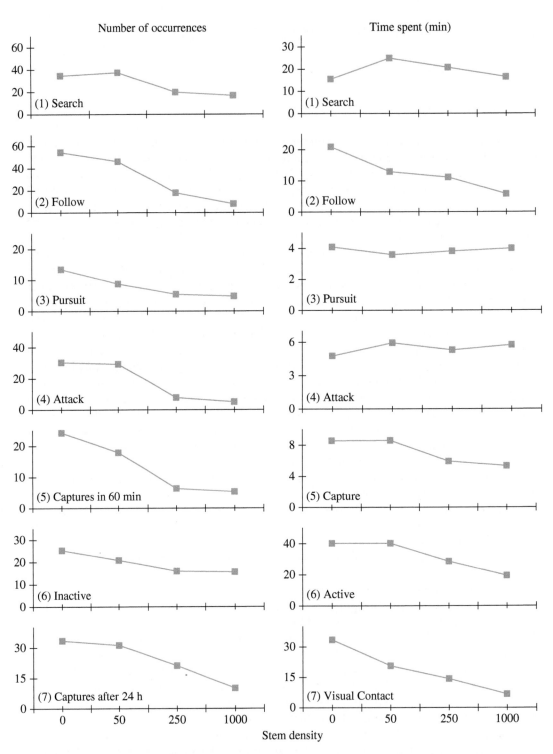

FIGURE 20-1   Number of occurrences (left) and time (right) in different behaviors at each stem density.
Source: Adapted from Savino and Stein (1982).

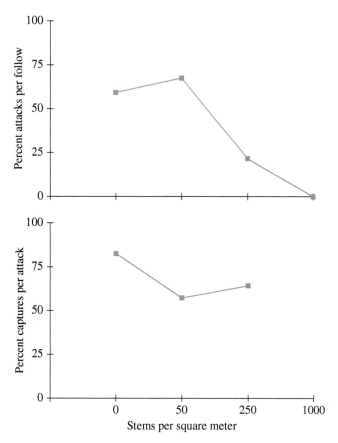

FIGURE 20-2   Percent of successful behaviors for bass at each stem density. *Source: Adapted from Savino and Stein (1982).*

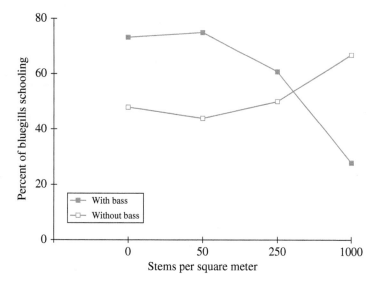

FIGURE 20-3   Percent of bluegill schooling in presence or absence of bass at each stem density.
Source: Adapted from Savino and Stein (1982).

position in the water column; small minnows usually located in the water column left the pools with bass and went into shallower water. Similarly, crayfish of all sizes, being a preferred prey, moved into shallower water. Small benthic fish like darters that are not preferred prey of bass and larger-bodied minnows responded less dramatically to bass presence.

Largemouth Bass, Optimal Foraging, and Vegetative Structure

The previous experiments evaluated changes in predator and prey behavior as influenced by vegetative structure. One important criterion of such behavior would be the ability of the predator to adapt to structure in its foraging behavior. The concept of optimal foraging and its influence on prey selection has already been presented. A follow-up question might be whether fish could change their foraging tactics to optimally forage in a variety of vegetated habitats or under any other changing environmental condition.

Anderson (1984) addressed this question for largemouth bass. He used laboratory experiments of bass foraging on either damselfly larvae or guppies and evaluated their foraging pattern related to optimal foraging predictions. The experiments tested bass in two structured environments: vegetation of low structure (plastic *Elodea* plants at 200 stems per m²) or high structure (670 stems per m²). In each 184-L arena, he added prey and bass and quantified the behavior of bass encountering prey. For prey abundance, he used 20 damselflies and 4 guppies, and he allowed the predators to forage for five minutes. He also did preliminary experiments to allow the bass to acclimate to the apparatus and learn handling and other characteristics for the two prey types. During all of these experiments, he measured the time spent in various behaviors as well as swimming speed of the bass and energy returns for various prey. One exception to the normal optimal foraging analysis was that he included search time in the handling time measurements to predict the optimal prey set using benefit–cost ratios.

In the preliminary experiments, Anderson found increased structural complexity (stem density) produced longer search times for both prey types. Predator behavior also changed with complexity as the swimming speed for bass searching for damselflies was higher at lower stem density. As guppy numbers declined, the search time for the remaining guppies increased considerably (Table 20-1). He could use these measures to predict energy return per search and handling time of guppies and damselflies as long as he corrected for the order of guppy capture.

TABLE 20-1	Search and Handling Times for Largemouth Bass Consuming Two Prey Types in Low or High Structured Habitats. Mean ± Se.

	Time Per Item	
Action	**Low Structure**	**High Structure**
Damselfly search	23.1 ± 3.1	37.5 ± 4.4
Damselfly handling	6.6 ± 0.5	8.9 ± 0.6
Guppy handling	5.9 ± 0.6	7.9 ± 0.7
Guppy search		
First captured	3.8 ± 1.0	15.8 ± 2.7
Second captured	2.4 ± 0.6	17.0 ± 4.4
Third captured	3.2 ± 0.9	113.1 ± 14.7
Fourth captured	12.0 ± 2.8	134.6 ± 13.8

Source: Data From Anderson (1984).

He then set out to evaluate optimal foraging predictions for mixed prey of guppies and damselflies in low and high structured habitats.

For the second experiment, Anderson evaluated learning by the predators by first exposing each to foraging in one type of structural complexity and then switching them to the other type. He continued an experiment for seven days in one density treatment, and then changed to the other and ran it an additional seven days. He compared behavior in each of these time periods to evaluate the learning pattern for bass in new habitats, and also whether they ended up with similar behavior regardless of the initial exposure type. For each habitat, he also predicted the optimal pattern of prey selection using cost–benefit ratios for guppies and damselflies determined during the preliminary experiment.

Optimal foraging predictions for low structural complexity were that bass should consume all four guppies first and then move on to eat damselflies. In highly structured habitat, where search time was much longer for guppies, the prediction was that bass should eat two guppies and then expand their diet to include damselflies. This pattern would result in the highest energy return per unit time compared to eating one or the other exclusively. For the low structure habitat, 32 of the 42 trials resulted in bass eating all four guppies first (the optimal predicted pattern; Figure 20-4). In high structure habitat, there was more variation, and the bass in these trials ate fewer guppies and more damselflies. In 23 of 42 trials, the fish ate two guppies first then damselflies, as predicted. Other trials varied from eating more guppies to eating damselflies first. Anderson also found as time in the apparatus increased, bass consumption was more similar to the optimal solution.

The changes encountered by bass in his experiment were fairly complex. Handling time for both species of prey increased considerably in dense structure, but under either habitat, guppies were about four times as valuable as damselflies in cost–benefit ratio (kJ/s). The major change in optimal foraging pattern resulted from decreased encounter rates for guppies in dense structure. Bass in these trials, with changing vegetation structure, learned the new patterns of prey availability, search, and handling times fairly quickly and appeared to forage

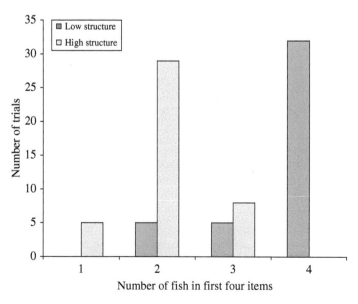

FIGURE 20-4   Number of fish out of the first four prey items consumed by bass during five-minute foraging bouts in low structure (blue bars) or high structure (yellow bars) habitats. *Source: Adapted from Anderson (1984).*

in an optimal manner by seven days. The behavior of these bass was much closer to foraging according to learned expectations for prey abundance and capture than it was for foraging by encounter. In fact, damselflies were avoided more than would be expected due to encounter alone during the times when density of vegetation was changed, indicating that some bass would continue to search for guppies that were not in view rather than eat a damselfly in view. Bass moving to high structure learned the new rules more slowly than those moving to low structure. This is not a surprise considering the fish in low structure could probably see all prey in the tank, while in high structure, a number of prey would not be visible at any given time. Anderson used these results to predict that bass should forage in low structured environments, when given the choice, but might have to alter this situation in the presence of predators. In this case, the increase in structure density produced better refuge for the prey, but the predator still seemed to be able to make foraging decisions based on maximizing the rate of energy intake.

Another similar study of pike interactions with perch and roach found results that differed as one would expect based on the behavior of pike (Jacobsen and Perrow 1998). In this study, an experimental pond 16 m² was set up with half of the pond having open water and the other half subdivided into three compartments with 25, 50, or 75% of the water volume infested with live *Elodea*. They used four treatments 5 with replicates each: 64 age-0+ roach alone, 32 roach with 32 age-0+ perch, 32 roach with 32 age-0+ perch and two yearling pike, and 32 roach with 32 age-0+ perch, 1 yearling pike, and 3 age-2+ perch. The normal roach distribution when alone was 13% in open water during daytime and 90% at night. They found that roach as prey were most vulnerable to pike predation in vegetation, and their habitat choice changed from 12% in open water during day without pike to 90% in open water during day for treatments with pike. Pike tended to forage and remain in the vegetation even when roach predominated in open water. There was no significant difference in distribution among various vegetation densities. They determined from these results that roach used cover in vegetation as a resting site when pike were not nearby, but shifted to open water as a refuge in daytime when pike were abundant. In spite of these behaviors, pike were quite successful consuming roach in this experiment, but there were no observations of pike predation, precluding details on the pike foraging method and location. The authors did believe that distance from the edge was an important factor in avoiding pike predation, and that previous experiments in smaller enclosures showed different results because sufficient space was not given for roach to avoid edge areas in daytime.

Field Experiments on Foraging in Structure

One problem with the previous experiments is the degree of comparability between results in aquaria and results in natural ecosystems. This concern was mentioned earlier in Werner and Hall's pond experiments; it is even more dramatic in laboratory systems. Continual vegetation (or ropes) at a certain density may explain behavior of bass and bluegill when they are in the midst of macrophytes, but it does not explain their selection of vegetation versus open water or illuminate the behavioral changes that might occur in the abrupt transition between two habitat types. Habitat incongruities may be very important, and the edge or interface between one kind of habitat and another may have a particularly critical effect. For example, the edge of a drop-off, where submergent and emergent vegetation declines and habitat becomes open, has a very clear interface, which may resemble a curtain. Edge effects occur in many habitats, such as the pool-riffle edge in streams, reef-flat edge in coral reefs, or mangrove-eel grass interface in tropical coastal areas. The effect of structure itself may not be as important as the interface of structure between one habitat and another.

Bass and Bluegill Interactions in Differing Distributions of Natural Vegetation

Kelley Smith, then a student at University of Michigan, did experiments to test this edge effect (Smith 1993). He used enclosures placed in natural lakes rather than the lab. These square enclosures were 15 feet per side, including fences buried into the sediments to exclude fish from burrowing out. He manipulated natural vegetation in each enclosure using four treatments: (1) removing all vegetation from one enclosure; (2) leaving two 4.5-foot wide strips of vegetation and a 6-foot wide strip of open water in the middle in another enclosure; (3) leaving three strips of vegetation (three feet wide each) with two strips of open water (also three feet wide) in the middle of another enclosure; or (4) leaving the last enclosure with complete cover. The two edge treatments (2 and 3) had the same aerial coverage of vegetation (135 square feet), but either 30 (one open strip) or 60 (two open strips) linear feet of edge. Using natural vegetation allowed fish to utilize vegetation as a food source or move into open water to forage on plankton. Thus, the bluegill's foraging decisions and cover seeking were both a part of the response. There can be an effect of the enclosure on this interaction if a predator crowds prey against a side. Smith did not observe fish regularly, so his measure of predation success was based on how many prey were eaten. These enclosures were much larger than the ones used in laboratory experiments, so the enclosure effect should be much less than in the previous studies.

Smith initiated an experiment by releasing 2 bass and 50 bluegill into each enclosure with a removable barrier to separate predators from prey until each had a chance to adapt to the habitats for 24 hours. He measured the number of bluegill that remained in each enclosure after three days and could then estimate the number of bluegill eaten given different vegetative cover or amounts of edge. There was a change in the number of bluegill eaten with increased cover (Figure 20-5), and enclosures with more open habitat had higher predation rates than those with more vegetated habitats. However, area of vegetation was not the only factor influencing predation rate because a doubling of edge with no change in vegetation area led to a

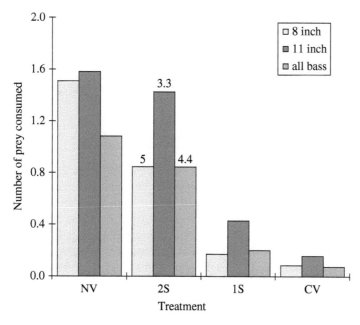

FIGURE 20-5   Predation rate (number of bluegill eaten) for each size group of bass in each treatment. Numbers above the 2S treatment are ratios of 2S:1S predation rates. *Source: Adapted from Smith (1993).*

4.4-fold increase in prey consumed. The amount of interface between vegetation and open water (edge) also had an effect. If vegetation's effects were a matter of area, both strip treatments should have the same predation rate. Treatment 1S had a lower number eaten than treatment 2S even though both had the same percent of the water covered by vegetation. This was not a simple increase in predation related to more edge, as the predation rate averaged 4.4 times greater in the 2S treatment compared to the 1S, even though edge only doubled between treatments. Since edge was doubled and open water strips were narrowed in the 2S treatment, a bass may have been able to search both edges and forage on both edges of the clearing at the same time. For the bass in 1S, the two edges were so far apart one fish could not forage both edges simultaneously. The 2S situation gave a bass twice as large a foraging rate, probably because it foraged along edges on both sides at the same time. This experiment indicates that edges gave an advantage to the predator when compared to complete vegetation, and the amount and configuration of the interface between vegetation and open water was important in determining predation rate.

Actual or simulated vegetation can be a refuge at very high density. In order to be a refuge, the vegetation area must be large enough to allow prey fish to hide in it. Complete vegetation in Smith's study was a refuge. Either the predator may have been able to penetrate through vegetation strips to forage in the treatments with strips or prey behavior changed because there was not a large enough area of vegetation. At low density or in patchy environments, vegetation actually may function to increase predation rate. Predators may take advantage of edges or the vegetation to distort their own silhouettes, mislead the prey, and achieve greater success than in complete vegetation.

Differences in predation rate between open, vegetated, or edge habitats can provide an opportunity to manage vegetation to increase predation rates. Olson et al. (1998) conducted such an experiment on 13 Wisconsin lakes. These lakes all had extensive growth of Eurasian milfoil, an exotic macrophyte that has become a pest in many North American lakes. The dense vegetation provided a large refuge for bluegill, increasing their survival and resulting in stunted populations. Four treatment lakes received macrophyte removals in strips of evenly spaced channels from shore through the littoral zone. Approximately 20% of the macrophyte cover was removed from each treatment lake. The year after macrophyte removal, bluegill exhibited significantly increased growth rates in treatment compared to control lakes, particularly for age-3 and age-4 bluegill. Largemouth bass did not exhibit a change in growth rate. This experiment was only one year in duration, and some of the effects of macrophyte removal might occur later in the lakes. However, it provided interesting implications for future experiments and for management of vegetation in lakes to alter predation rates and refuges.

Relative Predation Risk in Mangrove Habitats

Similar studies of predation risk and refuges have been conducted in a variety of marine habitats. Many of these studies use tethered prey fish as means to control where prey fish are most susceptible to predation. While tethered prey cannot use their full extent of predatory avoidance mechanisms, they can show where predatory encounters and risk are highest. Such a study was done in Biscayne Bay, Florida, by Hammerschlag et al. (2010) from University of Miami. The hypothesis driving this study was that many fishes shelter in mangrove habitat during the day but then move out to seagrass locations further offshore at night to forage, and the reason for this was high predation pressure during day in offshore habitats. To test this hypothesis, they tethered pinfish in various locations near mangrove locations and evaluated their consumption by predators based on location and time of day. To support their study, they did a number of tethering evaluations to show limited impact on the tethered prey as well as best methods for tethering their prey.

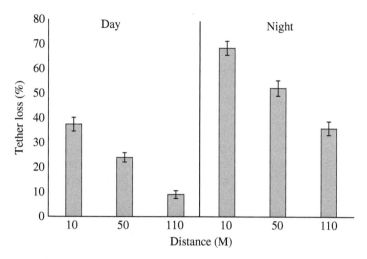

FIGURE 20-6   Percent predation of tethered pinfish related to mangrove proximity during day and night. Error bars indicate ± 1 SE. *Source: Adapted from Hammerschlag et al. (2010).*

Their results showed pinfish were more vulnerable to predators in locations at the interface between mangroves and eelgrass and that they also were much more vulnerable at night (Figure 20-6). The first result agrees well with the earlier study on bass in habitats with edge effects as well as a number of tethering studies in marine habitats. The latter result contradicts the commonly accepted ideas for behavior of prey fish in mangrove habitats. Nocturnal activity was believed to be an antipredator response, while in this experiment, it occurred at the time when predation risk was maximum. The authors believed that nocturnal movements were primarily driven by higher abundance of their prey at night, and that once the pinfish cross over the edge to open seagrass locations, they were relatively safe from predation at night. They also showed support once again for piscivores using dawn and dusk, or even night, as a time to focus foraging to take advantage of changing prey behavior as well as visual advantages predators might have, although neither were directly measured in this study.

Foraging and Predation Risk

The effect of predation risk is difficult to understand and evaluate. Fish may be able to assess and deal with food abundance by regular sampling. They can determine highest energy returns and shift habitats from vegetation to open water in order to maximize their rates of return. When fish are put into tanks with predators, they utilize predation escape behaviors. They school when they are in open water; they hide in vegetation when there is structure, but this is in the absence of food. When fish have to trade-off optimal foraging and predation risk, they are dealing with very different kinds of decisions. Optimal foraging measures food return and comfort in terms of feeding. Predation risk measures the chances of being captured by a predator. The presence or absence of predators can be sensed by prey, but if a predator is successful, that individual is no longer around to react; it cannot learn predation escape any better. As a consequence, predation is evaluated as a risk rather than as an actual occurrence. Of course, one individual fish might move into open water, and although there may be many predators in that habitat, the prey fish may not encounter them. Its concept of predation risk might be relatively low. Another individual might encounter many predators as it moves into open water, and its concept of predation risk might be very high. Differential experience may be very important in assessing predation risk, while in foraging, individual experiences probably average out among individuals.

Does an individual fish really have the ability to assess complex environmental conditions and respond in a predictable fashion? Gilliam and Fraser (1987) theorized foraging under risk of predation should result in animals minimizing the ratio of mortality rate (μ) to growth rate (g). This pattern has been labeled μ/g or Gilliam's rule (Mittelbach 2002) and could be used to predict optimal habitat use under foraging and predation constraints. Foraging patches of food and selecting sizes of food involve a fair amount of learning, but an even higher level of learning must be accomplished to trade-off feeding and predation risk. Are such costs and constraints as measured in the lab sufficiently realistic to predict behavior in nature?

Foraging and Predation Risk in Bluegill

It is a common observation that centrarchids such as bass, bluegill, pumpkinseed, and green sunfish demonstrate common behavior when they are small by remaining in macrophyte beds. They do not choose their habitats, as might be predicted, on the basis of optimal foraging, that is, they do not shift to different habitats because of their differential abilities to forage in those habitats. This may be a response to predation risk, and young centrarchids may be in vegetation because of the risk of predation in open water. Werner and his colleagues assumed in their work that open water was a risky habitat. That might be true in ponds in which open water habitat is limited, but it actually may be the transition of moving from one habitat to another that might be the biggest risk, rather than being in open water once that edge is crossed.

Werner et al. (1983) tested a foraging–predation risk hypothesis in ponds. They selected ponds purposely designed to have good forage in sediments and in open water but limited forage in vegetation. For these ponds, optimally foraging bluegill should utilize open water, where predation risk should be highest. In one experiment, they used a pond with a band of vegetation in shallow water and divided it in half with a net. On one side, they combined both predators and prey, and on the other, just prey as a control. This is actually a pseudo control because the two sides of the pond had very different progressions of plankton populations. In any event, optimal foraging would predict that bluegill should forage in open water, while predation risk would indicate that they should remain in the vegetation refuge.

On the experimental side of the pond system, they released eight bass and a mixed bluegill population with 500 small (1 1/2-inch), 300 medium (2-inch), and 100 large (3-inch) bluegill. The design allowed bass to easily consume the smallest bluegill and possibly eat a few intermediate bluegills, but the bass could not swallow the largest bluegill. Therefore, the design was fairly complex, including predation prone and predation free habitats, plus individuals in each habitat that were immune to predation or vulnerable depending on size. There were really two controls: behavior of the largest bluegill compared to the smallest and behavior of fish in predator or predator-free sides of the pond.

Once a week, ten bluegill from each size class were removed for stomach content analysis to assess where the fish foraged. These were replaced with ten more fish to keep density from decreasing artificially over time, which would influence what bluegill ate. A second challenge in this design was if bass foraged on bluegill, there would be fewer bluegill on the predator side of the pond over time. Since density has a strong effect on foraging, results from the two sides of the pond might differ in prey population structure because of bluegill density rather than because of bass effects on bluegill habitat selection. Therefore, bluegill were added to the predator side of the pond at a rate predicted to balance the number eaten by bass. Initially, 20 new fish were added per week, while toward the end of the experiment 10 were added per week. All added bluegill were of the smallest size (most vulnerable), and the number added was based on the assumption that a bass ate one bluegill every three days. As the experiment progressed, small bluegill grew, so bass would consume less in numbers, but about the same weight. Overall, 144 bluegill were added. The idea was not to end up with 900 bluegill, because some bluegill would die on the predator free side as well, but rather to end up with the same number of bluegill on both sides.

Mortality rate in the pond differed among size classes much as expected. Medium-sized bluegill had a 10% mortality rate on both sides of the pond, so the presence or absence of bass had no effect on their mortality rate. Large bluegill had an 11% mortality rate on the side with bass and 19% on the other side. Calculating mortality rate is complicated for the smallest size class because fish were added over time. At the end of the experiment, there were 348 small bluegill on the side with bass, and 359 on the side without bass, indicating the estimates for supplementing bluegill were reasonable. If the 348 small prey total is divided by 644 total fish on that side, the mortality rate for small bluegill was 46% on the predator side. Mortality of small bluegill was 28% on the control side, so mortality rate was almost double on the side with predators. The average bass ate 1 bluegill every 3.8 days, which is not very frequent compared to the common belief that fish predators such as bass are ravenous consumers that might eat many prey daily. Most studies of food habits demonstrate a similar slow rate of natural prey consumption. The only unexpected result for this part of the experiment was a higher mortality rate for large bluegill in the absence of predators.

Optimal foraging predictions indicated that all three sizes of bluegill should have initially foraged in the pelagic zone on zooplankton and then, as the season progressed, shifted to sediments. Werner et al. even predicted that the shift into sediments should occur in July. As predicted, the behavior of bluegill on the side without bass was to first to forage pelagically and then move to the sediment as the year progressed. In September, when plankton made a rebound, the smallest size groups switched back to feeding on plankton again. In the presence of bass, medium- and large-sized bluegill showed the same response, but behavior of small bluegill changed considerably.

In the absence of predators, 9% of the diet of small bluegill came from the vegetation (Table 20-2). That number increased to 34% with the presence of predators. In the absence of predators, 69% of the diet of small bluegill was benthos and the remainder plankton. For small bluegill in the presence of predators, a much lower fraction of their diets included benthos. These shifts in behavior, while significant, did not result in the choice of safe behavior by all small bluegill. Even in the presence of predators, small bluegill mostly foraged on benthos, either in an area where predation risk was still high or in sediments below the vegetation.

TABLE 20-2 Average Percent Composition of the Diet of Bluegill by Habitat for Three Size Classes.

Treatment	Size Class	Vegetation	Plankton	Benthos
No predator	Small[1]	9	19	69
	Medium	14	5	81
	Large	11	–	81
Predator	Small[1]	34	17	46
	Medium	9	16	74
	Large	14	6	78

Source: Adapted From Werner et al. (1983).
[1] Significant treatment effect.

While small bluegill in the presence of predators ate a higher fraction of prey inhabiting vegetation than small bluegill in absence of predators, they still ate mainly prey from other habitats.

Foraging responses of individual fishes did not necessarily reflect population averages but demonstrated several unique characteristics (Figure 20-7). Bluegill on the control side averaged about 10% prey from vegetation in their gut, and individuals were fairly consistent in diet composition. A few fish had no prey from vegetation, and not many had more than 20%. On the predator side, there was considerable variation among individuals. Some did not eat any prey from vegetation (they foraged entirely in open water), while some ate 100% vegetative prey. Werner et al. believed that some individuals might have been willing to risk moving into open water and foraging in spite of the presence of predators, while other individuals might have been conservative and spent their time hiding from predators. These data, with a fair degree of inter-individual variability in assessment of predation risk and foraging response, conform to that belief. Some individuals behaved as they would without predators, while others behaved quite differently. There is no indication of whether an individual that foraged in open water one day did so the next day, or whether individual risk assessment changed each day. However, clearly some individuals appeared to be risk averse, and others seemed to be risk

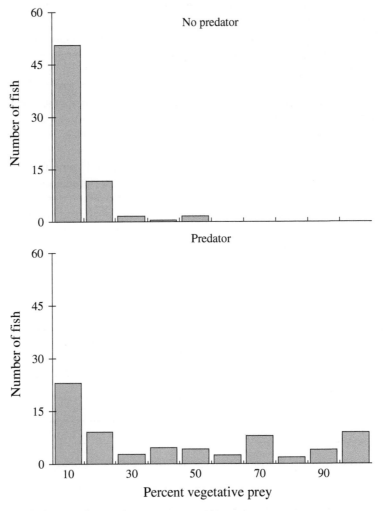

FIGURE 20-7 Number of small bluegill containing different percentages of prey derived from vegetation in presence or absence of bass. *Source: Adapted from Werner et al. (1983).*

prone. An individual's concept of predation risk may depend on its recent history – whether or not it has observed predators in open water. That history could vary among individuals. Bluegill appear able to assess risk of predation and changes in forage abundance and alter their behavior accordingly. However, small fish, which were more vulnerable to predation, did not demonstrate a predictable, all-or-none response.

Reduced growth rate has two effects on a prey population. The first is prey achieve a given size at a slower rate, which means they may have lower fitness, because it will take them longer to reach maturation size. Second, fish that grow slowly remain in a vulnerable size range for predation much longer. For small bluegill, halving their growth rate would be the same as doubling their mortality rate because they would be exposed to predators for twice as long. Therefore, growth and mortality are intertwined when evaluating the effect of a predator. Werner et al. tried to sort this out based on their measures. They found small bluegill in the presence of bass had a 37% reduction in survival compared to small bluegill in the absence of bass. Of that 37% reduction, 10% was due to a slower growth rate that exposed bluegill to predators for a longer time, 23% was due to the predators' direct consumption, and 4% was due to an interaction between these two factors.

One question remains, as in earlier experiments, regarding whether this response of fish in ponds also occurs in a larger environment. If predators use edges, prey are exposed to predators at higher rates in ponds than they might be in lakes in which they might move longer distances away from edges and remain there. Werner and Hall (1988) predicted the size at which bluegill should leave vegetation in natural lakes based on forage abundance and predation risk. These predictions matched well with observed size at which bluegill were found in open water, supporting the generality of their pond experiments.

Since vegetation is an important refuge from predation, many different species of centrarchids may remain in vegetation as young individuals. Competition may be quite severe for food resources or space in vegetation under those conditions. This competition may actually explain why bluegill dominate in abundance in most natural lakes although they do poorest of the bluegill–pumpkinseed–green sunfish combination in a pond. Bluegill are generalists and at small sizes are more capable of foraging on a variety of food than are the other species. Bluegill may dominate competition when all the species are restricted to vegetation and competing for food. Once bluegill are adults, they lose the competitive advantage, but if their size structure and number in each size class are dependent on surviving through the early refuge stage, adult competition may not matter as much.

Foraging and Predation Risk by Armored Catfish

Strong spatial gradients in predation risk and foraging gains also occur in streams. Mary Power from the University of California, Berkeley, has focused on this research topic for some time. She initially studied herbivorous armored catfish (family Loricariidae) in a Panamanian stream. These fish consume attached algae and periphyton. Such algal growth is more rapid with denser biomass in sunny reaches of streams, while areas with dense forest canopy have reduced algal biomass. Densities of catfish are negatively correlated to forest canopy, indicating that they forage more in sunny pools with higher algal production (Power 1984). The stream system in Rio Frijoles alternated between sunny, shallow riffles (5–40 cm deep) with cobble substrate, and pools (40–200 cm deep) with bedrock, cobble, or depositional materials.

Shallow riffles with cobble substrate provide areas of refuge for small catfish as these fish can hide in substrate and escape predation by birds. Larger catfish are seldom found in these riffles but move into deeper pools apparently to avoid avian predation. High standing crops of algae developed in the shallows of this stream due to limited herbivory by the mainly small catfish there. Power et al. (1989) tried to determine mechanisms for the distributions of fish and algae. They focused on two hypotheses: (1) grazing catfish limit algal abundance and (2) bird predators constrain catfish depth distribution.

For the first hypothesis, the effect of catfish removal on growth of attached algae in pools was evaluated. Three pairs of pools were manipulated by snorkeling and hand netting. Large numbers of catfish were removed from one pool, and the second pool of each pair remained as a control for six pools in three stream sections. Algal biomass above and below the 20-cm depth zone in each pool was then estimated. While some samples were lost in transport from Panama, there were enough to compare biomass in dark and half-shaded pools after 10 days. Control pools had 4–20 times higher algal biomass in shallow water than in deep water, while removal pools had similar algal biomass in both depth zones. The difference in algae with depth in control pools thus appeared to be correlated to catfish grazing.

For the second hypothesis, large catfish were placed in pens with open tops, and their numbers evaluated over time. Pens were all located in gravel substrates in 10, 20, 30, and 50 cm of water depth. Three replicate reaches of stream each had four pens installed, and reaches were chosen at locations where little blue herons had been observed feeding. Pens were stocked with similar numbers and sizes of two species of catfish and left for 12 days before re-census. The results for one stream reach (Figure 20-8) showed nearly complete mortality of *Rinelori-caria* from pens at 10 and 20 cm, with limited loss in pens from deeper water. Similar but less extreme trends were found for *Ancistrus* with losses in shallow pens of approximately half the fish. In the other two stream sections, no catfish were lost from any pens, indicating a lack of predation there. Still, results in the one reach supported the concept that predation was most intense for larger catfish in shallower water, and this process restricted catfish to deeper pools.

Overall, these studies indicated that armored catfish altered their foraging behavior under risk of predation, with larger fish foraging mainly in deep water to avoid avian predation while smaller fish could forage mainly in shallower water where they could also find refuge in substrate. In another study, Power et al. (1985) demonstrated that minnows grazing on algae showed similar avoidance of stream pools with predatory bass. In many stream systems with large predatory fish, it appears that small juvenile fish are restricted to shallow areas by predation pressure from larger fish in pools, while those larger fish do not commonly penetrate shallow areas due to predation pressure from fish-eating birds and mammals. Thus, both sizes of fish show foraging ranges restricted by predation pressure.

Assessment of Predation Risk by Sculpin

The previous experiments indicated bluegill were able to assess risk of exposure to predators and alter their behavior both in the lab and the field. The methods to assess this risk and the nature of the risk assessment were not determined. Chivers et al. (2001) conducted experiments to evaluate risk assessment in sculpin. The basis of their work was that animals should be able to determine the risk of exposure to a threatening predator and react to that but not to non-threatening animals. The reasons for this are fairly clear. If an animal reacted in fright or hiding to any larger animal, it would lose much valuable foraging time and would not grow well. Therefore, animals should be able to distinguish risks from threatening and non-threatening animals and alter their response accordingly. This is called threat-sensitive avoidance behavior.

Sculpin are common stream fishes that live in association with brook trout in many areas. Only large brook trout are capable of eating sculpin, so the risk of predation is dependent on the size of trout and sculpin. Sculpin may sense risk visually or chemically. Chivers et al. conducted field and lab experiments to test whether sculpin showed appropriate threat-sensitive responses to brook trout.

The first experiment involved cages in the field, where trout of various sizes were caged in a stream, and the distribution of different sized sculpin was assessed related to presence and size of

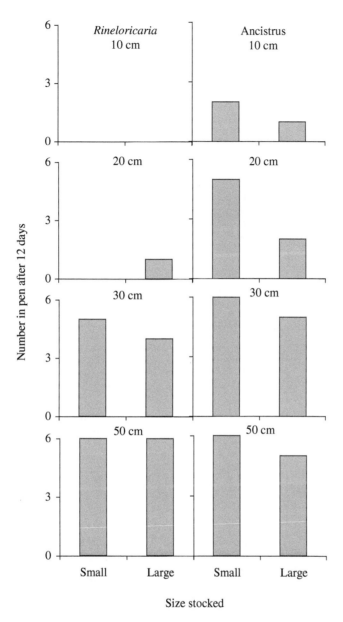

FIGURE 20-8 The number of *Rineloricaria* and *Ancistrus* remaining from six of each size originally stocked in pens at different depths after 12 days of exposure to predators in a Panamanian stream. *Source: Adapted from Power et al. (1989).*

trout. In this experiment, small sculpin were less common when large trout were caged in an area, but small sculpin did not avoid areas with small trout, and other sizes of sculpin did not react to presence of large trout. While this experiment includes several uncontrolled variables related to other factors influencing field distribution of fish, it at least demonstrates that sculpin distributions were in agreement with threat-sensitive predator avoidance under natural conditions.

The more interesting part of Chivers et al. experiments was the lab work exposing sculpin to visual and chemical cues of trout. For visual cues, the fish were put into adjacent tanks with visual contact, and barriers were put in place to test distributions of sculpin in visual presence or absence of trout. When small or medium sculpin were in visual presence of large trout, they showed a reduction in the number of short and long moves that would make them visually more obvious to the predator (Figure 20-9). Large sculpin, on the other hand, showed no differences in movement in visual presence or absence of trout. When smaller trout were used as predators, only the smallest sculpin showed a decrease in movement pattern.

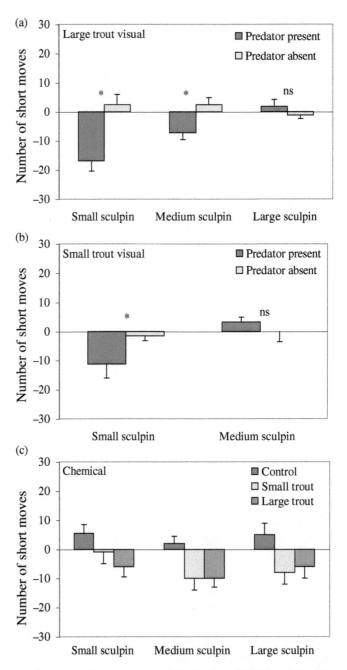

FIGURE 20-9 Changes (mean + SE) in numbers of short moves by different size classes of sculpin during exposure to various predatory cues: (a) presence or absence of visual cues for large trout; (b) presence or absence of visual cues for small trout; and (c) presence or absence of chemical cues for different sizes of trout. *Source: Adapted from Chivers et al. (2001).*

A final experiment was conducted using chemical cues. Trout were held in water for 24 hours without feeding, and then 60 ml of the water was put into tanks with sculpin. Two treatments were used: one with a single large trout and other with a number of smaller trout to produce an equal biomass. Sculpin in test tanks were exposed to cues from small trout, large trout, or control tanks. Once again, the sculpin reduced the frequency of short and long moves when the chemical cue for trout was present, but there was no difference among sizes of sculpin or to cues from large or small trout. In this case, all sculpin reacted equally to the chemical sense of trout and appeared to be unable to detect large from small trout or to use threat-sensitive reactions. The authors believe the chemical sense of trout probably occurs from a longer distance and makes sculpin aware of the potential presence of a predator, while visual cues are reacted to more strongly and in ways that are appropriate to the risk of predation posed by a trout that is seen. They also believe the chemical sense of trout size may be from chemicals egested from trout that have recently eaten sculpin, so this experiment may have eliminated that cue. In any case, these experiments provide interesting results regarding the ability of sculpin to react appropriately to risk of predation detected by both visual and chemical senses.

Summary

Predator–prey interactions are very important in aquatic communities, where predation may structure the abundance, behavior, and habitat selection of prey species. Various habitat elements, such as vegetation, substrate size, or other discontinuities, may alter the behavior and success of both predators and prey. Both predators and prey appear to learn to use such structures and adapt behavior to perform better in these structured environments. Such learning often involves complex trade-offs, such as the trade-off between larger foraging returns and increased risk of predation. Fish appear capable of reacting to such trade-offs in ways that benefit their long-term fitness.

Literature Cited

Anderson, O. 1984. Optimal foraging by largemouth bass in structured environments. Ecology 65:851–861.

Chivers, D.P., R.S. Mirza, P.J. Bryer, and J.M. Kiesecker. 2001. Threat-sensitive predator avoidance by slimy sculpins: Understanding the importance of visual versus chemical information. Canadian Journal of Zoology 79:867–873.

Gilliam, J.F., and D.F. Fraser. 1987. Habitat selection when foraging under predation hazard: A model and a test with stream-dwelling minnows. Ecology 68:1856–1862.

Hammerschlag, N., M.R. Heithaus, and J.E. Serafy. 2010. Influence of predation risk and food supply on nocturnal fish foraging distributions along a mangrove–seagrass ecotone. Marine Ecology Progress Series 414:223–235.

Jacobsen, L., and M.R. Perrow. 1998. Predation risk from piscivorous fish influencing the diel use of macrophytes by planktivorous fish in experimental ponds. Ecology of Freshwater Fish 7:78–86.

Magoulick, D.D., 2004. Effect of predation risk on habitat selection by water column fish, benthic fish and crayfish in stream pools. Hydrobiologia 527:209–221.

Mittelbach, G.G. 2002. Fish foraging and habitat choice: A theoretical perspective. Pages 251–266 in J.B. Hart and J.D. Reynolds, editors. Handbook of Fish Biology and Fisheries. Vol. 1. Fish biology. Blackwell Science Ltd., Oxford, UK.

Olson, M.H., and nine co-authors. 1998. Managing macrophytes to improve fish growth: As multi-lake experiment. Fisheries 23(2):6–12.

Power, M.E. 1984. Habitat quality and the distribution of algae-grazing catfish in a Panamanian stream. Journal of Animal Ecology 53:357–374.

Power, M.E., T.L. Dudley, and S.D. Cooper. 1989. Grazing catfish, fishing birds, and attached algae in a Panamanian stream. Environmental Biology of Fishes 26:285–294.

Power, M.E., W.J. Matthews, and A.J. Stewart. 1985. Grazing minnows, piscivorous bass, and stream algae: Dynamics of a strong interaction. Ecology 66:14480–1456.

Savino, J.F., and R.A. Stein. 1982. Predator–prey interaction between largemouth bass and bluegills as influenced by simulated, submersed vegetation. Transactions of the American Fisheries Society 111:255–266.

Smith, K.D. 1993. Vegetation-Open Water Interface and The Predator–Prey Interaction Between Largemouth Bass and Bluegills. Ph.D. dissertation, University of Michigan, Ann Arbor.

Werner, E.E., and D.J. Hall. 1988. Ontogenetic habitat shifts in bluegill: The foraging rate-predation risk trade-off. Ecology 69:1352–1366.

Werner, E.E., J.F. Gilliam, D.J. Hall, and G.G. Mittelbach. 1983. An experimental test of the effects of predation risk on habitat use in fish. Ecology 64:1540–1548.

PART 5

Reproduction and Life Histories

CHAPTER 21

Reproductive Traits

Fishes exhibit a diversity of reproductive traits both intra- and inter-specifically. They greatly differ in terms of the number and size of eggs they produce, their age and size when they first mature and reproduce, and whether they spawn once or multiple times per year. Most fish eggs are fertilized externally, but some species reproduce through internal fertilization, and while most fishes do not provide direct nutrition to developing eggs, some species provide placental support to developing offspring. Some fishes build nests and guard their developing offspring, while others broadcast their eggs and provide no additional parental support after egg deposition. For many fishes, initial sex of an individual is primarily determined genetically, but for some sex expression is related to environmental factors, such as temperature during development. And while many fishes remain a single sex throughout their lives, some species undergo sex changes later in life. Finally, some fish species spawn once and then die, while others spawn over several consecutive years.

The expression of these diverse reproductive traits is seemingly related to different fishes adapting approaches that allow for successful reproduction and increased fitness, given the constraints of their own morphology and physiology as well as the constraints of the environments they inhabit. In Chapter 23, we highlight some of the reproductive constraints placed upon fishes and present some frameworks to describe and explain the set of life histories expressed by fishes. First, we focus on describing the reproductive diversity exhibited by fishes.

Reproductive Differences

Fertilization

Key differences in fish reproduction relate to how eggs are fertilized and how nutrition is provided to developing embryos. The most common type of reproduction in fishes is ovipary, meaning these fishes lay eggs that are fertilized externally. As these fertilized eggs develop, they obtain nutrition from yolk reserves contained within each individual egg. At the other extreme, viviparity is a less common reproductive type among fishes including some sharks and some members of the family Poeciliidae, which involves internal fertilization and placental development of embryos. An intermediate reproductive approach involves internal fertilization of eggs, which are then held internally until young are born live, with little or no placental involvement in egg development. This reproductive approach was historically referred to as ovoviviparity, but this term has decreased in use because across animal species – and even among fish species – there are actually a variety of different reproductive approaches that may

Biology and Ecology of Fishes, Third Edition. James S. Diana and Tomas O. Höök.

all be grouped as ovoviviparity. For example, some internally fertilized and incubating fish eggs may obtain maternal nutrition without actual placental development. And some species of fish may be able to reproduce without fertilization as asexual parthenogenesis has been confirmed in several species of fishes, including bonnethead shark and smalltooth sawfish.

Sex Determination

Another highly variable aspect of reproduction among fishes is sex determination. For many species of fish, sex is primarily determined genetically, and once sex is determined, it remains fixed throughout an individual's life. However, environmental conditions during early development, in particular temperature, can also affect phenotypic sexual expression. Temperature-dependent sex determination has been documented for a number of fish species and was first documented and then broadly studied by David Conover and colleagues researching Atlantic silverside. Ospina-Álvarez and Piferrer (2008) reviewed temperature-dependent sex determination in fishes and highlighted that, for many species that exhibit genetically determined sex, exposure to certain temperatures during early development may disrupt this process. Therefore, they differentiated between "genotypic sex determination plus temperature effects" versus true temperature-dependent sex determination. They concluded that a subset of fish species actually exhibited true temperature-dependent sex determination, and for all of these species, the proportion of males increased with warmer temperatures during incubation. Regardless of whether sex is truly determined by temperature or whether environmental conditions simply disrupt otherwise genetically determined processes of sexual expression, the processes of environmental conditions influencing phenotypic sex expression can be complex. For example, phenotypic sex expression may not only be a function of mean temperatures experienced but also the variability of temperatures experienced. Coulter et al. (2015) demonstrated that when fathead minnow eggs were held consistently at 21, 25, or 29 °C, their phenotypic sex expression did not deviate from genotypic expression. But when eggs were incubated at temperatures fluctuating from 21 to 29 °C, some individuals reversed sex. The potential of temperature during early development to influence phenotypic expression is taken advantage of in some aquaculture practices to only produce male or female fish as the sexes can differ in terms of growth rates and maximum sizes. Changes in environmental temperatures could potentially alter sex expression in the future. For example, Ospina-Álvarez and Piferrer (2008) suggested that with plausible climate warming the sex ratio of fish species that exhibit true temperature-dependent sex determination could shift from 1:1 to 3:1 male to female.

Individuals of some species of fish will display characteristics of both males and females during their life. Most fish species exhibit gonochorism – where individuals maintain separate sexes throughout life and do not change sexes. However, some fishes are hermaphroditic and express both male and female gonads. There are different types of hermaphroditism. Simultaneous hermaphroditism – homogamy – implies that individuals simultaneously possess both male and female gonads. This type of hermaphroditism is less common among fishes but is evident among some families, including some species of hamlets and gobies. Sequential hermaphroditism is more common among fishes and involves a switch during life when individuals transition from expressing gonads of one sex to expressing gonads of another sex. Protandry involves transitioning from male gonad expression to female expression and is evident in species such as clownfish. Protogyny involves transition from initial expression of female gonads to express male gonads. Well studied examples of protogynous fishes include many species of grouper and wrasses. Avise and Mank (2009) provided an interesting review of hermaphroditism in fishes. The selective forces that have led to the evolution of hermaphroditism in fishes are not fully understood. Sequential hermaphroditism may be related to different size advantages when reproducing for male and female fish and may be triggered by social cues.

For example, the loss of females or males in a local area may signal other individuals to change sex. If size has a very strong positive effect on reproductive success for one sex but not the other, it is plausible that this could favor sequential hermaphroditism. Synchronous hermaphroditism may develop when densities are low and the probability of encountering an individual of the opposite sex would be low.

Number of Spawning Events

Fishes greatly differ in terms of the number of spawning events they undertake during their lives. Many fish species are iteroparous, meaning they have multiple reproductive cycles over the course of their lives. This implies that individuals will spawn over several consecutive years. However, in cases where there are limited resources and individuals need to build up sufficient energy for gonadal development, they may skip spawning some years. Skipped spawning is particularly evident among females in large, long-lived species such as sturgeon and lake whitefish. In contrast to iteroparity, semelparous fishes only spawn once and then die. Classic examples of semelparous fishes include many species of Pacific salmon and migrating eels. These fishes undertake long spawning migrations and then die shortly after reproduction. Annual species display a particular type of semelparity in which they only live a single year and reproduce once. Atlantic silverside is an example of an annual species, and this feature has made it very useful in multi-generational inheritance studies as it has a fixed generation time and complete population turnover each year.

Fish species not only differ in terms of how many years they spawn during their lives, but they also differ in the number of spawning events during a single spawning season. Many female fish deposit their eggs during a single event – synchronous spawning – but others are batch spawners – asynchronus spawning. Batch spawning fish deposit eggs on multiple occasions over an extended spawning season. These fish can be either determinate or indeterminate batch spawners. For determinate spawners, the number of eggs produced is fixed at the beginning of the spawning season and the eggs are simply not released at the same time. Indeterminate batch spawners have the potential to develop and release new eggs through the spawning season. Differences in the number of spawning events in a year (synchronous versus batch spawning) are often associated with when energy is allocated to reproduction – capital versus income strategies (see below).

Maturation

Regardless of the mechanisms of sex determination and reproduction, animals such as fish are considered immature before they reproduce and mature once they produce viable gametes and reproduce. Therefore, maturation is a binary process; once an individual fish becomes mature it remains mature throughout the rest of its life even if it skips spawning during a particular year or does not contain viable gametes during a given season. While factors influencing maturation differ among species, for many fishes, maturation is positively related to size and age. When individual fish become sufficiently large and old, they have potential to mature. However, size and age thresholds for maturation can vary both inter- and intra-specifically. Maturation thresholds generally vary by sex with females, often maturing at larger sizes and older ages than males, but the opposite occurs for some species. When one sex matures at a different size than the other sex, this is a type of sexual size dimorphism. Curtis Horne and colleagues (2020) explored patterns of sex-biased size at maturation across fish species. They found that female-biased size at maturation was more common and seemingly related to fecundity selection and benefits related to larger females having more energy available for reproduction. However, they also documented male-biased size at maturation for several species, including among species of wrasses. They suggested that relatively large size at maturation for males was more likely to occur for species in which larger males were better able to hold spawning territories, larger males were able to provide better parental care or when there was strong sperm competition for fertilization.

The size and age at maturation are seemingly controlled by both genetic and environmental factors. Populations of the same species often vary in terms of their maturation schedules, and maturation schedules within a population may change over time. However, the relative role of genetic differences versus environmental factors in influencing maturation schedules is generally unclear. Even if two populations were genetically identical, they could exhibit different maturation schedules. That is, if environmental conditions (e.g., temperature and prey availability) allow for faster growth in one population, individuals in this population may on average mature at a younger age and potentially greater size than the other population.

Maturation schedules have important implications for how fish stocks may respond to harvest or other types of perturbations, and therefore it is useful to be able to quantify maturation schedules. For example, if fish in a population are intensively harvested at small sizes before they have an opportunity to mature, such a population may be likely to severely decline due to limited recruitment success. There are many ways to describe maturation patterns of fish populations. The percent of mature individuals of a given age can be quantified. For anadromous, iteroparous populations – such as steelhead – the proportion of first-time versus repeat spawners in a river may be assessed. One of the most common approaches for describing maturation schedules involves estimation of age and length at 50% maturity. This approach involves fitting a logistic regression to data collected for individual fish (i), including individual age or length (independent variable, X) and maturation status (binary-dependent variable, Mat). A logistic regression involves fitting a sigmoid curve to data ranging from 0 to 1 to estimate Mat, the probability of maturity:

$$\text{Mat} = \frac{1}{1 + e^{-(\beta_0 + \beta_1 X_i)}} \qquad \text{(Equation 21-1)}$$

This model includes two parameters, β_0 and β_1, and the estimated values of these parameters can be used to estimate age and length at 50% maturity. Specifically, age and length at 50% maturity can be estimated as $-\beta_0/\beta_1$.

Hui-Yu Wang and colleagues (Wang et al. 2008) compared maturation schedules of lake whitefish in lakes Huron, Michigan, and Superior from the late 1970s to the early 2000s. During this time, growth rates of whitefish in lakes Huron and Michigan declined. Densities of *Diporeia* – an energy-rich amphipod that historically was the main prey for whitefish in these lakes – dropped precipitously in lakes Huron and Michigan, and whitefish switched to consume less energy-rich prey in these lakes. However, *Diporeia* did not decline in Lake Superior, and whitefish maintained similar growth rates in this lake over time. In estimating age and length at 50% maturity, Wang et al. (2008) found that female whitefish in all three lakes tended to mature at older ages and larger sizes than male whitefish. In addition, maturation schedules differed among the lakes, with lake whitefish in Lake Michigan having lower ages and lengths at 50% maturity compared to lake whitefish in lakes Huron and Superior.

Wang et al. compared maturation schedules between cohorts born before and after 1990; this year roughly corresponded to whether cohorts would have had access to high densities of *Diporeia* in lakes Huron and Michigan. They found that the age at 50% maturity for both male and female lake whitefish increased in lakes Huron and Michigan from pre-1990 to post-1990 cohorts. In contrast, they did not see a change in age at 50% maturity in Lake Superior.

They also found that lengths at 50% maturity did not change in any consistent manners. They attributed these temporal patterns to reduced growth rates in lakes Michigan and Huron, preventing lake whitefish from quickly reaching an appropriate size for maturation. This contributed to increased age at 50% maturity in lakes Huron and Michigan for both sexes.

Gonadal Investment

While maturation is binary, the amount of energy and time invested toward reproduction is not. Individual fish may greatly differ in terms of how much energy they invest in reproduction, including energy invested to gonadal development and gamete production. Gonadal investment can be measured in several ways, including fecundity. Fecundity is often used to refer to number of offspring produced by an individual organism over its lifetime. However, when describing fecundity of fish, it is common to index fecundity as the number of eggs produced by a female during a single spawning season. This is referred to as absolute fecundity, which differs from relative fecundity – eggs produced per unit mass of a female. Duarte and Alcaraz (1989) compiled information on fecundities of teleost fish species. Out of 230 studies – including 51 marine and 46 freshwater species, respectively – they reported a mean absolute fecundity of 1,836,000 with a range from 7 to 57,600,000.

An important aspect of gonadal investment and fecundity is when the investment takes place. Fishes are generally described as capital or income spawners. Capital spawners initially invest energy to storage tissue and to develop gonads and eggs, allowing them to spawn at a later stage. In so doing, capital spawners largely determine their fecundity well in advance of the time they actually spawn This is in contrast to income spawners, who allocate energy from food directly to reproduction. So, the fecundity achieved by income spawners may not be determined until just prior to spawning and will be closely related to the amount and quality of food consumed.

Gonadal investment can be expressed not only in terms of fecundity but also in terms of gonad size. The mass of fish gonads will differ with fish size, and therefore gonad size is often standardized to fish size. Gonadosomatic index (GSI) is the mass of an individual fish's gonad divided by the fish's total mass. Therefore, GSI can theoretically take values from 0 to 1 or 0 to 100%. GSI values are generally less than 10% for males and less than 35% for females. GSI values will differ across seasons, and for capital investing, synchronous spawners GSI values are generally greatest just prior to spawning (Figure 21-1). Seasonal patterns of GSI may look rather different when contrasting capital- versus income-investing species and when considering synchronous or batch spawners.

Similar to absolute fecundity, the relative size of gonads just prior to spawning will vary among species. GSI also often differs among individuals of the same species. As fish grow older and larger, their relative fecundity and GSI in many species will increase such that they dedicate a larger proportion of their body energy to gonad development and gamete production at larger sizes than they do at smaller sizes. Also, mean GSI values often differ among populations of the same species. Different populations will inhabit environments that may allow for variable energy acquisition and growth and may thereby facilitate differential gonadal development. Long-term conditions may also select for evolution of dissimilar levels of reproductive investment across environments leading to different GSI expressions among populations. So, inter-population differences in mean GSI likely reflect both environmental conditions experienced during the lifetime of individual fish as well as genetic differences among populations (Figure 21-2).

Most fish gonads can be viewed as vessels for developing and holding milt and eggs. Gonad size and fecundity are generally positively related within species, and GSI is strongly related to

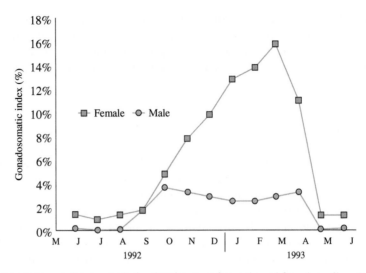

FIGURE 21-1 Monthly gonadosomatic index (GSI) values for male and female walleye in Lake Erie during 1992 and 1993. Note that the time series begins just after spawning in spring 1992. The GSI values remain low during summer and then begin to increase during fall, and values are near maximum just prior to spawning in spring 1993. *Source: Data from Henderson et al. (1996).*

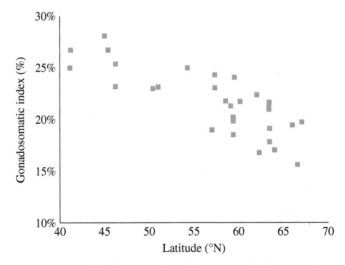

FIGURE 21-2 Patterns of female gonadosomatic index values for different populations of Eurasian perch distributed across a broad latitudinal gradient. *Source: Data from Heibo et al. (2005).*

absolute fecundity. Some species of fish do, however, produce and hold material other than just milt and eggs in their gonads. For example, the congeneric yellow perch and Eurasian perch also produce a gelatinous skein material that surrounds eggs in ovaries. When the eggs are released, the females drape skeins containing eggs over structures – such as rocks and woody debris – and the eggs are fertilized and develop within the skeins. It is unclear what the energetic cost of producing skein material is for perch, but there must be some cost, and it clearly takes up space in the ovaries, which could otherwise contain additional eggs. What then is the benefit of skein material? Zoe Almeida and colleagues (Almeida et al. 2017) analyzed the composition of yellow perch skein material and found that it contained chemicals with anti-freezing and anti-biotic properties. Skein material may therefore offer some protection from developing eggs from freezing and may reduce the likelihood of fungal or bacterial infection. Almeida and colleagues also experimentally demonstrated that while both crayfish and fish predators will readily consume yellow perch eggs if they are removed from the skein material, they are less willing to consume eggs in the skein material. The skein material is off-putting for potential predators, perhaps because it contains chemicals with known noxious properties, such as certain alkaloids. In short, perch appear to allocate some of their gonadal energy and space to material other than just eggs, but there do appear to be various benefits for the developing offspring for such allocation.

In addition to fecundity, another important aspect of gonadal investment relates to energy and size of individual gametes. Similar to fecundity, species of fish greatly differ in terms of the energy invested in individual eggs. While it is common for species of fish to produce quite small eggs, some species produce rather large eggs. Duarte and Alcaraz (1989) compiled information on not only fecundities of teleost fish species but also on egg sizes. Out of 305 reports, they found an average egg diameter of 2.3 mm with a range of 0.34–8.8 mm. As a particularly impressive example, the West Indian Ocean coelacanth (a non-teleost fish) produces individual eggs with a diameter up to 90 mm. Given constraints on energy for egg production and space in gonads, there tends to be a negative relationship between egg size and absolute fecundity. However, the specifics of such constraints will differ between capital-investing, synchronous spawners and income-investing, indeterminate batch spawners. Nonetheless, species that produce very large eggs tend to produce a relatively low number of eggs and invest a great deal of energy into each egg. Larger eggs generally yield larger larvae, and larger larvae tend to experience higher survival than smaller larvae.

An interesting inter-specific pattern among marine teleost fish species is the tendency for individuals to produce larger egg sizes in cooler environments relative to warmer environments. Anderson and Gillooly (2020) explored this pattern by developing equations to predict optimal egg size (mass) across different temperatures. Their predictions of optimal egg size incorporated considerations of how egg mass and temperature are related to various early life rates and time, including egg development time, egg mortality rate, larval growth rate, and larval mortality rate. Using such relationships, they calculated the egg mass that maximized maternal fitness across a range of temperatures (0–30 °C). Their resulting predicted relationship between temperature and optimal egg mass was negative and had a similar slope as an empirical relationship between temperature and mean egg size across 221 species; however, predicted optimal egg size was larger than mean egg size observed across species.

In addition to egg size varying among species, egg size often varies intra-specifically. Mean egg size may differ among populations of the same species. Similar to GSI, inter-population differences in mean egg size may partially reflect differences in environmental conditions and prey resources allowing some populations to obtain more energy and allocate more energy to individual eggs. Inter-population egg size may also partially result from long-term selection leading to the evolution of population-specific mean egg sizes. Within populations, mean egg size generally differs among females. Some individuals appear to produce relatively large eggs throughout their lives, and some individuals appear to produce relatively small eggs. This was demonstrated by Zach Feiner and colleagues who collected eggs from the same individual yellow perch over four consecutive years and showed that some individuals appeared to consistently produce relatively large eggs (Feiner et al. 2018). A common pattern within populations is a positive relationship between female size or age and the size of eggs produced by females. This positive maternal effect partially results from individuals increasing the size of eggs they produce as they become older and larger. Larger eggs are expected to yield larger larvae with better potential for survival. However, given energy and space constraints, production of larger eggs seemingly comes with a cost of producing a lower number of eggs. Therefore, positive maternal effects may lead larger, older female fish to sacrifice the number of offspring they produce and instead produce offspring with better chances for survival. This is a commonly studied pattern that has implications for fisheries management and the protection of large, old female fish as potentially important contributors to subsequent generations.

Multiple studies have examined egg size variation for walleye. Many walleye populations swim up rivers or gather on rocky reefs to spawn and it is therefore rather straightforward to collect walleye when they are ready to spawn. Several management agencies also collect eggs from wild walleye populations to fertilize and grow in hatchery systems for subsequent stocking into various natural systems; thereby facilitating collection of walleye eggs for size evaluations. Walleye express positive maternal size effects on egg size, but the strengths of such effects are rather weak. Walleye egg size tends to increase somewhat as females become older and larger, but there is a great deal of variation in walleye egg size that is not related to female size. Instead, walleye egg size appears to vary consistently among populations. Wang et al. (2012) and Feiner et al. (2016) demonstrated that after accounting for maternal size effects, walleye populations in the Laurentian Great Lakes differed in the mean size of eggs produced. Johnston and Leggett (2002) examined walleye eggs in smaller freshwater systems and showed that walleye egg sizes not only varied among populations but also tended to decrease with increasing latitude. Walleye egg sizes appear to respond to annual environmental condition. Feiner et al. (2016) examined egg sizes from walleye collected in the Saginaw Bay, Lake Huron system across seven consecutive years, and demonstrated year-to-year variation in mean egg sizes related to temperatures during the winter preceding spawning.

Parental Care

Parental care varies tremendously among fishes. Schooling, pelagic fishes, such as tuna, may broadcast and fertilize their eggs in open water, and the slightly buoyant or planktonic eggs drift off with the current. These species show virtually no parental care. Fishes such as salmonids might undergo long prespawning migrations, select particular spawning sites, dig nests, deposit eggs in the nests, and cover them; this demonstrates a high degree of prespawning parental care. Some fishes build nests, care for the nests until the eggs hatch, and continue to care for the young for some time. Good examples of this are bluegill and stickleback. Mouthbrooding or livebearing fishes care for the eggs until they are in very advanced stages of development. Some fishes, such as mouthbrooding cichlids, may retain eggs and larvae in their mouths to protect them, and even guard the young and allow them to hide in their mouths when danger approaches. These species demonstrate a high degree of postspawning parental care.

Patterns of parental care, from broadcast spawning to mouthbrooding or livebearing, are often associated with differences in both size and number of eggs produced. Mouthbrooding fishes usually have relatively few eggs, but because they care for those eggs, the mortality rate of young fish is relatively low. Species such as herring, which broadcast spawn, produce thousands of poorly developed young, nearly all of which die. There is obviously some relationship between mortality rate, egg size and number, and degree of parental care.

Once parental care is initiated for a given species, subsequent investments of energy in care generally become less costly. If a male builds a nest to attract a mate and spawn, the energy cost of staying on that nest for a short time to protect the eggs is low compared to the original cost of obtaining the territory and building the nest. The next behavioral steps – to stay on the nest, attract more females, and guard eggs – are fairly simple. In a progression of parental care levels, each subsequent step becomes less expensive than the last. Caring for the eggs is generally much more expensive for a parent than caring for young, because in caring for young, the adult can feed independently and need only remain near the young to protect them. The development of elaborate parental care is not surprising once the step of making a nest is taken because subsequent steps – if they improve survival of the young – are not as costly for the parent.

Summary

Wide variation in reproductive traits demonstrates the diversity of reproductive approaches employed by fishes. While external fertilization of large numbers of small eggs and limited parental care are common among fishes, there are also many species that exhibit internal fertilization, low fecundity, large egg sizes, and extreme parental care, including mouth brooding and even placental development. Individuals vary – both intra- and interspecifically – in terms of how many years they spawn throughout their life and how many times they spawn in a given year. Reproductive traits are generally related across fishes. For example, there are often trade-offs between fecundity and mean egg size, and internal fertilization and high parental care are often associated with lower fecundity and greater egg sizes. Different reproductive traits are often strongly associated with each other, and subsequent chapters describe frameworks developed to define how reproductive traits may co-vary.

Literature Cited

Almeida, L.Z., S.C. Guffey, T.A. Krieg, and T.O. Höök. 2017. Predators reject Yellow Perch egg skeins. Transactions of the American Fisheries Society 146:173–180.

Anderson, D.M., and J.F. Gillooly. 2020. Predicting egg size across temperatures in marine teleost fishes. 21:1027–1033.

Avise, J.C., and J.E. Mank. 2009. Evolutionary perspectives on hermaphroditism in fishes. Sexual Development 3:152–163.

Coulter, D.P., T.O. Höök, C.T. Mahapatra, S.C. Guffey, and M.S. Sepulveda. 2015. Fluctuating water temperatures affect development, physiological responses and cause sex reversal in fathead minnows. Environmental Science and Technology 49:1921–1928.

Duarte, C.M., and M. Alcaraz. 1989. To produce many small or few large eggs: A size-independent reproductive tactic of fish. Oecologia 80:401–404.

Feiner, Z.S., H.-Y. Wang, D.W. Einhouse, et al. 2016. Population and maternal influences on egg size and maternal effects in a freshwater fish. Ecosphere 7(5).

Feiner, Z.S., T.D. Malinich, and T.O. Höök. 2018. Interacting effects of identity, size, and winter severity determine temporal consistency of offspring phenotype. Canadian Journal of Fisheries and Aquatic Sciences 75:1337–1345.

Heibo, E., C. Magnhagen, and L.A. Vøllestad. 2005. Latitudinal variation in life-history traits in Eurasian perch. Ecology 86:3377–3386.

Henderson, B.A., J.L. Wong, and S.J. Nepszy. 1996. Reproduction of walleye in Lake Erie: allocation of energy. Canadian Journal of Fisheries and Aquatic Sciences 53:127–133.

Horne, C.R., A.G. Hirst, and D. Atkinson. 2020. Selection for increased male size predicts variation in sexual size dimorphism among fish species. Proceedings of the Royal Society B 287:20192640.

Johnston, T.A., and W.C. Leggett. 2002. Maternal and environmental gradients in the egg size of an iteroparous fish. Ecology 83:1777–1791.

Ospina-Álvarez, N., and F. Piferrer. 2008. Temperature-dependent sex determination in fish revisited: prevalence, a single sex ratio response pattern, and possible effects of climate change. PLoS ONE 3(7):e2837. doi:10.1371/journal.pone.0002837.

Wang, H.Y., T.O. Höök, M.P. Ebener, L.C. Mohr, and P.J. Schneeberger. 2008. Spatial and temporal variation of maturation schedules of lake whitefish (*Coregonus clupeaformis*) in the Great Lakes. Canadian Journal of Fisheries and Aquatic Sciences 65:2157–2169.

Wang, H.Y., D.W. Einhouse, D.G. Fielder, et al. 2012. Maternal and stock effects on egg-size variation among walleye *Sander vitreus* stocks from the Great Lakes region. Journal of Great Lakes Research 38:477–489.

CHAPTER 22

Reproductive Behavior and Spawning Migrations

Traits, such as internal versus external fertilization, maturation schedules, fecundity, egg size, and parental care are important for differentiating reproductive characteristics among fishes. However, they are not the only features of reproduction. Fishes express a variety of behaviors as part of the reproductive process, from broadcast spawning to detailed mate selection, and from local nest building to incredibly long spawning migrations. Reproductive behaviors vary among species but also differ among populations of the same species and even individuals of the same population. Here, we highlight some noteworthy examples of reproductive behavior and focus on fish spawning migrations in particular.

Generalized Reproductive Behavior

Site Selection

Spawning locations vary tremendously among fishes and are related to the number and vigor of young they produce. Pelagic fishes often spawn in open water and broadcast their eggs into the water column with no pattern of site selection. Many species spawn in vegetation and produce adhesive eggs that attach to vegetation and remain there. By attaching to vegetation, these eggs remain in areas with higher water quality compared to the muddy sediments below. Many other fishes, such as muskellunge and walleye, broadcast their eggs over rocky or woody substrates, and the eggs sink into crevices and develop there. Stream-spawning trout dig up small rocks or gravel and expel their eggs and then cover their eggs to protect them. Species with internal fertilization develop sophisticated patterns of spawning behavior and courtship leading to fertilization.

Spawning habitats are likely related to the tolerance of eggs for certain water quality characteristics – such as, pH, oxygen concentration, and temperature – although it is difficult to demonstrate a causal relationship between tolerance and spawning sites. A fish that spawns in muddy substrate or in vegetation generally has eggs that are tolerant of low water quality. Eggs placed on mud substrates are exposed to low oxygen content, high hydrogen sulfide, high ammonia, and other conditions. They also may be vulnerable to many benthic egg predators. Compared to this pattern, rainbow trout produce eggs that do not tolerate poor water quality, so trout actively select

Biology and Ecology of Fishes, Third Edition. James S. Diana and Tomas O. Höök.
© 2023 John Wiley & Sons Ltd. Published 2023 by John Wiley & Sons Ltd.

clean substrate and clean it further during prespawning activity. There is likely simultaneous adaptation of adult spawning locations and resistance of eggs to poor environmental conditions. One major cause of reproductive failure in fishes is the deterioration of spawning habitats due to sedimentation, pollution and cultural eutrophication so that conditions where adults spawn are much poorer in water quality than they were before human settlement.

Mate Selection

Many species of fish do not engage in direct mate selection. However, some species clearly do select mates, and for such species, a critical step is courtship – the attraction or selection of a mate and the stimulation of that mate to develop and release gametes in synchrony. In most fishes – indeed in most animals – the male attracts the female. Regardless of which sex does the attracting, several morphological and behavioral characteristics might be used as cues in the selection of a mate.

One of the most commonly accepted cues is large body size. Increased growth rate and large size lead to increased survival, higher fecundity, and larger territories in many fishes. Large body size may denote evolutionary fitness. A female, by selecting a large male, may ensure a higher degree of fitness for her eggs than a smaller male could provide. However, large body size is not selected in all species; some females, such as guppies, may select smaller males that are highly agile and adept at impregnation. The generality of the higher reproductive success for large males remains somewhat unproven.

Besides size, other cues used in mate selection might include characteristics, such as coloration patterns and appropriate behavior. Another characteristic of a good mate might be possession of a good territory with an appropriate breeding site. Appropriate timing – being in a condition to breed at the right time of year or time of day – should also be important. Both coloration patterns and reproductive timing are at least in part hormonally controlled and may be influenced by prespawning courtship. Males often tend to migrate to spawning sites earlier than females and remain there over longer periods to ensure they are ready with territories when females come to spawn. This higher energetic cost of spawning behavior often leads to higher mortality in breeding males.

It is very important for females to choose appropriate males as mates. Males of most species expend less energy in sperm production than females expend producing eggs. Since females invest much more energy prior to reproduction, they may be the limiting resource. A female's success could be limited not only by the number of eggs she can produce but also by fertility of her mate. An example of this is a study done on lemon tetra, which are common aquarium fish (Nakatsuru and Kramer 1982). The number of spawning acts – and subsequent development of fertilized eggs – declined with increased number of spawning acts by the male (Figure 22-1). In other words, males became less capable of fertilizing eggs after several breeding bouts. If a female selected a virgin male, she would have a high rate of fertility (about 80%) and a large number of eggs that developed successfully. If a female selected a male that had spawned many times, she would have a low rate of fertility (nearly 0%), and a low number of eggs would develop successfully. This rationale has been used by fish breeders to separate male and female fishes for about a week to maximize egg viability. While this has not been studied as well in many species, a similar pattern may exists across many species. One could hypothesize that females should select as mates males that had not bred before or during that spawning season.

The lemon tetra example indicates that a female should not select just any male but a male that has not recently spawned. Does a fish actually have this ability? Nakatsuru and Kramer (1982) set up experiments to test this selection using three compartments in a tank: a middle compartment with a fertile female, one with a male and fertile female, and one with a male and a spent female. The purpose was to determine whether the female in the middle compartment could sense the past

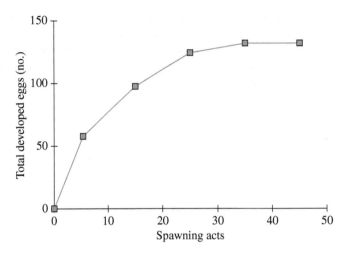

FIGURE 22-1   Relationship between total daily spawning acts by lemon tetras and cumulative number of eggs developing. *Source: Adapted from Nakatsuru and Kramer (1982).*

reproductive activity of males and select her mate accordingly. After some time, the other females were removed. The female in the middle compartment spent more time near the male that had been with the spent female and had not recently spawned. The female appeared to correctly choose her mate, or perhaps that male was more effective at courting than the recently spawned male.

Examples of Reproductive Behavior

Rainbow Trout

Stream-dwelling trout – which often reside in pools as adults – generally move to different habitats to spawn. Spawning locations may be nearby riffles where there are appropriate currents, or far away habitats, depending on habitat availability. Trout move into relatively fast water with gravel or rocky substrates to spawn. Since trout dig redds and then cover the eggs, high flow velocity is required to allow water to penetrate the substrate and maintain reasonable oxygen levels for the developing young.

Spawning behavior has been analyzed for many salmonines (see Keenleyside 1979 for review). The female selects a site in which to dig a redd in gravel substrate (Hartman 1970). The size of gravel that can be moved is dependent on the female's size. For example, while a large female chinook salmon might be able to move large rocks, a small female rainbow trout might only choose gravel. To dig the redd, the female lies on her side on the substrate and swims along the bottom, displacing the gravel with her caudal peduncle and fin. This behavior both excavates a depression and cleans sediment from the depression. The female measures the depth of the excavation with her anal fin and continues to dig until an appropriate depth is reached. Several males usually begin agonistic encounters for the right to breed with that female while she is still digging the redd. The female does not court the male rather the males compete for the right to breed. Once the female has dug the redd to her satisfaction, the male quivers next to the female and over the nest, which induces the female to spawn. The male may also nudge her abdomen to encourage spawning. She then lays her eggs in the redd, and the successful male fertilizes those eggs. The female chases off her mate, as well as any other males so they will not eat the eggs. She covers the eggs by moving the gravel back over the depression and then abandons the redd. Salmonines have elaborate pre-fertilization parental care but little post-fertilization care. This pattern makes good sense for adult semelparous salmon because these salmon do not survive long enough after spawning to protect the eggs. For iteroparous salmonines, limited post-fertilization care may allow adults to transition to feeding and increase likelihood of future successful reproductive events.

Mouthbrooding Cichlids

Another example of parental care is provided by the mouthbrooding cichlid, *Haplochromis burtoni* (Wickler 1962). There are many species of mouthbrooders in the family Cichlidae. A female *Haplochromis* lays eggs on a nest prepared by the male. These eggs may be fertilized by the male on the nest or taken into the female's mouth before they are fertilized. For oral fertilization, the male flashes his anal fin, which has a spot on it that resembles an egg. The female snaps at that spot in an attempt to take in the egg, and the male releases sperm into the water. The female ingests the sperm and fertilizes eggs in her mouth. This use of egg dummies – such as the spot on the anal fin – or other means to attract females to gulp sperm and fertilize eggs is common in cichlids (Keenleyside 1979). The female *Haplochromis* provides for the eggs, passing water across them during ventilation to maintain a high-quality rearing environment. Once the eggs hatch, the female digs a small depression in the sand and sets yolk sac fry there during the day, while she moves water across this nest with her pectoral fins. As the young absorb their yolk sacs and become free swimming, the female continues to protect them. She has several signals – the most common of which is a twitch of the body – to indicate danger is near. After this signal, she will open her mouth, and the young fishes will swim into her mouth to hide. When the young become able to fend for themselves, they are abandoned. Many species of fishes demonstrate similar parental care and guarding behavior, including the bowfin, which guards nests and schools of young, and Nile tilapia.

Bluegill

A very interesting and complicated pattern of reproductive behavior occurs in bluegill. Large male bluegill build nests, which are small circular depressions, usually in shallow areas. In late spring, their beds can be identified as many small, light-colored, round depressions in the midst of muddy or sandy substrate. The males defend the nests against other males. The females develop bright orange coloration on their ventral surfaces and are considerably smaller than the males. Once a nest is built, the male will regularly circle the nest as a means to attract females to spawn. Each male will attract as many females as possible to his nest to spawn with him, and females generally select larger males to breed with (Jennings et al. 1997). The guarding male intends for the eggs in their nest to comprise the offspring of themselves and many females.

Bluegill nests are located in a group, so a variety of males and females are congregated nearby. After spawning is completed, each large male drives off the females and any other fish, and guards the eggs in his nest. He may even fan them. Guarding males generally remain with the nest until the larvae hatch and become free swimming. Typically, guarding males do not feed during this time. Many studies have demonstrated that the energetic cost of sperm is less than the energetic cost of eggs, and GSI of males tends to be lower than for females. However, the entire reproductive process by large male bluegill may expend a great deal of energy. That is, if one factors in not only the energy of sperm production but also nest construction, guarding before and after fertilization, the difference in reproductive energy expenditure between large male and female bluegill may not be as great.

There are two additional breeding tactics for male bluegill. The first involves sneakers, males that mature at very small sizes (ages 1–3). These sneaker males remain near bluegill colonies, watch males and females spawn and, at critical moments, dart in, expel their sperm on the nests and leave. Scientists did not realize this behavior occurred until high-speed photography helped identify satellite males because they moved in so quickly. Another breeding tactic involves female mimicry, whereby males that are not as large as nest holders develop orange coloration and imitate female behavior (Gross 1982). A mimic may move into a nest and attempt to fertilize any previously unfertilized eggs. A mimic may even move in at the same time a male is courting another female. The mimics may be tolerated because they appear to be females and may steal some successful fertilizations from the territorial, guarding male. Sneaker males mature at relatively small sizes and young ages, female mimics are intermediate sizes and ages, and guarding males are generally older and larger. Thereby, sneaker males may potentially produce offspring at very young ages and allow for shorter generation times. The probability of surviving to reproduce at a young age is greater than the probability to survive to reproduce at an older age. Thus, even if a minority of males mature in early life, given the survival difference the majority of breeding males in a bluegill population may be sneakers or female mimics (Gross 1982)

This kind of elaborate reproductive behavior does not occur exclusively in bluegill, but because it is difficult to observe spawning behavior in most fishes, the complexity of their breeding behavior may be less well known. Cuckoldry has also been described for guppies, salmon (jacks), and wrasse (Gross 1991), and various other fish species display alternate male reproductive strategies (Taborsky 1994). Cuckoldry is the pattern of reproduction in which an animal uses deception to allow fertilization without normal provision of parental care. Both mimic females and sneakers follow this pattern by breeding deceptively and abandoning offspring to the care of nest-guarding males.

Sharks

The final example of reproductive behavior is for viviparous sharks. A few shark species have placental development, while most sharks have some form of internal fertilization. Males have claspers – modified pelvic fins – which are specialized to allow insertion in the female cloaca and transfer of sperm during copulation. The breeding behavior of sharks can be quite violent, and scars or wounds commonly develop. During instances of copulation between smaller sharks, the male typically has its clasper inserted and its tail wrapped around the female (Springer and Gold 1989). The male may also bite the female's pectoral fin to hold her in place. Lemon sharks have been observed mating while swimming in parallel, a behavior pattern that may be more common in large sharks. The female usually produces few eggs that develop slowly (often 9–22 months gestation). Perhaps the most unusual characteristic about shark reproduction is fetal cannibalism. As young sharks develop internally, they become cannibalistic, so a young shark attached to its yolk sac might eat its sibling or other undeveloped eggs. A female may start out with 10 or more sharks developing internally, and by birth have only one pup left alive. Fetal cannibalism results in larger offspring but reduced reproductive rates. Females may migrate to nursery areas to bear their young, and these lagoons or estuaries are separate from adult habitats.

Spawning Migrations

Large spawning migrations are spectacular events that have been well studied and comprise behaviors related to both spawning and general activity. Migrations are really a series of events that include not only locating places in which to breed but also returning to feeding grounds.

Young fishes born in remote environments must return to adult habitats later in life. There is a series of migrations involved, which can vary in extent among populations and species, from rainbow trout that might move a few hundred yards upstream to a riffle to spawn, to sockeye salmon that might move several thousand miles. Many of these migrations also involve homing.

Migrating fishes may migrate entirely within fresh water (potamodromous), moving from lakes to small streams (rainbow trout), from lakes to marshes (northern pike), or moving widely within lakes (lake trout) to spawn. In truly spectacular migrations, salmon move from the ocean to small streams. Diadromy is the process of moving across a salinity gradient during migration. Diadromous spawning migrations from the ocean to freshwater are termed anadromous; migrations from freshwater to the ocean are catadromous. An example of a catadromous migration is provided by eels, which live in freshwater streams and lakes in North America and Europe but migrate to the Sargasso Sea to spawn. Australian freshwater species like Galaxiids make amphidromous migrations, where young are hatched in streams, migrate to the ocean for a year, and then return to streams for adult life. Similar kinds of migrations take place within the ocean (oceanadromous), as when halibut spawn in shallow estuarine areas.

Examples of Migrations

Metcalfe et al. (2002) produced an excellent review of fish migrations related to fishery exploitation. They provided examples of a number of species besides salmon that undergo major migrations. Two important examples of marine migrations are covered here, those of eels and North Sea plaice. Later, salmon and lamprey spawning migrations are evaluated to describe homing and orientation mechanisms.

Eels from the family Anguillidae are catadromous, spawning as adults in the open ocean, rearing there for some time as larvae, and then returning to streams for adult life. Spawning occurs in tropical to subtropical seas, usually of great depth. Eels are semelparous, and adults die after spawning. Eggs develop into segmented larvae called leptocephalous larvae. These larvae are so different from adults they were originally believed to be a different species. Leptocephalous larvae live from one to several years at sea and then develop into juveniles that gradually move back to streams propelled by oceanic currents. In the Atlantic Ocean for example, spawning occurs in gyres like the Sargasso Sea, and young are transported by the Gulf Stream to nearshore locations. Young eels do not appear to home to the same stream system as their parents, nor if adults home to the same ocean location to spawn. However, groups such as North American and European eels appear to spawn in different locations and young eels show fidelity to continent of origin. While oceanic navigation for eels is aided by the Gulf Stream, it also undoubtedly involves other forms of orientation. A major difficulty in studying these migrations is the need to either implant transmitters to follow fish, or tag and later recapture them to determine their movements. This is particularly difficult in open ocean environments with small fish.

North Sea plaice are an oceanic species residing on the continental shelf and have been the target of a long-term and valuable fishery. Adults reside in dispersed feeding grounds in northern sections of the North Sea. In fall, they move south to spawning sites near the English Channel. Peak spawning occurs January to February, and then adults return to northern feeding grounds in late winter (Arnold and Metcalfe 1996). Males spend longer times in the spawning area than females. Immature fish also appear to follow mature ones on a false spawning migration, possibly to learn the location of the spawning grounds. Migrations are accomplished by selective tidal stream transport, where fish move up into the water column and swim with the current during the north- or south-directed tide (Buckley and Arnold 2001). They return to the ocean floor during the opposite tide. This behavior speeds up migration at considerably reduced costs of swimming. Additional orientation does not appear to be necessary, except in areas of slow tidal currents, where the fish show directed swimming near the bottom in the appropriate compass heading. Spawning and feeding sites appear to be widely distributed and require no precise form of homing. Many marine species undergo similar sorts of spawning movements.

A technique for evaluating migrations of marine and diadromous fishes involves the use of otolith microchemistry to determine prior residence. The otolith – inner ear bone – grows with a fish and contains chemicals related to the environmental experiences of the fish when the otolith was a smaller size. In other words, otoliths can be used to back calculate the conditions where a fish existed at a prior age in its life. For example, Zlokovitz et al. (2003) evaluated the strontium and calcium content in various locations from otoliths of striped bass to determine their previous occupancy of freshwater or marine environments. Certain ratios of Sr:Ca indicate freshwater residence, while others reveal marine residence. They used this to determine that striped bass showed differing migratory contingents with some fish residing long term in the Hudson River and others moving quickly after birth to coastal habitats. In another study, Dierking et al. (2012) used otolith microchemistry to evaluate habitat residence by marine flatfish. In their case, they used stable isotopes of more biologically active chemicals, including N, C, and O. These isotopes are related to prey consumed, as well as thermal conditions. They found distinct chemical patterns for sole residing in the open Mediterranean Sea or coastal lagoons and determined the timing of migrations based on these chemical signals. Overall, otolith microchemistry has been used for a variety of purposes and is an important technique for determining oceanic and diadromous migrations.

Salmon and Lamprey Migrations

Anadromous migrations are common in fishes with at least 150 species completing such migrations (Reide 2002, cited in Bett and Hinch 2015). The vast majority of these are philopatric, meaning that they home to their natal location to reproduce. Salmon spawning migrations have been the most studied and show the highest known degree of homing of any fish. Fisheries for salmon concentrate on their inshore or upstream migrations as the main times at which to catch them.

Spawning migrations – or any migration – allow fishes to utilize resources which are geographically isolated. Spawning in shallow areas where early survival and early growth are relatively high can be adaptive, and living as an adult in deeper areas where the food abundance may be suitable but predation risk is high may also be adaptable. Moving from one kind of an environment to another may maximize the benefits of both. Gross (1987) predicted that diadromy would be adaptive when fitness – the lifetime product of reproductive success times survivorship – is gained by diadromy. In this case, the gain in fitness from the utilization of the second habitat would be expected to exceed the migration costs of moving from one habitat to another plus the fitness in utilizing only the original habitat. Gross related increased fitness to higher availability of food in the ocean for temperate species (anadromy favored) or higher availability of food in freshwater for tropical forms (catadromy favored). While this may be one factor, clearly both eels and salmon overlap in geographic areas but demonstrate opposite patterns of migration.

In species with long migrations and separation of adult and young habitat, it is difficult for individual fish to use local cues to predict the suitability of a location for reproductive success. Homing – innate behaviors allowing fishes to return to their natal sites – may become more important the farther a population disperses. It is difficult for a salmon to judge the quality of a rearing environment in terms of flow consistency, predator density, and other characteristics during its brief spawning period in the fall. The farther a fish lives from its breeding habitat, the more homing allows some degree of predictability. The homing of salmon also allows stocks to adapt to local conditions. Reproductive strategies of fishes, such as American shad, show very specific adaptations from one stream to another (Leggett and Carscadden 1978), including semelparity in some rivers and iteroparity in others. These adaptations are related to the characteristics of the spawning sites, but can only occur if the stocks of animals are reproductively isolated. However, not all anadromous fishes home, and other means to detect suitable habitat are important. Lamprey use chemical cues on abundance of their young in spawning streams to determine where to select for spawning. This difference makes sense in ecological terms as some salmon species migrate to locations their young have already departed and therefore have little or no ability to use young density as a primary determinant of a suitable spawning site, while lamprey young reside in rivers for several years and therefore are able to produce signals the adults can identify as indicators of suitable spawning sites.

The Stream Phase of Migrations

The spawning migration of salmon can be considered beginning with the adults. The simplest phase to study is the stream phase after salmon find a river and move upstream. Art Hasler from the University of Wisconsin studied the migration of salmon for many years and reviewed these studies in several papers and books, including Hasler et al. (1978). Salmon can be captured more easily in this phase, and individuals can be identified when they are at the river mouth as well as at the spawning site. The first component of the upstream spawning migration of salmon is that salmon swim against the current or are positively rheotaxic.

Hasler and his colleagues hypothesized that salmon smell the odor of their home stream and use it for homing. They then conducted experiments to test the hypothesis of an olfactory mechanism of homing. The first series of experiments were done on the Issaquah River in Washington (Wisby and Hasler 1954). Hasler and colleagues placed traps in the river to capture upstream migrating salmon, and they marked these salmon with a distinctive pattern to identify whether they had been captured in the East Fork of the Issaquah or the mainstream. They also plugged the nasal passages on half of the fish, leaving the other half as controls. They then transported both groups of salmon several miles downstream and released them. If nasal plugging impaired the ability to sense their home, the test salmon would have swum back upstream and randomly gone up both forks of the river. If the ability to home was not impaired, both test and control fish would have moved to the same location and been trapped there again. If homing ability was impaired, since the East Fork was only about one-fifth the size of the mainstream, it seemed reasonable that 80% of randomly dispersing fish would be expected to go up the mainstream and only 20% up the East Fork.

The researchers recaptured 46 control fish from the main river, and 100% of them were recaptured in the original capture location (Table 22-1). In other words, they made the correct choice. Of the mainstream fish with plugged nasal passages, about 80% made the correct choice and 20% did not. For fish of East Fork origin, 71% of control fish chose the correct river, while 84% of occluded fish returned to the wrong river. These results are evidence that salmon used their nasal senses to determine the correct stream. However, there were a large number of strays (29%) of control fish from the East Branch, much higher than one might expect. Hasler believed this was simply an artifact – those fish might have continued upstream for some time and then recognized that they were in the wrong stream and returned downstream. Others questioned whether the trauma of the handling caused these results although sensory impairment has remained the most common way to determine olfactory homing.

In search of further proof of olfactory homing, Hasler conducted another series of experiments in Lake Michigan and some nearby tributaries. Exotic coho salmon were released at a particular location during stocking and were known to return to that same location a few years later to breed. Hasler believed he could influence their recognition of home by olfactory cues. He added an unusual organic chemical – morpholine – into rearing water to imprint the

TABLE 22-1 Recapture Data for Salmon Marked in the Issaquah River, Washington, and Released Downstream.

	Capture Location	Recapture Location	
Treatment	Issaquah	Issaquah	East Fork
Control	46	46 (100%)	0
Occluded	27	19 (77%)	8 (23%)
	East Fork		
Control	27	8 (29%)	19 (71%)
Occluded	19	16 (84%)	3 (16%)

Source: Adapted from Wisby and Hasler (1954).

salmon. He believed salmon learned the odors of their natal streams just prior to smolting and imprinted on those odors for later homing. Earlier studies had shown that salmon from a hatchery, stocked into a river prior to smolting, returned to the river rather than the hatchery. If stocked after smolting, salmon returned to the hatchery rather than the river, indicating that salmon learned the conditions of their stream of origin rather than this being genetically inherited, and the critical time for imprinting salmon was at smolt transformation. Hasler believed salmon learn the chemical cues of their home streams at smolting, and the smells are later recalled during homing migrations. Imprinting is the rapid establishment of a perceptual preference for some object or characteristic that takes place during a specific period in the life cycle.

During the morpholine experiments, there were other chemicals besides morpholine in the water that the salmon might recognize. If salmon did return home to morpholine, however, Hasler could clearly show this was a chemical response not a response to other characteristics of the home stream. Hasler released 8,000 fish that had been imprinted with morpholine and another 8,000 control fish. The fish were released directly into Lake Michigan after smolting at the mouth of Oak Creek in spring 1971. In fall 1971 and 1972, fish were recaptured at Oak Creek where morpholine had been added to the water to attract the fish. Out of the 8,000, 216 morpholine-imprinted fish homed, and only 28 of the non-imprinted fish homed. Later experiments without morpholine release indicated equal returns to Oak Creek for imprinted or control fish. This indicated that morpholine, released at this site, allowed the morpholine-imprinted fish to home better to that location than control fish.

Many questions remain about stream migrations of salmon. How can a fish that breeds in a small stream in the mountains but migrates thousands of miles to the Pacific Ocean sense water from that small home stream when diluted in the Pacific? The dilution problem may be overwhelming, not only in the ocean but also in larger rivers. Bett and Hinch (2015) from University of British Columbia completed an extensive review of migrations in Pacific salmon. Their evidence suggests that salmon imprint not on a single chemical in one location but on a series of waypoints during their migration (Oshima et al. 1969). This identification of a series of chemicals also makes it simpler to solve the question of straying because when salmon reach the mouth of their first river many stocks of fish may be imprinted on the same chemical cues. As they move farther upstream, the next chemical in the series may differentiate one stock from another. This would mean straying within a river system would be much more common than straying between rivers.

Chemical senses are also important in the stream phase of migration for sea lamprey although from a different perspective. The chemicals sensed by migrating salmon are likely caused by the geology of their natal stream and possibly from pheromones produced by young salmon. These pheromones may be stock specific or only indicate a species occurrence. Bett and Hinch proposed the Hierarchical Navigation Hypothesis, which relates instream salmon homing to three components: imprinted signals, presence of conspecifics, and non-olfactory environmental cues. The former are most specific to natal location, while the latter are more general but may also aid straying salmon in finding some suitable location to spawn. In the case of lamprey, they do not home to natal streams but use the intensity of pheromones released by juvenile lamprey in a stream to determine a suitable spawning site (Wagner et al. 2009). This results in successful reproduction for anadromous lampreys but not in local adaptation to stream conditions and locally resident stocks of lamprey. Once male lamprey migrate to breeding sites, they exude a strong pheromone to attract females to the site to spawn (Sorensen and Vrieze 2003).

Open Ocean Phase of Migration

It is unlikely that chemical sense of a home stream can allow salmon to migrate from the open ocean to the coast near their natal stream. This sort of migration must involve some other means of orientation. It begins when young salmon make their movement from streams to open ocean areas to forage as adults. Putman et al. (2014) provide evidence that young salmon possess an inherited ability to use information on the inclination and intensity of the Earth's magnetic conditions in any location to provide location information rather than a compass orientation.

While similar abilities may be used by adult salmon returning to their natal stream, there are other ideas about their ability to navigate over such distances in the ocean. One cue involves the Sun as a compass. Since the Sun is not always in the same compass direction, Sun compass orientation requires knowledge beyond simply following the Sun. Since the Sun rises in the east and sets in the west, early in the morning it is in the east, by noon it is in the south (in the northern hemisphere), and by afternoon it is in the west. To use the Sun as a compass, the Sun's position and the time of day must be known. Water currents, thermal fronts, and other oceanic conditions may also be used in navigation; visual landmarks such as islands may be important to some degree, especially in the nearshore migration. Tidal currents might also be used as in the earlier examples of plaice.

Hasler and his colleagues conducted several tests of direction finding ability in fishes (Hasler et al. 1958). First, pumpkinseed were trained to find food in an outdoor tank that had four doors (north, south, east, and west). The fish were trained to find food behind the south door. At 3:00 p.m., 7 out of 10 fish tested went south to find food. At 3:00 p.m., for the latitude of Madison, Wisconsin, the Sun would appear on the right during a southerly approach. When fish were tested at 9:00 a.m., 6 out of 6 still found the south door, even though the Sun was then to the left side when the fish were approaching south. In another experiment, bluegill were trained to escape to the north when frightened, in an arena with 16 potential escape directions. For tests on a sunny afternoon, most hiding was done in the north. If they were tested in the morning, the fish still hid to the north. On a completely overcast day, however, the fish could not recognize north as readily and moved in any direction. If a light was used as an artificial sun, the fish also oriented to the light as if it were the Sun, and located north in the morning but south in the afternoon because the light's position was not changed. It appears from these experiments that the Sun was the characteristic used to find north and fish could find directions with some precision.

In a previous chapter, it was proposed that schools may serve to reduce the costs of swimming, and they are common in migrating fishes. In oceanic migration, salmon appear to use magnetic location and Sun compass orientation cues to move close to their natal rivers, where chemical cues and olfaction come into play. To aid in locating their natal river, a school of fish with no obvious leader and continual swimming readjustments may navigate more precisely than isolated individuals. Basic statistical theory, such as the central limits theorem, would support this. Peter Larkin from University of British Columbia tested such an idea (Larkin and Walton 1969). They showed mathematically that larger schools should be more accurate in homing than smaller schools. For a low directional ability of 2 (from a range of 0–10), 5 fish could navigate within 95% confidence limits of ±48.6°, 30 fish to ±18.3°, and 100 fish ±9.5°. Stronger navigation abilities obviously gave improved direction finding. This improvement of navigation by schooling, coupled with orientation as shown earlier, makes it reasonable to expect fishes to readily find coastal areas near their natal rivers from oceanic sites even without precise orientation. From that point on, olfactory cues can be utilized.

Summary

The reproductive behaviors of fishes are some of the most interesting and stereotyped of all behaviors. They function to assist males and females to successfully find mates and select appropriate sites in which to breed. Some of these behaviors, such as migration, are spectacular indicators of the innate or learned capabilities of fishes. Natural selection should strongly act upon appropriate reproductive behaviors, including migrations, because they result in success or failure in genetic contributions to future generations.

Literature Cited

Arnold, G.P., and J.D. Metcalfe. 1996. Seasonal migrations of plaice (*Pleureonectes platessa*) through the Dover Strait. Marine Biology 127:151–160.

Bett, N.N., and S.G. Hinch. 2015 Olfactory navigation during spawning migrations: A review and introduction of the Hierarchical Navigation Hypothesis. Biological Reviews 91:728–759.

Buckley, A.A., and G.P. Arnold. 2001. Orientation and swimming speed of plaice migrating by selective tidal stream transport. Pages 263–277 *in* J. Sibert and J. Nielsen, editors. Electronic Tagging and Tracking in Marine Fisheries. Kluwer Academic Press, Dordrecht.

Dierking, J., F. Morat, Y. Letourneur, and M. Harmelin-Vivien. 2012. Fingerprints of lagoonal life: Migration of the marine flatfish *Solea solea* assessed by stable isotopes and otolith microchemistry. Estuarine, Coastal and Shelf Science 104:23–32.

Gross, M.R. 1982. Sneakers, satellites and parentals: Polymorphic mating strategies in North American sunfishes. Zeitschrift fur Tierpsychologie 60:1–26.

Gross, M.R. 1991. Evolution of alternative reproductive strategies: Frequency-dependent sexual selection in male bluegill sunfish. Philosophical Transactions of the Royal Society of London 332B:59–66.

Gross, M.R. 1987. Evolution of diadromy in fishes. American Fisheries Society Symposium 1:14–25.

Hartman, G.F. 1970. Nest digging behavior of rainbow trout (*Salmo gairdneri*). Canadian Journal of Zoology 48:1458–1462.

Hasler, A.D., R.M. Horrall, W.J. Wisby, and W. Braemer. 1958. Sun-orientation and homing in fishes. Limnology and Oceanography 3:353–361.

Hasler, A.D., A.T. Scholz, and R.M Horrall. 1978. Olfactory imprinting and homing in salmon. American Scientist 66:347–355.

Jennings, M.J., J.E. Claussen, and D.P. Philipp. 1997. Effect of population size structure on reproductive investment of male bluegill. North American Journal of Fisheries Management 17:516–524.

Keenleyside, M.H.A. 1979. Diversity and Adaptation in Fish Behaviour. Springer-Verlag, Berlin.

Larkin, P.A., and A. Walton. 1969. Fish school size and migration. Journal of the Fisheries Research Board of Canada 26:1372–1374.

Leggett, W.C., and J.E. Carscadden. 1978. Latitudinal variation in reproductive characteristics of American shad (*Alosa sapidissima*): Evidence for population specific life history strategies in fish. Journal of the Fisheries Research Board of Canada 35:1469–1478.

Metcalfe, J., G. Arnold, and R. McDowall. 2002. Migration. Pages 175–199 *in* P.J.B. Hart and J.D. Reynolds, editors. Handbook of Fish Biology and Fisheries. Vol. 1. Fish biology. Blackwell Science, Ltd. Oxford, UK.

Nakatsuru, K., and D.L. Kramer. 1982. Is sperm cheap? Limited male fertility and female choice in the lemon tetra (Pisces, Characidae). Science 216:753–755.

Oshima, K., W.E. Hahn, and A. Gorbman. 1969. Electroencephalographic olfactory responses in adult salmon to waters traversed in homing migration. Journal of the Fisheries Research Board of Canada 26:2123–2133.

Putman, N.F., M.M. Scanlan, E.J. Billman, et al. 2014. An inherited magnetic map guides ocean navigation in juvenile Pacific salmon. Current Biology 24:446–450.

Reide, K. 2002. Global register of migratory species. German Federal Agency for Nature Conservation, Project 808,05,081.

Sorensen, P.W., and L.A. Vrieze. 2003. The chemical ecology and potential application of the sea lamprey migratory pheromone. Journal of Great Lakes Research 29:66–84.

Springer, V.G., and J.P. Gold. 1989. Sharks in Question. Smithsonian Institution Press, Washington, D.C.

Taborsky, M. 1994. Sneakers, satellites, and helpers: Parasitic and cooperative behavior in fish reproduction. Advances in the Study of Behavior 21:1–100.

Wagner, C.M., M.B. Twohey, and J.M. Fine. 2009. Conspecific cueing in the sea lamprey: Do reproductive migrations consistently follow the most intense larval odour? Animal Behaviour 78:593–599.

Wickler, W. 1962. "Egg dummies" as natural releasers in mouth-brooding cichlids. Nature 194:1092–1093.

Wisby, W.J., and A.D. Hasler. 1954. Effect of olfactory occlusion on migrating silver salmon (*O. kisutch*). Journal of the Fisheries Research Board of Canada 11:472–478.

Zlokovitz, E.R., D.H. Secor, and P.M. Piccoli. 2003. Patterns of migration in Hudson River striped bass as determined by otolith microchemistry. Fisheries Research 63:245–259.

CHAPTER 23

Life-History Patterns and Reproductive Strategies

The preceding chapters focused on describing the reproductive diversity and behavior exhibited by fishes. In this chapter, we describe some approaches for classifying and understanding reproductive diversity and life-history trait variation. Researchers have sought to understand consistent ecological and evolutionary dynamics by identifying intra- and inter-specific patterns of reproductive strategies and developing frameworks for depicting life-history trait variation. Reproductive patterns interest ecologists because they differ so much among organisms and because Darwinian evolution should have such clear and strong influence on reproductive traits and decisions. Therefore, consistent patterns of reproductive life-history traits likely arise because they are adaptive. This has led scientists to try to understand the adaptive nature underlying these patterns.

Life-history traits are constrained, and such constraints lead to trade-offs among individual traits and behaviors. From a fitness perspective, an individual female fish would benefit by (a) maturing very early in life since they are more likely to survive to an early rather than late age and (b) reproducing many times throughout their lives. The reproducing female would also benefit by producing a huge number of eggs during each reproductive event. Ideally, each of these eggs would be very large, have large yolk reserves (or receive maternal nutrients during development), and experience high hatching success during a short incubation period, followed by a very high survival probability during early life. This high survival would also take place with no parental care so upon releasing eggs the female fish could immediately invest a high proportion of energy to the next reproductive event. Unfortunately (and quite obviously), all of these traits cannot be simultaneously optimized. The term, Darwinian demon, has been used to refer to an unrealistic organism that could maximize all fitness-related traits.

Individual fish are constrained by available energy, space, and time. There will be limited energy available for gamete development and limited space in gonads; therefore, there must be trade-offs between the number of eggs produced and the size of these eggs. It takes time for young fish to grow large enough to mature and produce eggs and milt. Energy invested in maturation at an early age may limit future size and energy available for future reproductive events. Increased parental care clearly has benefits for offspring survival, but investment in parental care may affect parental survival and limit ability to amass energy for future reproductive events. Due to these different trade-offs, fishes vary in the combination of reproductive life-history traits they express. Some may produce many small eggs, while others produce a lower number of larger eggs. Some mature at a relatively young age, but initially produce a somewhat limited number of offspring. Some may provide a high level of parental care, but this

Biology and Ecology of Fishes, Third Edition. James S. Diana and Tomas O. Höök.
© 2023 John Wiley & Sons Ltd. Published 2023 by John Wiley & Sons Ltd.

may contribute to higher parental mortality risk and limit production of offspring during future reproductive events.

Life-history theory provides "the analysis of the phenotypic causes of variation in fitness and exposes the pervasive tension between adaptation and constraint" (Stearns 1992). Since successful reproduction results in genetic contributions to future generations – and evolutionary fitness – analyses of these patterns often lead to both inter- and intraspecific evolutionary arguments. Combinations of reproductive life-history traits may vary not only among different species but also among populations of the same species and even among individuals within the same population. The purpose of this chapter is to overview reproductive strategies and present frameworks for describing life-history trait variation. We also focus on the American shad as an example species to demonstrate concepts in the evaluation of reproductive strategies.

Reproductive Strategies

Life-history strategies are the combination of evolved life-history traits expressed by a population. Gross (1987) defined strategy as a "genetically determined life-history or behavior program that has evolved because it maximizes fitness." Some populations are characterized as having simple life histories, where all individuals express essentially the same combination of life-history traits. In contrast, some populations express complex life histories where different individuals express divergent combinations of life-history traits. The different reproductive patterns of male bluegill are a good example of complex life histories. Gross (1987) suggested that simple life histories "evolve in the absence of frequency-dependent evolution," while the specific combination of traits expressed by individuals in populations with complex life histories are responsive to the frequency of life-history tactics present in the population. The success of female mimics and satellite male bluegill is dependent on their frequency relative to the frequency of dominant, nest-guarding males. These alternative life-history strategies are likely to be most successful when the proportion of large, nest-guarding males is high.

Species- or population-specific patterns may have arisen under frequency-dependent competition even though they are now common and comprise simple life histories. However, two species that each express simple life histories can differ in terms of the specific combinations of life-history traits they express. That is, fish species greatly differ in terms of combinations of life-history traits, allowing them to maximize fitness in their particular environment.

Life-History Patterns

Maturation Timing and Growth

A commonly evoked life-history trait trade-off relates to somatic growth and benefits to future reproductive success versus maturation and immediate gonadal growth. While the energetic investment in reproduction varies among species, populations, and individuals, there is always an energetic cost to reproduction. This energetic cost can be related to development of gonads, but can also be related to spawning migrations, courting behavior, nest construction, and antagonistic interactions among reproductive rivals. In many populations, reproduction can also increase the short-term risk of mortality due to energy depletion and moving to habitats with increased probability of predation, fisheries harvest, disease, or some other mortality factor. Reproductive success is often related to a fish's size as fecundity generally increases with fish size, and larger individuals may be more successful in attracting mates and competing with reproductive rivals.

The costs and risks of reproduction in addition to the positive influence of fish size on reproductive success set up a trade-off related to maturation timing. Maturation at a relatively young age is beneficial mainly because the probability that an individual will survive to this age and be able to reproduce is relatively high. However, the act of maturing and reproducing may increase short-term mortality risk and lead to slower growth and smaller size, lower fecundity, and potentially lower reproductive success if an individual survives to reproduce on subsequent occasions. In contrast, delayed maturation allows individuals to prioritize increases in somatic tissue and overall size. These attributes may contribute to greater reproductive success at the time of their first reproductive event. If maturation is delayed to an old age, there is a relatively high probability that an individual may die before they have a chance to reproduce.

Maturation timing varies among species, populations, and individuals and seemingly reflects balances among the risk of mortality and growth opportunities, in addition to the influences of fish size and age on reproductive success. The expression of maturation timing can be quite complex. Maturation timing is plastic for many species, and the growth experienced by an individual throughout its life may influence whether it matures and reproduces relatively early or late. Maturation timing is also influenced by evolutionary selective pressures, including both long-term and short-term selective pressures. For example, Chapter 32 describes how size-selective fisheries harvest can rapidly select for altered maturation schedules.

Egg Size Versus Fecundity

The relationship between mean size of eggs produced by female fish and their overall fecundity is further demonstrative of trade-offs between reproductive traits. While it may not always be possible to quantify constraints on reproduction, there must be some limitation to the amount of energy an individual fish can allocate to reproduction and, more specifically, to gamete development in their ovaries. Egg production must be limited by space in the gonads. By producing relatively small eggs, fish may be able to produce many offspring, but if these numerous small offspring experience very poor survival, the reproductive fitness of the fish may be very low.

A large number of studies have evaluated the influence of egg size on offspring performance. While findings across studies are not entirely consistent, in general, larger eggs tend to have greater hatching success, and they tend to yield larger larvae with greater yolk reserves. Such larger larvae appear to survive better for a variety of reasons. The switch from endogenous to exogenous feeding is a period of high mortality for many larval fishes because they may not develop rapidly enough to feed exogenously. Even if they are appropriately developed, they may not encounter appropriate prey. Larger larvae are generally more developed and therefore better able to make this transition. Their greater yolk reserves also allow them to survive longer without switching to exogenous feeding, which may be particularly important if they do not initially encounter appropriate prey. Larger larvae are also able to swim better and therefore may be able to feed more effectively, while also avoiding predators.

The trade-off between number of eggs produced and the energy invested to each individual egg was demonstrated by Smith and Fretwell (1974). Through theoretical quantitative models, they suggested that the effort or energy invested for each individual offspring was positively related to fitness of the offspring. However, this relationship may not be linear. When investment per offspring is already high, there may be minor increases in offspring fitness with increased parental investment per offspring. In such cases, parental fitness may actually be greatest at some lower, intermediate investment per offspring, as such lower investment would allow for production of more offspring. Smith and Fretwell demonstrated that the nonlinear relationship between investment per offspring and offspring fitness will lead to some intermediate investment per offspring that is optimal. This optimal investment – egg size – would be expected to differ among species and populations, given the long-term selective pressures of their environment. The specific conditions of a system should determine the appropriate egg size in that system. Houde (1994) suggested that differences in environments between large marine and smaller freshwater systems have contributed to generalized differences in larval size – and therefore egg size – between the two ecosystems types with marine larvae generally hatching at a much smaller size than freshwater larvae. Houde argued that more stable environmental conditions in small freshwater systems generally select for production of more robust larva, while highly variable, broad-scale physical processes in large marine systems favor production of a large number of small larvae.

The concept of optimal egg size was further evaluated by Einum and Fleming (2000). These authors conducted an empirical study with Atlantic salmon. They were able to rear female salmon to produce eggs of variable sizes. They took eggs from individual mature females, but in so doing, they separately grouped small and large eggs from the same female. They then fertilized these eggs with milt from one male salmon to control for parental effects and focus on egg size effects. Einum and Fleming then buried fertilized eggs in gravel nests and monitored survival of these different offspring as they emerged. They found that offspring from larger eggs survived better during 0–28 days post emergence but did not observe an effect of egg size on survival from 29 to 107 days post emergence. They were then able to calculate an optimal egg size by simultaneously considering how increasing egg size led to increased offspring survival but decreased number of eggs produced per unit investment. They predicted that salmon should produce individual eggs with an optimal mass of ~0.12 g, and this was a very close match with mean size of eggs actually produced by this salmon population.

Semelparity Versus Iteroparity

The number of lifetime reproductive events varies across individuals, populations, and species. A straightforward contrast is between populations displaying semelparity – a single spawning event per lifetime – versus those displaying iteroparity – multiple lifetime spawning events. Semelparous fishes tend to allocate a great deal of energy to their single reproductive event, which makes sense, given there is no benefit of retaining resources for future reproduction. However, semelparity is very risky if reproductive success varies greatly by year. If there is a reproductive failure during a single year, a semelparous individual would have no ability to contribute offspring during future reproductive years.

Cole (1954) produced a model predicting semelparity would be favored if an animal could produce at least one more viable offspring by increasing the amount of reproductive energy utilized and then dying. Cole evaluated many tactics related to semelparity or iteroparity, such as age at first reproduction and number of young in a litter. They estimated how much would be gained in future genetic fitness if an animal survived to breed again. Animals such as fishes with relatively large litter sizes (fecundity) and early maturation should be semelparous according to their model. However, this conflicts with a basic observation of fish life histories since iteroparity is very common among fishes.

Garth Murphy, a marine biologist from the University of Hawaii, tried to reconcile Cole's predictions with the fact that most vertebrates – particularly most fishes – are iteroparous. They wondered why iteroparity was so common if it was not advantageous to animals by natural selection. Murphy evaluated other factors that influenced iteroparity, particularly variability in survival of the young. For example, in sardines – a species studied extensively – Murphy (1968) estimated how much variation there might be in successful spawning or how much the relationship between reproductive success and reproductive lifespan tended to vary. There was a strong correlation between the two: more variation in success of spawning was associated with longer reproductive lifespan. Murphy predicted that spawning success was more important than simple fecundity in determining semelparity or iteroparity. This is much like the saying to not put all your eggs in one basket; a semelparous animal that chooses the wrong year in which to breed would make no contribution to the next generation because all of its young would die. With large variations in lifetime spawning success among individuals, over time, the frequency of genotypes with semelparous tactics would decline, while individuals that used iteroparous tactics would predominate.

Murphy characterized this trade-off in simple terms of steady reproduction and short or long life (Figure 23-1). This model indicates if reproduction was relatively certain, semelparity was possible. In other words, if survival of the young was certain, it could be adaptive to allocate extra energy into breeding and achieve better fitness. However, if reproduction was variable, it was necessary to have a long life as a hedge against variable years – to produce fewer young in each year – but eventually breed in a year when survival was reasonable and ultimately produce some offspring. Murphy believed it was possible to have steady reproduction and a long life, but not to have variable reproduction and a short life because the chances of gene survival over the long term were very limited under the latter conditions. From their perspective, semelparity should be favored only if there is reproductive certainty, and iteroparity should be favored if there is reproductive uncertainty.

Other traits are also involved in this semelparity/iteroparity trade-off. One cannot say whether these caused semelparity or iteroparity, or whether they arose as a result of semelparity or iteroparity, but at least they may be linked (Table 23-1). If there is normally poor adult

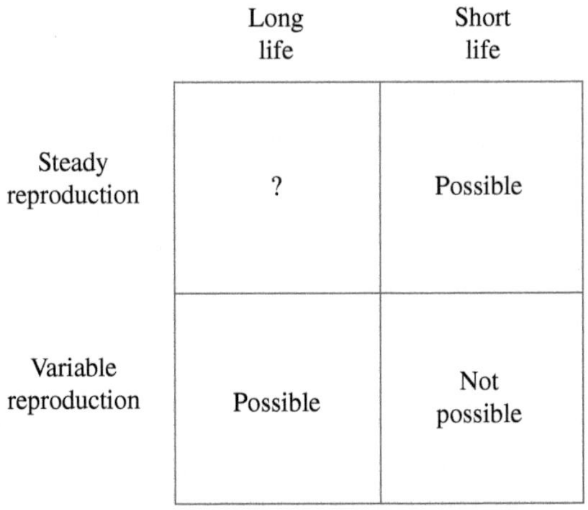

	Long life	Short life
Steady reproduction	?	Possible
Variable reproduction	Possible	Not possible

FIGURE 23-1 Summary of Murphy's arguments on life-history strategy for marine fishes. *Source: Murphy (1968)/The University of Chicago.*

TABLE 23-1 Factors Linked to Alternate Reproductive Life Histories in Fish Populations.

Factor	Semelparity	Iteroparity
Young survival	Constant or predictable	Variable or unpredictable
Adult survival	Low	High
Reproductive behavior	High energetic cost	Lower energetic cost

survival past spawning, semelparity should be favored. For example, for tropical fishes that live in wet/dry climates, where their habitats may become very limited after breeding, the chances of survival past breeding are very low. This would favor putting all surplus energy into reproduction. This might also be true in a heavily preyed upon species, made up of very small animals with many predators. Even if an adult survives breeding, its chances of living to the next year might be very low. In that case, semelparity should be favored. In contrast, animals with good potential to survive after spawning might be expected to develop iteroparity.

Animals that demonstrate reproductive behaviors that are energetically costly – particularly large migrations in the case of fishes – might also be semelparous. In these instances, the costs of migration are so high, they may limit potential adult survival. Allocating extra energy into reproduction may be a better strategy than conserving energy in somatic reserves for survival. For fishes such as salmon – which are semelparous – survival of the young is somewhat predictable. The streams they utilize for spawning will reach a carrying capacity, which can be exceeded if an excessive number of adults breed in a given year (in which case, density dependence occurs). The carrying capacity allows a consistent number to survive every year, and survival of young is predictable. It might not be constant, because a high density of adults may mean a lower survival of young per individual adult, but it is predictable. Salmon also have costly reproductive behavior (spawning migration), which greatly reduces their chances of survival past spawning.

Crespi and Teo (2002) of Simon Fraser University considered the evolution of semelparity in 12 species of salmonid fishes, salmon, and trout. These species, while taxonomically similar, differ in that some display iteroparity, while others are semelparous. Many of these species include both anadromous forms – that undertake long spawning migrations – and freshwater resident forms – that do not migrate long distances. Crespi and Teo sought to understand the evolution of semelparity by considering the taxonomic relatedness of these species, as well as patterns of their reproductive traits, including female body size, fecundity, gonadosomatic index (GSI), individual egg weight, and number of lifetime reproductive events. For the latter, they contrasted semelparity versus iteroparity, and for iteroparous fish they considered percent repeat breeding as an index of the degree of iteroparity. By comparing among species and anadromous versus resident forms, Crespi and Teo (2002) unsurprisingly found that female size was positively related to GSI, fecundity, and egg weight. More interestingly, they found that the degree of iteroparity was negatively related to GSI – a lower frequency of repeat breeding is associated with greater investment in gonads per reproductive event. They also found a strong association between semelparity and egg size, with semelparous species producing significantly larger eggs. Based on these patterns of reproductive traits and reconstruction of ancestral states, Crespi and Teo suggested that semelparity in salmonids developed in part because of long distance migrations contributing to high adult mortality. However, increased egg weight and associated increased juvenile survival were likely critical in the success and evolution of semelparity. Larger egg size increases likelihood of reproductive success, and this pattern for salmonids is generally consistent with Murphy's model, proposing that semelparity is possible if reproductive success is relatively certain.

Life-History Patterns

Specific life-history traits do not tend to vary independently among species or populations. Rather, life-history traits tend to covary. For example, fishes that mature at a young age tend to have relatively low fecundity, and fishes that have high parental care tend to produce offspring of relatively large size. Therefore, there appear to be some general patterns in how combinations of life-history traits vary. Several attempts have been made to differentiate species based on combinations of covarying life-history traits. While it may be attractive to classify species into distinct groups based on life-history traits, this is inappropriate because life-history traits tend to vary more continuously. A framework to differentiate species based upon life-history traits should instead be viewed more as a continuum with different species (or populations) falling along this continuum.

r–K Continuum

The most commonly invoked framework to describe life-history trait variation is the r–K continuum. This continuum has two endpoints, with r-selected species on one end and K-selected species on the other. The framework and its two endpoints build from the logistic population growth model with its two parameters: r (intrinsic growth rate) and K (carrying capacity). K-selected species may be limited more by the population's carrying capacity, while r-selected species may be limited more by their potential to rapidly increase in abundance. The fitness of individuals that are K-selected species may depend on producing offspring that are robust and can outcompete other offspring of the same species. In contrast, r-selected individuals may benefit by producing a large number of small individuals with short generation time and potentially allowing for rapid population growth.

While the r–K life-history framework is a continuum, it is worth considering traits that define the two ends of the continuum. A prototypical K-selected species is characterized by a relatively long lifespan and late age at maturation. K-selected species often grow to a relatively large maximum size. They may not produce a large number of offspring, but the offspring tend to emerge at a relatively large size and may be supported with a great deal of parental care during early life. In contrast, a prototypical r-selected species is characterized by a short lifespan, early maturation, high fecundity, with small offspring, and minimal parental care. r-selected species often have short generation times and their young may be dispersed broadly. Among aquatic organisms, some examples of K-selected species include whales and penguins, while jelly fish are an example of an r-selected species.

Equilibrium, Opportunistic, and Periodic Continuum

While several publications have described individual fish species along the r–K life-history trait continuum, this framework does not seem to be appropriate for many species of fish. As an example, consider sturgeon. These species reach a large maximum size, they have a long lifespan, and mature late – all characteristics of K-selected species. However, sturgeon are also very fecund. They produce a huge number of very small eggs and provide no parental care – all characteristics of r-selected species.

Kirk Winemiller and Kenneth Rose, at the time from Oak Ridge National Laboratory, evaluated overall patterns in the life history of freshwater and marine fishes from North America. They determined average values for each of 216 species for up to 16 traits. They then evaluated the patterns of these traits with univariate and multivariate methods. The result was three general groups of life-history patterns as well as some intermediate types. Their framework is based upon a triangular continuous plane of life-history patterns rather than the simple linear continuum of the r–K framework.

Winemiller and Rose (1992) condensed variation in freshwater fish life histories into three main groups: (1) periodic strategy – fishes with intermediate to large adult body size, delayed maturity, long life, large clutches, small eggs, and short reproductive seasons; (2) opportunistic strategy – species with early maturation, small adult size, small eggs, and long reproductive seasons with multiple spawning bouts; and (3) equilibrium strategy – species with small

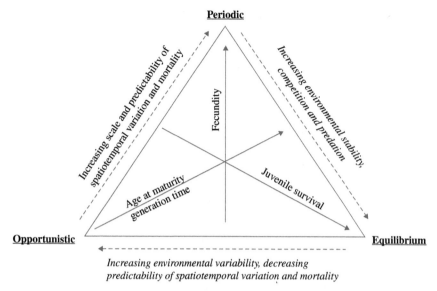

FIGURE 23-2 A model adaptive surface of fish life-history strategies based on demographic trade-offs selected under environmental variation. *Source: Winemiller (2005)/Canadian Science Publishing.*

or medium adult size, large eggs, small clutches, strong parental care, and long reproductive seasons. Winemiller and Rose proposed that these patterns arose because of the selection of life histories to environmental conditions and related their pattern to different environmental conditions (Figure 23-2). Periodic fishes should be favored in environments with large spatial or seasonal variation. Opportunistic fishes should be favored in colonizing situations, where survival may change often or stochastically over short time scales. Equilibrium species should be favored in environments with consistent habitat quality and strong biotic interactions. The distinction between opportunistic and equilibrium species is similar to the distinction between *r*- and *K*-selected species, although opportunistic species produce relatively small clutches and equilibrium species tend to be fairly small. Periodic species, however, do not neatly fit into the *r*–*K* continuum. This category includes many well-studied species exploited by fisheries, such as walleye and lake whitefish.

Life-History Invariants

The *r*–*K* framework and Winemiller and Rose's periodic-opportunistic-equilibrium framework seek to distribute species along a linear or triangular continuum based on multiple life-history traits. Another approach is to consider pairs of life-history traits that may vary together in a consistent manner. Such consistent relationships may seemingly create life-history invariants across species, which may allow one to predict the value of one life-history trait from the value of another. An example of such a potential invariant is the ratio of lifespan (E) to age-at-first reproduction (α) for mammals (Charnov 1993). For example, using E/α ratios, the life history of elephants and squirrels are very similar but quite different from fishes (Charnov et al. 2001). For mammals, E/α may be a life-history invariant, which is consistent across all species, even those not closely related.

The search for such invariants in fishes and other species with indeterminate growth has not been as clear as the example given for mammals. Jensen (1997) defined four invariants in fish populations: (1) the relationship between natural mortality (M) and age at maturation (x-m), where $M \cdot (\text{x-m}) = 1.65$; (2) the relationship between natural mortality and the Brody growth coefficient (K), where $M = 1.5K$; (3) the relationship between length at maturity ($L_{x\text{-}m}$) and asymptotic length (L_∞), where $L_{x\text{-}m} = 0.66L_\infty$; and (4) the relationship between growth coefficient and maximum length, where $L_\infty = K^{-0.33}$. They proposed that these invariants described the optimization of age at maturity and lifetime energetics for fishes. Also, note that these relationships are qualitatively similar to Pauly's (1980) model presented in Chapter 10 to predict natural mortality from estimated von Bertlanffy parameters, L_∞ and K. Charnov et al. (2001) hypothesized that the proportion of body mass allocated to reproduction annually (C) and adult lifespan (E) may be another invariant in fishes and lizards.

Establishment of invariants could allow for improved understanding, categorization, and modeling of animal life histories. Examination of data sets for a large number of species is one way to determine these patterns although this can also be considered for multiple populations of the same species. Of course, summarizing population level variations in life history – which can be quite large – into one general pattern for a species is necessary for such analyses but makes the end result appear more consistent than it really is. For example, such analyses would not reflect the variation among sneaker, female mimic, or parental male bluegill, as all male bluegill would be analyzed with one set of average life-history traits. In addition, life-history invariants reflect general patterns, and there are inevitably species and populations – even if they exhibit simple life histories – that do not match expectations based on such invariants.

Life-History Patterns of the American Shad

What species might be expected to show large differences in life history throughout its geographic range? Several characteristics immediately come to mind. One is that individual stocks should have reproductive isolation from each other. Each population should be isolated sufficiently, so the nature of the environment has a strong influence on survival and inheritance for both juveniles and adults. The life history of the American shad does not appear to follow this expectation because adults are not isolated. Shad spawn in rivers like salmon. The young develop slowly in rivers, migrate to the sea at the end of the first year, and become adults in the Gulf Stream. Adults from all areas remain in the Gulf Stream with a temperature of about 13–18 °C, which is located far north off Canada in the summer and farther south near Cape Hatteras, USA, in the winter. American shad from every river along the East Coast live together as adults in the Gulf Stream under similar environmental conditions. The only difference among the stocks occurs during spawning or early life in the stream. Since they live most of their adult lives together, one would not expect to see many differences among individual populations of shad, at least not as many as are found among individual populations of some fishes that are isolated throughout their entire lives. Yet, in fact, there are large differences in life history.

Different populations of shad spawn in rivers from Maritime Canada to Florida, USA. Obviously, the characteristics of those rivers vary considerably with latitude. Adult shad – moving in the Gulf Stream within the 13–18 °C isotherms – come in contact with local rivers when that water is near the coastline. Adult shad then enter the streams to spawn when the river temperatures are near 18 °C.

Bill Leggett, an ecologist from McGill University, evaluated populations, growth, and reproduction of shad in streams from various latitudes. Leggett and colleagues found the fish were semelparous at southern latitudes; none of the individuals survived to spawn again. In rivers farther to the north (above 35° latitude), there were individuals that survived to reproduce again (Figure 23-3). Similar to Crespi and Teo (2002) when studying salmonids, Leggett and Carscadden (1978) termed this increase in percent surviving to spawn again as degree of iteroparity. One may quibble with the term "degree of iteroparity" because semelparity and

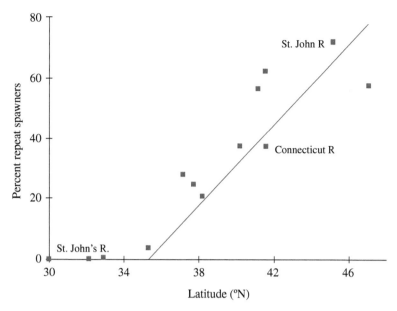

FIGURE 23-3 Relationship between percent repeat spawners and latitude for 13 populations of American shad. *Source: Adapted from Leggett and Carscadden (1978).*

iteroparity are actually either/or strategies. However, the important point is that among rivers shad differed in terms of displaying semelparity, iteroparity, and various levels of repeat spawning.

The life-history tactics of populations from each river varied dramatically. Individual growth rates demonstrated the two extremes mentioned above with fish in the northern rivers showing larger and more rapid somatic growth than fish in the southern rivers (Figure 23-4). By age 3 or 4, fish from northern rivers were considerably larger than fish from southern ones. If semelparity is an evolved trade-off between gonad and body growth, this pattern is exactly what would be expected – that semelparous fish in southern rivers would devote less energy to body growth and put the surplus into reproduction. The maximum age was also lower in southern populations, where fish bred once and then died. Not all individuals bred at the same age, so fish up to age 6 could be collected in southern rivers, but they died after breeding.

If the energy allocation between body and gonad growth followed the trade-off mentioned above, there would be higher egg production per unit body mass for southern populations than for northern ones. Relative fecundity – the number of eggs produced at a given weight – showed exactly that (Figure 23-5). The southernmost population in the St. John's River, Florida, had the lowest somatic growth rate, while the number of eggs produced per gram of body weight was much higher than in rivers that were more northern. On the other hand, age at first maturity did not differ much among populations. The average age at maturity was 3.8 years in the St. John's River compared to 4.2 years in the Miramichi River, New Brunswick. There appeared to be a difference in total reproductive lifespan, but semelparous populations did not breed at a much earlier age.

The pattern of energy allocation and relative fecundity for shad fits expectations of semelparity or iteroparity. However, the question remains: what are the differences between southern and northern rivers that cause such a large variation in life history for this one species of fish? The most obvious cause for these differences would be the thermal regime of rivers in the north and south, and the influence of thermal regimes on predictability of early life conditions. The seasonal distribution of temperatures through the year in Florida varies from a maximum around 30 °C to a minimum near 15 °C. The maximum temperature approaches

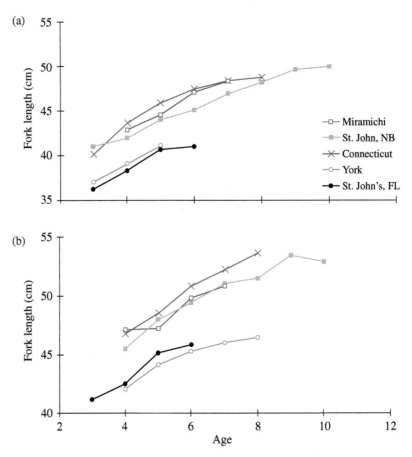

FIGURE 23-4 Size at age for male (a) and female (b) American shad from five rivers arranged on the legend in a north–south gradient. *Source: Leggett and Carscadden (1978)/Canadian Science Publishing.*

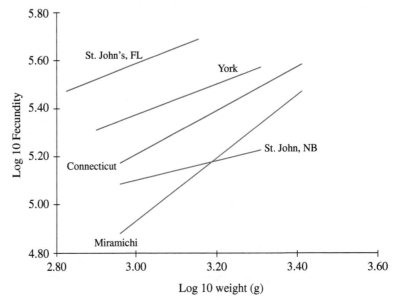

FIGURE 23-5 Relative fecundity at size for the same five shad populations. *Source: Leggett and Carscadden (1978)/Canadian Science Publishing.*

24 °C in Connecticut, and the minimum is less than 4 °C. There is a huge disparity between the thermal patterns of the rivers at these two extremes.

Since shad enter a river when 13–18 °C water from the Gulf Stream is near the coast, they enter the river in Florida from December to February (Figure 23-6). From that time, the temperature steadily climbs, so when shad spawn several weeks or months later, the temperature is near 18 °C. Young shad develop in water temperatures increasing from 20 to 30 °C. Spring pulses in primary production, caused by nutrient recycling and other factors, were mentioned in earlier chapters. These pulses indicate that spring is a good time in which to hatch young because food may be more abundant. Shad in the St. John's River can take advantage of this peak in food production. The best temperature for hatching of young is 13–15 °C, and the best temperature for the development of young is in the 15–18 °C range. The southern rivers have temperatures warmer than optimal for spawning, hatching, and developing, so they do not produce the best conditions for young development. However, the temperature regime is predictable from the time of spawning since the temperature will warm consistently over spring to summer.

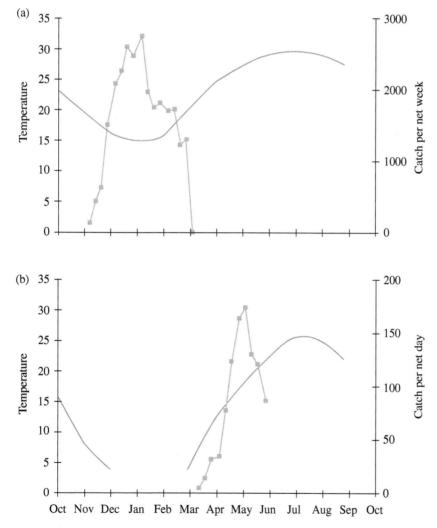

FIGURE 23-6 Shad catches (squares) and water temperature (line) for (a) St. John's River, Florida and (b) Connecticut River. *Source: Adapted from Leggett and Whitney (1972).*

Shad enter the river in April or May in Connecticut, again when 13–18 °C water is near shore (Figure 23-6). The spring pulse of food production that occurred there has passed before shad young emerge, and the time in which young shad can accumulate energy to survive the more severe winter is much shorter. Temperatures may continue to climb or stabilize after hatching, depending on the summer. The predictability of this thermal regime and food resources is much lower in Connecticut than in Florida, at least for American shad. It is probably even worse in Canadian rivers. Predictability of food and temperature appears to influence semelparity or iteroparity in shad populations. Actual temperatures may at times be better in Connecticut than in Florida for young rearing, but the temperature changes are less predictable. Temperatures may increase or stabilize after shad spawn, and any significant change may influence the survival of young. The difference between these populations is predominantly in terms of predictability of temperature and food resources rather than in average values of temperature and food resources.

Summary

As presented several times in this book, life-history patterns of fish populations are related to environmental characteristics. Reproductive patterns can vary dramatically within species and among species. These patterns vary from iteroparity to semelparity, and other important characteristics, such as fecundity, egg size, and age at maturation. Population-specific strategies appear to be correlated mainly with predictability of the survival of young, as exemplified by American shad. These patterns become important in understanding the ecology of young and adult fishes as well as in establishing some limits to fishery management.

Literature Cited

Charnov, E.L. 1993. Life History Invariants. Oxford University Press, Oxford, UK.

Charnov, E.L., T.F. Turner, and K.O. Winemiller. 2001. Reproductive constraints and the evolution of life histories with indeterminate growth. Proceedings of the National Academy of Sciences 98:9460–9464.

Cole, L.C. 1954. The population consequences of life history phenomena. The Quarterly Review of Biology 29:103–137.

Crespi, B.J., and R. Teo. 2002. Comparative phylogenetic analysis of the evolution of semelparity and life history in salmonid fishes. Evolution 56:1008–1020.

Einum, S., and I.A. Fleming. 2000. Highly fecund mothers sacrifice offspring survival to maximize fitness. Nature 405:565–567.

Gross, M.R. 1987. Evolution of diadromy in fishes. American Fisheries Society Symposium 1:14–25.

Houde, E.D. 1994. Differences between marine and freshwater fish larvae: Implications for recruitment. ICES Journal of Marine Science 51:91–97.

Jensen, A.L. 1997. Origin of the relation between K and L_∞ and synthesis of relations among life history parameters. Canadian Journal of Fisheries and Aquatic Sciences 54:987–989.

Leggett, W.C., and J.E. Carscadden. 1978. Latitudinal variation in reproductive characteristics of American shad (*Alosa sapidissima*): Evidence for population specific life history strategies in fish. Journal of the Fisheries Research Board of Canada 35:1469–1478.

Leggett, W.C., and R.R. Whitney. 1972. Water temperature and the migrations of American shad. Fishery Bulletin 70:659–670.

Murphy, G.I. 1968. Pattern in life history and the environment. The American Naturalist 102:391–403.

Pauly, D. 1980. On the interrelationships between natural mortality, growth parameters, and mean environmental temperature in 175 fish stocks. ICES Journal of Marine Science 39:175–192.

Smith, C.C., and S.D. Fretwell. 1974. The optimal balance between size and number of offspring. The American Naturalist 108:499–506.

Stearns, S.C. 1992. The Evolution of Life Histories. Oxford University Press, London, UK.

Winemiller, K.O. 2005. Life history strategies, population regulation, and implications for fisheries management. Canadian Journal of Fisheries and Aquatic Sciences 62:872–885.

Winemiller, K.O, and K.A. Rose. 1992. Patterns of life-history diversification in North American fishes: Implications for population regulation. Canadian Journal of Fisheries and Aquatic Sciences 49:2196–2218.

CHAPTER 24

Ontogeny and Early Life of Fishes

arly life stages of fishes – egg through juvenile – have very different characteristics from adults, and these characteristics affect their survival, growth, and abundance. It is clear that when many fishes hatch, many larvae do not at all resemble adults; larval fishes undergo a transition as they develop to juveniles. While the transition of larval fishes has previously been described as a metamorphosis akin to insect development, in reality, the development of larval fishes is far more gradual and less punctuated than metamorphism of insects. Larval development is related to habitat conditions and food acquisition since very small larvae of most fish species cannot thrive in the same habitats or eat the same kinds of food as adults. Given the effects of size and developmental stage on resource acquisition abilities, fish undergo ontogenetic shifts in food and habitat use in early life, therefore changing how they interact with other organisms, including prey, predators, and competitors.

While the larval stage is a relatively short period, the changes a fish experiences during this stage are dramatic. For example, from hatch to transition to juvenile, a larval northern anchovy may grow from 4 to 40 mm. This corresponds to an increase in length of 10 times during just 80 days and is equivalent to an even greater proportional increase in mass. While individuals that survive to late adulthood will gain greater length and mass, post larval growth generally will not be as dramatic on a relative basis. For example, as a juvenile and adult an anchovy may grow from 40 to 200 mm, corresponding to only a five times increase in length over multiple years (Figures 24-1 and 24-2).

Early Development in Fishes

There are some clear stages in the development of a fish from egg through adult. These stages have led a number of authors to categorize and name various stages of development. Eugene Balon from the University of Guelph adopted ideas from Kryzhanovsky (1947) to categorize development stages of fishes into five periods: embryonic, larval, juvenile, adult, and senescent (Balon 1975). These periods are related to developmental events, and fishes may spend a relatively long time in one period of development and then change to different characteristics common in the next period. An example of these changes can be seen for walleye and anchovy in Figures 24-2 and 24-3.

Biology and Ecology of Fishes, Third Edition. James S. Diana and Tomas O. Höök.
© 2023 John Wiley & Sons Ltd. Published 2023 by John Wiley & Sons Ltd.

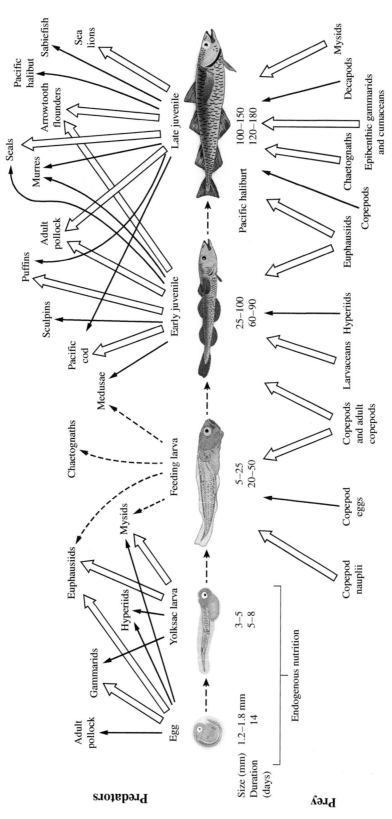

FIGURE 24-1 Prey and predators of early life-history stages of pollock in the Shelikof Strait region. *Source: Kendall et.al (1996)/John Wiley & Sons, Inc.*

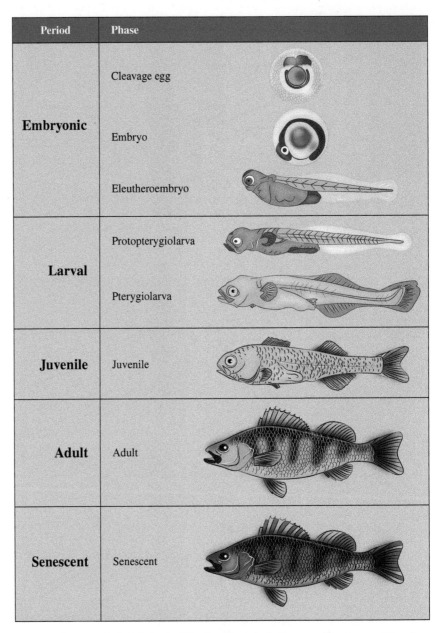

Period	Phase
Embryonic	Cleavage egg
	Embryo
	Eleutheroembryo
Larval	Protopterygiolarva
	Pterygiolarva
Juvenile	Juvenile
Adult	Adult
Senescent	Senescent

FIGURE 24-2 Developmental periods in the life of walleye, including specific phases of development used in larval fish descriptions. *Source: Reproduced with permission of Hunter and Coyne 1982 / CalCOFI.*

The embryonic period of a fish is considered from fertilization until hatching. Hatching occurs as enzymes degrade the chorion, and movements of the embryo result in emergence of a small fish. After hatching, many larvae are provisioned with a yolk sac, which contributes a considerable reserve of energy, as well as materials for building tissue. Balon actually considered the yolk-sac stage part of the embryonic stage, but most biologists now consider hatching the transition point from embryo to larvae. However, it is still common to differentiate between yolk-sac and post yolk-sac larval sub-stages. The yolk-sac sub-stage ends after complete absorption of the yolk. A yolk-sac larva utilizes yolk resources to build body tissue

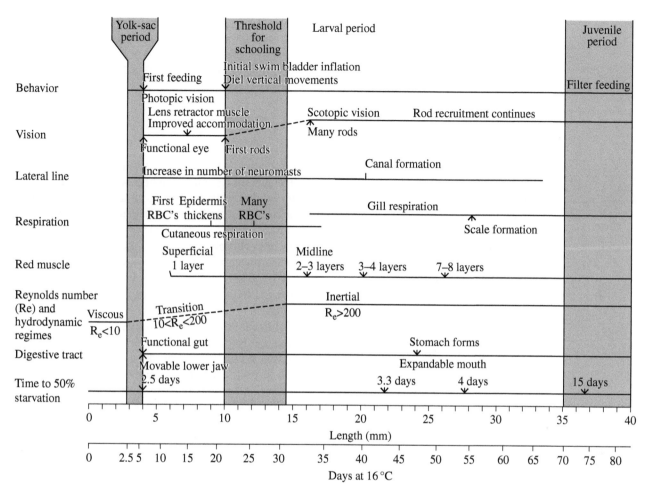

FIGURE 24-3   Developmental events in the early life of an anchovy. *Source: Reproduced with permission from Hunter and Coyne (1982).*

and may complete development of the neural system, digestive system, fins, and mouth during this period. Initially, a yolk-sac larvae will be entirely dependent on yolk reserves for nutrition, which is referred to as endogenous feeding. As the mouth develops, the yolk-sac larvae may begin actively feeding on prey, even before the yolk sac is entirely depleted. When the yolk sac is fully depleted, post yolk-sac larvae are forced to forage entirely for their own nutrition, which is referred to as exogenous feeding.

A fish in the larval stage must have most of its major organs developed to allow for food capture. Larvae may have fins that are not well differentiated at this stage, so the young fish is not completely developed, but at least it must capture and process food. The development that occurs during the larval period is considerable (Figures 24-1, 2 & 3). At transition to the juvenile period, the individual resembles the adult and has completed most development except gonadal maturation.

Species that produce large eggs – with much yolk – can undergo more development prior to yolk sac absorption. The time spent as yolk-sac larvae and post yolk-sac larvae as well as the degree of differentiation that occurs in the larval stage vary considerably among species. Some extreme examples of development include eels, which have leptocephalus – flat and transparent – larvae. Anguillid eels live as adults in fresh water and migrate to the ocean to spawn. Young eels are pelagic for a long time, and then metamorphose into juveniles when they migrate into rivers to live their adult lives. The larval stage is quite prolonged in species such as eels and lampreys. There are similar levels of metamorphosis in species like flatfishes, but over a shorter time interval. A flatfish larva has eyes on both sides of its head, but one eye rapidly migrates to the other side of its head, and the larva becomes benthic. Major morphological change such as this may occur long after hatching.

Precocial and Altricial Larvae

Survival of young fish is often strongly influenced by the success of transition from endogenous to exogenous feeding. This concept was introduced in Chapter 12 when describing the critical period concept put forth by David Cushing. There are two extremes in larval development – with gradations between them – forming a continuum of larval life histories. The extremes have been termed altricial and precocial, indicating the degree of advanced development at yolk absorption. Trout are precocial larvae, which complete much of their development prior to hatching. Altricial larvae hatch at a much less developed state. This is related to the amount of energy available for development. Altricial young undergo much development after yolk absorption. For example, there is generally much later development of the jaw, eye, gut, and fins for altricial larvae compared to precocial larvae.

A comparison of precocial and altricial fishes can reveal characteristics of the two extremes and examples of species whose larvae span the continuum. The more advanced kinds of reproductive behaviors in terms of parental care or investment usually produce much more precocial young (Table 24-1). The examples of species in each category are not necessarily taxonomically related, but they show similar characteristics because of their similar breeding habits. For example, taxonomically diverse pelagic spawners, such as cod or herring, produce poorly developed young at hatch, while live-bearing fishes, such as guppy and shark, may not be taxonomically similar but produce relatively well-developed young at hatch.

The growth patterns shown by each type of fish are related to their stages of early development (Noakes and Godin 1988). Precocial larvae are usually relatively good predators. When their yolk reserves are depleted, they have well-developed eyes, mouth, and fins. They are capable of using sit-and-wait tactics to feed on prey and can do well catching prey at a variety of prey densities because the larvae can swim effectively. These larvae also have well-developed guts that can extract more energy from their food (up to 80%). This combination of characteristics makes these young more robust and less susceptible to starvation at yolk absorption.

TABLE 24-1   Some Reproductive Characteristics of Adult Fishes that Produce Larvae that are Altricial or Precocial at Hatching, as Well as Species Examples of Fishes Across the Continuum.

	Altricial	Precocial
Development at hatching	Limited	High
Fecundity of adults	High	Relatively low
Yolk per egg	Low	High
Parental care	None	Variable to much
Examples	Herring Bluegill	Guppy
	Cod Trout	Shark

In contrast, many altricial species have just formed their mouths when they deplete their yolk sacs. The heads are not extensively developed, and the guts are linear tubes extending to the anuses. The amount of energy that can be extracted from food is low; the assimilation efficiency might be 20–30%. Many of them have less fin differentiation at yolk absorption and are less capable of swimming. They are less efficient at surprising and capturing prey due to their poorly developed fins. They can do well at relatively high prey abundance but cannot utilize distributed prey patches as effectively as precocial larvae. As a result, altricial development results in a much lower likelihood of survival than precocial development. Altricial young are most likely to show large mortalities due to a critical period because of their lesser development at yolk absorption. Altricial larvae tend to be associated with the life history of opportunistic and periodic species, while precocial larvae are often associated with equilibrium species.

The ontogeny of an altricial larva – the anchovy – demonstrates this late development (see Figure 24-3; from Hunter and Coyne 1982). About 40% of yolk in the egg is utilized prior to hatching. Remaining material declines exponentially, so the anchovy larva has completely utilized its yolk about 170 hours after spawning. The mouth is developed after about 90 hours, and there may be only 10–20% of yolk reserves left for development at that stage. These reserves might be utilized for metabolism rather than for growth if the temperature is high, in which case, mouth development might not be completed by yolk depletion, likely contributing to mortality. Critical stages in development of altricial larvae – acquiring the ability to move the lower jaw, stomach formation, and functional eye development – may not occur until very late. They may not take place at all if other circumstances – e.g., high temperature – reduce yolk resources available for growth or if early foraging is limited.

Unusual body shapes are common for fish larvae, and these shapes do not appear hydrodynamically sound compared to those of adult fishes. Adult fishes are generally streamlined with rare exceptions in fishes that do not perform much sustained swimming. An adult fish, viewed from above, has a shape like the cross section of an airplane wing. This is a good shape for minimizing inertial forces that occur as water moves past the fish. The adult body form is generally hydrodynamically sound for sustained swimming. Larval fishes do not often show similar shape characteristics – maximum width at one-third of the body length – which is one characteristic of streamlining. A streamlined object would also have a sharply tapered tail and a round head. Because of the protuberances of fins and other morphology, the same kinds of characteristics are not common in larval fishes or in zooplankton.

Is hydrodynamic advantage not as important to larval fishes as it is to adult fishes? Several characteristics of swimming make the forces affecting larval or adult fishes very different. One is the Reynolds number – *Re* – which indicates the kinds of forces that have to be overcome by a moving object. Reynolds number is a function of three terms:

$$Re = VL/v \qquad\qquad \text{(Equation 24-1)}$$

where *V* is swimming velocity of the fish, *L* is length of the fish, and *v* is viscosity of the water. An average adult fish has an *Re* value of approximately 10^4 to 10^8, and the major factor retarding swimming for this body form would be inertial forces. With such a preponderance of inertia, a resistance to water flow across the body makes swimming difficult, and evolution has resulted in fish shapes that reduce friction by being more hydrodynamic. Caudal fin propulsion is much more effective at high *Re* values when burst-and glide-swimming is most efficient.

In contrast, larval fish swimming occurs at *Re* values from 10^1 to 10^2. The viscosity of the water compared to fish length is quite high in larvae compared to adults. Larvae swim in a situation in which inertia is not that influential, but viscosity strongly acts on the individual larvae. Crustaceous zooplankton are often similar in size and have similar *Re* values as larval fishes, and one reason crustaceous zooplankton remain in the water column is their body shape and projections cause them to sink slowly. Larval fishes are intermediate between many crustaceous zooplankters and adult fishes, so frictional drag based on the shape of the fish is not as important as viscosity in terms of how larvae swim. This allows larvae to develop unusual shapes and sustain swimming with body flexures rather than use of the caudal fin. Caudal and medial fin development often occur later in the larval period.

Environmental Effects on Hatching and Emergence

The match/mismatch hypothesis proposed that reproductive timing is based on the matching of larval hatching and transition to exogenous feeding with a time when prey availability and other conditions are suitable for survival. When there is a mismatch between these, there is poor reproductive success; when there is a match, a strong year class develops. The match/mismatch hypothesis implies that development of egg and larval fish characteristics should ideally be synchronized with timing of environmental factors that might influence survival. A number of environmental factors come to mind: abiotic – temperature, dissolved oxygen, salinity, current/wind/wave action – and biotic – food and abundance of predators and/or competitors. Timing of hatch and development of young fishes may be synchronized to avoid predation. Many characteristics of spawning appear to be related to the avoidance of predation, including the movement of adults to benign environments to spawn, where predators are less abundant or eggs are protected. Some characteristics that influence hatching time have already been covered, such as temperature.

Temperature – specifically degree days – is the number of days times the °C above a base temperature or biological zero. Development of larvae can be estimated by the number of degree days required to achieve a specified amount of growth or reach a certain developmental stage (for example, see Chezik et al. 2014). A degree day can be calculated in various specific ways, but a unit of degree day is essentially exposure to 1 °C for one day. Eggs at 8 °C for one day or 4 °C for two days would both experience 8 degree days. Increased temperature increases the speed at which most ectotherms develop or the incubation time. Temperature also influences the size of young when they hatch through the allocation of energy to body growth or metabolism during embryonic development.

The following sections detail early development and hatching in several fish species (esocids, salmonines, grunion, capelin, and anchovy) and will review major ecological components that generally build from the match/mismatch concept.

Broadcast Spawning in Esocids

Northern pike and muskellunge are members of the same genus (*Esox*) and typical examples of freshwater fishes in hatching and early development. Northern pike have been introduced into lakes that had natural muskellunge populations, and in most cases, muskellunge declined in abundance or became locally extinct. Fishery managers have wondered why northern pike outcompete muskellunge in sympatry. In historic times, the two populations were simply not in the same inland lakes and streams. In the past several decades, pike have been introduced into areas by humans intent on improving angling or allowed access by inundation of geographical barriers during hydropower development and dam building.

Both esocid species show similar characteristics of maturation and development. Northern pike and muskellunge develop gonads early in the fall, before spawning the following spring. The males essentially complete all of their testes growth and maturation in the fall; females continue to grow ovaries over winter. Muskellunge are believed to have evolved as riverine-type fishes (Harrison and Hadley 1978). They use offshore spawning sites, often over open sediments. Northern pike (more lacustrine fishes) move into flooded emergent vegetation along the shorelines of lakes to spawn or even into marshes. Both species spawn in early spring under conditions of rising water level due to snow melt. Northern pike use the resultant flooded vegetation as habitat in which to spawn; the young hatch and persist there for a short time and then usually move back into lakes before water recedes completely. The rising water levels flood marginal areas that do not have many natural predators. Muskellunge spawn in shallow offshore sandbars, nearshore areas with woody debris, or reefs, where rising water levels and the resultant waves may clean sediments from the area, improving conditions for spawning.

Both species are broadcast spawners, that is, they lay eggs over selected areas usually with vegetation or sediments. Northern pike eggs are adhesive and attach to those materials. This is important for northern pike, which spawn in shallow, muddy, weedy areas. If their eggs settled into the muddy sediment, they probably would not survive since the mud–water interface often has very little dissolved oxygen. By attaching to vegetation, the eggs are suspended in the water column and are exposed to elevated dissolved oxygen conditions. The non-adhesive muskellunge eggs settle onto the sediments and may be moved from woody or hard substrates to soft sediments by wave action. Survival of muskellunge eggs in these sites is very poor, and the only eggs surviving may be ones laid on woody debris (Zorn et al. 1998). In both species, hatching occurs after a certain number of degree days. Both northern pike and muskellunge hatch in 150–200 degree days; at cold temperatures (8 °C) hatching might take 20 days, while at warm temperatures (15 °C), it would take 10 days.

Both northern pike and muskellunge eggs and young are subject to extensive predation. Northern pike that move into intermittent tributaries to spawn may be the only fishes in the areas, and survival of young could be high because of higher food availability and lower predator density than that which exists for individuals that spawn in lakes on emergent vegetation. Muskellunge spawning in open water may be subjected to even more predation than northern pike in littoral vegetation in lakes. Early development of both could be limited by low dissolved oxygen and predation.

The most important characteristics in the competition between northern pike and muskellunge are that pike spawn earlier, hatch earlier, and develop faster than muskellunge. For example, in one Saskatchewan lake, the average spawning date for northern pike was between April 14 and May 12 (Rawson 1932). The temperature was about 4 °C on 14 April and about 13 °C on 12 May. Muskellunge, on the other hand, spawn between mid-to-late May in Minnesota, as much as two to four weeks later than northern pike. The temperature during muskellunge spawning is between 9 and 14 °C. Northern pike would have spawned as much as two weeks earlier in that same lake, and even though it was colder, the accumulation of degree days would result in pike young hatching before muskellunge. Northern pike larvae, once they hatch, have adhesive glands on their heads; they attach to vegetation and remain there. Emergence could be considered the time when northern pike finish consuming their yolk, detach from the vegetation, swim away, and begin to feed. Northern pike begin with a zooplankton diet and quickly shift to benthic invertebrates and then to fishes. By the time northern pike are a little over a month old, they forage on other fishes – particularly larval fishes – and may even be cannibalistic.

Since muskellunge spawn over sediments, they remain there rather than on vegetation. They are several weeks behind in their development compared to pike. When pike and muskellunge are sympatric in a lake, pike emerge earlier, when zooplankton and phytoplankton populations may be rapidly increasing; this leads to rapid growth of pike. They may grow fast enough that by the time muskellunge hatch and emerge, and pike can consume them. The competitive superiority of pike is most commonly attributed to the earlier emergence of pike and the predatory effects of pike on muskellunge. In inland lakes, where there may not be good offshore habitats for spawning, muskellunge utilize similar habitats as northern pike for spawning, and young fishes coexist in these habitats. In the St. Lawrence River and Lake St. Clair, muskellunge utilize offshore and littoral habitats for spawning, while pike utilize marshes, so there is clear habitat segregation between young of the two species. In the latter case, muskellunge and northern pike coexist because the young of both species do not occur in similar areas.

Successful early survival for these species in similar areas appears to be related to the time of hatching, which is affected by the earlier spawning of northern pike. With the differences in growth rate, these factors may combine to allow northern pike to become predators on muskellunge. One might wonder what factors allowed ancestral muskellunge with an earlier spawning time to evolve a later spawning time. The advantage of later spawning may be related to the riverine origins of muskellunge. Temperature increases – especially in flooded areas – progress much more rapidly in lakes than in streams. The main body of a river – especially major rivers, such as the St. Lawrence – gradually warms and is much more resistant to temperature increase than the floodings where northern pike spawn. The timing of blooms of high densities of prey for larval fish may also differ between lakes and rivers, so the reproductive timing of muskellunge may be an evolved difference related to riverine development. This advantage is lost when both species occur in the same lake.

As in the esocid example, emerging earlier may be adaptive. Most temperate freshwater fishes spawn in the spring. Not only are muskellunge larvae available for northern pike to feed upon, but other fishes that spawn later than pike are also in larval stages. Northern pike tend to be one of the earliest species to spawn in a lake. Late-stage larvae and juvenile northern pike then have the opportunity to forage on other larval fishes upon their emergences. Early spawning – when a series of other individuals spawn slightly later – may be an advantage for a predator. However, emergence too early – before the plankton bloom – could result in reproductive failure.

Nest Spawning in Salmonines

The second example of emergence and hatching is for the species of salmonines, which show different timings of spawning. Some species spawn in fall and some in spring. All young emerge in the spring, so for a fall spawner such as Pacific salmon, young remain in the redd through the winter and emerge the following spring. Spawning in fall results in earlier emergence of young than spawning in spring. So, if there were an advantage to emerging earlier, that advantage would be for fall-spawning salmonines. However, the accumulated degree days over winter are limited, so hatching and emergence in the spring occur fairly close for both spawning times.

Development in salmonines is correlated with water temperature and dissolved oxygen (DO) concentration. If dissolved oxygen declines much, development slows down. Young brook trout larvae in the gravel – developing under low oxygen concentrations – will first suspend any development and maintain just enough oxygen exchange for maintenance. Therefore, reduced DO levels at nonlethal concentrations result in longer incubation times before hatching. These levels probably also result in smaller sizes at hatching since energy is used for metabolism alone. Mortality occurs under extremely low dissolved oxygen conditions or long periods of low oxygen.

Salmonines lay their eggs in a redd, which is dug out and covered over with gravel. Eggs remain in that redd for a long period – throughout hatching and yolk absorption. Once larvae have used most of their yolk, they emerge from the gravel into the water and begin foraging. The timing of swim up is not correlated with any particular environmental event; it is correlated with absorption of yolk. If early emergence is important to survival, fall spawning would be adaptive. However, conditions over winter may largely influence whether fall spawning will be successful. In small tributary streams with low water flow in winter, low DO could occur in the redd and limit development or cause mortality. Spawning in spring – with rising flows and water levels – may produce a survival advantage in more intermittent streams.

Beach Spawning of Grunion and Capelin

The next example – beach spawning in fishes – is another spectacular behavior and appears to be related to the match/mismatch concept. One marine species with a Pacific distribution – the grunion – spawns in the middle of the night at a full or new moon when the highest high tide occurs. During their spawning season – March to September – tides are higher at night than during the day. Adults move up the shore, dig holes in the sand, and lay their eggs, and then wiggle back out and are swept away by the next wave. Grunion eggs remain in the sand over the next two weeks and develop, but do not hatch until the highest high tide of the next tidal period occurs and they are re-immersed. At that time, young grunion hatch, move into the water, and return to the sea. For grunion, spawning and emergence of larvae from the sand are keyed to the highest tides each month. Re-immersion alone is not sufficient to stimulate hatching; wave action is also important (Griem and Martin 2000). If wave action is not sufficient, the eggs remain unhatched in the sand for a second or even third tidal cycle, depending on temperature (Smyder and Martin 2002). It is not entirely clear what the adaptive value of wave action is, compared to immersion under calm conditions, although reduced predatory efficiency in wavy conditions may be speculated.

Sand is an environment that contains few predators, limiting predation on young grunion. On the other hand, it is a very difficult physical environment. An adult grunion spawning on the beach has little chance to select habitats that will necessarily be flooded by water again before the larvae use up their yolk. The next tidal height – flooding of the sand – and turbulence also depend on wind and waves. The adults select a benign environment in terms of predators but a very physiologically challenging environment in terms of abiotic conditions.

Capelin are widespread in the Atlantic Ocean, where they also spawn on beaches. However, capelin do not spawn or hatch in synchrony with the tides. Capelin move into gravelly areas of beaches to spawn but not at the highest high tide each month. Their eggs become mixed by waves as deep as 15 cm into the sediments. Frank and Leggett (1981) studied emergence of capelin into Newfoundland waters. Emergence occurs from June and through August, and the number of hatched capelin in the nearshore zone does not peak at regular intervals. However, at several different sites, capelin do appear synchronously. The conditions for emerging are probably similar among these sites but are not tidal in nature, so other characteristics of the environment must function as emergence cues.

The nearshore area off the coast of Newfoundland is mainly exposed to offshore winds – winds blowing from shore. Occasionally, winds may shift and become onshore. This also influences temperature. When the wind blows offshore, there is upwelling – deep, cold water comes up near shore – and warm surface water moves offshore (for a simplified example of upwelling; see Figure 24-4). When the wind blows onshore, just the opposite occurs – warm water from the surface moves into the nearshore environment. Therefore, the nearshore water temperature is warmer during an onshore wind and cooler during an offshore wind. Waves may also be larger in an onshore wind than in an offshore wind.

Water temperature characteristics were correlated with onshore or offshore winds along the Newfoundland coast. Temperature increased slowly prior to an onshore wind, increased dramatically after the wind began, and then declined again once the winds shifted offshore (Figure 24-5). This happened repeatedly throughout the summer, and there was a strong

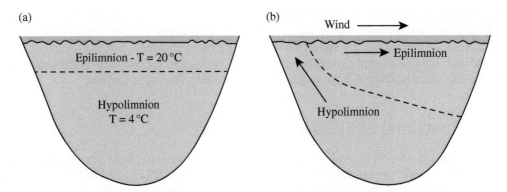

FIGURE 24-4   Two-dimensional, highly simplified representation of the effects of wind on a stratified lake: (a) stratified conditions and (b) upwelling after wind has blown.

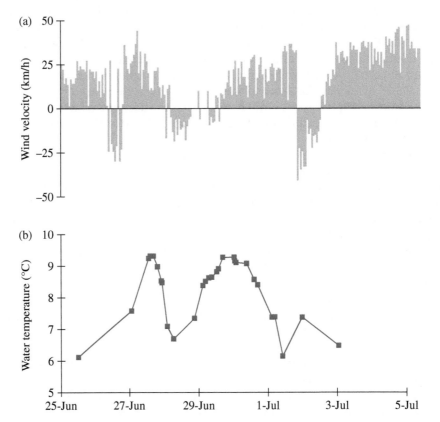

FIGURE 24-5   Wind speed and direction (a) and nearshore water temperature (b) for Bryant's Cove, Newfoundland. Positive wind speeds: offshore; negative wind speeds: onshore. *Source: Adapted from Frank and Leggett (1983).*

correlation between water temperature and wind direction. The thermal environment was largely modified by the direction of wind and its influence on temperature conditions.

The density of capelin larvae in the beach sediments was strongly correlated with the amount of time since the last offshore wind and with wind intensity (Figure 24-6). Stronger periods of offshore winds resulted in more larvae accumulating in the sediments. Onshore

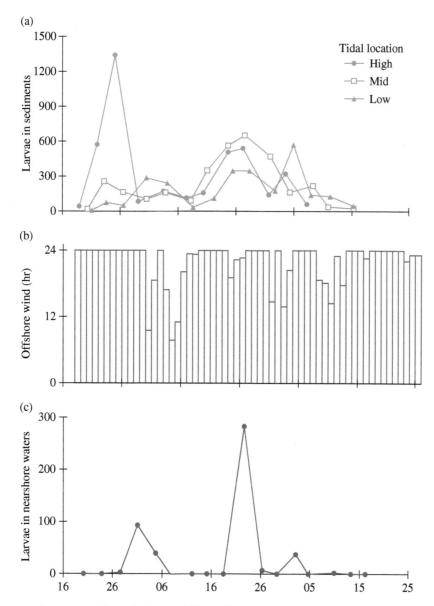

FIGURE 24-6   Occurrence of capelin larvae (a) in sediments at three locations designated as high, mid, or low tide, and (c) in nearshore waters, as well as (b) the duration of offshore winds for Bryant's Cove, 1978. *Source: Frank and Leggett (1981)/Canadian Science Publishing.*

winds resulted in fish emerging from the gravel. Larvae emerged when there were longer and stronger onshore winds and remained in the gravel when there were offshore winds. There was also a correlation of more young capelin in nearshore waters when offshore winds were less frequent (Figure 24-6). Frank and Leggett used their own, as well as historic data, to show that the density of capelin larvae in sediments prior to an onshore wind was correlated with the density of young capelin one day later in the nearshore zone (Figure 24-7). This emergence appears to be a well-established pattern.

Laboratory experiments on capelin demonstrated that the key characteristic for larval emergence was exposure to rapidly increasing temperatures, which are characteristic of

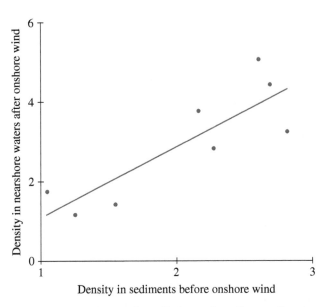

FIGURE 24-7   Relationship between density of capelin larvae in sediments (log thousands of larvae/m³) before an onshore wind event and larval density in the intertidal zone (log larvae/m³) one day after an onshore wind event. *Source: Frank and Leggett (1981)/Canadian Science Publishing.*

onshore winds. One might propose that passive means – increased wave action during onshore winds – could cause capelin to be moved from the sediments rather than assume that capelin actually select onshore wind conditions as a preferred time to emerge. However, experiments in absence of wave action showed that increased swimming by larvae in sediments occurred after exposure to rapidly increasing water temperature, and this actively moved larvae into locations where wave action could return them to the ocean.

Capelin larvae remained in the sediment until onshore winds brought warm water into the nearshore zone rather than emerge during offshore winds. Fish larvae in the sand could detect an offshore wind by sand temperatures and by water exposure; they then emerged. Fish eggs that remained in the gravel hatched; larval fish used yolk reserves and then used endogenous energy stores during their residence in the gravel. Unlike grunion, their hatching was not dependent on water immersion and turbulence (Frank and Leggett 1983). Since emergence was not tidally synchronized, fish larvae may have experienced several inundations before emergence. As expected, capelin larvae became debilitated during their residence in the sediments and use of endogenous stores. From the time of yolk absorption on, swimming ability declined in accordance with increased time in the sediments (Figure 24-8). The frequency of aberrant twitching movements also increased with time in the sediments. So over time, after yolk was absorbed, fish larvae declined in condition while they remained in the sediments.

Beyond temperature, other characteristics differed between onshore and offshore wind conditions. There is not much difference in salinity of water during an offshore or onshore wind event. However, there are substantial differences in the presence of other species in the nearshore waters during these wind events. During offshore winds that force cooler water nearshore, there is a high density of chaetognaths, jellyfishes, and ctenophores, and a low density of small zooplankton. This predator-rich environment – at least for capelin larvae – developed in the cooler surface waters during offshore winds. The nearshore zone was also a food-poor environment at that time with few small zooplankton. The reverse conditions developed during

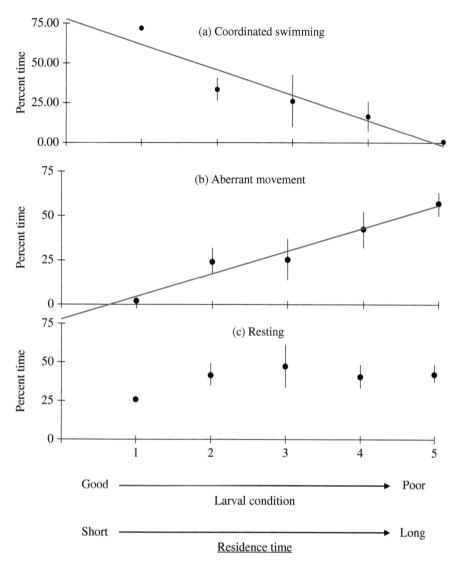

FIGURE 24-8   Percent of time (mean + 95% CL) capelin larvae spent in each behavior with progressive residence times in the sediments, which are correlated with declining physical condition of the larvae. *Source: Frank and Leggett (1982)/Canadian Science Publishing.*

onshore winds. Therefore, emergence of capelin appeared to be correlated with the presence of water containing better food resources and lower predation.

Capelin larvae evolved behaviors that caused them to remain in the sediments and die rather than disperse into water during offshore winds. The predictability of onshore winds is very low. Capelin do not all breed synchronously, but more or less continually, and fish larvae in the sediments appear to wait for onshore winds to emerge rather than risk emergence into predator-rich water. The correlation between onshore or offshore winds and the abundance of other species of fish larvae follow exactly the same pattern as capelin. Nearshore abundance of fish larvae in general is highly correlated with onshore winds in the region. This appears to be a predator avoidance and food-seeking behavior, which agrees well with the match/mismatch hypothesis.

Pelagic Spawning of Anchovy

A more typical example of the early life of marine species is the anchovy, a pelagic spawner. The general pattern is similar for clupeids and probably many other marine pelagic fishes. Reproductive behavior of the anchovy is not elaborate and involves broadcast spawning of eggs and mass fertilization over pelagic waters. The eggs tend to remain pelagic and develop in the pelagia to the larval and juvenile stage. This developmental ontogeny is shown in Figure 24-3.

Anchovy is a species suspected of having intense critical periods at the stage of transition to exogenous feeding. Houde (2002) and Cowan and Shaw (2002) both presented excellent reviews of the concepts involved in these early life histories. The Critical Period Hypothesis of Hjort (1914) and the match-mismatch concept of Cushing (1969) identify a paradox – that larval fish need a lot of food but live in areas with less food available (on average) than needed.

Reuben Lasker and colleagues at The National Marine Fishery Service Lab in San Diego focused on this issue for anchovy over many years (Lasker 1981). Lasker developed the Stable Ocean Hypothesis, which proposes that larval anchovy survival is enhanced when winds are calm. Calm conditions produce stratified waters, stronger layers of chlorophyll maxima, and very patchy food resources for anchovy. Lasker believed anchovy could forage more effectively under these patchy distributions and calm conditions. Peterman and Bradford (1987) were also able to demonstrate correlations between calm conditions and good survival of anchovy off the coast of California (Figure 24-9).

While winds leading to more turbulent conditions appear to be detrimental to anchovy survival, upwelling events are believed to enhance survival. Upwelling brings cold, nutrient-rich waters to the surface where primary and secondary production is enhanced, also producing more potential food for anchovy. Upwelling can be generated by wind or current events and can vary in intensity. One hypothesis (Cury and Roy 1989) is that intermediate levels of upwelling enhance larval survival, while weak or strong upwelling events limit survival (Figure 24-10). This is consistent with Lasker's hypothesis regarding turbulence and describes the trade-off between increased food production and widely distributed food resources. It also becomes clear from this work that numerous events may enhance or reduce larval survival of anchovy, which should be highly variable under such constraints.

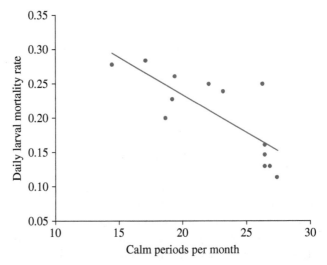

FIGURE 24-9 Daily mortality rate of anchovy compared to the number of calm periods per month in the California Current over a 13-year period *Source: Peterman and Bradford (1987)/American Association for the Advancement of Science.*

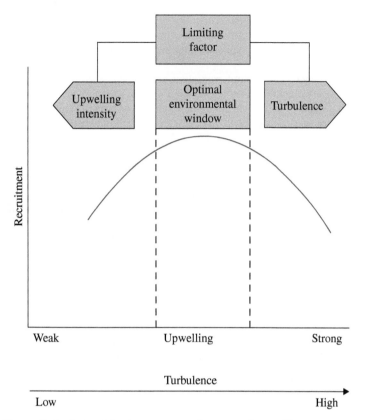

FIGURE 24-10 Theoretical relationships between recruitment and upwelling *Source: Cury and Roy (1989)/ Canadian Science Publishing.*

Summary

Larval fishes vary dramatically in their development at hatching and metamorphosis. Some of this variance is due to adult investments in egg size and quality, while some is due to hatching and emergence cycles. The early development of fishes can be strongly influenced by abiotic conditions – temperature, dissolved oxygen, mechanical action – as well as by biotic conditions – presence of prey, predators and competitors. Larval fishes are highly susceptible to mortality. Processes that affect their vulnerability to predators and overlap with suitable prey during key developmental transitions can strongly influence their survival and ultimate recruitment success.

Literature Cited

Balon, E.K. 1975. Terminology of intervals in fish development. Journal of the Fisheries Research Board of Canada 32:1663–1670.

Chezik, K.A., N.P. Lester, and P.A. Venturelli. 2014. Fish growth and degree-days I: Selecting a base temperature for a within-population study. Canadian Journal of Fisheries and Aquatic Sciences 71:47–55.

Cowan, J.H., Jr., and R.F. Shaw. 2002. Recruitment. Pages 88–111 *in* L.A. Fuiman and R.G. Werner, editors. Fishery Science: The Unique Contribution of Early Life Stages. Blackwell Science, Malden, Massachusetts.

Cury, P., and C. Roy. 1989. Optimal environmental window and pelagic fish recruitment success in upwelling areas. Canadian Journal of Fisheries and Aquatic Sciences 46:670–680.

Cushing, D.H. 1969. The regularity of the spawning season of some fishes. ICES Journal of Marine Science 33:81–92.

Frank, K.T., and W.C. Leggett. 1981. Wind regulation of emergence times and early larval survival in capelin (*Mallotus villosus*). Canadian Journal of Fisheries and Aquatic Sciences 38:215–223.

Frank, K.T., and W.C. Leggett. 1982. Environmental regulation of growth rate, efficiency, and swimming performance in larval capelin (*Mallotus villosus*), and its application to the match/mismatch hypothesis. Canadian Journal of Fisheries and Aquatic Sciences 39:691–699.

Frank, K.T., and W.C. Leggett. 1983. Survival value of an opportunistic life-stage transition in capelin (*Mallotus villosus*). Canadian Journal of Fisheries and Aquatic Sciences 40:1442–1448.

Griem, J.N., and K.L.M. Martin. 2000. Wave action: The environmental trigger for hatching in the California grunion, *Leuresthes tenuis* (Teleostei: Atherinopsidae). Marine Biology 137:177–181.

Harrison, E.J., and W.F. Hadley. 1978. Ecologic separation of sympatric muskellunge and northern pike. American Fisheries Society, Special Publication 11:129–134.

Hjort, J. 1914. Fluctuations in the great fisheries of northern Europe viewed in the light of biological research. International Council for the Exploration of the Sea, Rapports et Proces-Verbaux des Reunions 20:1–228.

Houde, E.D. 2002. Mortality. Pages 64–87 *in* L.A. Fuiman and R.G. Werner, editors. Fishery Science: The Unique Contribution of Early Life Stages. Blackwell Science, Malden, Massachusetts.

Hunter, J.R., and K.M. Coyne. 1982. The onset of schooling in northern anchovy larvae *Engraulis mordax*. Reports of the California Cooperative Oceanographic and Fishery Investigations 23:246–251.

Kendall, A.W., Jr., J.D. Schumacher, and S. Kim. 1996. Walleye pollock recruitment in Shelikof Strait: Applied fisheries oceanography. Fisheries Oceanography 5(Suppl. 1):4–18.

Kryzhanovsky, S.G. 1947. Classification of the Cyprinidae. Zool. Zhurn 26:53–64 (in Russian)

Lasker, R. 1981. Marine Fish Larvae, Morphology, Ecology and Relation to Fisheries. University of Washington Press, Seattle.

Noakes, D.L., and J.G.J. Godin. 1988. Ontogeny of behavior and concurrent developmental changes in sensory systems in teleost fishes. Pages 345–396 *in* W.S. Hoar and D.J. Randall, editors. Fish Physiology, Vol. XI Part B. Academic Press, Inc., San Diego, California.

Peterman, R.M., and M.J. Bradford. 1987. Wind speed and mortality rate of a marine fish, the northern anchovy (*Engraulis mordax*). Science 235:354–355.

Rawson, D.S. 1932. The pike of Waskesiu Lake, Saskatchewan. Transactions of the American Fisheries Society 62:323–330.

Smyder, E.A., and K.L.M. Martin. 2002. Temperature effects on egg survival and hatching during the extended incubation period of California grunion, *Leuresthes tenuis*. Copeia 2002:313–320.

Zorn, S., T. Margeneau, J.S. Diana, and C.J. Edwards. 1998. The influence of spawning habitat on natural reproduction of muskellunge in Wisconsin. Transactions of the American Fisheries Society 127:997–1007.

PART 6

Fish Communities in Aquatic Ecosystems

CHAPTER 25

Description and Measurement of Fish Communities

Focus on community ecology generally involves studying patterns of occurrence and abundance of multiple species or functional groups. This is in contrast to population ecology, which focuses on a single population or a small number of interacting populations. The term community is often used in a generalized way to refer to groups of species or functional groups. However, various other terms also have been used to describe groups of species, and there is a history of inconsistent and confusing terminology in describing species groups. An ecological community is comprised of populations of two or more species inhabiting the same system or location. Communities may change over time, and therefore the time window used to define a community may be influential. Also, the spatial extent within which a community is measured will influence its composition. A community contained within an entire river network will likely be more specious than a community contained within a single headwater stream of the same river network. Depending upon how groups of species are defined, they can incorporate a breadth of taxonomic levels (e.g., both plants and animals) or be more restrictive. Grouping can be defined for specific taxonomic groups, such as the populations of fishes inhabiting a particular estuary, lake, or stream.

The inconsistent and confusing terminology used in community ecology was recognized by Fauth et al. (1996), who proposed a framework to make the terminology for community ecology consistent. They advocated for the use of different terms to describe groups of populations based on whether the groups have different phylogeny, co-occur geographically, or exploit similar resources. They developed a Venn diagram to visually communicate their terminology framework (recreated in Figure 25-1). They define communities as species that co-occur in a location at a particular point in time. Under their definition, members of a community do not need to use similar resources or have similar phylogeny. The term guild refers to species that exploit similar resources. However, members of a guild do not necessarily need to have shared phylogeny or co-occur in a system. Groups of species that co-occur in a system and have shared phylogeny are termed an assemblage. All fish populations that inhabit a particular lake may collectively be considered an assemblage.

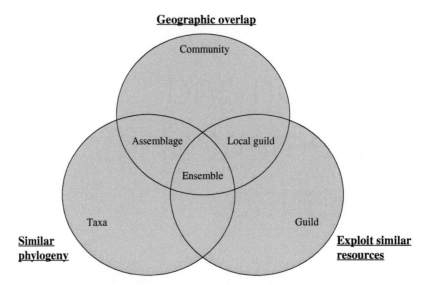

FIGURE 25-1 Venn diagram depicting terminology framework for groups of species with shared geography, resource used, and phylogeny. *Source: Fauth et al. (1996)/The University of Chicago.*

Groups of species that co-occur in a system and exploit similar resources are termed a local guild. For example, all benthivorous organisms – both vertebrates and invertebrates – in a lake would constitute a local guild. Groups of species that co-occur in a system, exploit similar resources, and have shared phylogeny are termed ensembles. All the benthivorous fishes in a lake would be considered an ensemble. In considering the ecology of fishes, it is important to clearly define the criterion for including species when defining a community, assemblage, local guild, or ensemble.

The populations comprising a community (or other community ecology grouping) have been traditionally defined taxonomically at the species level, and the diversity of communities has been assessed based on occurrence and proportional abundances of species. However, diversity is increasingly assessed not simply based on species occurrence and abundances but also based upon functional, morphological, and phylogenetic characteristics. If species are functionally very similar – for example, in terms of prey they consume and habitats they occupy, they may be somewhat functionally redundant. Several recent studies have focused on assessing the functional breadth of groups of organisms rather than only taxonomic breadth.

Community ecology involves describing the patterns of composition and exploring the processes contributing to the structure of groups of organisms. As described above, organisms can be grouped in a variety of ways – for example, based on location, time, resource use, taxonomy, morphology, and function. Rather than continuously listing the different possible groupings of organisms, we will use the term community in a generalized way to consider groups of organisms when we focus on taxonomic diversity. There are a variety of approaches for describing and comparing communities, and there are various processes that contribute to community structuring. Subsequent chapters describe attributes of fish communities (and assemblages and ensembles) in different types of aquatic systems. Here, we describe basic approaches for describing and comparing fish communities and introduce some processes contributing to community structure.

Fish Richness and Abundance

There are many approaches to describe communities and to quantitatively compare communities. The most common approach is to measure richness. Richness is defined as the number of species in a community or assemblage – for example, the number of fish species in a reservoir. Richness is dependent on the community one considers. In addition to total fish species richness, one can consider richness of distinct groups of fish, such as the richness of darters in a stream (the number of darter species), richness of fish species classified as sensitive to eutrophic conditions, richness of species classified as preferring cold water, or richness of primarily piscivorous fish species. One can also consider richness at coarser taxonomic levels, such as genus richness or family richness. Regardless of what aspect of richness is considered, in most cases, it is impossible to obtain a true measure of richness from the community as not all individuals from the community can always be assessed. Rather, richness tends to be assessed from a subset of individuals sampled from the greater community. This implies that any sampling biases that make it difficult to collect certain species will influence indices of richness. Even if only considering fishes, different gear and sampling methods will effectively collect different species. The amount of sampling conducted and the number of individual organisms assessed can also strongly affect richness quantification. By exerting more sampling effort and assessing more individuals, one is likely to encounter more total species.

Abundances of different species in a community are generally not uniform. Some species tend to be rather abundant, while others tend to be much rarer. In fact, percent dominance is a common index to capture such distributions and simply involves quantifying the percent of all assessed individuals categorized as one of the most abundant species. Percent dominance is quantified for the one, two, or three most abundant species. Given such relatively high abundance of a few species, when one collects individuals from a community, one is most likely to encounter these dominant species. Therefore, if one only collects a few individuals from a community, it is probable that the richness of species encountered will be relatively low. However, as more and more individuals are assessed, the rarer species are also likely to be encountered.

The pattern of accumulating more species with additional sampling of a community can be described as a species accumulation curve (Figure 25-2). If the true number of species in a

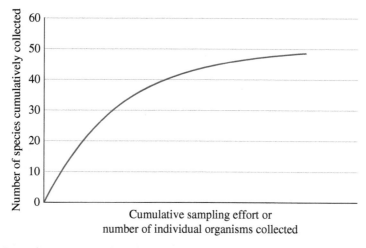

FIGURE 25-2   Curve demonstrating the accumulation of additional species with more and more sampling of a community and inspection of additional individual organisms.

community is Y, then with increasing sampling and collection of individuals from that community, one should accumulate more species – collect an ever-increasing proportion of Y. Even with limited sampling of a community, one would expect to rapidly accumulate the more abundant species. However, the rate of accumulating new species will decrease as one approaches Y. This is because the likelihood of capturing very rare species is low, and a great deal of sampling may be necessary to accumulate these species.

The differential abundances of species in a community and patterns of species accumulation with more sampling and inspection of additional individual organisms collectively have important implications for comparing richness measures among communities. If one community of fishes is sampled more intensely than another, or if researchers inspect a greater number of individual fish from one community than another, it is highly likely that more fish species will be observed for the community with greater sampling and where more individuals are encountered. Such an effect makes it less straightforward to compare changes in richness over time or among spatially distinct communities. It is important that measures of richness are based on consistent sampling methods, as well as consistent levels of sampling effort or consistent numbers of individual organisms. If such consistency is impractical, it is possible to use a technique called rarefaction to standardize measures of richness to a fixed amount of effort or fixed number of individuals. This method involves determining what the richness of a sample would be if the entire sample is not considered, but only a smaller proportion of the sample is considered. For example, suppose 10 seine hauls from a stream collect a total of 215 individual fish and 11 different species, but another stream is only sampled with 5 seine hauls and yields 145 individual fish and 7 species. It is not appropriate to directly compare the richness values from these two samples. However, one could use a rarefaction approach to estimate what the richness of the first stream's sample would be if 5 (rather than 10) seine hauls would have been conducted or if 145 individual fish (rather than 215) would have been inspected.

Fish populations – and the broader aquatic food webs they occupy – are structured through a variety of processes, including biotic interactions, colonization, local extirpation, and adaptation and acclimation to environmental conditions. Patterns of fish species richness are related to relative rates by which systems gain new species and lose existing species. An aquatic system may gain new species, either through colonization from another system or through speciation within the system. Systems lose species when they become locally extinct. Factors that affect these three processes (colonization, speciation, extinction) are thought to strongly influence patterns of species richness across gradients. For example, conditions that favor colonization from external systems and conditions that have facilitated high rates of speciation in the past are expected to lead to higher levels of richness. The following chapters present some of these processes and their influence in structuring fish communities in various aquatic systems.

Diversity and Evenness

In addition to the number of species present, communities are often described in terms of the relative abundances of different species. While there is a tendency for abundances of species to be skewed, with some species being very abundant and others being quite rare, the extent of such skewness will differ among communities. In some communities, the distribution of species abundances will be far more equitable. Various measures have been developed to simultaneously index species richness and abundance equity. Such diversity indices are based on the notion that communities are more diverse when they are not only more rich with species, but when the abundance of the species constituting a community is more evenly distributed. The two most common diversity indices are the Shannon–Wiener index, H', and the Simpson index, $1 - D$:

$$H' = \sum_i^R p_i \ln p_i \qquad \text{(Equation 25-1)}$$

$$1 - D = 1 - \sum_i^R p_i^2 \qquad \text{(Equation 25-2)}$$

For both indices, p_i is the proportion of species i in the community and R is the number of species in the community (richness). The Shannon–Wiener index, H', can take values from 0 (if there is only one species in the community) to infinity. For example, if a community has 10 evenly distributed species, $H' = 2.3$. The Simpson index, $1 - D$, can take values from 0 to 1. If a community has 10 evenly distributed species, $1 - D = 0.9$.

Both of these diversity indices simultaneously account for the richness of the community as well as the relative abundances of the species constituting the community. In some cases, it may be relevant to consider the equity of species abundances independent of the actual number of species in the community. To this point, both diversity indices facilitate calculation of evenness indices:

$$J' = \frac{H'}{\ln R} \qquad \text{(Equation 25-3)}$$

$$V' = \frac{1 - D}{1 - \dfrac{1}{R}} \qquad \text{(Equation 25-4)}$$

Here, J' is the Shannon–Wiener evenness index and V' is the Simpson evenness index.

Alpha-, Beta-, and Gamma-Diversity

In evaluating richness or diversity in a system or across a region, it is common to collect organisms from multiple locations or sampling sites. In so doing, it is plausible to examine richness or diversity within an individual site, as well across all sites. Whittaker (1960), at the time working at Brooklyn College, introduced a framework to describe these different spatial levels of diversity. Whittaker defined α-diversity as the mean diversity value at a single site and γ-diversity as the diversity when data across all sites are combined. Whittaker introduced the concept of β-diversity as a means of comparing α and γ. β-diversity was introduced as γ-diversity divided by α-diversity. Since γ-diversity is usually greater than α-diversity, β-diversity will generally take values greater than 1. A system with homogeneous species composition will tend to have relatively low β-diversity with values close to 1.0, while a system with very distinct compositions and diversity values among sampling subunits will express relatively high β-diversity.

Similarity of Fish Communities

Measures and indices such as richness, diversity, and evenness are useful to describe a community in that they integrate information for a variety of species to describe the entire community. However, as with any index, there is a trade-off between summarizing a lot of data into a single index, which allows for a holistic measure, but in the process loses the more refined information used to create the index. In considering communities, it may be insightful to know the actual species present in the community or the relative abundances of strongly interacting species, such as predators and prey. Indices of richness, diversity, and evenness are species independent. Therefore, two different communities may be characterized by near-identical richness, diversity, and evenness indices but not have a singles species in common. When comparing among spatially distinct communities or evaluating how a community changes over time or along environmental gradients, these indices may be used along with other approaches to compare the composition of communities.

There are various methods to compare communities based upon species composition and relative abundances of species. These methods are generally based on either evaluating how similar the communities are or how dissimilar – or distant – they are from one another. With indices of similarity, greater values imply that paired communities are more alike. In contrast, with distance measures, greater values imply that paired communities are more distinct.

The most straightforward index of similarity is Jaccard's coefficient, $C_{j,k}$, which is simply based on species presence and absence. When comparing two communities, j and k, the Jaccard's coefficient describing their similarity is simply based on the number of species present in both communities, p, versus the number of species only present in one of the communities, m:

$$C_{j,k} = \frac{p}{p+m}$$

(Equation 25-5)

Jaccard's coefficient can take values from 0 – if no species are shared between the communities – to 1, if they have identical species compositions.

Another approach involves indexing percent similarity, $PS_{j,k}$. Unlike Jaccard's coefficient, percent similarity is based on the percentage abundances of species in each community, $P_{i,j}$ and $P_{i,k}$. For each species, i, the percentage abundance is compared between the two communities and the minimum value is selected. These minimum percentages are then summed including all species present across both communities (n) to calculate percent similarity between the two communities, j and k:

$$PS_{j,k} = \sum_{i}^{n} \text{minimum}\left(P_{i,j}, P_{i,k}\right)$$

(Equation 25-6)

Note that percent similarity does not consider actual abundances of different species but is based solely upon percent abundances. Two communities can be vastly different in terms of actual abundances of species, but if the percentage composition is similar by species, their overall percent similarity will be high. If they include identical species and express the same percentage composition, their paired percent similarity will be 1.0 or 100%. In contrast, two communities that have no shared species will have a percent similarity of 0.

There are various distance measures to evaluate the distinctness of fish communities. Two of the most common distance measurements are Euclidean distance, $d'_{j,k}$, and Bray–Curtis distance, $b_{j,k}$. These measures are based upon actual measures of abundance, X, of different species, i, in the communities being compared. Therefore, when using these measures, it is appropriate to have a similar sampling effort when assessing different communities for comparison:

$$d'_{j,k=} \sqrt{\frac{\sum_{i}^{n}\left(X_{i,j} - X_{i,k}\right)^2}{n}}$$

(Equation 25-7)

$$b_{j,k} = \frac{\sum_i^n \left| X_{i,j} - X_{i,k} \right|}{\sum_i^n \left(X_{i,j} + X_{i,k} \right)}$$

(Equation 25-8)

Here, $X_{i,j}$ and $X_{i,k}$ are the measured abundances of species i in community j and k, and n is the total number of species present across both communities. If two communities are identical in terms of species abundances, both distance measures will equal 0. Euclidean distance can range from 0 to infinity, while Bray–Curtis distance measures can take values from 0 to 1, with 1 being completely distinct communities.

Evaluating Fish Communities Among Indiana Lakes

Having described various indices of fish communities and methods to compare among fish communities, it may be useful to demonstrate these analyses using example data. Bass Lake (568 ha), Bruce Lake (99 ha), and Nyona Lake (43 ha) are three proximate glacial lakes in northwest Indiana, USA. Fish communities of these lakes were sampled by the Indiana Department of Natural Resources during June and July using standard assessment gear. Methods included overnight sets with trapnets and experimental, multi-mesh gillnets as well as along shore, nighttime electrofishing. Table 25-1 presents the total catches of fish in these lakes through these multi-gear assessments along with lake-specific community indices. While some species were collected in all three lakes, others were only caught in one or two of the lakes. Note that the abundances of specific species shared among lakes are quite different. Bruce Lake stands out as having a very large number of bluegill and largemouth bass. This lake also contained the largest number of species with a richness of 19. However, due in large part to the high dominance by these two species (81%), diversity, and evenness, indices for the Bruce Lake fish community are lower than for Bass Lake and Nyona Lake.

Table 25-2 presents similarity indices and measures of distance among fish communities in Bass, Bruce, and Nyona lakes. Based on Jaccard's coefficient (and the data presented in Table 25-1), it is clear that Bruce and Nyona fish communities have a relatively large number of species in common. The Bass Lake fish community, however, has a relatively large number of unique species, such as channel catfish, walleye, and white crappie. When comparing the Bass Lake community to each of the other two lakes, not only is Jaccard's coefficient relatively low, so are percent similarity indices. In considering distance measures, there is a bit of inconsistency in patterns based on Euclidean versus Bray–Curtis distances. Bray–Curtis distance is lowest (more similar) between Bruce and Nyona communities, which is consistent with the larger number of shared species and high percent similarity. However, Euclidean distance is lowest when comparing Bass and Nyona communities and comparison of Bruce Lake community with either Bass or Nyona fish communities yields higher Euclidean distances. This disparity between Bray–Curtis and Euclidean distances is simply a reflection of how the indices are calculated. Euclidean distance measures are strongly influenced by disparities in very abundant species. Therefore, the very large numbers of bluegill and largemouth bass in Bruce Lake lead to large Euclidean distances when comparing with communities in the other two lakes.

	Bass Lake	Bruce Lake	Nyona Lake
Black crappie			3
Bluegill	5	552	60
Bluntnose minnow	3	1	
Bowfin		2	
Brown bullhead	2	5	11
Channel catfish	72		
Common carp	7	1	
Gizzard shad	20	40	19
Golden shiner		8	
Hybrid striped bass			4
Lake chubsucker		1	
Largemouth bass	2	102	86
Longear sunfish		11	
Longnose gar		1	
Muskellunge		4	
Pumpkinseed	1	7	1
Redear sunfish		3	47
Quillback	13		
Spotfin shiner	2		
Spotted gar	5	14	13
Striped shiner	6		
Walleye	15		
Warmouth		10	4
White bass	22		
White crappie	44		
White sucker		4	8
Yellow bullhead		3	1
Yellow perch	4	36	20
Species richness	16	19	13
Number of individuals	223	805	277
Percent dominance (top two species)	52%	81%	53%
Simpson diversity	0.83	0.51	0.81
Shannon–Wiener diversity	2.13	1.25	1.96
Simpson evenness	0.88	0.54	0.88
Shannon evenness	0.77	0.43	0.76

TABLE 25-1 Counts of Different Fish Species Collected by the Indiana Department of Natural Resources During Standardized Assessment of Three Glacial Lakes Using Experimental Gillnets, Trap Nets, and Electrofishing. Below the Counts are Values of Community Indices Calculated Based Upon these Counts.

TABLE 25-2 Paired Measures of Similarity and Distance Based on Comparisons Between Fish Community Assessments Conducted in Three Indiana Lakes (Data Presented in Table 25-1).

	Bass and Bruce	**Bass and Nyona**	**Bruce and Nyona**
Jaccard's coefficient	0.35	0.32	0.52
Percent similarity	13%	15%	49%
Euclidean distance	110.80	30.82	108.08
Bray–Curtis distance	0.92	0.85	0.60

Changing Fish Communities in Lake Erie and Saginaw Bay

As demonstrated for the three lakes in Indiana, fish communities will differ among locations. In addition, fish communities can change over time. The Laurentian Great Lakes have been affected by a wealth of external stressors over time that have collectively altered the abiotic and biotic environments, including various invasive species, overfishing, habitat degradation, pollution, and nutrient loading. In the 1960s and 1970s, environmental regulations were implemented in the United States and Canada, which over time have collectively led to reductions in pollution and nutrient loading. However, since the 1960s, extant and new invasive species have continued to alter Great Lakes food webs. Although some degraded habitats have been restored, habitat degradation has continued to be an issue. While fishing pressure has declined for some stocks, commercial and recreational fisheries continue to exploit Great Lakes fish stocks. Considering these various changing stressors, it is interesting to consider how fish communities may have changed over decadal time scales. Given that stressors have varied concomitantly, it may not be feasible to precisely determine which specific stressors may have led to specific changes in fish communities. However, the type of changes observed in a community may offer insights as some likely drivers of changes.

Lake Erie is the smallest – and on average, shallowest – of the five Great Lakes, but it is also the most productive. The western and central basins of Lake Erie are highly productive and have historically supported some of the greatest fish biomass concentrations in the Great Lakes. Saginaw Bay is a large embayment in western Lake Huron, and similar to Lake Erie, Saginaw Bay, and in particular inner Saginaw Bay, has historically supported high biomass of fishes. Being relatively shallow, both Lake Erie and Saginaw Bay are classified as eutrophic or mesotrophic, and both are strongly influenced by discharge from large watersheds. Relative to deeper and less productive Great Lakes, Lake Erie and Saginaw Bay tend to support species of fish tolerant of warmer temperatures and more eutrophic conditions. These fish communities have been monitored consistently over time, with the Ohio Department of Natural Resources and Michigan Department of Natural Resources, conducting bottom trawling surveys in Lake Erie and Saginaw Bay since 1969 and 1970, respectively. In two similar but separate analyses, Ludsin et al. (2001) and Ivan et al. (2014) used these long-term data collected by state natural resource agencies to evaluate how fish communities changed in western and central Lake Erie from 1969 to 1996 and in inner Saginaw Bay from 1970 to 2011.

Ludsin et al. examined temporal patterns of species richness, separately for the western and central basins of Lake Erie. They found that overall species richness declined in the western basin but increased in the central basin. They also separately examined signal species richness and defined a signal species as species that (a) were captured during more than three years but less than 28 years of the survey and (b) were not a primary prey of walleye, the main piscivore in Lake Erie. They used the latter caveat in selecting signal species to limit the influence of predation in driving shifts in signal species richness. Finally, they separately considered signal species richness for groups of species either tolerant or intolerant of eutrophic conditions. They found that richness of tolerant signal species in the western basin of Lake Erie declined from 1969 to 1996, but intolerant signal species richness increased in both the western and central basins of the lake during this same time frame.

Ivan et al. used somewhat similar approaches to examine changes in the inner Saginaw Bay fish community. First, they conducted a rarefaction analysis to standardize fish community analyses across years during which sampling effort varied. They then examined trends in not only total species richness, but richness of signal species and native species, as well as richness of species tolerant, intolerant, and moderately tolerant of eutrophy. They found that total species richness, native species richness, and richness of moderately tolerant species all strongly increased over time, while richness of tolerant species moderately decreased and richness of intolerant species moderately increased (Figure 25-3a).

Both Ludsin et al. and Ivan et al. used ordination methods to examine how species compositions had changed in Lake Erie and Saginaw Bay. Ordination methods are multivariate statistical and mathematical techniques that build on the concepts of multivariate similarity and distance to compare communities. However, unlike the paired comparisons described above, ordination methods allow for simultaneous comparison of more than two communities to evaluate which communities – or in these cases, which year-specific communities – were more similar or more distant in terms of species compositions. Ordinations are often depicted graphically with communities with more similar species composition located in close proximity in ordination space and communities with very distinct species composition located far apart in ordination space. Ludsin et al. used a detrended correspondence analysis to show that the fish communities present in western and central Lake Erie during the 1970s were distinct from communities present during the 1990s. For example, in the 1970s, tolerant species, such as brown bullhead (*Ameiurus nebulosus*) and common carp (*Cyprinus carpio*), were relatively abundant in the western basin, while during the 1990s, relatively intolerant species, such as burbot (*Lota lota*) and lake whitefish (*Coregonus clupeaformis*), increased in the central basin. Ivan et al. used a nonmetric multidimensional scaling ordination based upon Bray–Curtis distances to demonstrate that Saginaw Bay fish community compositions during the 1970s and early 1980s were distinct from community compositions during the 1990s and 2000s. They found that relative abundances of the majority of species increased from the 1970s to 1990s and 2000s. However, some tolerant species, such as black crappie (*Pomoxis nigromaculatus*) and pumpkinseed (*Lepomis gibbosus*), did decline over time, while more moderately tolerant and intolerant species, such as lake whitefish, increased (Figure 25-3b).

It is difficult to unequivocally determine which environmental drivers contributed to fish community changes in Lake Erie and Saginaw Bay, given that multiple potential drivers trended over time. However, considering the type of changes observed in both Lake Erie and Saginaw Bay fish communities – as well as the timing and direction of fish community changes – both Ludsin et al. and Ivan et al. reached the conclusion that the main drivers of fish community changes were likely reductions in nutrient loads and shifts from more eutrophic to more mesotrophic conditions. In both systems, the richness of fish species tolerant of eutrophic conditions decreased, while the richness of species intolerant or moderately tolerant of such conditions increased. These patterns were accompanied by decreases in tributary nutrient loads into the systems as well as decreases in nutrient and chlorophyll concentrations in Lake Erie and Saginaw Bay.

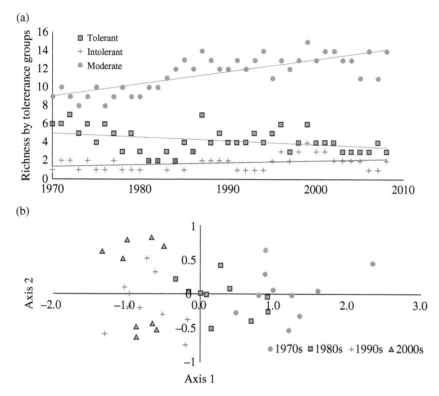

FIGURE 25-3 Temporal patterns of the Saginaw Bay fish community from 1970 to 2008 based on annual trawling surveys conducted by the Michigan Department of Natural Resources. (a) Median fish richness per 10 bottom trawls with richness calculated separately for species grouped as tolerant of eutrophic conditions, intolerant of eutrophic conditions, and moderately tolerant of eutrophic conditions. Note that Saginaw Bay is considered a mesotrophic to eutrophic system, and improved water quality from 1970 to 2008 seemingly contributed to a slight decline in the number of tolerant species in the system and a slight increase in the number of intolerant system. However, the greatest change was the more pronounced increase in the number of moderately tolerant species. (b) A nonmetric multidimensional scaling (NMDS) ordination to compare the Saginaw Bay fish community among sampling years. In the NMDS ordination, each point represents a year and points (years) closer to each other in ordination space have more similar fish community composition based on shorter Bray–Curtis distances. Note that the ordination suggests that the fish community changed markedly from the 1970s to the 1980s and into the 1990s; however, the community was more similar during the 1990s and 2000s. These data analyses were conducted by Lori Ivan and are similar to analyses presented in Ivan et al. (2014). *Source: Adapted from Ivan et al. (2014).*

Assessing Rarity

Some fish species seem to be essentially ubiquitous in broad regions. They are not only present in every local fish community but abundant in most systems. Bluegill and largemouth bass are examples of such species in Indiana lakes. While such common species may be influential, there are several reasons to focus on the rarer species, similar to Ludsin et al.'s focus on signal species. Rare species are numerically at greater risk of extirpation, and there may be ethical, ecological, and social reasons to justify identifying rare species to allow for protection of such species. The loss of rare species may have cascading effects throughout a food web as the loss of even weak biological interactions (e.g., predator–prey, competition) may modify the strength of extant biological interactions. Rare species may serve important ecosystem level functions, and there are ethical, pragmatic, intergenerational, and economic rationales for preserving rare species (Moyle and Moyle 1995; Olden et al. 2000).

Given the potential value of rare species, it is important to be able to identify such species. Rabinowitz (1982) presented a framework to consider three aspects of species rarity: the species' geographic range, the specificity of habitat needs, and the size of local populations. The most severely rare species would have a small range, very specific habitat requirements, and small population size. Pritt and Frimpong (2010) built on this framework to evaluate rarity of North American stream fishes based on quantitative criteria. They analyzed two datasets of fish distributions: one compiled by the US Geological Survey that included 1040 stream sites from 44 US states and another compiled by the US Environmental Protection Agency that included 308 stream sites from the US mid-Atlantic region. They developed a range measure that simultaneously considered a species areal range (in km²) as well as the species' latitudinal and longitudinal range. They generated a measure of habitat specificity by assessing how species associated with 19 different habitat types. They also assessed whether local population sizes were small, based upon indices of numerical and proportional abundances. For each of these three aspects of rarity, they established thresholds for rarity and determined for different species whether they met any two or all three of the thresholds for rarity. They compared their assessments of rarity with the American Fisheries Society's (AFS) listing of species as endangered, threatened, and vulnerable. While many of the species they identified as rare were listed by the AFS, they identified 30 species not listed by the AFS but were considered rare based on all three aspects of rarity. They suggested that these species should be evaluated for conservation needs.

Summary

Fish communities and assemblages are collections of populations of different species that co-occur in space and time. Communities are multivariate and distinct indices are necessary to describe and compare these constructs. Communities vary among systems and will vary within a system over time. There are many processes that can influence community composition. However, there are some general gradients across which community richness tends to vary. In particular, richness tends to vary if gradients lead to differences in rates of species colonization, extinction, and speciation. Subsequent chapters explore these concepts further for different types of aquatic systems.

Literature Cited

Fauth, J.E., J. Bernardo, M. Camara, et al. 1996. Simplifying the jargon of community ecology: a conceptual approach. The American Naturalist 147:282–286.

Ivan, L.N., D.G. Fielder, M.V. Thomas, and T.O. Höök. 2014. Changes in the Saginaw Bay, Lake Huron, fish community from 1970–2011. Journal of Great Lakes Research 40:922–933.

Ludsin, S.A., M.W. Kershner, K.A. Blocksom, R.L. Knight, and R.A. Stein. 2001. Life after death in Lake Erie: Nutrient controls drive fish species richness, rehabilitation. Ecological Applications 11:731–746.

Moyle, P.B., and P.R. Moyle. 1995. Endangered fishes and economics-intergenerational obligations. Environmental Biology of Fishes 43:29–37.

Olden, J.D., J.R.S. Vitule, J. Cucherousset, and M.J. Kennard. 2000. There's more to fish than just food: Exploring the diverse ways that fish contribute to human society. Fisheries 45:453–464.

Pritt, J.J., and E. Frimpong. 2010. Quantitative determination of rarity of freshwater fishes and implications for imperiled-species designations. Conservation Biology 24:1249–1258.

Rabinowitz, D. 1982. Seven forms of rarity. Pages 205–217 *in* H. Synge, editor. The Biological Aspects of Rare Plant Conservation. John Wiley and Sons, Chichester, England.

Whittaker, R.H. 1960. Vegetation of the Siskiyou Mountains, Oregon and California. Ecological Monographs 30:279–338.

CHAPTER 26

Aquatic Food Webs

Populations in an ecosystem are connected to each other through various interactions, including the direct interaction between predators and prey. Some predators only consume a single type of prey, and some prey are only consumed by a single type of predator. However, the majority of predators consume multiple prey types, and the majority of prey are vulnerable to be consumed by various predators. Therefore, the numerous connections among predators and prey in an ecosystem can be quite complex, forming a network of connections termed a food web. Fish populations are components of broader aquatic food webs, and in many cases, fish are components of food webs that span across aquatic and terrestrial systems. The size, complexity, and stability of food webs are quite variable, but describing and understanding food web structure and dynamics can be important for understanding how fish populations may respond to perturbations in different parts of the food web.

Food Web Structure

The simplest food webs are actually food chains, i.e., where predators consume a single prey type and prey are only consumed by a single predator. The simplest food chain is a two-level food chain with a single prey and single predator. For example, there may be a single primary producer (prey) and a single primary consumer (predator) that preys upon the primary producer. If another consumer (secondary consumer) preys upon the primary consumer, this would then form a three-level food chain. If another consumer (tertiary consumer) preys on the secondary consumer, this would then form a four-level food chain. The different levels of food chains can also be described as trophic levels, starting with primary producers (trophic level 1) and then primary consumers (trophic level 2), secondary consumers (trophic level 3), tertiary consumers (trophic level 4), and so on (Figure 26-1). Again, most systems where fish are found are composed of many different species with few species consuming only a single type of prey. For study purposes, one might consider a simple food chain within a larger food web. For example, in a pond one might consider a food chain consisting of phytoplankton (like green algae), herbaceous zooplankton (such as *Daphnia* spp.), a planktivore like bluegill, and a piscivore, such as largemouth bass.

Highly interconnected food webs exist when predators consume multiple prey, and prey are consumed by multiple predators. The type of prey fish consume changes with ontogeny. First-feeding larval fish are likely to consume very small planktonic organisms, and as fish grow, they will consume larger and larger prey. Some fish eventually becoming piscivorous and consume other fish, and some fish species even grow to consume other vertebrates, such as

Biology and Ecology of Fishes, Third Edition. James S. Diana and Tomas O. Höök.
© 2023 John Wiley & Sons Ltd. Published 2023 by John Wiley & Sons Ltd.

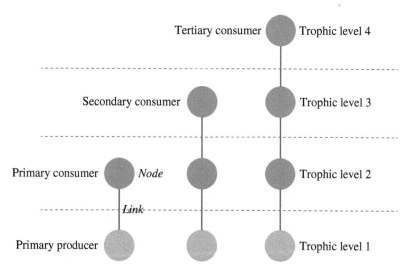

FIGURE 26-1   Simple two, three, and four trophic level food chains. Primary producer nodes are depicted as green circles, consumers as blue circles, and links between food chain components as lines.

reptiles, birds, and mammals. A species like yellow perch may feed on very small zooplankton, like copepod nauplii, during initial exogenous feeding, and then as they grow switch to larger and larger zooplankton like adult copepods and cladocerans. Eventually, many yellow perch transition to feed benthically on larger macroinvertebrates, such as midges and mayfly larvae, and several individuals will become large enough to become partially or exclusively piscivorous. Such ontogenetically variable prey consumption can contribute to complex food webs. Independent of ontogenetic effects, species may differ in terms of being more specialized or generalized foragers. That is, while most fishes are opportunistic, euryphagous predators (have the potential to consume a wide variety of prey), populations and species may nonetheless display a relatively narrow or broad range of prey consumed.

Reliance on multiple types of prey becomes even more evident in considering entire populations of fish, not just an individual fish. Some individuals may target certain prey types within a population, and other individuals may target other prey types. Such differential prey targeting may be related to ontogeny and size differences, but even independent of size effects, individuals within a population may differ in the type of prey consumed. For example, individuals occupying different habitats within a system may be exposed to different prey. There are often seasonal shifts in availability of different prey. Invertebrates may seasonally emerge from the sediment and become available to fish in the water column. Young fish may be of a size where they are vulnerable to fish predators early in the season and may then grow past this vulnerable size range later in the season. The result of all of these variable predator–prey connections can be a complex web of trophic connections.

Depicting Food Webs

Food webs are generally presented as networks with nodes representing the different components of the food web (e.g., different populations or trophic groups) and paths representing the predator–prey connections among nodes. It is possible to include not only the connections among food web nodes but also the strength of these connections. For example, it may be appropriate to include the proportion of predators' diets consisting of different prey. Therefore, one can more appropriately consider the flow of energy through the food web.

While food web networks tend to represent consumptive predator–prey interactions, there are other important biological interactions that influence food web constituents (e.g., positive, competitive, and apparent competitive interactions). Some of these other interactions may actually act through predator–prey dynamics – for example, two predators competing for the same prey – but other interactions are not directly related to predator–prey connections. Predators may also affect prey through non-consumptive means. Several studies have highlighted the notion of the ecology of fear (Brown et al. 1999), where the presence of predators can alter behavior – and even physical development – of prey. For example, in the presence of predators, prey may occupy inferior environments (e.g., move to deeper, cooler, darker depths in the water column). Prey may be less active to avoid detection and therefore feed less and grow slower. Some prey will divert energy from growing larger to develop defensive structures (e.g., some crustacean zooplankton will develop longer spines in the presence of predators). These predator-induced changes in behavior, development, and ultimately growth can come at a cost to biomass production of the affected prey population. The reduced growth and production may be even greater than the biomass that would have been lost through direct consumption by the predator population, which induced altered behavior and development. Although such ecology-of-fear effects can be quite influential and affect predator–prey dynamics, they do not tend to be represented in food web presentations because these effects do not actually lead to the direct transfer of energy from prey to predator.

Quantifying Connections within Food Webs

The above descriptions of connections within food webs and proportional reliance on different prey may seem rather straightforward. However, generating data to describe entire food webs is very involved. Given ontogenetic, inter-individual, seasonal, and spatial variation in prey consumption, diets of a large number of individuals across sizes, seasons, and habitats should be considered in order to fully describe what a single predator population consumes. Expanding this to all the different populations that may constitute a food web can be quite daunting. In fact, for most food web analyses, diet data for some food web components are lacking. Researchers may make assumptions about food web connections based on studies in other systems or data on related species. Groups of species may often be considered as a single food web node rather than multiple nodes. For example, calanoid copepods may be considered as a single food web component, while in reality this order of crustacean zooplankton may be represented by a large number of different species in a system. If species have similar trophic roles in a food web (i.e., consume similar prey and are predated on by similar predators), this type of consolidation in food web analyses may be appropriate.

Methods for describing prey consumption and rations were presented in Chapter 19. These methods are the basis for elucidating food web connections and are briefly reviewed here. Historically, the most common method for describing diets involves direct examination of stomach contents. This can be accomplished by dissection and analysis of stomach contents after mortality, or through gastric lavage, in which fish are forced to regurgitate their stomach contents. This latter approach may not be as effective in recovering all stomach contents but does have the advantage of not requiring mortality and sacrificing individual fish. Either approach allows for direct examination of prey consumed, potentially including species and size, therefore providing detailed consumption information and high taxonomic resolution of diets. However, if prey differ in their digestibility and energetic value, stomach contents may not accurately and proportionally represent how different prey are assimilated and support the growth of predators. Stomach contents only represent prey consumed over the past day or so and would need to take place throughout the year to account for seasonal variation. Given that different prey items may digest at different rates (e.g., soft tissue larval fish versus gastropods) and some prey may be more readily identified than others, stomach content analysis may provide a somewhat biased perspective on prey consumed by fish. To overcome this latter issue, various methods have been used to identify prey in stomachs, including identifying fish by undigested hard parts (e.g., otoliths or vertebrae) and using genetic methods to identify prey consumed by predators, even if there is limited structural evidence.

Other methods for describing diets overcome some of the limitations of stomach content analysis but may have their own biases. The differential transfer and assimilation of stable isotopes from prey tissue to predator tissue is well established. Carbon isotopes are assimilated into consumer tissue roughly consistent with their proportion in consumed prey tissue, while assimilation varies between different types of nitrogen isotopes. By measuring isotope ratios of predators together with information on isotopic composition of potential prey, it is possible to quantitatively estimate the type of prey predators consumed to achieve observed isotopic ratios. Unlike stomach contents, isotopic ratios reflect prey tissues actually assimilated by consumers and tend to reflect prey consumed over a longer time horizon (as long as the past several months). Therefore, stable isotopes may not necessitate as frequent collection of diet information as stomach content analysis. However, stable isotopes are unlikely to provide the same level of taxonomic resolution as stomach content analysis. This is especially the case if (1) there are a large number of potential prey types available to consumers, and it is difficult to isotopically discriminate among different prey, and (2) one is only relying on two types of stable isotope ratios, such as the commonly used $\delta^{13}C$ and $\delta^{15}N$. Various other molecular markers of prey consumed and assimilated have been developed and may complement or replace $\delta^{13}C$ and $\delta^{15}N$ analysis. Other types of isotopic ratios, such as isotopes of hydrogen and sulfur, may also provide insight into the type of prey consumed. Most consumers are unable to synthesize all of the fatty acids in their tissues and are therefore dependent on prey consumption to obtain certain fatty acids. Prey generally differ in relative fatty acid composition, and analysis of fatty acid composition in consumer tissues can help elucidate prey consumed. More recently, methods have been developed to analyze isotopic composition of specific biomolecules, i.e., compound-specific isotopic analysis of amino acids and fatty acids, which should provide even greater resolution for the prey types assimilated by consumers.

Network Analysis and Trophic Levels

Regardless of precisely how food web connections are estimated, once these are established for a food web the entire network can be described in various ways. Networks are often described in terms of their number of connections and overall complexity. There is potential for more connections if there are more species or nodes in a food web. It is common to standardize for number of species in quantifying connectivity of food webs. A common index of food web connectivity is connectance (c) – a measure of the number of paths (or links, L) relative to the number of nodes (or species, S). Given a certain number of nodes in a food web, the maximum number of nodes is equal to

$$S(S-1)/2 \qquad\qquad \text{(Equation 26-1)}$$

For example, if a food web contains three nodes, the maximum number of links is 3, and if a food web contains six nodes, the maximum number of links is 15. The index of connectance is simply the number of observed links divided by the maximum number of links for that number of nodes:

$$c = \frac{L}{S(S-1)/2} \qquad\qquad \text{(Equation 26-2)}$$

This index can take values from 1, if the food web is maximally connected, to 0, if there are no connections in the food web.

A similar index of connectivity is linkage density (D – an index of the realized number of connections in a food web):

$$D = L / S \qquad \text{(Equation 26-3)}$$

Unlike connectance, linkage density will tend to increase as the size of food webs increase. However, both of these indices will be relatively low for simple food chains and will increase with complexity (see Figure 26-2). These are rather simple indices to summarize potentially complex networks of connections within food webs and allow for comparisons. These indices do not incorporate the relative strength of connections among different food web components; they simply reflect whether or not connections exist. A variety of other indices and analytical tools exist to consider the structure of networks – such as food webs – and some of these tools incorporate both presence of connections and strength of such connections.

A common analysis of food webs is to calculate the trophic level of individual nodes that constitute food webs. Calculating trophic levels is quite easy in simple food chains because as one moves from considering primary producers to primary consumers to secondary consumers, the trophic level simply increases by one integer level (from 1 to 2 to 3). However, in more complex food webs, different species may be feeding on prey from different trophic levels. For example, a planktivorous fish may consume both herbaceous zooplankton (trophic level 2) and predatory zooplankton (trophic level 3). The planktivorous fish would not be considered at a trophic level below 3 because it is feeding on prey from trophic level 3. It would not be considered as high as trophic level 4 because it is also feeding on prey from trophic level 2. Instead, the planktivorous fish would be considered some intermediate trophic level between 3 and 4. If 50% of its diet is from herbaceous zooplankton, and 50% is from predatory zooplankton, the trophic level of the piscivorous fish would be 3.5. With information on the strength of food web connections (the proportion of predators' diets consisting of different types of prey), it is

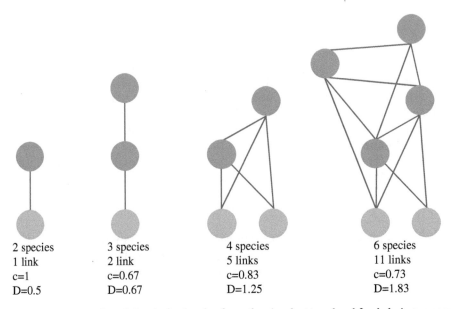

2 species	3 species	4 species	6 species
1 link	2 link	5 links	11 links
c=1	c=0.67	c=0.83	c=0.73
D=0.5	D=0.67	D=1.25	D=1.83

FIGURE 26-2 Four examples of simple food webs, from the simplest two-level food chain to more complex food webs. Symbology is the same as in Figure 26-1. In addition, connectance (c) and linkage density (D) are presented for each food web.

possible to calculate the trophic levels – including intermediate trophic levels – of different components of food webs. This involves starting with primary producers (trophic level 1) and working up the food web. For a predator consuming a number of different prey, its average trophic level (TL$_{pred}$) would simply be

$$TL_{pred} = 1 + \sum_{i=0}^{n} TL_i \times Prop_i$$ (Equation 26-4)

Here, TL$_i$ is the trophic level of prey i and Prop$_i$ is the proportion of the predator's diet that consists of prey i. Most fish in food webs will be characterized by non-integer trophic levels, as most fishes are omnivorous and display varied diets, especially at the population level (Figure 26-3).

Food Web Patterns

Indices allow for the characterization of different components of food webs as well as holistic food web summaries. Moreover, analyses of individual food webs can facilitate comparisons among food webs and consideration of recurring patterns. For example, the number of trophic levels or "food chain length" of food webs is of interest because of influences on processes such as community dynamics and potential bioaccumulation of contaminants by top predators. Post et al. (2000) considered how the trophic level of top predators varied among lake systems. Specifically, these researchers considered 25 lakes of different sizes and different productivity. Past studies had posited that food chain length would increase with system productivity. That is, food webs with greater primary production would be expected to support greater overall biomass, including more trophic levels. Alternatively, the physical size of a system may have strong influence on number of trophic levels with larger systems potentially containing greater overall biomass, more habitats, and greater functional diversity among species. Post et al. (2000) quantified C and N stable isotopes of primary consumers and top predatory fish in the different lakes as a means of estimating the trophic level of the top predators, including lake trout, largemouth bass, northern pike, and walleye. They then compared the maximum trophic level observed in a lake – the estimated trophic level of top predatory fish – with the

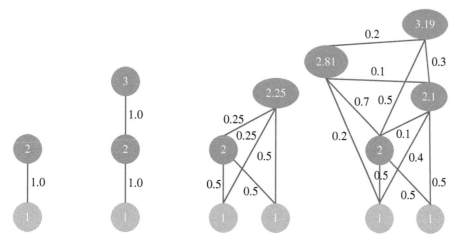

FIGURE 26-3   Four examples of simple food webs. The numbers next to each link represent the proportion of each consumer's diets that comes from various prey. The values in each node represent the calculated trophic level for that food web component.

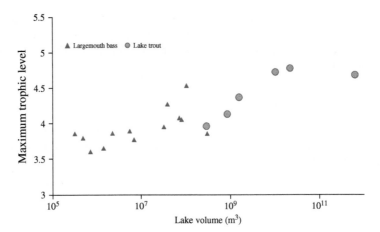

FIGURE 26-4   Relationship between the estimated trophic level of top piscivores (largemouth bass or lake trout) in lakes and the corresponding volume of a lake. Note logarithmic scale of lake volume axis. *Source: Adapted from Post et al. (2000).*

size of a lake and the productivity of the lake. Lake size was indexed as lake volume, and productivity was based on concentration of phosphorous, the limiting nutrient in most freshwater systems. They found that maximum trophic level was consistently positively associated with lake volume, while lake productivity did not have an association with maximum trophic position (Figure 26-4).

The increase in maximum trophic level with system size observed by Post et al. (2000) was partially related to the addition of species in large lakes. For example, lake trout expressed the greatest overall trophic level across lakes and this species was absent in the smallest lakes. However, the addition of a top predatory species to large lakes was not the only mechanism contributing to greater maximum trophic level. That is, even if just considering lake trout in the subset of lakes where this species was present, lake trout trophic level increased with lake volume. This pattern suggests that in larger lakes the energy and biomass initially generated by primary producers passes through a relatively large number of trophic levels before supporting production of lake trout. Post et al. (2000) specifically suggested that larger lakes were characterized by more intermediate trophic level predators and lower levels of trophic omnivory, both of these features could extend trophic levels and lead to greater maximum trophic level.

Flow of Energy through Food Webs

Food web indices like connectance and linkage density are useful for describing the structure of a food web, but they do not incorporate the relative strengths of different linkages to describe the overall flow of energy or matter through a food web. More holistically describing food web flow has been a goal of ecologists for almost a century. However, as discussed above, it remains no small feat to quantify the flow of energy to a single consumer population, much less an entire food web. Seasonal, spatial, ontogenetic, and individual diet variation can contribute to difficulty in estimating overall annual energy flow. And all methods to describe diets and flow to consumers have some limitations as described above. However, there have been various studies to quantitatively describe flow food webs in aquatic systems. For example, Wyatt Cross and colleagues (2013) considered food web flow patterns in the Grand Canyon region of the Colorado River.

Constructed flow food webs allow for a variety of considerations regarding food web functioning. Such food webs are often constructed at an annual time step but could be considered at finer or coarser temporal resolutions. Flow food webs can also be constructed for different systems or for different habitats within a larger system. Flow food webs can be compared among different time periods or among different locations to examine how food webs function differently in space and time. For example, Cross et al. (2013) considered food webs in six different locations – both above tributaries and below large tributaries draining into the Colorado River.

Flow food webs generally involve examination of production of various food web components – the increase in energy or biomass of a component over time. One can then estimate the amount or fraction of a consumer's production attributable to different consumed prey. For organisms that serve as prey, one can compare overall production versus overall consumption by predators. Consider an invertivorous fish population like fathead minnow in the Colorado River system. One can estimate annual production of fathead minnow in a certain region of the Colorado River system as well as estimate the overall amount of fathead minnow consumed by the various predators that consume fathead minnow in the same region. If production exceeds consumption by predators, the fathead minnow population likely increases. However, if consumption by predators exceeds production, the population declines. One can similarly consider the balance between production and predation from the perspectives of predators. This is termed ecotrophic efficiency, which represents the percentage of a prey's production consumed by its predators.

Another interesting aspect of flow food webs is the ability to estimate the trophic basis of production – or the ultimate contribution of different resources and primary producers to production of different food web components. By tracing energy or matter flows from the base of the food web, it is possible to estimate the relative importance of primary producers, such as different phytoplankton, macrophytes, bacteria, and fungi. For riverine systems in particular, it is often of interest to consider the relative contribution of autochthonous primary production from within the system versus allocthonous resources supporting consumers.

Perturbations and Stressors of Food Webs

Food webs may be depicted as static networks of trophic connections, but in reality, they can be quite dynamic. Environmental conditions may change and affect predator foraging abilities and prey vulnerabilities, and population abundances may change pronouncedly through natural and anthropogenic-induced causes. As populations become more or less abundant, their connections with other populations may change. For example, if a predatory fish population's primary prey were to decline precipitously in abundance, the predator population would likely switch to consume other prey to a larger extent. Therefore, the strength of connections between the predator population and alternative prey would increase. Potentially, abundances of these alternative prey would decrease, which could negatively affect other predators and lead these other predators to switch to exploit alternative prey to a greater degree. If the alternative prey are themselves consumers, their decline in abundance could release their prey from predatory control, which could have additional cascading effects on various other components of the food web (see Chapter 30 for additional discussion of trophic cascades). The main point here is that due to various connections across a food web, changes to one component of the food web can translate to various other effects on food web components both directly and indirectly connected.

Numerous perturbations can affect food web structure. Some may seem rather minor, while others are more dramatic. Natural perturbations may include phenomena such as severe winter conditions or drying of small systems during summer, both of which could lead to high mortality and extreme abundance declines of certain populations. Changes in oceanic circulation patterns and coastal upwelling conditions can strongly alter marine food webs.

As described in Chapter 12, fish recruitment is naturally highly variable from year to year, and this can lead to large changes in abundances of populations with potential cascading effects throughout the food web. New species may invade food webs and contribute to new and altered connections. While species invasions are a natural phenomenon, they have become far more common through anthropogenic activities (see Chapter 33). Other anthropogenic-induced food web perturbations may be brought about by fisheries harvest, habitat alteration and destruction (e.g., dam construction and wetland loss), nutrient loading, other pollution, and climate change. Some of these stressors may directly affect a single component of the food web with cascading effects to other components, while other severe stressors may directly affect almost the entire food web.

Stability

Several researchers have considered the issue of stability when food webs are perturbed. That is, whether a perturbation will have minor effects on food web structure and composition or whether a perturbation will fundamentally alter the food web. Stability can actually be conceptualized in a variety of ways. One can consider both system resistance (the ability of the food web to resist fundamental restructuring in the first place) and resilience (the ability of the food web to quickly return to the structure in place prior to perturbation). Locally stable food webs are stable in the face of single or multiple small perturbations, while globally stable food webs are able to resist and be resilient, even when exposed to very large perturbations. Measuring food web stability is also not straightforward and can include consideration of whether or not species simply persist in food webs or whether species' abundances remain stable.

Most studies of food web stability have relied upon modeling approaches and mathematically examined the attributes of food web, which may influence stability. In particular, studies have focused on how food web complexity affects stability. Food web complexity relates to the number of species in food webs and number of links between species. Complexity can be indexed a variety of ways, including using indices such as connectance; for a food web with a given number of species, greater values of c would be indicative of greater complexity. The prevailing previous understanding was that greater food web complexity was related to greater stability. This understanding was in part related to Robert MacArthur's (1955) and Charles Elton's (1958) arguments that in systems with more paths for energy to flow, there is greater potential for the system to overcome spikes or declines of one or more food components. The idea that more complexity begets more stability was challenged by mathematical models. Mark Gardner and Ross Ashby (1970), as well as Robert May (1972), used constructed food web models to ask the question, "Will a large complex system be stable?" They concluded that as webs became more complex (measured by number of species, connectance, and interaction strengths), they should eventually become unstable. Gardner and Asby (1970) pointed to a threshold level of connectance, beyond which systems will transition from some level of stability to instability. Debates over how food web structure contribute to stability continue. Many more recent analyses have reverted back to provide support for the notion that increased complexity promotes stability. Studies have pointed to various factors that influence model stability, such as (a) whether model food webs are constructed randomly or more realistically, (b) the extent to which food webs are supported by allochthonous production, i.e., subsidies from outside of the food web itself, (c) the presence of compartments within food webs, and (d) the presence and locations of particularly strong connections within food webs. Ultimately, a diversity of food webs – including some rather complex food webs – clearly do persist in a variety of aquatic ecosystems. While some mathematical models may suggest that stability is unlikely, there appear to be certain food web features that promote stability even when networks are complex.

Hysteresis and Regime Shifts

When food webs are faced with a perturbation or when an external influence on a food web changes, the response of the food web may in part depend on its history. This is referred to as hysteresis – a system being dependent on its past. The concept of hysteresis is by no means unique to food webs. For example, when rain falls on a landscape, the amount of precipitation that transfers to the soil will depend on previous precipitation and evaporation events influencing whether soil is already saturated with water. In physics, a magnetic system can display hysteresis as can economic and immunological systems. A well-described example of hysteresis for aquatic systems and food webs is the response of lakes to changes in nutrient loading.

Hysteretic processes affecting how lakes – in particular shallow lakes – respond to changes in nutrient loading have been described by a team of researchers, including Steve Carpenter from the University of Wisconsin, Erik Jeppsen from Aarhus University, and Martin Scheffer from Wageningen University. These researchers worked both independently and collaboratively, and through a series of publications (e.g., Scheffer et al. 1993, 2001) described responses of lakes to changes in nutrient loadings. In general, phosphorous is the nutrient limiting primary production in lake systems. With increased phosphorous loading, primary production and overall biomass will increase. However, different types of primary producers may respond differently to changes in nutrient loading, and history of the system may determine these responses.

Several aquatic systems experienced a prolonged period of increased nutrient loading during the 20th century. Through various abatement programs, many systems have more recently experienced a period of reduced nutrient loading. Prior to intense nutrient loading, a shallow lake may have been characterized by stands of rooted macrophytes and relatively clear water. As nutrient loading increases, densities of both macrophytes and phytoplankton may increase and water clarity may decline a little, but to a large extent, a lake may be able to maintain its clear water state. With initial nutrient loading increases, existing macrophytes sequester much of the additional nutrients and therefore limit the potential for phytoplankton to increase in response to higher nutrient loading. Limited densities of phytoplankton and maintained water clarity allows light to penetrate to the bottom of a greater proportion of the lake. Such light penetration is necessary for rooted macrophytes to be able to grow from the bottom of the lake. The presence of macrophytes also limits the resuspension of sediment from the bottom of the lake, further maintaining water clarity. The presence of macrophytes resists the ability of phytoplankton to take advantage of higher nutrient loads. However, if nutrient loading becomes sufficiently high, macrophytes will not be able to resist phytoplankton production, and these algae will become far more abundant. Phytoplankton will then contribute to reduced water clarity and limit light penetration, confounding the ability of macrophytes to grow from the bottom of the lake. As macrophytes become less abundant, resuspension of sediments will increase, further limiting macrophyte growth. Carpenter, Jeppsen, Scheffer, and others have referred to lakes as having two alternate stable states – a clear water state with high densities of macrophytes, and a turbid water stage with high densities of phytoplankton and low densities of macrophytes. These alternate stable states have also been referred to as different equilibria or regimes.

When a lake system is in a particular state, it resists transitioning to another state. The presence of many macrophytes limits the ability of phytoplankton to grow and promotes the maintenance of the clear water state. In contrast, the presence of high concentrations of phytoplankton limits the ability of macrophytes to establish and allows for the maintenance of a turbid water state. If one plots the relationship between nutrient loading and lake state (or water clarity), an abrupt transition from clear water to turbid water state may be expected. However, the exact nutrient loading level that triggers this transition will vary if nutrient loading is increasing (transitioning from clear water to turbid water) versus if nutrient loading is

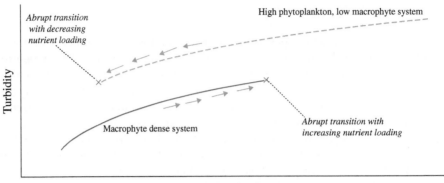

FIGURE 26-5   Conceptual plot of the relationship between changing nutrient loading to a lake and turbidity. As nutrient loading increases to a macrophyte dense system (solid curve), the state of the system resists a dramatic shift in turbidity until nutrient loading becomes very high (indicated by × in plot above). In contrast, when nutrient loading decreases to a high phytoplankton, turbid system (dashed curve), the lake state resists transition to low turbidity state until nutrient loading becomes very low (indicated by × in plot above). This is an example of hysteresis. *Source: Adapted from Scheffer et al. (1993, 2001).*

decreasing (transitioning from turbid water to clear water). This is an example of hysteresis. In fact, with nutrient abatement programs leading to reduced nutrient loading to lakes, many systems have not transitioned to a clear water, macrophyte abundant state. Rather, nutrient loading must decline very low before such a transition takes place (Figure 26-5).

Fish assemblages in shallow lake will respond to transitions between the clear water, macrophyte state and the turbid water, phytoplankton state, and some fish may contribute to the resistance of the system to transition from one state to the other. Certain fish species are able to tolerate fairly turbid conditions and high phytoplankton concentrations (e.g., common carp), while other species – such as northern pike – perform well in clear water and systems with high macrophyte densities. In addition to tolerating turbid conditions, common carp will resuspend sediment as they forage for prey and may also root and displace macrophytes as they move into nearshore vegetated areas ahead of spawning. Thereby, species like carp can contribute toward maintaining a turbid water state and limit the establishment of macrophytes.

Endogenous Feedback Loops among Macrophytes, Invasive Crayfish, and Native Sunfish

The above example of alternate lake states demonstrates hysteresis and how the state of a system (A: clear water, macrophyte-abundant lake versus B: turbid, phytoplankton-abundant lake) can determine how a system responds to an external driver or stressor. There are several examples of how internal food web feedbacks can influence how a system responds to external influences. For example, one such endogenous feedback loop was proposed by Brian Roth (then of University of Wisconsin and now with Michigan State University) and colleagues to describe how lake food webs respond to invasion by crayfish (Roth et al. 2007). Roth and colleagues studied lakes in northern Wisconsin, many of which were invaded by rusty crayfish. While some lakes that were invaded by rusty crayfish were seemingly strongly altered by this invasion, other lake systems invaded by crayfish experienced minimal changes. In particular, some lakes invaded by crayfish were characterized by a dramatic decline in littoral macrophyte coverage. In addition, some lakes where rusty crayfish became established were characterized by low numbers of native sunfish (bluegill and pumpkinseed). Roth and colleagues hypothesized that destruction of macrophytes by rusty crayfish negatively affected sunfish who rely on macrophyte beds for foraging and protection from large piscivores. They explored relationships among crayfish, macrophytes, and sunfish through long-term records for Trout Lake (annual records from 1981 to 2004), surveys of 57 northern Wisconsin lakes (surveyed from 2001 to 2004), and directed surveys of sunfish and crayfish in four lakes, intended to assess whether sunfish predation of crayfish could limit crayfish populations.

Roth and colleagues (2007) found a negative relationship between crayfish and macrophyte biomass in Trout Lake. They also found that as rusty crayfish became more abundant in Trout Lake, catches of sunfish declined, indexed through both beach seine sampling and fyke net sampling. When looking at patterns across the 57 northern Wisconsin lakes, they found a negative association between rusty crayfish catches (based on catches in baited minnow traps) in a lake and electrofishing-based catches of sunfish. They also found that across these lakes, the percent of the littoral zone covered by macrophytes was positively associated with sunfish catch rates. In the four lakes they intensively surveyed, Roth and colleagues found that sunfish consumed crayfish, in particular targeting juvenile crayfish (less than 15 mm carapace length). They estimated that sunfish consumed a large proportion of the crayfish population in two lakes (Arrowhead and Wild Rice) but consumed a very small percentage of the overall crayfish population in two other lakes (Trout and Big). Of note, Arrowhead and Wild Rice contained very low densities of rusty crayfish, while Trout and Big contained large crayfish populations.

Roth and colleagues proposed an endogenous feedback loop to explain the observation that some lakes invaded by rusty crayfish were fundamentally altered, while food webs in other lakes invaded by crayfish were largely unaffected. The crux of the proposed feedback loop relies on whether sunfish were abundant in a lake at the time of crayfish invasion. If sunfish are abundant, they exert strong predatory control over juvenile crayfish and therefore never allow adult crayfish to become abundant. In turn, macrophyte densities remain high, providing foraging and refuge habitat for juvenile sunfish and leading to high numbers of adult sunfish capable

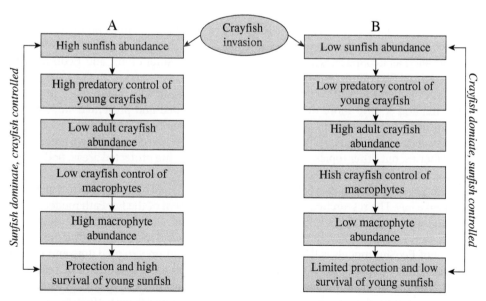

FIGURE 26-6 Reciprocal feedback loops proposed by Roth et al. (2007), with figure adapted from Roth et al. (2007). When crayfish invade lakes where sunfish abundance is high (A), sunfish prey on juvenile crayfish, which feeds back to maintain abundant macrophyte beds, large numbers of sunfish, and limited numbers of crayfish. In contrast, in lakes where sunfish are less abundant (B), they are unable to control invading crayfish, which leads to decreased macrophyte numbers, poor survival for young sunfish, and high numbers of crayfish.

of controlling crayfish. In contrast, if sunfish densities are low when crayfish invade, there is limited predation pressure on crayfish, allowing for expansion of crayfish. This leads to large numbers of crayfish, which decimate macrophytes, leading to poor survival for young sunfish and ultimately low numbers of adult sunfish, incapable of controlling crayfish (Figure 26-6).

Ecopath and Ecosim to Explore Fisheries Effects

Various effects of fisheries harvest are described in Chapter 32, including potential effects on food webs. Here, we focus on one of the main tools used to consider the effects of fisheries harvest on food webs – the modeling software, EcoPath. EcoPath was initially developed during the 1980s by Jeffrey Polovina (1984) of the National Oceanic and Atmospheric Administration and then expanded on by a team principally from the University of British Columbia which included Villy Christensen and Daniel Pauly (Christensen and Pauly 1992). It is based on principles of mass balance as energy moves through a food web. Briefly, it is a path model in which food web components and connections among these components must be specified. The biomasses of food web components and strength of connections can be initially estimated using existing abundance and diet data. The model allows for quantitative adjustment of biomasses and strengths of connections to meet the requirement of mass balance. EcoPath on its own is therefore a tool for developing a model of a balanced, static food web, and among other metrics, allows for calculation of fractional trophic levels of individual food web components. EcoPath was extended upon through the development of EcoPath with EcoSim (EwE; led by Villy Christensen and Carl Walters), which allows for temporally dynamic consideration of an ecosystem as it is perturbed through fisheries harvest. With EwE, one can explore how reduced biomass of one or several components of a food web can translate to affect other components of the food web. The framework has been further expanded on through additional development of components such as EcoSpace, which allows for spatial considerations, and Ecotracer, which allows for tracing of contaminants in a food web.

In 2015, Colléter et al. (2015) documented that there were 433 EwE models in the scientific literature, and several more models have been developed and presented since this time. Most EwE models are for marine systems, but EwE has also been developed and applied for freshwater systems and even terrestrial systems. Most EwE research topics have focused on the effects of fisheries, but other topics, such as the effects of marine protected areas, pollution, and aquaculture, have also been explored.

An example of the development of an EwE model is presented by Christensen et al. (2009) for the Chesapeake Bay. The researchers described the development of the EcoPath model, including data used and workshops to inform construction of the model. Ultimately, they developed a food web configuration representing Chesapeake Bay during 1950 and including 45 functional groups. A key aspect of EcoSim models is the vulnerability of different prey to predators and in particular how vulnerability to predation increases for prey when predator abundance increases. These parameters can have strong influence in EcoSim models and are difficult to directly estimate from field data or laboratory data. Rather, these vulnerability parameters have often been developed based on quantitative agreement with time series data. Christensen et al. (2009) calibrated their EcoSim model to time series of component biomasses over the period 1950–2002.

While the emphasis of Christensen et al.'s (2009) effort was to develop and validate the EwE model, they also used the model to consider various management scenarios. For example, planktivorous Atlantic menhaden are preyed upon by striped bass in Chesapeake Bay. The team

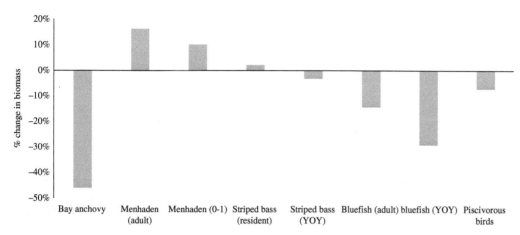

FIGURE 26-7   Estimated change in biomass of selected Chesapeake Bay food web components from EcoPath with EcoSim analyses by Christensen et al. (2009), with figure adapted from Christensen et al. (2009). The percentage change values represent how much the component biomass was simulated to change from baseline conditions with bay anchovy harvest at 2002 levels compared to after 50 years of increased fisheries harvest of bay anchovy. Note that the most direct negative effect was on the harvested species, bay anchovy. Species that relied on bay anchovy as prey (e.g., bluefish and piscivorous birds) also decreased, while species that competed with bay anchovy for food (e.g., menhaden) increased.

explored how changes in the commercial fishery for menhaden might affect striped bass and ran the EcoSim model 25 additional years to simulate effects of either (a) retaining the menhaden fishery at the level of 2002, (b) decreasing fishing of menhaden by 50%, or (c) ceasing harvest of menhaden. These simulations suggested that striped bass biomass would decrease if menhaden fishing remained at 2002 levels, would remain fairly stable if menhaden fishing was reduced by 50%, and would increase if menhaden harvest was to cease. The team also considered various fishing options for bay anchovy by either (a) maintaining 2002 fishing levels or (b) increasing fishing of bay anchovy. They simulated these two scenarios 50 years beyond 2002. Obviously, bay anchovy declined with higher fishing pressure, but some groups, including menhaden, increased as bay anchovy declined, presumably because they were released from resource competition with bay anchovy. Striped bass also increased slightly in the high anchovy harvest scenario, which was related to increased availability of menhaden as prey. Other predators that heavily relied on bay anchovy as prey were also expected to decline under this high harvest scenario (Figure 26-7). It should be noted that various studies subsequent to Christensen et al. (2009) have made use of the model developed by this team to consider other management questions for Chesapeake Bay.

Summary

Fish populations exist within interconnected food webs. Such food webs vary in structure and complexity, and these attributes seemingly influence how food webs respond when perturbed. A number of both natural and anthropogenic phenomena can perturb a food web. Due to the variety of connections within a food web, the perturbation (increase or decrease) of a single food web component will affect other food web components. However, food webs differ in terms of their vulnerability to perturbations and their ability to return to their previous state after a perturbation. These types of effects have been demonstrated in a variety of aquatic food webs, and the cascading effects of fisheries harvest have in particular been explored with dynamic network models.

Literature Cited

Brown, J.S., J.W. Laundré, and M. Gurung. 1999. The ecology of fear: Optimal foraging, game theory, and trophic interactions. Journal of Mammalogy 80:385–399.

Christensen, V., and D. Pauly. 1992. ECOPATH II – a software for balancing steady-state ecosystem models and calculating network characteristics. Ecological Modelling 61:169–185.

Christensen, V., A. Beattie, C. Buchanan, et al. 2009. Fisheries Ecosystem Model of the Chesapeake Bay: Methodology, Parameterization, and Model Explanation. U.S. Dep. Commerce, NOAA Tech. Memo. NMFS-F/SPO-106, 146 p.

Colléter, M, A. Valls, J. Guitton, et al. 2015. Global overview of the applications of the Ecopath with Ecosim modeling approach using the EcoBase models repository. Ecological Modelling 302:42–53.

Cross, W.F., C.V. Baxter, E.J. Rosi-Marshall, et al. 2013. Food-web dynamics in a large river discontinuum. Ecological Monographs 83:311–337.

Elton, C.S. 1958. The Ecology of Invasions by Animals and Plants. London. Methuen.

Gardner, M.R., and W.R. Ashby. 1970. Connectance of large dynamic (cybernetic) systems: Critical values for stability. Nature 228:784.

MacArthur, R.H. 1955. Fluctuations of animal populations and a measure of community stability. Ecology 36:533–536.

May, R.M. 1972. Will a large complex system be stable? Nature 238:413–414.

Polovina, J.J. 1984. Model of a coral reef ecosystems. I. The ECOPATH model and its application to French Frigate Shoals. Coral Reefs 3:1–11.

Post, D.M., M.L. Pace, and N.G. Hairston Jr. 2000. Ecosystem size determines food-chain length in lakes. Nature 405:1047–1049.

Roth, B.M., J.C. Tetzlaff, M.L. Alexander, and J.F. Kitchell. 2007. Reciprocal relationships between exotic rusty crayfish, macrophytes, and *Lepomis* species in northern Wisconsin lakes. Ecosystems 10:74–85.

Scheffer, M., S.H. Hosper, M.L. Meijer, B. Moss and E. Jeppesen. 1993. Alternative equilibria in shallow lakes. Trends in Ecology and Evolution 8:275–279.

Scheffer, M., S. Carpenter, J.A. Foley, C. Folke, and B. Walker. 2001. Catastrophic shifts in ecosystems. Nature 413:591–596.

CHAPTER 27

Temperature and Fish Distributions

Temperature is one of the many factors that influence the geographic ranges of fish species. Temperature has a strong influence on distribution, both on a global and regional scale. Earlier in this book, the concept of thermal guilds was presented in relationship to fish species in inland lakes (Chapter 1), fish metabolism (Chapter 5), and growth (Chapter 6). The importance of thermal guilds in understanding fish distributions is emphasized in this chapter.

There are significant differences in the impacts of temperature on fishes in oceans, inland lakes, and streams. The following chapters review factors influencing fish communities in temperate streams, tropical rivers, lakes, and oceans. Fishes in those ecosystems are all impacted by thermal conditions and histories, and these impacts will be further evaluated in those chapters. Here, we cover basic thermal capacities of fishes and how the landscape and local conditions influence ecological processes for fishes.

Factors Influencing Geographic Ranges of Fishes

A map of the geographic range for any inland fish species shows distinct boundaries. The native range of largemouth bass, for example, extends to the coasts in the south and east of North America, passes west to the Mississippi River and beyond (but not to the Rocky Mountains), and continues north to the Canada–US border (Figure 27-1). That pattern of distribution is typical for a variety of warmwater fish species. The northern extremes of the ranges for some sunfishes are farther south, and a few ranges extend slightly farther north. However, the general pattern is fairly consistent.

There are two major questions related to geographic ranges: (1) what is the relationship between temperature conditions and north–south ranges and (2) what limits east–west distributions? North and south boundaries seem easier to understand because they coincide with climatic differences influenced by latitude. The northern extreme of a range is often believed to be limited by temperature, in that cold summer or spring temperatures may limit reproduction or rearing success of fishes. Southern bass populations spawn as early as February, when the temperature warms above 16 °C. Bass begin to spawn in late May in northern

Biology and Ecology of Fishes, Third Edition. James S. Diana and Tomas O. Höök.
© 2023 John Wiley & Sons Ltd. Published 2023 by John Wiley & Sons Ltd.

FIGURE 27-1 Native and naturalized ranges of largemouth bass. *Source: Robbins and MacCrimmon (1974)/ Biomanagement and Research Enterprises.*

areas, when temperature exceeds 16 °C, and they may continue to spawn in June at temperatures near 20 °C. The northern limit of largemouth bass distribution (Figure 27-1) appears at least casually related to temperature too cold for successful reproduction. However, the southern extreme for bass – at least within the United States – is as far south as geographically possible. It is doubtful that the southern extreme of the range for bass meets a warm lethal temperature limit. Most bass that survive in Florida and do not penetrate far into Mexico probably are not limited by summer temperatures reducing survival, but the actual causative agent for this limit is unclear. It is possible bass may simply be poorer at competing with other fishes existing in that area although various stockings of bass in southern lakes and reservoirs would not bear that out.

In theory, fish populations near the extremities of their range are often limited by abiotic conditions, such as poor temperature for reproduction or growth, which may cause density-independent factors to dominate population dynamics. Fish populations in the middle of their geographic range, where conditions are good for survival, are generally believed to be limited more by density-dependent effects, including abundance of food and abundance of predators or competitors. One important effect of climate change will be changes in the geographic ranges of many fish species historically limited by thermal conditions. This is covered in more detail in Chapter 35.

Much of the historical distribution of fishes appears limited by temperature. However, current distributions present some problems to such an interpretation. The current ranges of fishes are problematic because people like to import their favorite fishes for sport, food, or aesthetic purposes, so the current distributions are strongly influenced by human interference. Largemouth bass populations in states west of the Rocky Mountains were introduced; the native range never extended there. However, this current distribution also indicates that fish populations were not only limited by temperature because they often survived once they were introduced into unique areas.

East–west separations of inland fish species within the temperate zone are at least partly due to existence of glacial refuges during the Pleistocene Era or earlier. During these times, there were refuges where fishes could exist in water even though much of the continent was covered in ice. If individual species existed in only one or two of these refugia, their distributions after the glaciers retreated were limited by dispersal from the refugium. Northeastern North America had main refuges in the Mississippi basin and Atlantic drainages, while other northern refuges existed in the west. Northern pike in the Great Lakes region apparently migrated from refuges in the Mississippi and in Alaska. Various populations of pike have somewhat different characteristics, which appear to indicate that the stocks arose from these different refuges. Northern pike are circumpolar, continuing east and west of the Rocky Mountains and into Eurasia, and other populations likely had more refuges as well. In contrast, muskellunge appear to have had the Mississippi as their sole refuge, and from there they were able to expand only throughout the midwest US region. Natural muskellunge populations are currently limited to that part of the world. East–west limits to the ranges of fishes – at least in North America – may be due more to glaciation effects than to modern geographic conditions. The north–south orientation of major mountain ranges proves to be a strong barrier to wider fish distribution on an east–west gradient in North America during post-glacial times, a factor that differs in Europe and other continents.

The above description of geographic ranges applies in part to marine organisms as well. Temperature certainly has a major influence on distribution of marine fishes, particularly in the north to south latitudinal gradient. However, thermal conditions in the ocean are also influenced by currents, and ranges of fishes are altered by nutrient availability and a variety of other characteristics. Stuart-Smith et al. (2017) conducted a meta-analysis of distributions for 1790 coastal marine species and found that distribution was largely defined by similar temperature limits as found in freshwater fishes, with tropical fishes having the largest ranges and temperate fishes smaller ranges. Marine fish distributions have been evaluated on the basis of Marine Biogeographic Realms (MBRs), which are larger regions defined by similar characteristics in geological history, temperature, and the effects of currents. There is potential for widespread distribution of any species in oceans because of continuity, but most species reside within a single MBR and very few are cosmopolitan. The influence of MBRs on fish distribution in the ocean is covered more thoroughly in Chapter 31.

Thermal Guilds of Fishes

There appear to be three groups of fishes in relation to evolution of thermal characteristics: tropical, temperate, and polar. Tropical fishes are believed to exist in a warm and relatively narrow range of temperatures. This classic perception of thermal conditions in the tropics is not necessarily true for all fishes because many may undergo wide ranges of temperature in their natural range due to elevation differences, upwelling, and other effects. However, tropical fishes may still be considered warm stenotherms when compared to temperate or polar fishes. For this definition, recall that therm is used to denote tolerance of temperature; steno indicates a narrow range of temperature; and warm indicates that the narrow temperature range is in the warm extreme. These fishes generally prefer temperatures from 25 to 35 °C and cannot exist at temperatures less than 10 °C.

Thermal conditions are extremely variable in the temperate zone. The range of largemouth bass exemplifies this. Although water may never freeze in southern Florida, winter conditions in the middle of the bass range may exist for a month or less, while in the northern extreme of the range, winter conditions may last for as long as six months. Inland temperate fishes must deal with these temperature variations in order to survive, while marine species can also migrate to remain within a preferred temperature condition. Strong seasonal effects differ in accordance with latitude and altitude. Individual populations undergo variable temperatures, and so temperate fishes can be considered eurytherms, which means they tolerate a wide range of temperatures. The most consistent characteristic of these eurytherms is they can survive prolonged exposure to temperatures near freezing.

The polar fauna appears even more adapted to very specific temperature conditions than tropical fishes. There are species of marine fishes in polar regions that can survive at −1.5 °C (below the freezing point of freshwater but not seawater) because peptides and glycopeptides in their blood depress its freezing point. However, some polar fishes cannot survive at temperatures much above 5 or 6 °C, so they are limited to very cold conditions. Recall the earlier examples for metabolic compensation, in which the adaptations to temperature were described. Fishes in the polar regions showed the best compensation and are truly cold stenotherms.

The evolution of temperate and tropical fishes differ in terms of the way each group deals with temperature even though these temperature ranges overlap. Temperate fishes are very robust in their tolerance of temperature. These species have abilities to tolerate temperature conditions that differ with latitude. For example, southern populations of largemouth bass in Texas are quite different from populations of bass in Ontario. These differences in metabolic rate and lethal temperature limits appear to be recent adaptations – probably since glaciation – or they may represent evolved differences among populations that survived in different refuges during Pleistocene glaciation. The adaptations of some fish species to less seasonal conditions through acclimation and acclimatization are very common; these populations appear to have become more specialized rather than more robust. The Arctic fishes that can live at −1.5 °C also specialize tremendously to adapt body functions to those cold and constant temperatures. This adaptation is costly; however, they cannot tolerate much variation in temperature. Many tropical and polar fishes do not have the ability to tolerate widely varying temperatures; overwinter temperatures are more commonly the problem for tropical fishes.

The recent work of Stuart-Smith et al. (2017) described earlier also evaluated the thermal influence on biogeography for marine coastal fishes. In their study, they defined a realized thermal niche, which – unlike the fundamental niche – involves the extreme maximum and minimum temperatures influencing a species rather than the average maximum and minimum influencing an individual or population. They found that coastal marine fishes were either in the tropical or temperate guild, with very few subtropical fishes found near 30° latitude. Realized thermal niche differed dramatically between temperate and tropical marine fishes, with tropical fishes having a much narrower niche. The difference in niche was predominantly affected by differences in minimum temperatures tolerated, while maximum temperatures were more similar among thermal guilds. They also found that the geographic ranges of tropical coastal fishes were much larger than those for temperate fishes, with tropical fishes having twice the range size in latitude and 1.3-fold the range size in longitude. They described two hypotheses for existence of thermal guilds: the climate variability hypothesis that fishes adapted evolutionarily to local temperature conditions and the metabolic scaling hypothesis that fishes with a certain metabolic capability have prospered under local temperature regimes. While both of these concepts may help to understand current fish distributions, the distributions themselves do not answer the question of which (or both) is the causal mechanism for thermal guilds. However, local adaptation of thermal performance of isolated fish populations gives support to the evolution to temperature differences.

There may also be human influence on the thermal tolerance of fishes. For example, northern pike's geographic range is circumpolar in the north temperate zone. Northern pike demonstrate predictable thermal capabilities for a circumpolar distribution. They are not very tolerant of warm water; generally, the maximum temperature at which they can survive is about 27 or 28 °C. However, they can survive – and even grow – under cold conditions. Their growth over winter can give them the capability of living in areas with very short summers as long as there is enough of an open water season in which to breed and allow young to grow sufficiently to survive winter. Because of human changes in habitat conditions, pike have been spread throughout much of North America, including into more southerly reservoirs, where they can move into cold, deep water and survive. Pike were taken into hatcheries in locations south in their native range, such as Ohio, and used to stock these reservoirs. The populations of pike were sometimes maintained as brood stock and kept in ponds or lakes, and those locations may have become very warm during summer.

It appears that over the course of recent years, the thermal capabilities of pike from brood stocks in Ohio are quite different from those of pike in their native range. Optimum temperature for growth for northern pike in their native range is considered to be approximately 20 °C (Casselman 1978). Offspring of Ohio fish, where brood stocks were maintained at higher temperatures, have an optimum temperature for growth at 25 °C (Bevelhimer et al. 1985). It appears the Ohio stock has increased the ability to tolerate warm temperatures by five degrees in recent years, while being artificially maintained as a brood stock population. Genetic selection, although unintentional, appears to have occurred in the Ohio hatcheries. Surviving fish in the brood stock were probably more tolerant of warm conditions than those that died. Over time, through selection of surviving pike as brood stock, there was apparently selection for fish tolerant of warmer water. Introductions and hatchery selection appear to have had a strong influence in the thermal tolerance of pike.

Temperate Guilds

Within the temperate zone, there are also different thermal guilds of fishes; earlier these thermal guilds were defined based on their metabolic rates. Three groups were described: temperate eurytherms (warmwater fishes), mesotherms (coolwater fishes), and stenotherms (coldwater fishes). These temperate groups are different from the earlier stenotherms (tropical or polar) and eurytherms (temperate) because all three of these temperate guilds undergo seasonal temperatures near 0 °C. Temperate fishes have to exist under cold conditions during the winter, and even though temperate warmwater fishes appear to have adapted to growth in summer, they undergo the largest temperature range of all temperate guilds. They thrive in the warm temperatures, but they also survive over winter. Temperate coolwater fishes may have the best ability to deal with cold to moderately warm water, but they have a narrower temperature range. Each guild is defined by the range of temperatures it can tolerate. These guilds were defined earlier based on optimum temperature for growth or other laboratory characteristics. One question may be: is the classification based on lab data artificial, or do these thermal guilds in the temperate zone sort out ecologically based on their selections of habitats and temperatures in nature?

John Magnuson, Larry Crowder, and Patricia Medvick from University of Wisconsin evaluated the generality of these thermal guilds (Magnuson et al. 1979). They examined temperature selection in the lab as well as temperature occupation by fishes in the field to see if these characteristics were correlated to the thermal guilds (Figure 27-2). Temperature preference in the lab was determined in a test tank in which fish were offered a variety of temperatures; their final selection of a temperature was considered the thermal preferendum. Determining temperature preference in the field was more difficult. Obviously, if fish lived in an isothermal lake at 18 °C, there would be no fish selecting 25 or 10 °C at that time even if they preferred to be at those temperatures. This discrepancy points out the importance of the realized thermal niche with extremes for a species being more predictive of guilds than the average conditions influencing individuals in a natural setting. A stratified lake with a variety of temperatures and good oxygen conditions throughout might be a good place to test field temperature preference since there is the potential for fish to select a preferred temperature. Temperature selection may be determined there by evaluating fish distributions in relation to temperature or temperatures occupied by fishes fitted with transmitters.

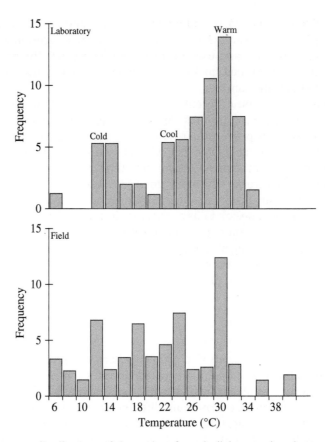

FIGURE 27-2   Frequency distributions of thermal preferenda (laboratory) and actual temperature occupation (field) for temperate freshwater fishes. *Source: Adapted from Magnuson et al. (1979).*

The frequency distribution of temperature preferences by fishes in the lab indicates one clear group of fishes that prefer temperatures of 10–15 °C. Another group peaks in frequency near 20–25 °C, but this latter group is merged with the final peak for fishes preferring 30 °C. There is a fairly clear separation between cold and coolwater fishes but not between cool and warmwater fishes. These analyses provide some evidence in the lab for existence of thermal guilds based on temperature selection, but the overlap is considerable. This separation is even more difficult using field distributions rather than lab preference. The frequency patterns of natural populations indicate a cold (6–11 °C) group and a warm (30 °C) group, with a broad coolwater group between (10–25 °C). Based mainly on lab data, coldwater fishes select 10–15 °C as their preferred temperature, coolwater fishes 20–25 °C, and warmwater fishes 27–32 °C. However, their field distributions are not as distinct as would be predicted on the basis of lab temperature preferences alone.

Another problem in evaluating field distribution is that a fish may face a choice between remaining in its preferred temperature or moving to a less desirable temperature where there is more food. With that choice, the fish might not stay at the preferred temperature. Several other behaviors might influence temperature selection. For example, bluegill by themselves will remain at temperatures around 31 °C. For bluegill placed together in a temperature selection tank, a dominant fish selects 31 °C, and a subordinate fish selects 27 °C (Magnuson et al. 1979). There may be a large shift in temperature selection based on presence of dominant fish. The presence of competitors or predators could equally influence temperature occupation.

Optimal foraging might predict a location for fishes to forage based on abundance of preferred prey. Optimal temperature selection might predict a location to remain based on temperature preferendum. Those two locations may not always coincide. Fishes might select an environment that would maximize their growth under the combination of temperature and food availability, and not optimize either food or temperature. Such selection of optimal growth conditions based on both temperature and food availability is described earlier in Chapter 8 based on work from Brandt et al. (1992). This habitat selection could be to remain at some temperature lower than preferred, but that offers higher food abundance.

Vertical Distribution of Fishes in Lakes and Oceans

Temperature also strongly influences the distribution of fishes within a lake, stream, or ocean. For example, all three of these temperate guilds may coexist in a lake because of the separation between temperature conditions with vertical stratification. In a stratified lake in summer, the epilimnion may vary from 20 to 30 °C, the metalimnion from 4 to 20 °C, and the hypolimnion would be 4 °C. Hypolimnetic temperatures depend on local conditions and might be as high as 8–12 °C. A deep oligotrophic lake might have fish distributions including bass and bluegill (temperate eurytherms) in the epilimnion during summer; pike, perch, and walleye (temperate mesotherms) in the metalimnion during summer; and lake trout (temperate stenotherms) in the hypolimnion during summer. This exact pattern can be seen for fish distributions in the Great Lakes. Habitat may be separated, and competition may be reduced or eliminated by vertical stratification in lakes. Even largemouth bass and pike – which might be expected to compete since they feed on the same kinds of food – probably do not compete strongly in such a stratified lake because they occur in very different temperature zones or are active in different seasons, so competition during summer may be reduced because of habitat segregation.

For many temperate fishes, competition during winter might be very intense because there are no thermal barriers to limit fish distribution. For example, young perch and bluegill both feed on zooplankton, and zooplankton abundance is highest in spring, when temperature resources are not vertically stratified. Both perch and bluegill may occur in the same habitat in spring, but may not compete for food because food is fairly abundant then. As summer progresses, zooplankton abundance declines, the lake stratifies, and although there is less food available, perch and bluegill may not interact because they become vertically segregated in the lake. In winter, thermal resources are similar throughout the lake, and demand for food is lower than in summer, but the abundance of food is also considerably lower. Bluegill and perch overlap in distribution at this time, and the highest degree of interaction might occur then. The combination of thermal resources and food distribution may alter the timing and extent of competition among fishes.

Highly eutrophic lakes often become depleted of hypolimnetic oxygen in summer. If this depletion occurs regularly, coldwater fishes cannot exist in the lake since they cannot survive in an anoxic hypolimnion and they cannot tolerate the warmer surface temperatures. The relationship between anoxia and temperature selection was studied by Headrick and Carline (1993) for northern pike. In Ohio reservoirs, the hypolimnion and much of the metalimnion becomes anoxic in summer, and only the epilimnion has water saturated in oxygen. However, the epilimnion of these lakes approaches 30 °C, and northern pike cannot tolerate those high temperatures. Telemetry studies on northern pike indicated that pike utilized a very narrow column of the water in mid-summer (often less than a meter in depth), where the temperature was as cool as possible and where there was enough oxygen in the water for survival (Figure 27-3). Pike habitat became extremely limited in summer. It is not difficult to visualize a pike suspending its body into the colder anoxic water and extending its head into the oxygenated warmer water to breathe. In this case, pike selected the coolest available temperature, where dissolved

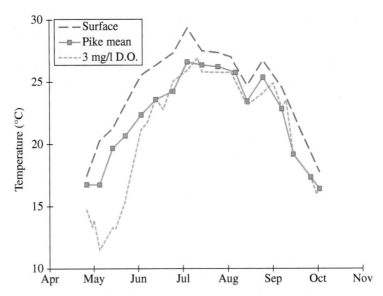

FIGURE 27-3   Mean temperatures of the lake surface, the depth at the 3.0 mg/l oxygen isopleth, and the location selected by pike from Rupert Lake, Ohio. *Source: Adapted from Headrick and Carline (1993).*

oxygen was at least 3.0 mg/l for much of the summer. This habitat compression affected their food consumption and growth compared to fish from other lakes.

This same concept has been evaluated to describe the ecological conditions under which a variety of coldwater and coolwater species persist in the northern US. Pete Jacobson and colleagues from Minnesota DNR and University of Minnesota have evaluated abundance of cisco, lake trout, burbot, and lake whitefish in inland lakes. Earlier studies had shown that cisco mortality occurred under a combination of low oxygen and high temperature, and as oxygen levels declined, the temperatures that cisco could tolerate also declined. They then used these ideas to propose a value of TDO3, which is essentially the temperature of a lake in late summer in locations where the dissolved oxygen reaches 3 mg/l, as a predictor of presence in coldwater and coolwater fishes (Jacobson et al. 2010). They defined TDO3 and other limnological conditions for 1623 lakes and evaluated presence or absence of each fish species in 997 of those lakes. Overall, they found that the TDO3 threshold differed for species in lakes based on factors like P concentration, depth, and presence of stratification. Increasing P concentration leads to higher productivity and lower oxygen content in deeper waters due to higher rates of decomposition. Overall, they found that lake trout had the lowest TDO3 at 5.7 °C, cisco had highest at 18.2 °C, and the other two species were similar in values to cisco. They were able to use TDO3 values and other measures to predict the presence of these species in lakes with reasonable accuracy. They also used these values to predict the distribution of all four species under climate and eutrophication changes in the future, and in all cases species presence in lakes declined under the human-impacted alterations.

While there is clear evidence that some fishes occupy locations in a water body based on optimal temperature for growth, and others avoid areas due to poor oxygen or temperature conditions, one should not assume that the vertical segregation of temperature in a lake or ocean results in limiting the distribution of all fishes. Diel vertical migrations have been described for some time as potential ways that fishes can forage more effectively, avoid predators, or optimize metabolic expenditures while foraging and growing by utilizing the entire water column. The behavioral and physiological components of diel vertical migration were evaluated in the second edition of this book (Diana 2005) in Chapter 5. However, good field and lab comparisons of diel vertical migrations were not available at that time.

The same vertical variation in water temperature occurs in oceanic environments, and similar distribution patterns exist there. David Sims and colleagues from the Marine Biological Association of the United Kingdom conducted a very comprehensive study of the diel vertical migrations in dogfish sharks (Sims et al. 2006). The study utilized acoustic telemetry, data storage tags, behavioral arenas in the lab, prey distribution, and predator avoidance measures to determine the basis for diel vertical migrations in dogfish. Sims and colleagues also conducted energetic modeling under different thermal regimes to determine whether the strategy resulted in savings in energy and better growth. Studies were done in Lough Hyne, which is a shoreline estuary attached to the Atlantic Ocean off southern Ireland.

Their initial studies showed that in late summer there was strong vertical stratification of temperature with surface temperatures around 16 °C and bottom temperature is approximately 14 °C. Their measures of prey abundance show that the common prey for dogfish (prawns, gobies, and swimming crabs) were 18–80 times more abundant in shallow and warm areas than deep areas and cooler areas. While grey seals were common predators of dogfish, none were found in the area during the summer. The authors determined that dogfish showed vertical migrations moving to shallow and warm areas in the evening, staying there through the night, and then moving back into deep water in the morning (Figure 27-4). They generally occurred over small home ranges within the estuary. While fish did move into shallow water at night,

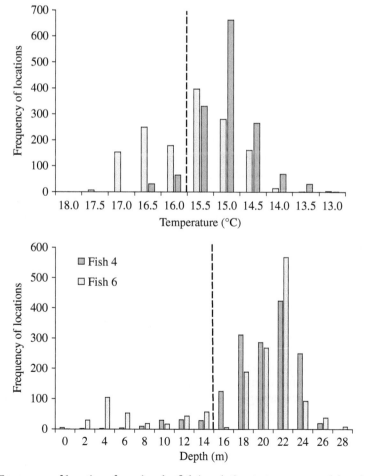

FIGURE 27-4   Frequency of locations for spiny dogfish in relation to temperature (above) and depth in July and August 2003. Dotted lines: locations of the thermocline. *Source: Adapted from Sims et al. (2006).*

they did not continually stay there but moved back and forth between shallow and deep waters. Their measures of activity were significantly higher when dogfish were in shallow water above the thermocline than when they were below it.

In the laboratory, dogfish generally moved into the warmer side of the choice compartment only to obtain food and chose to remain in the cooler side during most of the time observed.

The authors had three hypotheses regarding the distribution of dogfish: (1) that dogfish would move to deep waters during the day to avoid predation by seals, (2) that they would move to shallow water during the night to forage more effectively, and (3) that they would move to deep water during the day to reduce metabolic rate. Hypotheses 2 and 3 were supported by the data. They finally hypothesized that dogfish sharks hunt during warm and rest during cool periods, which allows for better growth rates and lower metabolic rates than constant occupation of either warm or cold conditions. They also believe that this may be common for many fishes showing diel vertical migrations. Their data supported such a concept, even though the temperature difference between warm and cool habitats was fairly small (only about 2 °C).

Conditions in streams make distribution of fishes there somewhat different than the vertical pattern described for lakes. Many temperate streams have water sources from snow runoff in the mountains and are coldest at the headwater areas and warmest at downstream locations. Conversely, some streams might be groundwater fed and have cold conditions throughout. The fish communities in temperate streams have often been defined based on warm or cold water conditions, and thermal guilds segregate longitudinally based on the temperature conditions as one moves downstream. These patterns are described much more thoroughly in the following chapters on fishes in temperate streams and tropical rivers.

Thermal Habitat and Population Biomass in Lakes

The previous example of pike behavior in Ohio reservoirs points out that behavior and population processes are probably strongly affected by the thermal habitat volume in a given ecosystem. The dimensions of this volume may well be set in critical time periods as in summer stratification for pike. Gavin Christie and Henry Regier from the University of Toronto evaluated correlations between commercial fish yield and thermal habitat volume for four commercial fish species (lake trout, lake whitefish, walleye, and northern pike) from a variety of Canadian lakes. These fishes are temperate stenotherms (former two) or mesotherms (latter two). Christie and Regier (1988) believed the stable yield (a function of fish biomass) of these species in inland lakes should be correlated to the amount of thermal habitat available to each in the lakes.

Christie and Regier defined thermal habitat for each fish based on temperature preferenda. They evaluated existing data to determine mean preferendum and optimal temperature for growth; both had similar values for each species. They considered the optimal thermal niche to be ± 2 °C of the temperature preferendum (or optimal temperature) and then calculated the volume of each lake that had such thermal characteristics. Summing this over time gave a thermal volume available as cubic hectometers per 10 days. For all four species, sustained yield was strongly correlated with thermal habitat volume ($r^2 = 0.25$–0.86; Figure 27-5), and yield showed a better correlation to habitat volume for coldwater fishes than for coolwater fishes. This analysis indicated that not only fish behavior – but also population abundance and biomass – were related to thermal habitat. Other lake factors correlated with sustained yield in this study included lake area and total dissolved solids. Similar analyses have also been used by Magnuson et al. (1990) to forecast effects of global warming on fish populations.

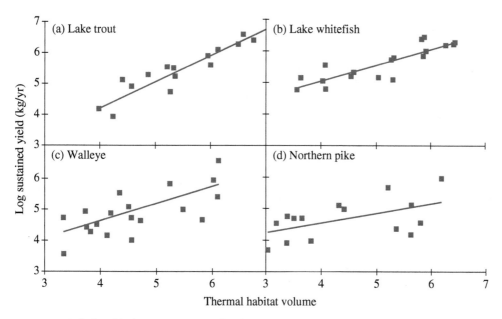

FIGURE 27-5   Relationship between sustained yield and thermal habitat volume for four fish species.
Source: Christie and Regier (1988)/Canadian Science Publishing.

Summary

The thermal capabilities of fishes strongly affect their distributions, particularly on a north–south gradient. Thermal guilds of fishes might be expected to show some similarities in their distributions as well as in their behavior in natural systems. However, even the temperature preferences of fishes within a thermal guild may differ, and these differences may affect distributions and interactions of fishes. Obviously, thermal capabilities are not the only characteristics influencing fish distributions; several other characteristics are covered later. However, the importance of temperature in determining distributions of fishes will remain in the future as we face issues such as global warming and its influence on fish populations.

Literature Cited

Bevelhimer, M.S., R.A. Stein, and R.F. Carline. 1985. Assessing significance of physiological differences among three esocids with a bioenergetics model. Canadian Journal of Fisheries and Aquatic Sciences 42:57–69.

Brandt, S.B., D.M. Mason, and E.V. Patrick. 1992. Spatially-explicit models of fish growth rate. Fisheries 17(2):23–31.

Casselman, J.M. 1978. Effects of environmental factors on growth, survival, activity, and exploitation of northern pike. American Fisheries Society, Special Publication 11:114–128.

Christie, G.C., and H.A. Regier. 1988. Measures of optimal thermal habitat and their relationship to yields of four commercial fish species. Canadian Journal of Fisheries and Aquatic Sciences 45:301–314.

Diana, J.S. 2005. Biology and Ecology of Fishes, 2nd edition. Biological Sciences Press, Traverse City, Michigan.

Headrick, M.R., and R.F. Carline. 1993. Restricted summer habitat and growth of northern pike in two southern Ohio impoundments. Transactions of the American Fisheries Society 122:228–236.

Jacobson, P.C., H.G. Stefan, and D.L. Pereira. 2010. Coldwater fish oxythermal habitat in Minnesota lakes: Influence of total phosphorus, July air temperature, and relative depth. Canadian Journal of Fisheries and Aquatic Sciences 67:2002–2013.

Magnuson, J.J., L.B. Crowder, and P.A. Medvick. 1979. Temperature as an ecological resource. American Zoologist 19:331–343.

Magnuson, J.J., J.D. Meisner, and D.K. Hill. 1990. Potential changes in the thermal habitat of Great Lakes fish after global climate warming. Transactions of the American Fisheries Society 119:254–264.

Robbins, W.H., and H.R. MacCrimmon. 1974. The Blackbass in America and Overseas. Biomanagement and Research Enterprises, Sault Ste. Marie, Ontario, Canada.

Sims, D.W., V.J. Wearmouth, E.J. Southall, et al. 2006. Hunt warm, rest cool: Bioenergetic strategy underlying diel vertical migration of a benthic shark. Journal of Animal Ecology 75:176–190.

Stuart-Smith, R.D., G.J. Edgar, and A.E. Bates. 2017. Thermal limits to the geographic distributions of shallow-water marine species. Nature Ecology & Evolution 1:1846–1852.

CHAPTER 28

Fish Communities in Temperate Streams

The classical concept of stream zonation described in Chapter 1 has long been established in the literature and proposes that fish communities in streams follow a longitudinal zonation pattern. The temperate zone extends roughly from 30° to 60°N and S latitude, encompassing the areas where winter cold conditions are common and temperate fish species persist. Upstream areas in temperate streams have communities of coldwater fishes, including mainly salmonids; downstream areas have warmwater fishes; and intermediate areas have cool-water fishes, such as grayling. The concepts supporting zonation were that physical and thermal conditions change as water progresses downstream, which results in orderly replacement of species in a longitudinal pattern. A second major process that also influences fish assemblages in streams is one of continual addition of species downstream as streams increase in size. With this pattern, as stream conditions change in a downstream direction, there is not necessarily a dramatic shift in species composition but an addition of species tolerant to the new conditions. Eventually, there may be replacements, as well as additions, because tolerances of some species may not allow them to extend downstream – or competitors and predators may exclude them – but the pattern of change for continual addition is much more gradual than for zonation.

There is opportunity for continuous migration throughout a stream because lotic systems are connected from headwater to mouth. This means immigration and extinction processes are less important to community structure in streams compared to inland lakes as local extinctions can be overcome regularly by migrations. Fish species at a stream site are part of a metapopulation, connected directly to the genetic resources of many nearby populations. The zonation concept asserts the landscape position is important in regulating stream characteristics and fish assemblages. In addition, stream gradient is an important determinant of fish assemblage structure as streams with high power produce different habitat types than those with lower gradient and power. The purpose of this chapter is to overview ecology of temperate streams and relate fish assemblages to factors such as zonation, addition, and gradient. Subsequent chapters evaluate tropical rivers and fish assemblages in these ecosystems as well as fish assemblages in lakes and marine systems.

Stream Ecology

The major difference between lotic and lentic environments is the linkage of habitats within the lotic system with a longitudinal change in habitat characteristics from upstream to downstream sites. Changes occurring from headwater to mouth also depend on local geological conditions.

Biology and Ecology of Fishes, Third Edition. James S. Diana and Tomas O. Höök.
© 2023 John Wiley & Sons Ltd. Published 2023 by John Wiley & Sons Ltd.

Landscapes with high topographic relief have large changes in gradient between headwater and floodplain streams. Streams in landscapes with less topographic relief may also have large variations in flow stability and temperature. The conditions in the latter streams depend more on topography, geology, and precipitation than altitude and gradient. Since landscape position of streams also involves changes in elevation, stream power, and average air temperature in areas with high topographic relief, conditions in streams do not vary independently but are strongly linked to landscape position. In areas with low topographic relief, landscape position also strongly affects stream conditions through changes in geology, slope, and land cover.

Stream order defines the location of a stream in a watershed with the smallest permanently flowing stream being a first-order stream (Strahler 1957). When two first-order streams merge, downstream waters become a second-order stream, and the convergence of two second-order streams results in a third-order stream, etc. It is also important to visualize stream positions in the landscape as a network. Avenues for species colonization of a stream system are considerable in locations where a low-order stream is a tributary to a much larger stream. This concept of a linkage classification based on proximity to streams of different order incorporates the view that streams are not only being influenced by their own order but also by the order of the streams to which they connect.

The River Continuum Concept

While zonation of streams has been recognized for some time, knowledge of the pattern of connectivity between stream systems was slower in development. Vannote et al. (1980) first proposed the River Continuum Concept as a way of better understanding the linkage among different habitats in river systems. They based their analyses on stream order and how changes in energy processing and animal communities vary with stream order. Energy and nutrients do not cycle in a stream system in a similar manner to a lake but rather are transported downstream as they are processed. Stream ecologists have termed this a spiral, where, as materials are processed and moved downstream, they become available and reprocessed by other organisms in a downstream manner. Vannote et al. used this concept to evaluate the major sources of energy and the major types of communities that should be expected in streams of different order. A schematic of the relationship between stream size and the structure and function of communities (Figure 28-1) shows very large differences between streams of different order. Headwater streams – order one to three – are theorized to be strongly influenced by riparian vegetation, and terrestrial material in the form of coarse particulate organic matter (size greater than 1 mm) is the major input of energy to upstream reaches. Therefore, benthic invertebrates are predominantly shredders and collectors utilizing this coarse particulate organic matter and breaking it down into smaller particles. There is a higher rate of respiration than photosynthesis in these upstream reaches because of shading and relatively low nutrient conditions, and the stream is dependent on allochthonous materials (materials from the watershed).

The pattern of community structure and energy change differs in higher order streams. Medium-sized streams – order four through six – have higher nutrient levels due to processing of materials that occurred upstream and often include photosynthetic rates higher than respiratory rates, indicating primary production by periphyton (algae attached to substrates like rocks). This production increases energy available for grazers and collectors that utilize internally produced material. As coarse particulate organic matter is transported downstream, it becomes further broken down into fine particulate organic matter (less than 1 mm), which is predominantly fed upon by collectors rather than shredders. Therefore, the community has moved from one dominated by heterotrophy to autotrophy depending on the degree of shading and light in the stream reach.

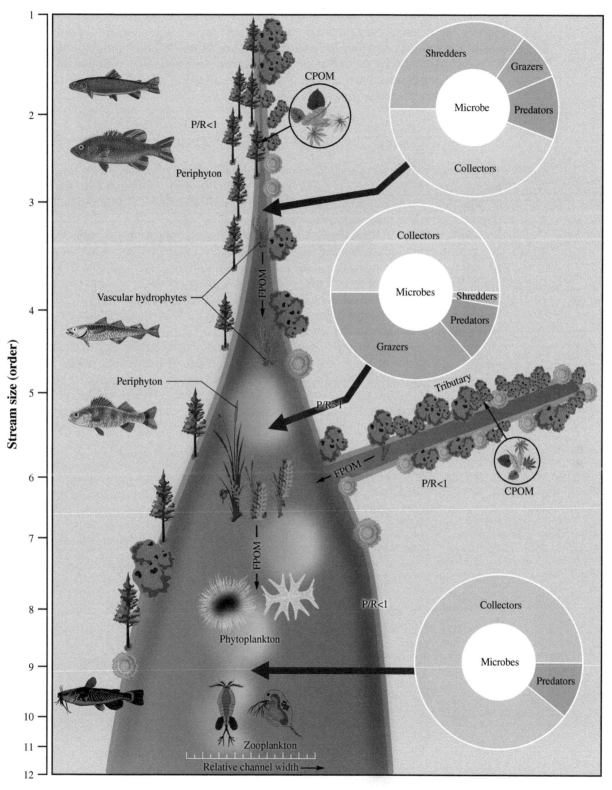

FIGURE 28-1   Relationship between stream order and structural or functional attributes of stream communities as hypothesized by the River Continuum Concept. *Source: Reproduced with permission of Vannote et al. (1980)/Canadian Science Publishing.*

Large rivers – order greater than six – continue to receive larger quantities of fine particulate organic matter from upstream processing. Primary production is often limited due to turbidity and depth, so the stream may be heterotrophic once more. As the conditions in the floodplain reduce current velocity, phytoplankton and zooplankton become more abundant rather than periphyton and benthic invertebrates. These large river systems are predominated by collectors and predators with few grazers or shredders.

The Importance of Connectivity to Stream Fishes

The connection between stream habitats is important not only from an energy flow perspective but also from the perspective of fish life histories. Schlosser (1991) emphasized the importance of migration between different types of habitat in the life history of stream fishes. This concept (Figure 28-2) is that many fish species show a spawning migration to a habitat with good conditions for incubation of eggs, then feeding migration to locations with seasonally favorable food and growth conditions, wintering migration to locations with poorer growth but good habitat for survival in winter, migration back to feeding habitats, and eventual spawning migrations again. The importance of this concept is that the life history of fish in streams requires mobility to capitalize on areas of excellent habitat quality. Connection of these types of habitats is very important in population dynamics. If fish cannot migrate to adequate habitat for one of these processes, habitat quality will limit the population.

Schlosser emphasized the importance of longitudinal heterogeneity much in the same manner as Vannote et al., except he focused on communities and populations, rather than whole stream processes. One important concept Schlosser emphasized is the relationship between lateral habitats and stream communities. In floodplain rivers – streams of high order – lateral habitats may extend large distances from the stream channel. These lateral habitats can be seasonally wet during floods or include oxbow lakes and other habitats isolated from the

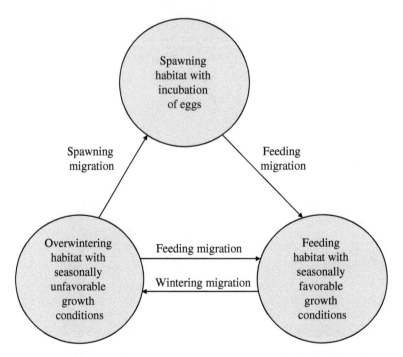

FIGURE 28-2 Schematic of the life history of habitats and migrations in stream fishes. *Source: Redrawn from Schlosser (1991).*

river channel but still wet. Much terrestrial vegetation and material exists within the floodplain channel and can enrich the river system during flooding.

In addition to the importance of specific habitats and migrations, Schlosser evaluated human-influenced changes within the watershed and their potential effects on fish. In particular, human land use and other disturbances have a strong influence on sources and patterns of nutrient and material addition to streams. River systems show dramatic alteration due to human development of the landscape, including the direct effects of dam construction, channelization, bank alteration, and flow divergence for irrigation or domestic use, as well as indirect effects of changes in the landscape that influence stream ecosystems. Construction of dams fragments connectivity of habitats and migration of fishes emphasized by Schlosser. Dams not only create barriers to movement but also change the flow dynamics of a river, especially when peaking hydropower is employed, which releases water at high discharge during times of day when energy need is greatest and reduces flow at times when energy need declines. Such changes in flow can actually exceed the annual seasonal dynamics of flow in a river. Dams also alter temperature downstream based on the pattern of water release through the dam, produce an impoundment that destroys riverine habitat, and may provide a refuge or a barrier for native fishes as well as predators or competitors not normally abundant in the river system and reduce sediment and woody debris delivery to locations below the dam.

Landscape alterations, such as urbanization, deforestation, and agricultural development, dramatically alter delivery of water and materials to the river, which can result in changes to river habitats and flow regimes, loss of connectivity, alterations in water quality, and impacts on fish assemblages. Human-induced alterations of landscapes not only have specific impacts, but many are combined to produce multiple stresses on aquatic systems. While we understand that human-induced alterations have a dramatic impact on river systems, it is difficult to predict impacts based solely on knowledge of which alterations have occurred (Infante et al. 2019). Schlosser proposed five effects that human-induced land use changes would have on fish communities in headwater streams, including reducing the diversity of adult and juvenile fishes because of lower habitat heterogeneity, decreasing the complexity and size structure of fish populations because of the absence of large fish often found in pool habitats, increased abundance of juvenile fish because of larger areas of shallow water refuges, higher growth rates of juvenile fish due to elevated primary production during the summer, and greater temporal variability in fish abundance because of absence of structural complexity.

Fish Assemblage Studies in Streams

Streams in Areas with High Topographic Relief

Frank Rahel and colleagues at the University of Wyoming have studied communities of fish in Wyoming streams for a considerable time. In particular, Rahel and Hubert (1991) attempted to evaluate the importance of zonation or continual addition of species downstream as processes affecting fish assemblages in Wyoming streams. They evaluated fish distributions in Horse Creek – a small tributary to the North Platte River – in order to test these two processes. They sampled 20 sites from headwater to mouth of the stream. They determined fish species composition at each site by electroshocking and also evaluated habitat, including depth, substrate, vegetation, type of habitat (pool, riffle, run), current velocity, and bank features. One important point in their analysis of fish assemblages was they estimated relative abundances as well as species richness (number of species in a location). Rahel and Hubert estimated the percent composition of different species in order to evaluate the additive hypothesis by determining importance of different species at a site rather than simply presence or absence.

The longitudinal pattern of change was fairly complicated, including species that showed evidence of zonation and downstream addition. Headwater sites were characterized by a simple assemblage dominated by trout, while farther downstream sites had more complex assemblages dominated by minnows and suckers. The transition between headwater and middle-stream sites appeared to be dramatic and an indicator of zonation with the fish species composition changing dramatically in a short linear distance. In comparison, middle stream sites seemed to have gradual species change caused by downstream addition, where minnows and suckers dominated the central reaches, and larger-bodied species – such as carp and sunfishes – were added in farther downstream reaches. Additionally, four species – stonecat, suckermouth minnow, central stoneroller, and Johnny darter – were absent from headwater sites, reached high levels in mid-stream sites, and then became rare or absent in downstream sites, further supporting zonation.

To evaluate these community changes, Rahel and Hubert utilized detrended correspondence analysis (DCA) as a statistical technique. DCA analyzes data on abundances of each species at each site with no emphasis on site location. DCA ranks and groups species and sites simultaneously and evaluates which species contribute to the major change in assemblages across sites. Their results with DCA (Figure 28-3) showed the three trends described above. The principal axis of ordination was based on upstream and downstream locations, with the upstream, middle, and downstream communities clearly segregated from one another. The ordination distance between headwater and mid-reach locations was so dramatic to indicate zonation. The segregation of mid to downstream areas was not as clear, particularly for site number 7, which did not group well related to its actual location. The second axis of ordination separated mid-stream from downstream sites, where downstream sites categorized into two groups, with presence of carp and green sunfish in one group or stoneroller and common shiner in another. This patterning on this axis was not as clear as axis one.

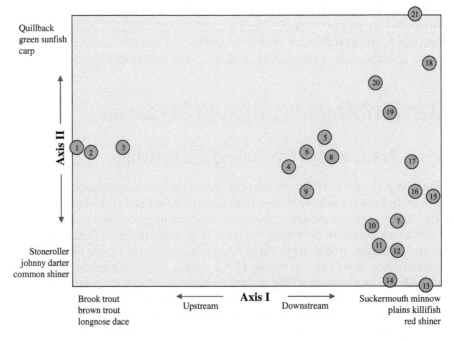

FIGURE 28-3 Detrended correspondence analysis of 20 study sites (numbers increase in a downstream pattern) from Horse Creek, Wyoming. *Source: Rahel and Hubert (1991)/Taylor & Francis.*

Rahel and Hubert also evaluated environmental gradients by performing principal component (PC) analysis on habitat characteristics. This would determine the main variation in habitat characteristics among sites. The first principal component described 74% of the variation in habitat characteristics, and a low score was associated with high elevation sites that were shallow, narrow, and clear. High scores in this component were low elevation sites that were wide, deep, and turbid. The second principal component characterized conditions of the stream bank and riparian zone. Low scores on this axis indicated areas with bare conditions on the bank and little undercut or overhanging vegetation, while high scores indicated the opposite. These values demonstrated that the longitudinal pattern shown in communities were in part related to upstream and downstream gradients in habitat in the stream itself and, in part, correlated to differences in bank vegetation.

In order to further evaluate the effect of zonation, Rahel and Hubert did a community similarity analysis for pairs of sites along stream longitude. This index calculated a similarity between fish assemblages at adjacent stream sites. Adjacent locations with low similarity would indicate dramatic changes in fish fauna, similar to the transition from one zone to another. Since presence and abundance of different species at each site were somewhat variable, community data for two upstream sites were averaged and compared in similarity for the next two downstream sites in order to remove some of this variation. The end result (Figure 28-4) showed a strong transition among the upstream sites but then a fair degree of similarity among farther downstream sites. It appears that zonation may describe differences in fish communities found in the headwater sections, but addition of species is probably more important in mid to high order streams. In fact, the studies reviewed by Rahel and Hubert indicated that workers who have supported additive patterns for stream fish assemblages generally studied systems lacking major thermal or topographic transitions. For example, streams in the southern US with limited longitudinal temperature differences, or streams in areas of low topographic relief with little change in physical conditions downstream, were locations that supported the additive downstream hypothesis. In the end, Rahel and Hubert believed a combination of both zonation and additive patterns described the variation of fish assemblages found in Horse Creek.

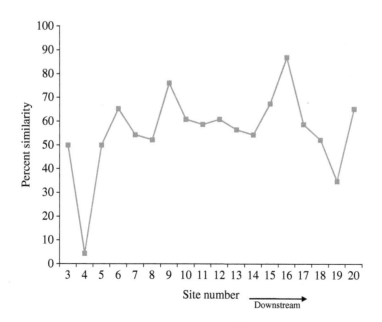

FIGURE 28-4 Assemblage similarity for pairs of adjacent sites in a downstream pattern for Horse Creek. *Source: Rahel and Hubert (1991)/Taylor & Francis.*

There have been a number of additional studies on river systems with high topographical relief in temperate and tropical locations, and in these, temperature seems to be the main driving factor in the change in fish assemblages longitudinally. Jonathan Mee, Geneva Roberts, and John Post from the University of Calgary conducted an interesting study on branches of the South Saskatchewan River in Alberta. They evaluated fish assemblages on three branches of the river over about 800 km from the Rocky Mountains to the plains with an elevation drop of 1,400 m. Their data on fish distributions in the system were taken from collection records of a number of organizations rather than by their own sampling, and this can lead to problems with different sampling methods, timing, etc. They hypothesized that temperature conditions would strongly affect fish distributions in the system, so they estimated the mean water temperature in July as an indicator of this factor. They also estimated a number of other characteristics of the habitat, such as gradient, water velocity, and chemical conditions.

Mee et al. (2016) estimated there were 33 species in these rivers, with 4 rare or atypical species removed from analysis, and only 2 exotic species found (rainbow and brook trout). They found the expected results of fish diversity increasing in downstream areas with warmer temperatures. They also found evidence for major changes in fish assemblages occurring in locations with increasing mean July temperatures, with a major change in locations at 15 °C and another in the area of 19–21 °C. They used the Czekanowski coefficient, which estimates the percentage of similarity between two locations, and found major reduction in assemblage similarity for each river in locations within those temperature ranges (Figure 28-5). They observed similar communities in each river across their longitudinal reaches and indicated that both additions and zonation influenced the fish assemblage in any location. The similarity indices remained fairly high even in locations of major changes with the upper two zones reflecting the presence of coldwater and coolwater fish and the final zone reflecting warmer tolerant fish but never reaching conditions for a true warmwater fish assemblage. One other interesting interpretation was that the present distribution and abundance of bull and cutthroat trout were altered by the exotic rainbow and brook trout in the watersheds.

Several other studies have found zonation on rivers in warmer climates, such as the Rio Grande River (McGarvey 2011), and in the tropics where elevation still results in major temperature changes with elevation (Jaramello-Villa et al. 2010). For the Columbian Andes study, Jaramello-Villa et al. found one major zonation occurring at about 1250 m altitude, and additions – as well as deletions – occurring at lower elevations. They also found a fairly large number of exotic species among the 62 species they collected for this study. McGarvey, in his study of the Rio Grande River, found three zones of fish assemblages and many more exotic species that probably interfered with native fish distributions.

Fish Communities in Streams with Low Topographic Relief

Streams in areas of low topographic relief have smaller changes in gradient and ambient air temperature in a longitudinal pattern compared to mountain streams. However, these streams differ throughout the drainage network in physical conditions and fish assemblages. For many years, streams in the midwestern United States have been managed as either coldwater or warmwater streams based on temperature conditions and fish populations. Coldwater streams were those capable of sustaining trout populations, while warmwater streams had a more mixed assemblage of species. In some cases, coldwater streams might reflect headwater areas of river systems, while warmwater systems could be lower in the catchment. However, this was not always the case; instead, the categorization of streams usually required on-site evaluation.

A number of recent studies have attempted to understand geographic patterns of warmwater and coldwater streams in temperate locations with low topographic relief as well as fish

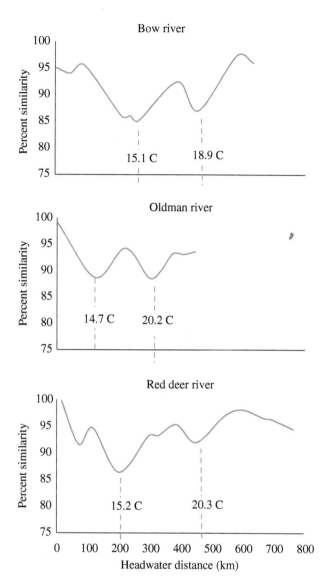

FIGURE 28-5   The Czekanowski coefficient (percent similarity) of the fish assemblage for sliding 50 km sections of river at 5 km intervals using 29 species of fish at each section of river. The temperatures listed at zone boundaries are mean July temperature for that river at that location. *Source: Adapted from Mee et al. (2016).*

assemblages contained in them. Some have worked with the basic assumption that warmwater and coldwater are a good starting point for fish assemblages, while others have re-evaluated the whole concept of fish assemblages in these streams.

Michigan Streams

Troy Zorn, Paul Seelbach, and Mike Wiley from the Michigan Department of Natural Resources and the University of Michigan have done considerable research on the landscape patterns of stream conditions and the fish communities contained therein. Zorn et al. (2002) evaluated streams in the Michigan Rivers Inventory, a database that includes a large number of sites with consistent data collection for fish and invertebrate assemblages

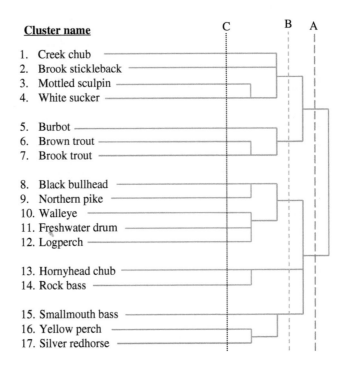

FIGURE 28-6 Dendogram for a cluster analysis of stream fish clusters in Michigan based on LFY and CA, with names based on dominant fish species in each cluster. *Source: Zorn et al. (2002)/ Taylor & Francis.*

as well as physical and chemical conditions. They evaluated distribution and abundance patterns for 69 fish species found at 226 stream sites. They used cluster analysis to detect groups of fish that shared similar abundance patterns at each site. The analysis resulted in 17 clusters of species with common overlapping distributions. The cluster analysis described 39% of the total variance in the distribution of these species, indicating that considerable variation still existed. The cluster analysis itself (Figure 28-6) could be evaluated at several different levels to understand differences among fish clusters and the patterns of streams that influence those differences.

The first split in these clusters ((A) in Figure 28-6) separated groups in small streams from those of larger downstream reaches. The second cut (B) distinguished five groups that reflected both stream size and temperature requirements of species. In these clusters, fishes from coldwater streams – brook trout, brown trout, and burbot clusters – were distinguished from those of smaller coolwater streams – hornyhead chub and rock bass clusters – and from those in medium to large warmwater streams – walleye, freshwater drum, logperch, northern pike, and black bullhead clusters. However, the main point of the analysis was the 17 clusters distinguished at level (C). These clusters do not reflect fish assemblages, that is, they do not include all of the species found at a given site but rather a subgroup of species commonly found together. The assemblage at a given site might include several of these clusters depending on local conditions.

In order to evaluate conditions in the streams associated with each cluster, Zorn et al. did an ordination analysis utilizing the low-flow yield (LFY) and catchment area (CA) conditions of the watersheds (Figure 28-7). LFY is an indicator of groundwater influence in a stream and reaches its highest level in basins with highly permeable surficial geology and relatively steep topography. Under these conditions, groundwater mainly influences flow of the streams,

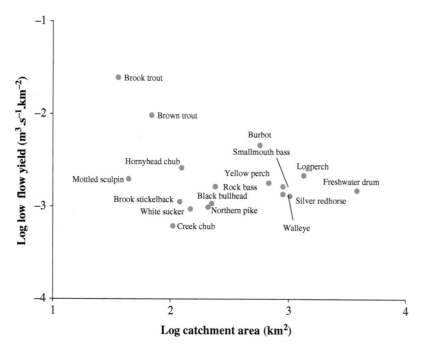

FIGURE 28-7   Mean LFY and CA conditions of sites where each cluster was most abundant in Michigan streams. *Source: Zorn et al. (2002)/Taylor & Francis.*

which is consistent and cool. LFY is also associated with discharge gradients, in that streams with high LFY have relatively constant flow conditions – not much flashiness or variation in flow over time. The second stream characteristic used by Zorn et al. was the area of the watershed upstream from the sampling site (A). This is a measure of stream size. They believed these two characteristics defined much of the variation in fish habitat in Michigan streams and did an ordination evaluating distribution of clusters between these two conditions.

The results of this ordination had two main sub-groupings. Coldwater clusters were mainly restricted to small CA and high LFY streams. Clusters typical of warmwater streams were found in small CA and low LFY sites, while mainstream species were found in moderate LFY and large CA sites. There did appear to be a separation of these clusters into coldwater, coolwater, and warmwater sites, as well as into main river and small sites. However, evaluating only the mean location for each cluster did not offer much detail, while using variation in each cluster over LFY-CA characteristics showed much overlap. Since these clusters were not the species assemblage, Zorn et al. believe sites in locations with intermediate conditions of LFY and CA would include both clusters in an assemblage.

One interesting result of their analysis was that Michigan streams did not often display typical longitudinal patterns of fish assemblages or stream conditions. In some cases, headwater areas of river systems emerged from wetlands or lakes and had warm and slow-flow conditions, and then moved through mid-stream channels in valleys, with high groundwater input and steeper gradient. This caused a change to coldwater conditions. As streams approached the Great Lakes and lake plain geology, they exhibited mainstream and warmwater conditions again. Therefore, coldwater to warmwater species did not occur in a longitudinal basis consistently in Michigan streams. For this reason, the combination of LFY and CA axes provided data to predict assemblages. Cold or warm segments (and the associated fish clusters) still occur predictably in Michigan streams but are based on landscape position – geology, stream order, and topographic slope – rather than longitudinal location.

Kevin Wehrly and colleagues at Michigan DNR and the University of Michigan also evaluated stream characteristics and fish assemblages based on data from the Michigan Rivers Inventory. Their focus was on temperature effects on fish assemblages, and they used thermal conditions in July as the variables predicting fish assemblages (Wehrly et al. 2003). They found that mean temperature and temperature flux had strong influences on fish assemblages, and these conditions varied, based on channel characteristics, riparian forest cover, and groundwater input. The fish assemblages they found segregated well into coldwater, coolwater, and warmwater guilds, with species compositions similar to Zorn et al. (2002). The thermal limits had coldwater fishes in streams with July weekly mean temperatures of 10–18 °C, coolwater fishes in 19–21 °C, and warmwater fishes in 22–26 °C. The latter distributions were undoubtedly truncated by the lack of very warm stream conditions in Michigan. In addition, they found these same assemblages were also affected by temperature fluctuation. Coldwater fishes were only found in locales with stable temperature (<5 °C) or moderate (5–10 °C) fluctuations, while coolwater and warmwater fishes were mainly found in areas of moderate fluctuations. Few cases of extreme fluctuation existed in Michigan streams. Also, species richness increased with both mean temperature and stream size, and these two attributes were correlated in most streams.

Little Tennessee River

Joseph Kirsch and James Peterson from the University of Georgia conducted a similar study to the Michigan group on the Little Tennessee River. Kirsch and Peterson (2014) collected 36 fish species and habitat data from 525 channel units within 48 study reaches of the river system. Once again, fish species diversity and abundance was negatively correlated to elevation, and temperature was an important determinant of residency by different species. They proposed that the overall assemblage across stream reaches was mainly influenced by topography, stream topology, and temperature. In comparison, distribution of fishes within a stream reach was controlled more by channel habitat types and urban land use. Warmwater intolerant species were largely found in headwater areas, while warmwater tolerant species were widely distributed. Higher stream linkage and size of the nearest downstream reach were also correlated with higher species diversity likely due to better dispersal capability. Finally, they showed a strong effect of urban land use on lower species diversity and fish abundance. While they did not evaluate zonation or addition as the means for species change longitudinally, their results support additive processes throughout most of the watershed.

Kansas Streams

Thornbrugh and Gido (2010) from Kansas State University analyzed fish assemblages in the Kansas River system with the intent to relate species abundance to connections of small tributaries with major river systems on a local as well as landscape scale. They first used data collected by the Kansas Department of Wildlife and Parks at 413 locations to evaluate larger scale processes (up to 200 km distant). There were 68 species of fish collected in the overall basin, dominated by cyprinids, centrarchids, and percids. The number of fish species was positively correlated with drainage area, a large-scale regional effect, and also with stream size. Larger bodied fish, such as freshwater drum, flathead catfish, and gizzard shad, were mainly associated with mainstem sites (order >5), while smaller fishes, such as redbelly dace, orangethroat darter, and common shiner, were mainly found in smaller tributaries. There were few significant differences among tributaries near mainstem sites compared to sites more distant from the mainstem, indicating little large-scale network dispersal.

Thornburgh and Gido also conducted collections on their own, on sites in three local tributaries. They collected 39 species in different seasons and locations. They found two distinct assemblages in this work, with fish in the mainstem showing strong differences from fish in the tributaries, while the fish assemblages in the smaller tributaries were similar to each other. Locations near the confluence of a small tributary to the mainstem tended to have fewer small tributary species and few fishes from mainstem assemblage. The difference in the small tributary assemblage was hypothesized to be due to habitat differences in these locations reducing suitability for some of the small river species rather than predation from large river species. This tended to revert with distance, and by 20 km upstream, the normal small tributary assemblage prevailed. Assemblages in small order streams (1–3) near confluence with larger order streams (2–4) tended to have more species than isolated locations, indicating network or confluence exchange among the smaller components of the drainage system. They demonstrated differences in mainstem and small tributary assemblages with limited exchange between the two but with significant enrichment of assemblages in smaller tributaries from the network of larger tributaries nearby. One might hypothesize that the small tributary to mainstem changes indicate zonation, while the confluence effects within smaller tributaries support addition.

The Index of Biotic Integrity

Earlier, we covered various issues with human-induced degradation of river systems and their potential effects on stream fishes. The Index of Biotic Integrity (IBI) is an analysis method developed by Jim Karr and colleagues in Illinois to determine degree of degradation of river systems. Karr et al. (1985) were interested in how various degradation processes had influenced a stream system and potentially the biotic communities in streams. IBIs have been used widely in the United States to conduct such evaluations and require understanding of the conditions in minimally impaired rivers as well as rivers of degraded quality. IBIs are not applied to all river systems with similar metrics, but rather they are developed to fit the conditions and species found in a region, and in particular often differentiate among wadeable streams, non-wadeable rivers, and small coldwater rivers. An interesting component of an IBI is it not only reflects species richness, but also other characteristics of communities more functional in nature, such as the percent of total omnivores and percent of tolerant species. Therefore, such indices do not necessarily reflect species assemblages but rather functional aspects of assemblages.

IBIs are established by producing metrics for various conditions of the community and scoring a gradient of poor through good criteria based on local conditions. John Lyons from the Wisconsin Department of Natural Resources has focused on the development and validation of the Index of Biotic Integrity for Wisconsin rivers. Lyons et al. (2001) used 10 metrics that described species abundances as well as characteristics of the fish community (Table 28-1). The conditions for poor to good streams were developed using a database of warmwater river conditions in sites with few human impacts compared to sites that were heavily impacted. Karr and Chu (1999) recommended that metrics should include at least six characteristics of the community: species richness and composition, presence of indicator species, trophic function, reproductive function, abundance, and fish condition.

Once IBI metrics have been determined and scores calculated, changes that have occurred in different stream reaches can be evaluated based on relative IBI scores. Lyons' analysis of Wisconsin streams (Figure 28-8) indicated that least impacted streams had mainly excellent conditions for IBI scores, while streams impacted by peaking hydropower, non-point source pollution, or multiple stresses generally had fair or poor scores. The distribution and clustering of these scores give some indication of the degree of impairment based on the source of degradation. It can be used as evidence of stream damage done by different conditions as well as

TABLE 28-1   Metrics and scoring criteria (points) for an IBI of warmwater Wisconsin rivers.

Metric	Geographic Location	Scoring Criteria Rating		
		Poor (0)	Fair (5)	Good (10)
Biomass/effort	All	<10	10–25	>25
Number of native species	North	<8	8–9	>9
	South	<12	12–15	>15
Number of sucker species	All	<3	3–4	>4
Number of intolerant species	All	<2	2	>2
Number of riverine species	North	<2	2–3	>3
	South	<5	5–6	>6
Percent diseased/injured fish	All	>3	0.5–3	<0.5
Percent riverine species	North	<11	11–35	>35
	South	<11	11–20	>20
Percent lithophils	North	<45	45–69	>69
	South	<25	26–40	>40
Percent insectivores (by weight)	North	<11	11–60	>60
	South	<21	21–39	>39
Percent round suckers (by weight)	North	<11	11–60	>60
	South	<11	11–25	>25

Source: adapted from lyons et al. (2001).

targets on which stream improvements might be judged. Therefore, IBI scores are commonly applied methods to determine conditions of rivers compared to some standard but not necessarily to predict the fish assemblage that might exist in a given river system.

Summary

The assemblages of fishes in streams have been largely defined based on zonation, temperature, and gradient patterns. Stream zonation, primarily affecting temperature, holds strong predictive value for fish communities in areas of high topographic relief. In areas with lesser relief, temperature is still a primary determinant of fish assemblages, but the influence of groundwater and other conditions are more important than longitudinal zonation in defining temperature and assemblages. Under these conditions, coldwater, coolwater, and warmwater groupings still appear to remain although considerable overlap and uncertainty exist in actual species assemblage. Species assemblage has been predicted based on landscape characters or stream habitat characteristics with some degree of accuracy, but considerable unexplained variation remains. Finally, an Index of Biotic Integrity is a means to test degradation of stream communities against some standard of less degraded habitat and use in evaluation of the impacts of different human influences on stream systems.

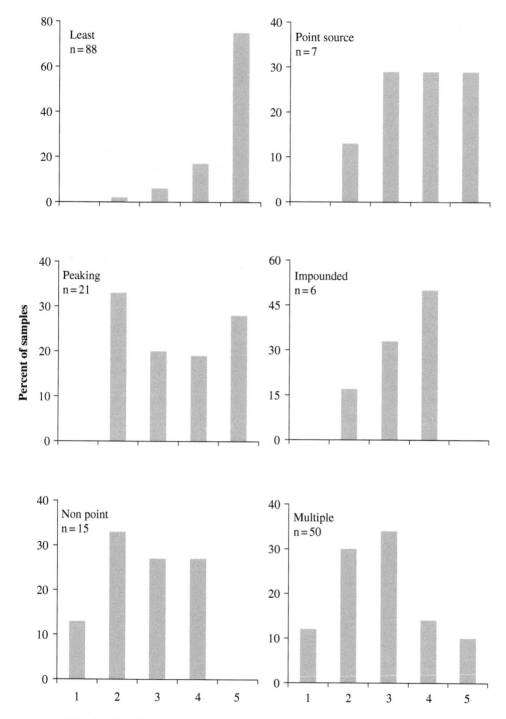

FIGURE 28-8 Distribution of IBI scores among rivers in various impact categories for 187 river sites in Wisconsin. IBI score: 1 is very poor, 2 poor, 3 fair, 4 good, and 5 excellent. *Source: Lyons et al. (2001)/ Taylor & Francis.*

Literature Cited

Infante, D.M., L. Wang, R.M. Hughes, K. Chen, and B.F. Terra. 2019. Advances, challenges, and gaps in understanding landscape influences on freshwater systems. American Fisheries Society Symposium 90:463–495.

Jaramello-Villa, U., J.A. Maldonado-Ocampo, and F. Escobar. 2010. Altitudinal variation in fish assembly diversity in streams of the central Andes of Columbia. Journal of Fish Biology 76:2401–24.

Karr, J.R., and E.W. Chu. 1999. Restoring Life in Running Waters: Better Biological monitoring. Island Press, Washington, D.C.

Karr, J.R., L.A. Toth, and D.R. Dudley. 1985. Fish communities of Midwestern rivers: A history of degradation. Bioscience 35:90–95.

Kirsch, J.E., and J.T. Peterson. 2014. A multi-scaled approach to evaluating the fish assemblage structure within southern Appalachian streams. Transactions of the American Fisheries Society 143:1358–1371.

Lyons, J., R.R. Piette, and K.W. Niermeyer. 2001. Development, validation, and application of a fish-based index of biotic integrity for Wisconsin's large warmwater rivers. Transactions of the American Fisheries Society 130:1007–1094.

McGarvey, D.J. 2011. Quantifying ichthyofaunal zonation and species richness along a 2,800-km reach of the Rio Chama and Rio Grande (USA). Ecology of Freshwater Fish 20:231–242.

Mee, J., G.L. Robins, and J.R. Post. 2016. Patterns of fish species distributions replicated across three parallel rivers suggest biotic zonation in response to a longitudinal temperature gradient. Ecology of Freshwater Fish 27:44–61.

Rahel, F.J., and W.A. Hubert. 1991. Fish assemblages and habitat gradients in a Rocky Mountain-Great Plains stream: Biotic zonation and additive patterns of community change. Transactions of the American Fisheries Society 120:319–332.

Schlosser, I.J. 1991. Stream fish ecology: A landscape perspective. Bioscience 41:704–712.

Strahler, A.N. 1957. Quantitative analysis of watershed geomorphology. Eos, Transactions American Geophysical Union 38:913–920.

Thornbrugh, D.J., and K.B. Gido. 2010. Influence of spatial positioning within stream networks on fish assemblage structure in the Kansas River basin, USA. Canadian Journal of Fisheries and Aquatic Sciences 67:143–156.

Vannote, R.L., G.W. Minshall, K.W. Cummins, J.R. Sedell, and C.E. Cushing. 1980. The river continuum concept. Canadian Journal of Fisheries and Aquatic Sciences 37:130–137.

Wehrly, K.E., M.J. Wiley, and P.W. Seelbach. 2003. Classifying regional variation in thermal regime based on stream fish community patterns. Transactions of the American Fisheries Society 132:18–38.

Zorn, T.G., P.W. Seelbach, and M.J. Wiley. 2002. Distributions of stream fishes and their relationship to stream size and hydrology in Michigan's lower peninsula. Transactions of the American Fisheries Society 131:70–85.

CHAPTER 29

Tropical Rivers

Many ecological studies confirm the presence of high species diversity in the tropics for both plant and animal communities. Explanations for this higher diversity are numerous, often speculative, and have been based largely on terrestrial organisms, particularly rainforest plants and birds. However, the same pattern is also borne out in aquatic ecosystems. This chapter briefly overviews the explanations for higher diversity in the tropics, tropical freshwater fishes and their species diversity, and then specifically describes fish assemblages in the Amazon River. Currently, there is much concern about clearing and burning of tropical rainforests and their effects on species diversity; often this issue focuses on the Amazon Basin. However, there has been little public concern about the interaction between fishes and the forest in the Amazon ecosystem. In fact, two tropical fish communities (coral reefs and tropical rivers) probably exceed any other vertebrate communities in species diversity and potential for loss of vertebrate genetic diversity – they are indeed special ecosystems.

Species Diversity in the Tropics

The flora and fauna of the tropics is much more diverse than that of the temperate zone. This is obvious to ecologists in spite of the bias that there has been much more effort in the temperate zone to identify and categorize species than there has been in the tropics. For example, Nelson et al. (2016) reviewed biodiversity and distribution of fish species. They considered the number of valid named species of extant fishes to be over 38,900, over half of all vertebrate species. About 43% of this fauna is freshwater resident and less than 1% diadromous. While Mooi and Gill (2002) estimated 25,000 species at that time, they provided relative species numbers for each biogeographic region (Table 29-1), which indicated a very large proportion of these fishes are tropical, including 14,350 from primarily tropical areas, and 90% of the freshwater fauna.

One example of tropical diversity can be seen by comparing the relative abundance of fish species in several watersheds. The fish fauna of the Amazon River basin includes more than 3,000 species of fishes (van der Sleen and Albert 2017); in the Panama Canal Zone of Central America, there are 456 species, while in the Laurentian Great Lakes there are only 172 species of fish present. This pattern, termed the Latitudinal Diversity Gradient, has been shown for a wide variety of organisms and indicates a clear decline in the number of species in communities on a tropical to temperate gradient.

There are many hypotheses as to why there is much higher species diversity in the tropics than in the temperate zone. These hypotheses are not all mutually exclusive ideas but often involve similar concepts. Some contradict each other and some seem almost identical, but they

Biology and Ecology of Fishes, Third Edition. James S. Diana and Tomas O. Höök.
© 2023 John Wiley & Sons Ltd. Published 2023 by John Wiley & Sons Ltd.

TABLE 29-1	Estimated Number of Fish Species for Freshwater and Marine Biogeographic Regions.

Region	Number of Species
Freshwater	
Nearctic (North America)	1,060
Neotropical (South and Central America)	8,000
Palearctic (Europe)	360
Ethiopian (Africa)	2,850
Oriental (Southeast Asia)	3,000
Australian (with New Guinea)	500
Marine	
Western North Atlantic	1,200
Mediterranean	400
Tropical Western Atlantic	1,500
Eastern North Pacific	600
Tropical Eastern Pacific	750
Tropical Indo-West Pacific	4,000
Temperate Indo-Pacific	2,100
Antarctic	200

Source: Mooi and Gill (2002)/John Wiley & Sons.

comprise different hypotheses that have been forwarded to explain why there are more species in the tropics than in the temperate zone.

The first concept has been called the time hypothesis. The tropics are warm and humid with little change in climate and a long history during which these conditions have existed. Scientists have suggested that this long history of relatively consistent conditions provided animals greater potential to evolve into specialized species. Support for the time hypothesis can be found among tropical fish communities but also from Lake Baikal in Russia. Lake Baikal is a very old lake, much older than lakes that have a geological history since the last glacial period – approximately 10,000 years. Lake Baikal houses a variety of endemic species, particularly sculpin. The large number of endemic species has been hypothesized to result from sympatric speciation in Lake Baikal due to relatively constant environmental conditions over a long period of time. This observation supports the concept that time is important in allowing evolution of an endemic fauna. Since the tropics have not been interrupted by glaciation, there has been a much longer time horizon over which evolution could occur.

The second hypothesis on tropical diversity is spatial heterogeneity. There may be more diversity of habitat structure in the tropics, which allows more niche specialization and therefore more species evolution. This can be a circular argument when such speciation actually caused that spatial heterogeneity, at least in ecosystems such as tropical rainforests and coral reefs. This hypothesis proposes that the diversity of habitats allows specialization of animals to utilize those unique habitats. This would result in more species utilizing a similar surface area of habitat than in areas with generalist species. The basic idea is that the more complex an environment, the greater the opportunity for species to specialize to use distinct habitats.

The third hypothesis on tropical diversity is competition. This hypothesis proposes that competition – and competitive displacement – is strong in the tropics and has driven evolution of a variety of species because the conditions have been relatively stable. It is obviously linked to the time hypothesis and provides a mechanism for that concept to function. Chapter 14 addressed the question of whether current competition is most important in species interactions or whether species have evolved through character displacement to avoid competition. While most people favor the latter argument, in either case, competition would result in character displacement and speciation. If competition is (or was) strong in the tropics because of the number of species present, then competition and time have allowed a larger amount of evolution. López-Fernández et al. (2013) found that neotropical cichlids largely showed most diversification early in their evolutionary history, but this likely varies in different systems, such as the African rift lakes, where much of the diversification appears to be quite recent (Albertson et al. 1999).

The fourth hypothesis on tropical diversity is predation. This idea proposes that there are more predators to reduce abundance, which results in less competition and more coexistence. The predation hypothesis opposes the competition hypothesis. In the Amazon River, many biting predators have evolved that do not swallow prey whole (such as piranha), and there is a high abundance of piscivorous fishes. Adaptations of biting mouth parts in predatory fishes are not as common in other freshwater environments. Predation, at least in some tropical habitats, is intense; whether that predation drives higher species will be covered later in this chapter.

The fifth hypothesis for tropical diversity is stability. This stability hypothesis is very similar to the time concept – that the tropics have existed over evolutionary time with stability of habitats found there. However, the tropics are not as stable seasonally, particularly in aquatic habitats. Temperature conditions in tropical freshwater habitats may fluctuate considerably day to night. As elevation increases, temperatures fluctuate and reach cold conditions seasonally or daily. Similar variations occur in wet/dry seasonal areas. This idea of stability, which has been used much to support evolutionary arguments for terrestrial communities, applies to aquatic habitats in much of the tropics but not necessarily to annual conditions within a habitat.

The final hypothesis commonly applied to tropical diversity is high productivity. This concept proposes that many characteristics drive high productivity in the tropics. Some characteristics include the biotic community itself, such as layering of tropical rainforests. Others include the high degree of insolation for much of the year and the warm conditions that can drive higher primary production. All of these conditions stimulate primary productivity, which encourages more effective use of energy by specialization. It may be difficult to be a specialist if food consumed is not relatively constant in distribution throughout the year. A specialist can persist better if productivity is high and food is relatively constant throughout the year. High productivity is hypothesized to drive specialization and evolution in the tropics. However, some tropical systems like blackwater rivers have very low natural productivity yet high species diversity of fishes, particularly piscivores (Winemiller and Jepsen 2004). Within the Amazon, there are blackwater rivers that contain much humic acid from their drainage and low nutrient concentration, whitewater rivers that are turbid and nutrient rich, and clearwater rivers with limited turbidity or humic acid. These latter two river types generally have high productivity of aquatic macrophytes as well as periphyton, high fish productivity, and diversity. Large detritivorous fishes from blackwater rivers migrate during the wet season to higher productivity whitewater rivers where they forage in the floodplains and spawn. They migrate back in the dry season, with added numbers and size due to their foraging in the more productive rivers, and become the main prey for piscivorous fishes in the blackwater rivers. This is an example of food web subsidy – similar to nutrient subsidies described earlier – and occurs in a number of low productivity aquatic systems.

All of these different ideas, and probably more, have been invoked to explain why the tropics claim more species than the temperate zone. Much of this argument is based on birds or mammals, yet there are more species of fishes than any other vertebrate, so such concepts should also apply to fishes. Examining diversity of fish species in the tropics might support or refute these different hypotheses because the species diversity of fishes is so high. It may also aid in the understanding and management of tropical fish communities as they are impacted by increasing human demands.

Tropical Freshwater Fishes

In different zoogeographic areas of the tropics, there have been different species groups from which adaptive radiation proceeded. In most locations, there is a diverse freshwater fish fauna in the tropics, but very different families form the basis of that diversity. The African fish fauna is an interesting example of such adaptive radiation. First, there are a number of ancient or relict forms of species in Africa, such as lungfish. Mooi and Gill (2002) determined there were more than 2,850 species of freshwater fishes in Africa, a very diverse fish community. Two-thirds of this diversity was in seven families: Cyprinidae (475 species); Characidae and Citharinidae (208 species); Clariidae, Clarioteidae, and Mochokidae (345 species); and Cichlidae (870 species). Undoubtedly there are many more species that have been described since that time, but the main groups are likely consistent. The fauna has many cichlids; tilapia is one example of a cichlid mentioned several times in this book. One interesting component of the tropical fish fauna in Africa is the importance of the great rift lakes: Lake Malawi, Lake Tanganyika, and others. These are mainly deep lakes because they occur along an earthquake fault line; several are also very old (two million years or more old, similar to Lake Baikal). The African rift lakes have had considerable evolution of endemic species, particularly cichlids. Chapter 2 covered the concept of dispersal evolution or speciation occurring among populations that were sympatric. Such sympatric speciation requires behavioral or physiological characteristics to serve as isolating mechanisms for evolution rather than geographical barriers. Adaptive radiation of the species flocks of cichlids in the great lakes of Africa is the basis for much of the unique fish fauna of Africa.

In South America, there are more than 8,000 species of obligatory freshwater fishes. These are minimal estimates because the numbers and the kinds of species existing in some areas have not been totally documented and described. One interesting fact about South America is there are no cyprinids (minnows or carp); this is the only region in the world where cyprinids are not a major component of the freshwater fish fauna. Mooi and Gill described five major groups of fishes that dominate the fauna of South America: Characiformes, with 1,280 species, which include piranhas and *Colossoma*; Siluriforms (catfishes, with over 1,400 species); Gymnotiformes (knifefishes, 100 species); Cyprinondontiformes (375 species); and Cichlidae (450 species). Again, these numbers are underestimates of the currently described fauna. Siluriforms and Characiformes have dominated the evolution of fishes in South America.

Feeding Adaptations of Tropical Fishes

One interesting characteristic of the freshwater fauna in the tropics is the elaboration of specific feeding mechanisms. Winemiller (1991) compared temperate and tropical food webs and found that, compared to temperate riverine species, tropical rivers have more herbivorous, detritivorous, and omnivorous species. In addition, invertebrate feeders and piscivores show greater niche specialization in the tropics. Some feeding adaptations that have occurred in the tropics have no parallel in the temperate zone, including fishes that feed on large particles of vegetation, chewing leaves or roots. The grass carp is probably the best example of this, and many species of tropical fishes eat macrovegetation, often even terrestrial vegetation. Another specialization that occurs but is not limited to the tropics includes fishes that scrape rocks, leaves, or other surfaces for periphyton (attached algae).

The evolution of biting predators is one of the more unusual developments for feeding in South American fishes. The size limit to prey consumption by most freshwater predatory fish is gape size, at least for fishes that swallow prey whole. Most temperate predatory fishes do this and are limited in the size of prey they can handle and swallow. There are many biting predators in tropical South America; probably the best example is piranha, which can consume very large items. Groups of fishes might consume prey considerably larger than each individual fish could eat itself. In addition to predators of large animals, there are biting predators that specialize on unusual items, such as scales or pieces of fins. Some of these fin eaters or scale eaters may mimic cleaner fishes, but rather than remove ectoparasites, they bite pieces from hosts and then depart.

Another unusual feeding method in the Amazon is seed and fruit eating by fishes. Many fishes in the Amazon River seem to have functionally replaced fruit eating mammals in the temperate zone, at least in terms of dispersing seeds. Similarly, the wet and dry seasons flood land areas at times and make terrestrial insects, such as termites, vulnerable to fishes. The regular flooding of terrestrial areas gives fishes in the Amazon access to terrestrial production, especially for fruits. The coupling of most fruit production by trees with the flooded season also makes fruit an abundant food source in these inundated areas. Detritivory may also be more important in the tropics than in the temperate zone due to high levels of primary productivity and large inputs of terrestrial materials to the aquatic ecosystem during flooding. Many tropical fishes consume detritus.

Other Adaptations in Tropical Freshwater Fishes

Several other unusual physical or physiological adaptations occur in tropical freshwater fishes. One physiological characteristic that is often very important in the tropics is air breathing. Species of fishes that can extract oxygen from air as well as water are largely found in the tropics. Air breathing appears to have evolved several times and has occurred in species of lungfish, catfish, perch, and others. For example, some species of *Clarias* (the walking catfish) are capable of surviving with no oxygen in water by breathing air but are not capable of surviving in well-oxygenated water with no access to air. *Clarias* extract oxygen from air by a modification of their gills. Their gills are unusually club shaped, and the fish gulp air at the surface, hold it in their branchial chambers, and use that air bubble to enhance oxygen levels near the damp gills.

Other fishes use their skin as a breathing organ (*Clarias* probably does this), and these fish have soft, unscaled skin to allow gas exchange. Air breathing is one adaptation to the wet and dry climate common in the tropics as well as an adaptation to poor water quality. The stagnant, turbid waters of many lowland tropical areas become low in oxygen due to high temperature and much plant decomposition. During the dry season, fishes may become stranded in stagnant pools with declining water levels, and air breathing allows dispersal from these pools or survival under the poor conditions there. Many fishes can crawl on land from one pond to another, breathing air while on land. The walking catfish is not the only air breather; in fact, it is common to fence aquaculture ponds used for catfish production in Southeast Asia, not to keep the walking catfish in but to keep other predators that disperse over land – like snakehead – out.

Salinity and temperature tolerance may also be similarly important for certain tropical fishes. For example, one common group in Central America is cyprinodonts or pupfishes. Cyprinodonts are the most salinity-tolerant fishes known, occurring at times in drying pools in the desert with salinities in excess of 55–60 ppt – nearly double the salinity of sea water. These fishes also are very temperature tolerant, surviving in temperatures in excess of 43 °C. The ability to tolerate desiccation of ponds, which involves increased salinity and high temperature, is common in more arid parts of the tropics.

A third adaptation to similar conditions includes specialized kinds of reproduction, often reproduction to avoid low oxygen or declining water levels. Lungfish can bury their eggs in sediments, and the eggs will survive dormantly in dried mud for years. The eggs hatch once the mud is flooded. Lungfish themselves can burrow into mud and become dormant. Other specialized reproduction, such as bubble nests (Siamese fighting fish), attachment of eggs to terrestrial vegetation, and spraying eggs with water (archer fish), has also occurred. These are adaptations of reproduction to the relatively poor water quality and high predator abundance that often develop in temporary water bodies as they desiccate.

Seasonality in the Tropics

Obviously, climate in the tropics may vary as much among sites as in the temperate zone. Temperature and rainfall patterns are variable, and seasonal changes in these patterns may be strong or weak. However, one widespread characteristic of the tropics is presence of seasonal wet or dry conditions. For example, the climate in Thailand has distinct wet and dry seasons in most years (Figure 29-1). The dry season usually occurs from December through April. Neither the beginning nor the intensity of the wet season is always predictable. In 1983, there was rainfall of nearly 1 m in the month of October and total rainfall more than 4 m in that year. The Bangkok area became a large flooding for much of that year. Average conditions would have produced 20–30 cm of rainfall per month, and a total of about 1 m during the wet season. In 1984, an "El Niño year," there was no peak rainfall in the Bangkok area and no wet season. Zeng (1999) evaluated effects of seasonal and

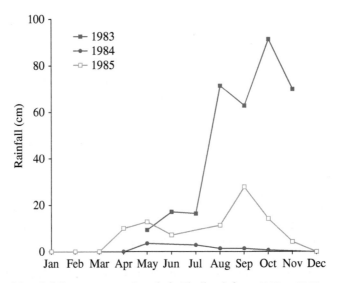

FIGURE 29-1 Monthly rainfall pattern near Bangkok, Thailand, from 1983 to 1985.

annual variations in rainfall on discharge of the Amazon River from 1986 to 1993. Annual variation in maximum and minimum monthly rainfalls in the Amazon was 20–30% and was affected by El Niño events. In Thailand, which is considered to be in the wet tropics, there were large variations in the existence, timing, and intensity of wet and dry seasons. This also differs among regions as the monsoon climate is influenced by ocean patterns, land temperatures, and circulation. El Niño is the change in water circulation and temperature over the Equatorial Pacific Ocean that regularly (on two- to four-year cycles) alters climate and rainfall throughout much of the world by changes in global circulation patterns in the atmosphere.

During the wet season, much of the floodplain region is inundated, and food for fishes may be produced by terrestrial nutrients and terrestrial plants. During the wet season, many fishes move into the floodplain to forage. Most plants also have their major growth period at this time, so productivity is seasonally high. One might expect fish migrations to be more common in the tropics compared to the temperate zone because of the regular changes of the water level in the tropics (Winemiller and Jepsen 1998). There is flooding in the temperate zone as well, particularly seasonal flooding in the spring, due to snow melt. This flooding might have similar effects by removing nutrients from the watershed and depositing them into the water. The time duration and extent of flooding are much lower in most temperate locations, indicating a reduced magnitude of flooding in most upstream areas. However, Schlosser (1991) believes migrations to special habitats are common for different life history stages of stream fishes in tropical or temperate areas, so inundation of floodplains, connectivity of the system, and these longitudinal and lateral exchange processes are probably important in all river systems. Winemiller and Jepsen (2004) even attribute the existence of very large-bodied species in many tropical river systems to the importance of these long migrations and trophic web subsidies.

Rivers and Flooding

Wet and dry seasons in the tropics produce extreme flooding and secession in some locations although flooding cycles are common in most river systems. The importance of this flood pulse was recognized by several scientists who have studied tropical and temperate rivers (Junk et al. 1989). Floods affect many riverine processes, including sedimentation, channel morphology, and organic matter recycling. The regularity and predictability of floods appear to influence use of the floodplain by animals (Figure 29-2), resulting in much fish migration. Many tropical fishes show most of their growth during seasonal inundations and harvest the productivity of terrestrial crops. The floodplain area is generally more productive than its riverine counterpart so that use of such a habitat may largely influence animal growth and production. The actual river channel can be relatively unproductive, serving as an avenue for migration but only producing limited fish growth. All of these characteristics led Junk et al. to conclude that main nutrient and energy exchange may occur laterally in large river systems rather than in a downstream direction as is more commonly believed in temperate rivers.

This concept opens some interesting areas of study. Rivers vary dramatically in the likelihood of their flooding. Groundwater rivers, such as many in sandy areas of the Midwestern United States, have relatively consistent flows regardless of season. Montane rivers are often dramatic in their flood and drought cycles brought about by seasonal snow melt. The faunas of such rivers may be as much influenced by flow regularity as by other local processes. Many human influences may reduce flooding (such as flood control dams or reservoir projects) or may change its periodicity (such as peaking hydropower dams). The flood pulse concept is very useful in understanding the dynamics of river systems.

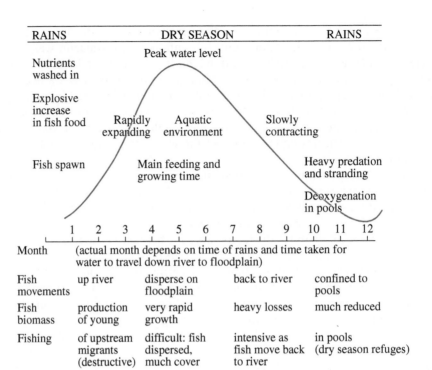

FIGURE 29-2 The seasonal cycle of events in a tropical floodplain river. *Source: Adapted from Lowe-McConnell (1977).*

Fish Communities of the Amazon River

One major example of tropical freshwater fishes in a seasonally flooded location is the Amazon River ecosystem. This massive habitat has been studied by Michael Goulding from University of California, Berkeley, as well as many other scientists, but our understanding of this area is still limited compared to more temperate rivers. The Amazon is a huge habitat; it is the largest river system (in volume flow) in the world (Goulding 1980). The Amazon River, under average flow, discharges 6.6 thousand cubic kilometers of water per year. In comparison, the second largest river in the world (Congo River) has an average discharge of 1.3 thousand cubic kilometers (1/5 of the Amazon), and the Mississippi River has less than 0.6 thousand cubic kilometers (1/11 of the Amazon). The Amazon River has a seasonal habitat that receives much rainfall (between 1,500 and 3,000 mm per year) predictably during the wet season. There is regular flooding of the Amazon, but due to the size of the drainage that flooding pattern differs with location. The southern reaches of the Amazon (as far as 25°S) have seasonal temperatures typical of the temperate zone, while the main channel of the Amazon is along the equator. This strongly influences annual discharge patterns. For example, two tributaries of the Amazon – the Rio Madeira (a southern tributary) and the Rio Negro (an equatorial tributary) – differ in discharge pattern (Figure 29-3). Peak discharge occurs in the southern river in January through March, and the wet season extends from November through March. The equatorial river, with a wet season from about March through July, has a peak discharge much later in the year (June and July). The timing of flooding may be predictable in the Rio Madeira or Rio Negro but may not be the same from one river to the other.

The work of Goulding emphasizes the influence of these changes in runoff, where water level fluctuates as much as 7–13 m between the wet and dry seasons. An additional 13 m of

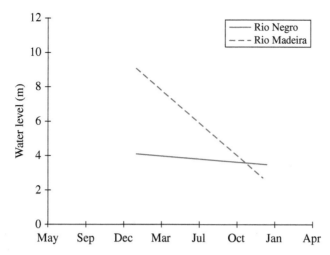

FIGURE 29-3   Monthly changes in the water level of two Amazon tributaries. *Source: Adapted from Goulding (1980).*

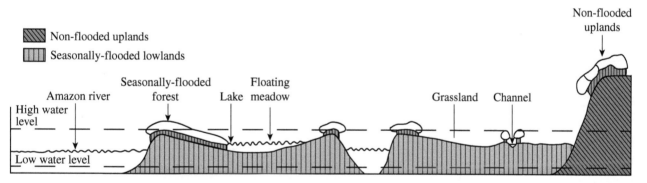

FIGURE 29-4   Cross section of the Amazon floodplain. *Source: Adapted from Smith (1981).*

water in the wet season is indeed a dramatic change. As water level rises, it covers terrestrial vegetation in the floodplain and actually may completely submerge much of the terrestrial vegetation (even large trees) (Figure 29-4). Isolated pockets of water in the floodplain during the dry season become exposed to the entire fish fauna during the wet season. This regularly opens a tremendous area of new habitat to fish movement.

One interesting characteristic about the Amazon is both plants and fishes have adapted to this regular seasonal flooding. Many trees can survive and do well in inundated conditions, in which their roots or even the entire tree might be under water for some period. The Amazon has a very low gradient for much of its watershed; the main reach (more than 3,000 km) drops approximately 80 m. It is more a large flooded plain rather than rapids running through the forest. Up to 70,000 km² of forests around the Amazon are seasonally flooded and available for fish habitation during the wet season (there are also upland terra firma forests that do not flood seasonally), while flooded forests become dry during the dry season.

What are some adaptations of fishes to this ecosystem? Probably the first and most unusual is fruit eating. A variety of fishes in the Amazon River eat fruit, especially the large characins (Table 29-2). Some species chew fruit and eat it all (including seeds); some chew fruit only to remove seeds and then eat the seeds; and some crack nuts and eat the flesh of the nut. All of these habits involve mouth parts that develop strong chewing and crushing

TABLE 29-2　Foods of Adult Characins of the Rio Madeira Basin.

	Fruit/seeds	Leaves	Flowers	Wood	Allochthonous Invertebrates	Feces	Arboreal/terrestrial Vertebrates	Aquatic insects	Crustacea	Mollusks	Zooplankton	Fishes	Algae	Detritus
CHARACIDEA														
Tambaqui														
Colossoma macropomum	X				*	*		*			*	*		
Pirapitinga														
Colossoma bidens	X	X			*	*				*		*		
Jatuarana														
Brycon sp.	X	*			*	*	*							
Pacu toba														
Mylossoma cf. duriventris	X	X	*		*	*		*						*
Pacu vermelho														
Mylossoma cf. albiscopus	X	*	*		*								*?	
Pacu mafura														
Myleus sp. A	X	X												
Pacu mafura														
Myleus sp. B	X	X												
Sardinha comprida														
Triportheus elongatus	X	*	*	*	X			*				*		*
Sardinha chata														
Triportheus angulatus	X	*	*	*	X									X?
Piranha preta														
Serrasalmus rhombeus	*	*	*		*			*	*			X		*
Piranha encarnada														
Serrasalmus serrulatus	X				*							X?		
Piranha mafura														
Serrasalmus cf. striolatus	X													
CYNODONTIDAE														
Peixe Cachorro														
Rhaphiodon vulpinus												X		
Pirandira														
Hydrolycus pectoralis												X		

TABLE 29-2   (Continued)

Pirandira					
Hydrolycus scomberoides				X	
ANOSTOMIDAE					
Aracu comum					
Schizodon fasciatus	X?	X	*		*
Aracu amarelo					
Leporinus fasciatus	X				
Aracu cabeca gorda					
Leporinus friderici	X	X			

*Minor Food Item.
X, Major Food Item.
Source: Goulding (1980)/University of California Press.

capability (Correa et al. 2007). By fruit eating, the fishes also aid in the dispersal of seeds; when they eat fruit, many seeds pass through the digestive system intact and are defecated elsewhere. Fishes may function as seed dispersers in the Amazon like birds and mammals do in the temperate zone. There are more than 40 species of fruits eaten by *Colossoma*; some of them are swallowed whole, and some chewed up. Goulding even hypothesized that consumption of fruit by fishes may result in a higher diversity of plants in the rainforest. In temperate forests, climax communities are established with few species predominating, shading out the others, and dominating the landscape. In the Amazon, tree reproduction may be limited by fishes selectively eating plants, fruits, and seeds. Instead of one plant becoming predominant in an area, fish removal of seedlings, seeds, and fruit may offer the opportunity for other plants to become established at that site, and a more diverse plant community to develop. This community may become more dependent on fishes, especially consumption of fruits by large fish and dispersal of seeds. In fact, Anderson et al. (2011) demonstrated that Amazonian fishes disperse seeds farther than known terrestrial species due in part to their large size, long displacements, and slow digestion rates of seeds. They also found a high degree of seed germination after passage through the guts of *Colossoma*. In this view, seasonal habitation by fishes provides regular disturbance and results in nonequilibrium processes influencing plant community structure.

A fruit-eating fish must be large to handle its food; as a result, there are many large-bodied characins in the Amazon River. The same problem occurs for predators, which usually are similar or larger in size than their prey to overcome them. This large body size of fish is quite different from some of the tropics, such as Central America, where most fish species (cyprinodonts and poeciliids) are small. The adaptation of fruit eating and large size must have been a selective force in evolution of piranhas. This selection for chewing predators would occur because large herbivorous fishes that grow rapidly are difficult prey, and chewing predators, rather than predators that swallow their prey whole, are more capable of utilizing these large-sized prey. Chewing carnivores are another adaptation of fishes in the Amazon with biting mouth parts to remove pieces of flesh. These fishes often use group feeding behaviors.

Goulding also hypothesized that the discharge characteristics of the Amazon River results in a high diversity of fishes there by nonequilibrium processes. The fish community that may remain in an isolated water body during dry season can vary tremendously. For example, a large school of piranhas might happen to become entrapped in the floodplain. These piranhas may consume most other species also in that lake, including large frugivores. Intense hunger of piranhas during the dry season, and intense crowding that may occur in isolated lakes, may result in much predation in those water bodies. Other areas might become isolated without many predators, and these would serve as refuges for less predator-tolerant species. In a way, this is similar to earlier descriptions of island biogeography, except this isolation occurs annually. At least, the chance occurrence of each species in a given pond or lake can be viewed as stochastic, resulting in sporadic, nonequilibrium events occurring locally in the floodplain. Both refuge and predator areas become intermixed again when water level rises. This seasonal change may prevent common species of fishes from becoming dominant, much the same way that fruit eating keeps common trees from becoming dominant, and probably contributes to the diversity of the fish fauna.

Are nonequilibrium processes due to seasonal flooding and drying the dominant factors influencing community assemblages in tropical rivers? Rodriguez and Lewis (1994) produced results that contradict the importance of nonequilibrium processes, at least in Venezuelan rivers. They studied fish assemblages in 20 floodplain lakes and compared assemblage patterns between lakes and years. Their basic premise was that equilibrium processes could be important in these systems if the characteristics of individual lakes outweigh the potential for stochastic recruitment during the flood period. Their results showed patterns supporting the importance of equilibrium processes in floodplain lakes. The lakes differed dramatically from one another in fish assemblage structure, and while they were disturbed by annual flooding and species reassortment, they showed considerable internal stability in assemblage composition as well as considerable variety between lakes. The assemblage properties were reestablished in each lake and occurred soon after the annual flood disturbance, and the lakes appeared consistent in species assemblage over multiple years.

Apparent equilibrium processes that helped structure fish assemblages in floodplain lakes also caused differences among lakes. The assemblage of turbid lakes was dominated by catfishes and knifefishes that are nocturnal and adapted to poor visibility conditions. In comparison, Characiformes and cichlids dominated more transparent lakes. While the species assemblage differed between lakes, trophic composition did not, because each trophic role was taken up by another species. This adds support to the equilibrium concept for floodplain lakes.

Rodriguez and Lewis hypothesized that fish assemblages in the Orinoco River were regulated by three mechanisms: (1) site-dependent culling of species by predators, (2) habitat selection by young-of-year and yearling fish in receding waters, and (3) selection of spawning sites by mature females. This supports the concept of Goulding that piscivory modifies relative abundance of species during the dry season, but in this view, piscivory does not cause stochastic change in assemblages but rather eliminates surplus individuals or species less capable in a particular floodplain lake.

A similar evaluation was done by Fitzgerald et al. (2017) using different methods. They conducted fish collections along a 400 km stretch of the Xingu River, a tributary to the Amazon. They then determined the functional trait distributions (FTDs) of fish assemblages throughout the region. Functional traits are characteristics of organisms that influence their fitness, and changes in these traits of an assemblage indicate ecological processes that may influence assemblages. In this study, Fitzgerald et al. determined the functional traits of 41 species from nine families which represented most of the fishes present in these river reaches. They evaluated 19 traits related to habitat use and swimming ability, such as fin dimensions and body depth, as well as 23 morphological traits related to feeding behavior, including mouth width and gut length. They then used the FTDs to evaluate whether wet season assemblages are more similar than dry season ones, and whether FTDs were different from random. In this context, more similar FTDs indicate assemblages more likely structured by equilibrium mechanisms related to local ecological conditions, while less similar FTDs may indicate the importance of stochastic processes in assemblage structure.

Fitzgerald et al. found that FTDs differed between the wet and dry seasons for both habitat and trophic traits (Figure 29-5). Wet season assemblages were more tightly clustered compared to dry season assemblages when comparing FTDs of the mean nearest neighbors (how close species are in niche space), standard deviation of the nearest neighbors (how evenly species are spaced), or functional dispersion (position relative to the mean niche space). They also found that the wet season assemblages had significantly lower values of functional richness (overall niche volume). They believe that the wet season assemblages show more equilibrium

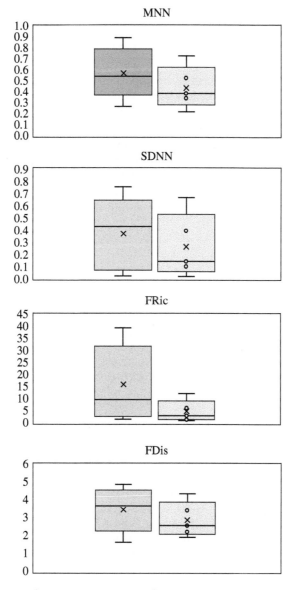

FIGURE 29-5 Comparison of observed FTD of local fish assemblages during dry and wet seasons based on traits related to trophic strategy. FTD metrics include mean nearest neighbor distance (MNN), standard deviation of nearest neighbor distance (SDNN), functional richness (FRic), and functional dispersion (FDis). *Source: Adapted from Fitzgerald et al. (2017).*

processes with habitats being connected and open due to flooding that occurs then and fish accessing their preferred habitats. In the dry season, when habitats become fragmented and movements among habitats limited, they hypothesize stronger stochastic influences on distribution or nonequilibrium processes. This FTD analysis demonstrates that characteristics of the whole assemblage are affected by both density-dependent (competition and predation) and density-independent (flooding) mechanisms, and these vary during different seasons.

The equilibrium–nonequilibrium argument continues for species assemblages in tropical rivers. Winemiller and Jepsen (1998) present evidence supporting both processes, and it is likely that the characteristics of the river, lake, and flooding cycle cause different mechanisms of control in different tropical systems.

Summary

This chapter on tropical rivers emphasizes the importance of equilibrium and nonequilibrium processes in establishing a diverse tropical fauna. Equilibrium processes of evolution and specialization of feeding niches have led to higher species diversity in the African rift lakes and other tropical fish communities. Similar adaptations of fruit-eating and biting predators have led to diversification of the Amazon fish fauna. However, seasonal flooding and recession of water in the Amazon basin also provide nonequilibrium processes by chance events. Isolation of differing fish groups in floodplain lakes may allow persistence of some groups and reduce dominance of others, further leading to increased species diversity. However, habitat conditions in these systems may also result in more equilibrium processes and stability in assemblage structure over time.

Literature Cited

Albertson, R.C., J.A. Markert, P.D. Danley, and T.D. Kocher. 1999. Phylogeny of a rapidly evolving clade: The cichlid fishes of Lake Malawi, East Africa. Proceedings of the National Academy of Sciences 96:5107–5110.

Anderson, J.T., T. Nuttle, J.S. Saldaña Rojas, T.H. Pendergast, and A.S. Flecker. 2011. Extremely long-distance seed dispersal by an overfished Amazonian frugivore. Proceedings of the Royal Society B 278:3329–3335.

Correa, S., K. Winemiller, H. López-Fernández, and M. Galetti. 2007. Evolutionary perspectives on seed consumption and dispersal by fishes. BioScience 57:748–756.

Fitzgerald, D.B., K.O. Winemiller, M.H. Sabaj Pérez, and L.M. Sousa. 2017. Seasonal changes in the assembly mechanisms structuring tropical fish communities. Ecology 98:21–31.

Goulding, M. 1980. The Fishes and the Forest. University of California Press, Berkeley.

Junk, W.L., P.B. Bayley, and R.E. Sparks. 1989. The flood pulse concept in river-floodplain systems. Pages 110–127 in D.P. Dodge, editor. Proceedings of the International Large River Symposium. Canadian Special Publications of Fisheries and Aquatic Sciences 106, Fisheries and Oceans Canada, Ottawa.

López-Fernández, H., J.H. Arbour, K.O. Winemiller, and R.L. Honeycutt. 2013. Testing for ancient adaptive radiations in Neotropical cichlid fishes. Evolution 67:1321–1337.

Lowe-McConnell, R.H. 1977. Ecology of Fishes in Tropical Waters. Edward Arnold Ltd., London.

Mooi, R.D., and A.C. Gill. 2002. Historical biogeography of fishes. Pages 43–68 in P.J.B Hart and J.D. Reynolds, editors. Handbook of Fish Biology and Fisheries, Vol. 1 Fish Biology. Blackwell Science, Ltd., Oxford, UK.

Nelson J.S., T.C. Grande, and M.V.H. Wilson. 2016. Fishes of the World, 5th Edition. John Wiley & Sons, New York.

Rodriguez, A., and W.M. Lewis, Jr. 1994. Regulation and stability in fish assemblages of neotropical floodplain lakes. Oecologia 99:166–180.

Schlosser, I.J. 1991. Stream fish ecology: A landscape perspective. Bioscience 41:704–712.

Smith, N.J.H. 1981. Man, Fishes, and the Amazon. Columbia University Press, New York.

van der Sleen, P., and J.S. Albert, editors. 2017. Field Guide to the Fishes of the Amazon, Orinoco, and Guianas. Princeton University Press, Princeton, New Jersey.

Winemiller, K.O. 1991. Ecomorphological diversification of freshwater fish assemblages from five biotic regions. Ecological Monographs 61:343–365.

Winemiller, K.O., and D.B. Jepsen. 1998. Effects of seasonality and fish movement on tropical river food webs. Journal of Fish Biology 53(A):267–296.

Winemiller, K.O., and D.B. Jepsen. 2004. Migratory neotropical fish subsidize food webs of oligotrophic blackwater rivers. Pages 115–132 in G.A. Polis, M.E. Power, and G.R. Huxel, editors. Food Webs at the Landscape Level. University of Chicago Press, Chicago.

Zeng, N. 1999. Seasonal cycle and interannual variability in the Amazon hydrologic cycle. Journal of Geophysical Research 104:9097–9106.

CHAPTER 30

Fish Communities in Lakes

Inland bodies of water are generally differentiated as being either lotic or lentic. Fish communities in lotic environments – systems characterized by running waters – were discussed in the previous two chapters. This chapter focuses on fish communities in lentic environments – systems characterized by standing waters. There are a number of different types of lentic systems, including wetlands, ponds, and lakes. Here, we specifically focus on fish communities in lakes (relatively large, deep lentic systems) – given the large percentage of surface freshwater contained in these systems, the large number of fish species in lakes, and the long history of researchers studying lake dynamics and the fish that occupy them.

Lakes are highly variable. They differ in surface area, depth, geological age, position in the landscape, and whether they contain fresh or saline water. Most lakes are formed naturally through glacial, tectonic, or volcanic activities, portions of lotic systems becoming isolated (e.g. oxbow lakes), landslides damming a valley, as well as dissolution of underground rock material or meteor impacts creating new basins. However, lakes can also be formed artificially, and humans have created reservoirs for various purposes. Not all lakes contain fish; for example, some high altitude lakes are fishless. However, most lakes either naturally contain fish or support fish that were introduced by humans.

The composition and dynamics of fish communities in lakes are influenced by the same general processes described in Chapters 25 and 26, including biotic interactions, colonization, local extirpation, speciation, and adaptation and acclimation to environmental conditions. Lakes are likely to contain more fish species if they are larger, connected to other aquatic systems, experience intermediate levels of disturbance, and are geologically older, as more time allows for greater colonization and speciation. Many of these concepts are described elsewhere in this book, and we do not revisit all aspects of lake fish community structuring in this chapter. We highlight some key attributes of lakes that influence their abilities to support different fish species, discuss the influence of bottom-up and top-down processes in structuring lake fish communities, and emphasize the attributes and connectivity of lakes influencing colonization and extinction rates determining richness and species composition.

Lake Zones

Lakes consist of various zones with distinct structural, physical, and chemical attributes and with differential access to prey and susceptibility to predators. The depth at which light penetrates into a lake has important implications for primary producers as well as visual foraging

Biology and Ecology of Fishes, Third Edition. James S. Diana and Tomas O. Höök.
© 2023 John Wiley & Sons Ltd. Published 2023 by John Wiley & Sons Ltd.

fish species. The euphotic zone of a lake is the top portion of a lake, from surface to a depth at which light penetrates. This is often defined as the depth at which photosynthetic active radiation is 1% of that at the surface. The nearshore portion of the euphotic zone is known as the littoral zone and is often characterized by aquatic plants and algae, including emergent, floating, and submerged macrophytes. Terrestrial debris, such as coarse woody structures, also is common in the littoral zone. This highly structured nearshore zone is used for reproduction by many fish species, and many young fish take advantage of the nearshore structure, both as a source for prey and cover from potential exposure to predators. In contrast to the littoral zone, the more offshore, open water volume of the euphotic zone is called the limnetic zone. Certain species of fish thrive in this open environment, but there is potentially more exposure to predators. Also, foragers do not have access to prey that inhabit bottom sediments in the limnetic zone. Therefore, relatively strong swimming fish – planktivores and piscivores – tend to dominate in the limnetic zone. Below the limnetic zone is the profundal zone, where light levels are much lower, and at the bottom of the profundal zone is the benthic zone, where water and bottom sediments interface. The benthic zone often provides some structure in the form of boulders and rocks. Fish species that are not particularly strong swimmers can essentially rest on bottom sediments in the benthic zone. Some benthic fishes have very poorly developed or entirely absent swim bladders. Also, in the benthic zone, piscivorous fish may forage on smaller, benthically oriented fish species, while invertivores can access benthic invertebrate prey that are often much larger than the relatively small zooplankton found in the limnetic zone.

In addition to lake zones defined by light penetration, lake zones are often differentiated thermally. Vertical stratification related to temperature is described in Chapter 1 and only briefly reviewed here. Lakes that are sufficiently deep in temperate regions will thermally stratify during the warm growing seasons. The timing of stratification will vary among lakes and years but may last from late spring to early fall. Stratification generally develops and persists due to density differences of water at different temperatures. During summer stratification, warm, less dense water rests on top of cold, more dense water. The warm, top layer is known as the epilimnion, while the cold bottom layer is called the hypolimnion, and the transition layer is referred to as the metalimnion. The thermal characteristics of these layers, in addition to their light and structural attributes, imply that different fish species will perform better in the different layers. For example, species with unsustainable metabolic demands in warm temperatures may only be able to persist in the hypolimnion, while species with very low activity potential in cool temperatures may be limited to the epilimnion. Deep, stratified lakes may contain both warm and cold water species. Through thermal stratification, cold water species are able to persist in temperate regions. That is, without maintenance of cold, hypolimnetic waters, many cold water species would be restricted to much higher latitudes.

Some fish species in lakes display diel vertical migration (DVM), moving vertically between or within depth zones on a daily basis. This behavior generally involves moving up in the water column at dusk, remaining high in the water column during night and returning to deeper depths during dawn. Several mechanisms leading to the development of DVM behavior have been proposed, but the mechanisms underlying DVM are not always clear. Thomas Mehner from the Leibniz-Institute of Freshwater Ecology and Inland Fisheries reviewed some of the processes that may lead to DVM in fishes (Mehner 2012). They considered both proximate triggers of DVM and the ultimate benefits that may lead to this type of behavior. The main cue that appears to lead fish to initiate upward or downward migration is light and changes in light level at depth during different parts of the day. Other factors, such as water temperature and pressure, may also guide these vertical migrations, in particular influencing when fish may stop their upward or downward movement. Interestingly, Mehner's exploration of the benefits of DVM focused on growth potential as well as predator avoidance. Fish are not the only aquatic organisms that display DVM; many species of micro- and macrocrustaceans also move up and down in water column diurnally. Therefore, fish DVM may partially reflect tracking of prey movement, as planktivorous fish track their zooplankton prey, and piscivorous fish track fish prey. Fish may also gain a bioenergetic growth advantage by moving between depth layers. Essentially, this is based on the notion that fish that feed in warmer temperatures – higher in the water column – may be able to eat more, and fish that occupy cooler temperatures – lower in the water column when they are not feeding – may burn less energy through metabolism. DVM may allow fish to increase prey consumption potential without a huge cost in terms of metabolic energy losses (see earlier chapters on bioenergetics). Vulnerability to visual predators is directly tied to light levels. During the day, visibility will be lower in deeper waters, and occupation of such waters at this time may decrease risk of predation. However, at night there is less risk of moving higher in the water column as visibility will also be low at shallow depths at this time. Many fish that display DVM likely do so as a balance between growth potential and predation risk. While decreased light levels may decrease potential foraging success for both planktivorous and piscivorous fishes, planktivores may be able to maintain relatively effective foraging at lower light levels than piscivores. By moving higher in the water column at dusk, small, planktivorous fish may be able to feed on their migrating invertebrate prey without an equivalent increased risk of predation.

While DVM is by far the diurnal migration behavior that has received the most attention for fish in lakes, there is evidence that fish also display other types of migration behavior. For example, fish may display diurnal horizontal migration (DHM), where they move between the limnetic and littoral zones during day and night. There are examples of fish migrating from the littoral zone during the day to the limnetic zone at night, and vice versa, moving from the limnetic zone during day to the littoral zone at night. For example, golden shiner have been documented to migrate from the littoral zone during the day and then move to the limnetic zone at night (Hall et al. 1979). There are also examples of fish in lakes displaying diurnal bank movement (DBM), where they move along the benthic layer from deep hypolimnetic waters to more shallow metalimnetic or epilimnetic waters while remaining along the bottom of a lake. Thereby, DBM is simultaneously an example of DVM and DHM. Species such as burbot, lake whitefish, and lake trout have all been documented to display DBM in lakes (Gorman et al. 2012; Cott et al. 2015.). Similar to DVM, DHM and DBM are seemingly related to enhancing foraging and growth opportunities, while minimizing risk of predation.

The different zones of lakes provide distinct habitat conditions and may support different species and life stages of fish that thrive in specific conditions. Strong, cruising swimmers that forage on plankton or smaller fish may thrive in the limnetic zone. In the more structure-rich littoral zone, one may be more likely to encounter fish able to effectively maneuver through structure, ambush predators, and small or young fish that require cover from predation. The thermal layers of a lake also create environments where species of distinct thermal guilds (warm water, cool water, and cold water) are differentially likely to inhabit. These zones are not entirely disconnected and, as demonstrated by DVM, DHM, and DBM behaviors, several species of fish will move between zones during a diurnal cycle. Given that distinct lake zones represent environments where fishes with different traits may thrive, lakes with more distinct zones and larger zones may on average support larger populations and a greater number of fish species.

Lake Productivity

Lakes are frequently differentiated in terms of their productivity levels. Lakes are naturally expected to become more productive and shallower over time as they gradually accumulate more material from the atmosphere and surrounding landscapes. Naturally, these processes would be expected to progress very slowly, but in many lakes around the world, human activities have greatly accelerated the transition of lakes to more productive states. Through natural and artificial processes, lakes exist along a continuum of productivity, from more productive, eutrophic lakes to less productive, oligotrophic lakes. At the middle of this continuum, mesotrophic lakes are intermediately productive, and at the extreme ends of the continuum, lakes have been described as hyper-eutrophic or hyper-oligotrophic. Through natural processes, lakes would be expected to transition very gradually between these states, but extreme nutrient and sediment loading humans have accelerated the lake transitions toward more eutrophic conditions – cultural eutrophication. More recently, with nutrient control programs and enhanced filtering by native, non-native, and anthropogenically propagated filter feeders, some culturally eutrophied lakes have experienced a transition in the opposite direction – a phenomenon known as oligotrophication.

There is not a single variable that defines where lakes fall along the eutrophic–oligotrophic continuum. Eutrophic lakes generally have relatively high inputs and concentrations of nutrients, such as phosphorous. So, they are more productive (greater chlorophyll concentrations and higher rates of photosynthesis per unit volume), and due to high algal densities and potentially high sediment loads, they have relatively low water clarity. The high productivity of eutrophic lakes may be accompanied with depleted oxygen levels. In contrast, oligotrophic lakes have lower nutrient inputs and concentrations, lower primary productions rates, greater water clarity, and are usually well oxygenated. Oligotrophic lakes are often deeper than eutrophic lakes, and lakes at higher latitudes and greater altitudes are more likely to be oligotrophic.

Differential conditions in eutrophic and oligotrophic lakes strongly influence the organisms that dominate in these different types systems. Eutrophic lakes do not only contain greater densities of planktonic algae, they are often dominated by certain types of algae, such as bluegreen algae – cyanobacteria. Low nutrient concentrations in oligotrophic lakes may limit both planktonic algae and rooted macrophytic plants, while low water clarity in eutrophic lakes may somewhat limit densities of macrophytes as sufficient light may penetrate to the bottom of only a small portion of such lakes to allow for rooted plant growth. The types of invertebrates in the water column and on the bottom of lakes will also vary between eutrophic and oligotrophic lakes, as eutrophic lakes may favor organisms able to feed in poor visibility and tolerate low oxygen levels (e.g. midge larvae), while oligotrophic lakes may favor invertebrates that are better swimmers and those that require higher oxygen levels (e.g. mayfly larvae). The composition of primary producers and invertebrates has implications for how fish forage in these different types of systems.

Similar to invertebrates, physical, chemical, and biological conditions in lakes will directly affect the relative performance of different fish species. Some fish species are sensitive to poor water quality as water is often described for eutrophic lakes. Fish, such as most members of the Salmonidae family (trout, char, and whitefish), are unable to tolerate low oxygen conditions and do not forage well under poor visibility. In contrast, other species, such as many members of the North American families Centrarchidae (freshwater sunfish and bass) and Ictaluridae (catfish), can tolerate lower oxygen levels and/or forage effectively under minimal visibility.

The potential for low oxygen (hypoxia) to affect fish communities in lakes is related to lake zones described above. During summer stratification, the top, warm epilimnion is separated from the bottom, cool hypolimnetic layer. The algae produced during massive surface blooms, characteristic of eutrophic lakes, will ultimately die and settle to the bottom of lakes. This organic material will then decompose in the bottom layer. Decomposition will use up oxygen, and due to the density separation between the hypolimnion and surface atmosphere, depleted oxygen in the bottom layer cannot readily be replenished. Therefore, surface algal blooms in productive lakes can ultimately lead to hypoxic (or even anoxic, i.e. the complete absence of dissolved oxygen) conditions in cool bottom layers. Lakes that would normally be able to support both warm and cold water species during the summer stratified period may lose cold water species as they are unable to survive in the warm epilimnion or the oxygen-deprived

426 CHAPTER 30   Fish Communities in Lakes

hypolimnion. Such losses of cold water species have been particularly evident though cultural eutrophication. For example, at the beginning of the 20th century, cold water cisco inhabited at least 50 glacial lakes in the US state of Indiana. However, by 2012, they were only present in six of these lakes, and these losses appear to be directly related to cultural eutrophication leading to hypoxic conditions in lake hypolimnia and local extirpations of cisco (Honsey et al. 2016).

Bottom-Up and Top-Down Control

As mentioned in Chapter 26, when describing food webs, aquatic ecologists have highlighted the importance of both primary productivity and predatory control in structuring the biomass and productivity of different food web components and trophic levels. This is especially true for food webs in lakes. Various studies have emphasized the extreme influence of bottom-up processes – relative rate of primary production – in determining biomass and productivity of higher trophic levels in lakes. However, a multitude of other studies point to top-down predatory control as the primary determinant of biomasses and size structures at different trophic levels. In the past, researchers debated whether bottom-up or top-down control is more important and which process should be the focus of research and management. This is seemingly a false dichotomy. Both bottom-up and top-down influences are important for structuring food webs. The relative importance of these different processes may vary depending on the time and space scales being considered and what aspect of a food web is being considered. For example, if comparing total fish biomass across a range of lakes with vastly different nutrient loads and rates of primary productivity, bottom-up processes may trump within-system top-down influences. However, within an individual lake, and especially over short time frames, the biomass of a particular trophic level may be strongly affected by top-down predatory control by the trophic level directly above.

Bottom-up Effects

Limnologists traditionally believed that resource limitation was the main process affecting productivity of lakes. The availability of nutrient resources, such as phosphorus or nitrogen, is believed to set the rate of primary production and biomass of phytoplankton. The availability of phytoplankton limits herbivorous zooplankton, and such interactions pass on through each trophic level. The resource limitation concept was formalized by Lindeman, a limnologist from University of Minnesota, as the trophic dynamics of aquatic ecosystems. In Lindeman's view, trophic pyramids of lakes were controlled by resource quantity, and a lake ecosystem was ultimately limited by primary production based on availability of nutrients and light (Lindeman 1942). In particular, primary productivity of most lakes is limited by phosphorous concentrations. This has been demonstrated across a variety of systems and was convincingly demonstrated by David Schindler (Schindler 1974), while working in the Experimental Lakes Area of western Ontario, Canada. One of the lakes in this area (Lake 226) has two basins separated by a narrow section. Schindler separated the two basins by having a nylon curtain installed in this narrow section. They then added carbon and nitrogen (sucrose and nitrate) to one basin of Lake 226 and nitrogen, carbon, and phosphorous to the other basin. While productivity did not change in the basin receiving only C and N, the basin receiving C, N, and P experienced a rapid increase in primary production.

Resource limitation has found much support in limnology and fisheries management. Regional scale studies of lake ecosystems usually indicate positive correlations among phosphorus concentration, and the biomass of phytoplankton, zooplankton, and fish (Figure 30-1).

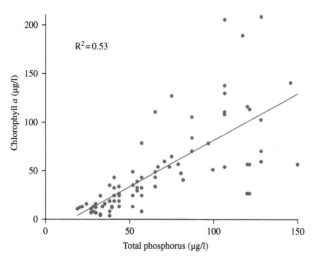

FIGURE 30-1 Relationships between chlorophyll a and total phosphorus concentrations during summer in lakes. *Source: Adapted from Shapiro (1979).*

Such correlations, while not proving cause and effect, indicate that increased abundance of limiting resources (nutrients and food) are often correlated to increased fish biomass. In fact, relating fish yield to indices of lake productivity, such as Ryder's morphoedaphic index (total dissolved solid concentration ÷ mean depth of lakes), build from this bottom-up perspective and are often successful as a first tool for predicting and managing fish yield in lakes (Ryder 1965). These correlations between nutrients and upper trophic level productivity, while significant, are often weak, indicative of the high degree of variation inherent in aquatic ecosystems.

Predatory Effects in Lake Communities

In the early 1960s, several scientists examined the effects of fish predators on the community of plankton that developed in a lake. John Brooks and Stanley Dodson from Yale University compared species composition and size of zooplankton in Crystal Lake, Connecticut, prior to 1942, when no alewife were present, and in 1964, after alewife were introduced (Brooks and Dodson 1965). Prior to alewife introduction, the average size of zooplankton was approximately 0.8 mm in length, and the community was dominated by *Diaptomus*, *Daphnia*, and *Mesocyclops* – all large-bodied zooplankton (Figure 30-2). By 1964, after alewife introduction, the zooplankton community of Crystal Lake showed a large shift. Average body size declined to 0.3 mm, and the species composition shifted to *Bosmina* and *Tropocyclops* (small-bodied plankton) instead of the larger *Daphnia*. Brooks and Dodson also compared zooplankton size distributions across multiple Connecticut lakes with or without landlocked populations of alewife. The lakes with alewife essentially lacked large zooplankton, such as *Daphnia*. These patterns were seemingly related to alewife predation of large-bodied zooplankters, and Brooks and Dodson's surveys demonstrated the potential of fish predation to have a top-down influence on size structure of a lower trophic level. While Brooks and Dodson's study has received more attention and has been more widely cited, it is worth noting that work by Jaroslav Hrbáček and colleagues from Czechoslovakia Academy of Sciences' Hydrobiology Laboratory documented similar patterns and interactions between fish and plankton in Czech lakes (Hrbáček et al. 1961) a few years earlier than Brooks and Dodson's study.

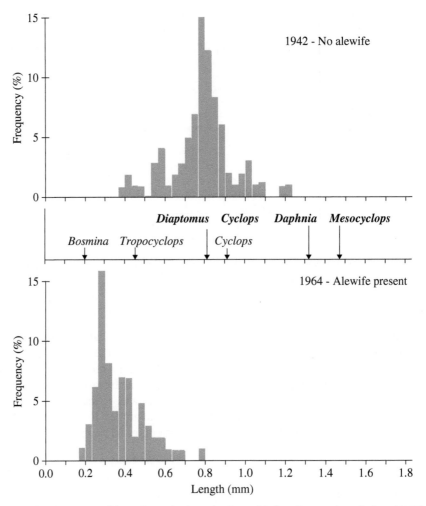

FIGURE 30-2 The size composition of zooplankton in Crystal Lake, Connecticut, before (1942) and after (1964) alewife establishment. The arrows and names represent the sizes of each dominant species in the plankton; bold lettering: dominants in 1942; normal lettering: dominants in 1964. *Source: Adapted from Brooks and Dodson (1965).*

The Trophic Cascade

The fact that predators cause changes in their prey populations, both in abundance and in species composition, led Steve Carpenter, Jim Kitchell, and others in a research group from the University of Wisconsin and Notre Dame to the next step in hypothesizing how those changes could affect community structure and flow of energy in the entire aquatic ecosystem (Carpenter et al. 1985). They used surplus production (described earlier) as the basis of their model, which indicates that the highest level of production occurs at an intermediate biomass. Cropping – either by harvesting or predation – releases compensatory effects and results in a higher production than would occur in absence of cropping.

Carpenter et al. extended this argument to the entire trophic pyramid. They reasoned that an increase in a piscivore should result in a decrease in biomass of planktivorous fishes because of predatory consumption. While this might result in increased production at intermediate prey biomass, higher piscivore biomass should generally result in lower biomass of planktivores, and higher biomass of herbivorous zooplankton, because there are fewer or smaller zooplanktivores foraging on herbivorous zooplankton. Ultimately, the abundance of phytoplankton should be lower too. With this alternating series of steps, increase in biomass at one trophic level is hypothesized to result in a decline at the next lower trophic level, an increase at the next trophic level, and this continues down through the entire food web. This concept is called trophic cascade and may be considered a top-down perspective (consumer controlled rather than resource controlled). The interesting part of this concept is not the effect of predators on their prey, which had already been established, but the effect of predators on trophic levels below that of their prey.

Based on the trophic cascade, Carpenter et al. made predictions of changes in an aquatic community. For example, a strong year class of piscivores becoming established in a lake would be planktivores early in life and then have a gradual effect on piscivore biomass as they grew (Figure 30-3). The long-term effect would be to gradually reduce vertebrate planktivores, increase invertebrate planktivores and large herbivorous zooplankton (prey of

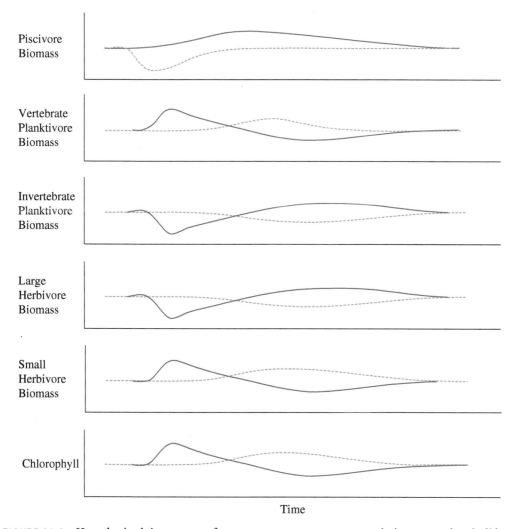

FIGURE 30-3   Hypothesized time course of ecosystem response to a strong piscivore year class (solid line) or a partial winterkill of piscivores (dashed line). *Source: Redrawn from Carpenter et al. (1985).*

the vertebrate planktivores), and decrease small herbivore zooplankton and phytoplankton biomass (prey of the invertebrate planktivores) over time. The system would then return to its initial state as the piscivores decline. As a second example, winterkill could remove large piscivores and have a reverse effect that would occur much more rapidly in the piscivores but more slowly on other trophic levels. Each alteration would have a different time course and a different result. One interesting part of this prediction is that vertebrate planktivores should have their strongest influence on large zooplankton. As demonstrated by Hrbáček et al. and Brooks and Dodson, certain fishes may selectively feed on large plankton and affect the balance between large and small plankton. In comparison, invertebrate planktivores selectively feed on small plankton, so changes in vertebrate or invertebrate planktivore density will likely differentially affect the abundance of large and small zooplankton not just total zooplankton.

A trophic cascade pattern may be more likely to be evident in relatively simple food webs where predators do not feed across multiple trophic levels. However, aquatic food linkages are generally not linear food chains but web like, without a straight-line relationship between one planktivorous fish eating one species of zooplankton, which eats one species of phytoplankton. High degrees of omnivory and prey switching (e.g. through ontogeny) may greatly diminish the potential for trophic cascades. In most lakes, there is not a simple one-to-one direct relationship among different trophic levels. A piscivore might eat vertebrate zooplanktivores, vertebrate herbivores, and occasionally invertebrate herbivores or carnivores. Moreover, virtually all top carnivores eat invertebrates at young ages and later switch to fishes. Thus, while strongest effects of a piscivore may be evident for its main fish prey, some effects may also be evident for invertebrates. There is an obvious sequential relationship between the abundance of young and adult predators, with a strong cohort of young predators likely subsequently leading to large numbers of adult predators. As young and adult predators may crop different trophic levels in a lake, this can further dampen potential trophic cascades.

While trophic cascades may not be evident in all lake systems, the hypothesis has obvious and important implications for lake management and fisheries. The use of fishes to manipulate natural lakes for management purposes is an interesting concept, and at least three related ideas on this manipulation have been proposed: biomanipulation, top-down management, and the trophic cascade (see Carpenter and Kitchell 1992; DeMelo et al. 1992 for more details). Tests of the trophic cascade concept have attempted to determine the extent of the cascade in natural or artificial settings. If adding piscivorous fishes and/or reducing planktivorous fishes could dramatically reduce algal biomass in a lake, it would be a very attractive technique for lake management in many systems burdened by cultural eutrophication.

Trophic Cascade in Lakes Peter and Paul

Carpenter, Kitchell, and colleagues conducted much research manipulating the ecosystem of several lakes at the University of Notre Dame Environmental Research Center in Wisconsin and Michigan. This research produced information on changes in manipulated lakes and resulting variability in the community of organisms at different trophic levels. The result of these experiments has been evaluated in Carpenter and Kitchell (1993) and particularly by the synthesis of Kitchell and Carpenter (1993).

Experiments adding or removing predators have shown that piscivores had rapid and massive effects on planktivores in these lakes. This effect was both direct (consumption) as well as indirect (predator avoidance by small fishes). In the latter case, introduction of a large predator not only resulted in the predatory removal of some soft-rayed species but also migration of some prey species out of the lake and into tributary streams where a predation refuge existed (He and Kitchell 1990). These changes in planktivorous fish resulted in changes in zooplankton and phytoplankton populations. Sometimes these changes reflected expectations from the trophic cascade, and other times they did not. The overall result of these studies demonstrated that food webs of lakes are complex and, on occasion, behave in ways predictable by theories such as trophic cascade. Some unexplained variance in the relationship between lake productivity and fish biomass was attributable to food web effects. Since phosphorus was the major limiting nutrient in these lakes, examination of the amount of phosphorus bound in organisms from different trophic levels gave some idea of the possible influence of organisms on phosphorus dynamics. Most phosphorus in these lakes was tied up in organisms rather than in water and sediments. As much as 50% of the total phosphorus was contained in predatory fish and invertebrates. Changes in abundance of such predators made a large difference in the nutrient budget of the lake and the resulting phytoplankton communities.

Biomanipulation of Shallow European Lakes

The practice of manipulating planktivorous fish biomass to affect herbivorous zooplankton and ultimately algal biomass has been attempted in many systems, including shallow, temperate European lakes. Lars-Anders Hansson from Lund University and collaborators from other European institutions provided a synthesis of some of the perspectives gained from such manipulations (Hansson et al. 1998). While these authors suggest that reduction of phytoplankton through trophic cascade processes is possible, they highlight the complexity of such effects and the conditions most likely to lead to successful biomanipulation. They emphasize that biomanipulation is unlikely to affect only the sequential food chain from fish to zooplankton to phytoplankton, but will also have other effects, and these other effects should also be considered. For example, reduction of adult fish biomass can allow for increased production and recruitment of young fish, which can confound biomanipulation over both the short and long terms. They emphasize that for many successful biomanipulations, not only did phytoplankton biomass decrease, but macrophyte biomass and cover increased. As described in Chapter 26, phytoplankton and macrophytes can act to counter each other. Lakes with high densities of phytoplankton are generally turbid, leading to low light penetration, which may limit macrophyte establishment. However, if macrophytes become established, they may limit phytoplankton biomass by making nutrients less available to phytoplankton and by contributing to increased water clarity. Macrophytes serve to limit the resuspension of sediment leading to increased water clarity and further establishment of macrophytes. Therefore, biomanipulation may be most successful if the environment allows for establishment of macrophytes.

Hansson and colleagues also point to conditions that may limit successful biomanipulation. First, if nutrient loading is very high, it is unlikely that manipulation of fish biomass will have demonstrative effects on phytoplankton biomass. With very high nutrient loading, phytoplankton production is expected to remain high and not possible to control through predation. Second, benthic feeding fish may upset intended biomanipulation. Trophic cascades are generally considered as a set of pelagic processes, where planktivorous fish affect herbaceous zooplankton, which in turn affect phytoplankton. However, high numbers of benthic feeding fishes may disrupt such intended effects by contributing to increased sediment resuspension, reduced light penetration, and inability for macrophytes to establish. In addition, benthic feeding fish can serve to transfer nutrients contained in the sediments to the water column, where they are more available to phytoplankton. Hansson et al. argue that biomanipulation to control eutrophic conditions is most likely to be successful if macrophytes can readily establish, recruitment of young-of-year fish is considered, external nutrient loading is limited, and benthic feeding fish are also controlled.

Is Bottom-up or Top-down Regulation More Influential in the Laurentian Great Lakes?

While it is now accepted that both bottom-up and top-down processes can contribute to trophic structure in lakes, studies still assess which of these processes is more influential. David "Bo" Bunnell from the US Geological Survey collaborated with a number of other researchers to consider the relative influence of these processes in the Laurentian Great Lakes (Bunnell et al. 2014). They compiled annual data (1998–2010) collected by various agencies across multiple trophic levels from each of the five Great Lakes. During the time frame they considered, the Great Lakes had experienced a pronounced decrease in nutrient loading. In addition, during the 1980s and 1990s, invasive zebra and quagga mussels established and proliferated in the Great Lakes. Intensive filtering by these bivalves targeted phytoplankton and other lower trophic level organisms. The combined reduced nutrient input and mussel filtering led to a reduction in primary producer biomass and potential for evidence of bottom-up effects. Many of the Great Lakes experienced an increase in top predator biomass due to stocking, increased successful natural reproduction, control of parasitic sea lamprey, and changes in harvest. Therefore, top-down predator control could become more intense during the period Bunnell and colleagues considered.

Within each lake (and for three separate basins in Lake Erie), Bunnell and colleagues examined annual correlations (1998–2010) between adjacent trophic levels (phosphorous to chlorophyll concentrations to zooplankton densities to prey fish biomass to piscivore biomass). If bottom-up processes were influential, they expected to see positive associations between adjacent trophic levels – consistent with notion that increased biomass at lower trophic level supports increased biomass at higher trophic level – whereas they expected to see negative associations between adjacent trophic levels if top-down effects are more influential – consistent with potential for increased biomass of higher trophic level to contribute to decreasing biomass of trophic level below. They found that most associations between adjacent trophic levels were positive. Eleven significant correlations were consistent with bottom-up regulation, and only one correlation was consistent with top-down regulation. Evidence for bottom-up influence was particularly strong for Lake Huron – a lake that experienced particularly strong reductions across all trophic levels during this period.

It is inappropriate to consider lake food web regulation as dichotomous – either top-down or bottom-up regulated. Over the period Bunnell and colleagues considered, they found a positive association between the annual biomass of lower trophic levels and upper trophic levels. This is demonstrative of the importance of bottom-up transfer of energy and carbon. However, this does not mean that predatory control is entirely unimportant. In fact, a multitude of studies in the Laurentian Great Lakes has pointed to both direct and indirect effects of predators on prey. Again, bottom-up and top-down processes occur simultaneously, and the relative influence of these different processes may vary, depending on the time and space scales being considered and what aspect of a food web is being considered.

Immigration and Extinction in Lake Communities

Island Biogeography

The concept of island biogeography was developed for oceanic islands, but any kind of isolated habitat can be viewed as an island. Lakes – especially those isolated from one another due to the absence of connecting streams or rivers – are similar to islands. Many ecosystems become fragmented and can be viewed as islands, at least in their geographic distributions. The key component of island biogeography is that these habitats are cut off from other similar areas by dissimilar habitats.

One can generalize on the fauna of islands, and indeed, island biogeography originated with such generalizations (MacArthur and Wilson 1967). The farther an island is from shore, the lower number of species is usually found there. This will not necessarily result in fewer individuals within a species on the island because abundance also depends on the productivity and diversity of habitats on the island. Also, larger islands generally have more species than smaller islands. The community on an island today might also be very different than it will be in 100 years. The proportions of species may shift and new species may colonize, or old ones may become extinct. Islands often do not appear to be in equilibrium, and their faunas are likely to continually change.

These three observations led to the concept of island biogeography. The animal community on an island is not hypothesized to reach an equilibrium in which those species remain there indefinitely, but dynamic changeovers of species are believed to occur. If there is a continual turnover on an island, there is a balance between new species immigrating to an area and the local extinction of existing species in the area.

MacArthur and Wilson proposed how an island becomes inhabited and believed that, over time, the number of species on the island would increase toward an equilibrium number of species (S). MacArthur used a series of simple drawings to illustrate the factors influencing S (Figure 30-4). These relationships have been depicted as straight lines and curves, but for our purposes, this matters little because the concept is more important than the shape of the curve. Since a population colonizes an island by establishing individuals on the island, immigration and extinction rates for species should determine equilibrium number of species. Once a species becomes established on an island, it can no longer colonize, so rate of colonization should decrease with increasing number of species in the community. Colonization (or "immigration" in MacArthur's terms) should decrease with number of species present or with time. However, animals cannot become extinct until they are established, so the number of species becoming locally extinct should increase with the number of species present. As species colonize, local extinction rates should increase because there are more species present. When the number of species becomes very high, competition, competitive displacement, and predation (or trophic cascade) might drive a species to extinction even though that species persisted on the island prior to arrival of its predators or competitors. The equilibrium number of species on the island would be the point where colonization and extinction rates cross.

Characteristics of islands influence these rates of colonization and extinction. Islands farther offshore tend to have lower numbers of species than islands closer to the mainland because of the greater distance over which immigration must occur. Thus, colonization rates are expected to be directly related to island distance. Island size also influences the number of species present. Large islands generally have more species than small islands, even if they are the same distance from shore. Larger islands would be expected to be more stable and able to support larger populations. The risk of extinction should be relatively lower for large populations compared to small populations. Also, larger islands usually have more habitats to occupy, and by having more habitats, they offer more opportunities for different species to coexist once they become established. A smaller island would be expected to have a much higher extinction rate than a larger island because there are fewer available habitats and it would generally have populations of smaller sizes which are also more likely to become extinct.

While the isolation of islands affecting colonization rates and the size of islands affecting extinction rates may be thought of as the primary drivers of island biogeography theory, this can be extended to include additional effects of island size and isolation. The size of islands can also affect colonization rates. Larger islands are more likely to be randomly encountered by migrating animals or propagules from plants. In a lake setting, larger lakes are more likely to be connected to

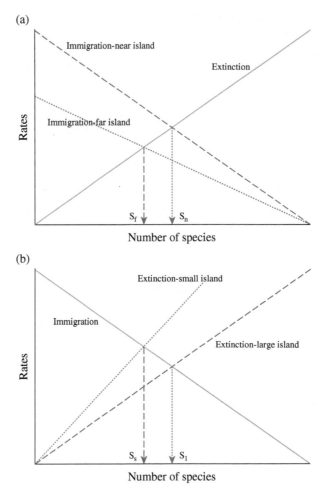

FIGURE 30-4 Hypothetical changes in immigration or extinction rates for (a) near and far islands or (b) small and large islands, indicating equilibrium number of species (S) for each condition.

other aquatic systems, either permanently or intermittently (e.g. through flooding). Also, during a time step (e.g. year) a population on an island may go extinct, but at the same time the species persistence on the island may be rescued by colonization from elsewhere. The idea of a rescue effect was developed by Ilkka Hanski of the University of Helsinki when considering meta-populations, but the concept is also relevant to island biogeography theory (Hanski 1982). Rescue from local extinction is most likely to occur for islands more connected to the mainland or other islands. Larger islands (and lakes) are likely to contain more species, not only because their size means that extinction rates will be lower and their habitat diversity may support more species, but also because their size implies that colonization rates are potentially higher. More proximate islands (and lakes) may support more species, not only because colonization rates are likely to be relatively high, but also because the risk of extinction will be lower due to potential rescue effects.

Importance of Colonization and Extinction in Lakes

Lakes can be viewed as islands in seas of land. Fish populations in lakes can go extinct, and species can colonize new lakes. Colonization may occur during flooding events as otherwise unconnected lakes become connected. Many lakes are characterized by streams draining into and out of them, and several lakes exist in a network of lakes and streams. Streams may be intermittent and not allow for fish passage during normal conditions. However, after heavy precipitation and during flooding events, such streams may become passable, allowing fish to potentially colonize new lakes. Human activities have greatly enhanced colonization rates of fish among lakes. Construction of canals has connected lakes that would otherwise not be connected. Movement of watercrafts among lakes and the release of captive individuals have contributed to accidental introductions. And, purposeful introduction of sport and forage species has been widespread across many lake regions.

Fish populations in lakes also routinely go extinct through natural and anthropogenic related processes, such as disease, intense predation by a dominant piscivore, over fishing, and loss of critical habitat. A common source of mass mortality relates to anoxic conditions. Anoxia may develop during summer and lead to mass mortality for cold water fish or during winter and lead to mass mortality for various species regardless of thermal guild. Lakes may become covered by ice during winter, greatly limiting the flux of oxygen from the air. This can lead to declining concentrations of dissolved oxygen in a lake, which can be particularly severe in shallow productive lakes. High productivity contributes to high respiration and oxygen consumption rates. Normally, oxygen consumption may be countered by diffusion of oxygen from the atmosphere and through high primary production. Ice cover not only limits contact with air, but shading by snow covered ice can also dramatically reduce photosynthetic activity and limit this source of dissolved oxygen. These effects can be particularly harsh in shallow lakes because the limited volume in such lakes implies there is less overall oxygen to deplete than in deeper lakes.

Historically, another common source of mass mortality in lakes relates to acidification. Some lakes are naturally acidic, but acidic conditions in lakes have been greatly exacerbated by loading of acids from runoff and deposition of acids from the atmosphere, often referred to as acid rain. Acid rain comes about due to gaseous pollutants, such as nitrogen oxides and sulfur oxides, reacting with water in clouds leading to precipitation having very low pH. Due largely to geology, some lakes have greater buffering capacity and can tolerate some acidic deposition, while other lakes are very sensitive to acidification. Aquatic organisms differ in their tolerance of low pH. Therefore, acidification can lead to diverse responses among species, including persistence of some and local extirpation of others.

Richness Patterns in Lakes in Wisconsin and Finland

Building on ideas related to the concept of island biogeography, John Magnuson from the University of Wisconsin and colleagues evaluated the species richness and composition of lakes and the roles of lake size and connectedness in influencing local species extinction and colonization. Magnuson et al. (1998) evaluated the isolation and extinction mechanisms in small lakes from Wisconsin, USA, and Finland. They evaluated 169 lakes from 0.2 to 87 ha in size. Related to island biogeography, they first determined there was a strong positive species-to-area curve for the lakes, and lakes from Wisconsin and Finland had no significant difference in their species–area curve (Figure 30-5).

These authors were interested in the relative role of factors related to extinction versus factors related to colonization in affecting species richness and colonization. Extinction variables included pH, conductivity, and lake area. Isolation parameters involved horizontal distance between lakes, vertical distance over land to the next lake, horizontal water distance, stream gradient, area of the next lake downstream, and distance to a road. They used statistical analyses (classification and regression tree and discriminant analysis) to determine which type of variables did the best job describing patterns of species richness and composition in these lakes. They found expected relationships between variables and fish species richness.

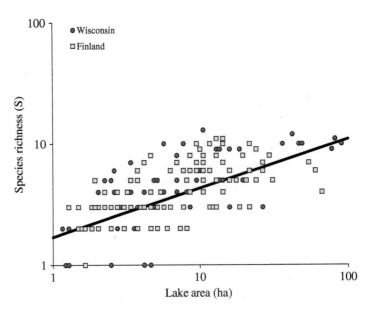

FIGURE 30-5 Species–area relationship ($\log x + 1$) for 169 small lakes in Finland and Wisconsin.
Source: Magnuson et al. (1998)/John Wiley & Sons.

For example, larger lakes, higher conductivity, and less extreme acidity were all associated with greater richness. Extinction variables were much more important than isolation variables in predicting species richness in both Finnish and Wisconsin lakes. Extinction-related variables were also more important in describing species composition than isolation variables. However, isolation variables appeared to be somewhat more influential in contributing to describing species composition patterns than they were in describing species richness patterns.

This data set demonstrated the importance of lake conditions in determining fish species assemblage. Extinction processes – mainly operating through limnological conditions or presence of predatory fish – were more important than colonization processes. Magnuson et al. believed this was due to the probability of local extinction being much higher than that of colonization, at least in North American lakes. However, they also concluded that the influence of barriers to movement and colonization on community richness and composition differed between Wisconsin and Finnish lakes due to geomorphic differences of the two lake regions.

Landscape Position

Magnusson et al.'s study of isolation and extinction processes generally demonstrated that limnological conditions strongly influenced fish assemblages through extinction processes. This led the Wisconsin researchers to evaluate the importance of geomorphic conditions on fish communities in lakes. In particular, a lake's landscape position – location in both elevation and in connection to upstream and downstream lakes – has a strong influence on the fish communities and possibly on lake conditions. Kratz et al. (1997) considered the landscape position to be the hydrologic position of a lake within the regional flow system. Lakes at higher elevation in a watershed receive more water from precipitation, while lakes at lower elevation have a greater proportion of water inputs from groundwater. Lakes higher in the landscape should be more responsive to drought, both by changes in nutrient concentration and in water level. Lake area and fish species richness may be characteristics of landscape position, with lower elevation lakes being larger and more species-rich, due to higher immigration rates and lower extinction rates. Kratz et al.'s concept was that the landscape position of a lake has a strong influence on all characteristics and may be used in a predictive sense to understand fish communities, as well as limnological conditions. They supported these trends with data from Wisconsin lakes, including strong associations among landscape position, lake area, and littoral fish species richness.

Landscape position can have even more dramatic effects on fish communities in more extreme regions. For example, in mountainous regions, lakes have formed at very different altitudes. High mountain lakes are often isolated and naturally fishless, while lakes at intermediate altitudes may contain a very small number of species, simply because of a very small upstream species pool capable of colonizing such lakes. Human introduction of fish into otherwise fishless or species-poor lakes has increased species richness in many mountain lakes, but overall fish species richness remains low in such systems and is generally negatively associated with altitude.

Summary

The species composition of fish in lakes is affected by multiple processes. Lakes contain a variety of zones defined by light and thermal properties. These zones may differentially favor distinct species of fish although some species migrate among zones. The physical, chemical, and biological features of lakes have strong influence on composition of fish in lakes. Specifically, the productivity of lakes and where they fall along the eutrophic–oligotrophic continuum may strongly affect the performance of different species. The relative influence of bottom-up versus top-down processes in structuring lake food webs has been debated in the past. However, both processes are important, with bottom-up or top-down control being potentially more influential, depending on the time and space scales across which systems are compared. Fish communities in lakes become established and are maintained through colonization and extinction processes. Similar to islands, attributes such as size and connectivity of lakes will influence species composition and richness.

Literature Cited

Brooks, J.L., and S.I. Dodson. 1965. Predation, body size, and composition of plankton. Science 150:28–35.

Bunnell, D.B., R.P. Barbiero, S.A. Ludsin, C.P. Madenjian, and 17 other co-authors. 2014. Changing ecosystem dynamics in the Laurentian Great Lakes: Bottom-up and top-down regulation. Bioscience 64:26–39.

Carpenter, S.R., and J.F. Kitchell. 1992. Trophic cascade and biomanipulation: Interface of research and management — A reply to the comment by DeMelo et al. Limnology and Oceanography 37:208–213.

Carpenter, S.R., and J.F. Kitchell. 1993. The Trophic Cascade in Lakes. Cambridge University Press, Cambridge, UK.

Carpenter, S.R., J.F. Kitchell, and J.R. Hodgson. 1985. Cascading trophic interactions and lake productivity. Bioscience 35:634–639.

Cott, P.A., M.M. Guzzo, A.J. Chapelsky, S.W. Milne, and P.J. Blanchfield. 2015. Diel bank migration of Burbot (*Lota lota*). Hydrobiologia 757:3–20.

DeMelo, R., R. France, and D.J. McQueen. 1992. Biomanipulation: Hit or myth? Limnology and Oceanography 37:192–207.

Gorman, O.T., D.L. Yule, and J.D. Stockwell. 2012. Habitat use by fishes of Lake Superior. I. Diel patterns of habitat use in nearshore and offshore waters of the Apostle Islands region. Aquatic Ecosystem Health and Management 15:333–354.

Hall, D.J., E.E. Werner, J.F. Gilliam, et al. 1979. Diel Foraging Behavior and Prey Selection in the Golden Shiner (*Notemigonus crysoleucas*). Journal of the Fisheries Board of Canada 36:1029–1039.

Hanski, I. 1982. Dynamics of regional distribution: The core and satellite species hypothesis. Oikos 38:210–221.

Hansson, L.-A., H. Annadotter, E. Bergman, et al. 1998. Biomanipulation as an application of food chain theory: Constraints, synthesis and recommendations for temperate lakes. Ecosystems 1:558–574.

He, X., and J.F. Kitchell. 1990. Direct and indirect effects of predation on a fish community: A whole lake experiment. Transactions of the American Fisheries Society 119:825–835.

Honsey, A., S. Donabauer, and T.O. Höök. 2016. An analysis of lake morphometric and land use characteristics that promote persistence of Cisco *Coregonus artedi* in Indiana. Transactions of the American Fisheries Society 154:363–373.

Hrbáček, J., M. Dvorakova, V. Korinek, and L. Prochazkova. 1961. Demonstration of the effect of the fish stock on the species composition and the intensity of the metabolism of the whole plankton association. Verhandlugen — Internationale Vereinigung fur Theoretische und Angewandte Limnologie 14:192–195.

Kitchell, J.F., and S.R. Carpenter. 1993. Synthesis and new directions. Pages 332–350 *in* S.R. Carpenter and J.F. Kitchell, editors. The Trophic Cascade in Lakes. Cambridge University Press, Cambridge, UK.

Kratz, T.K., K.E. Webster, C.J. Bowser, J.J. Magnuson, and B.J. Benson. 1997. The influence of landscape position on lakes in northern Wisconsin. Freshwater Biology 37:209–217.

Lindeman, R.L. 1942. The trophic-dynamic aspect of ecology. Ecology 23:399–417.

MacArthur, R.H., and E.O. Wilson. 1967. The Theory of Island Biogeography. Princeton University Press, Princeton, N.J.

Magnuson, J.J., W.M. Tonn, A. Banerjee, et al. 1998. Isolation vs. extinction in the assembly of fishes in small northern lakes. Ecology 79:2941–2956.

Mehner, T. 2012. Diel vertical migration of freshwater fishes – proximate triggers, ultimate causes and research perspectives. Freshwater Biology 57:1342–1359.

Ryder, R.A. 1965. A method for estimating the potential fish production of north-temperate lakes. Transactions of the American Fisheries Society 94:214–218.

Schindler, D.W. 1974. Eutrophication and recovery in experimental lakes: Implications for lake management. Science 184:897–899.

Shapiro, J. 1979. The importance of trophic-level interactions to the abundance and species composition of algae in lakes. Pages 105–116 *in* J. Barica and L.R. Mur, editors. Hypertrophic Ecosystems. Junk, The Hague.

CHAPTER 31

Marine Ecosystems

The oceans cover about 70% of the world's surface, and their massive volume houses a large diversity of fishes. One interesting difference from freshwater systems is the presence of sharks, skates, and rays (Class Chondrichthyes), which have only a few freshwater representatives but are very common in the ocean. Some patterns of oceanic environments were described in earlier chapters: recall tidal patterns and intertidal fishes, the effect of wind on capelin emergence, and the relationship between circulation and larval fish survival. However, the physical nature of marine ecosystems has not been described. This chapter begins with a discussion of marine biology and oceanography, and then reviews the characteristics of fish assemblages in pelagic waters, coral reefs, the deep sea, and estuaries.

Oceanographic Patterns

There is a very predictable major circulation pattern in the ocean, which has much influence on distribution of fishes. There are also minor current patterns influenced by local weather and wind conditions and have influence on growth and survival of fishes. Most upper ocean waters are poor in nutrients, and nutrient input occurs by processes of upwelling or by runoff from continents. An offshore area with limited currents and no contact with a continent generally has low nutrient concentration and low productivity. There are some exceptions, mainly coral reefs, which are often in areas of low nutrient concentration but high productivity. The major currents in the Atlantic and Pacific oceans are repeatable and generally predictable (Figure 31-1). In semi-equatorial regions, there are limited currents and the water is relatively stagnant, although there are major currents that run along the equator as well.

Nutrients and Primary Productivity

Tropical regions have low nutrient concentrations in oceanic waters because of the lack of currents. The predominant pattern of circulation has strong cross-oceanic currents north and south of the equator, a weak equatorial counter current, and major return currents at mid-latitudes. In the Atlantic, the Gulf Stream crosses from North Africa through the Caribbean, moves up along the eastern coast of North America, and then eastward again to Europe. A similar pattern is demonstrated by currents in the South Atlantic, North Pacific, South Pacific, and Indian Ocean. Gyres of currents are relatively strong in the Atlantic because of its smaller size; they become more dispersed in the Pacific because of its huge size. The middle

Biology and Ecology of Fishes, Third Edition. James S. Diana and Tomas O. Höök.
© 2023 John Wiley & Sons Ltd. Published 2023 by John Wiley & Sons Ltd.

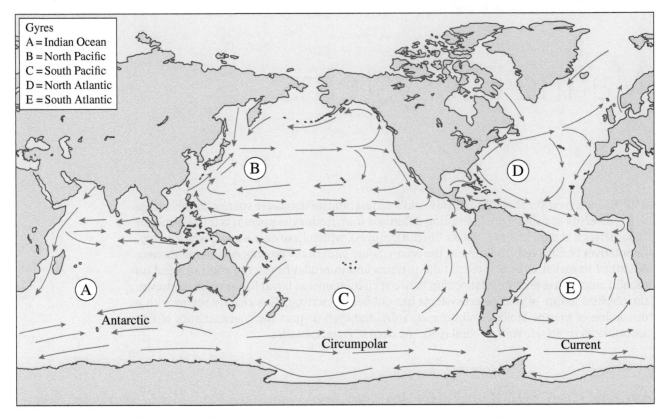

FIGURE 31-1   Major surface currents of the oceans. *Source: Reprinted with permission from Morrissey et al. (2018).*

of these gyres is usually an area of fairly stagnant water; the Sargasso Sea is one example in the middle of the Gulf Stream gyre. These offshore areas have very low nutrient availability because the main ocean current blocks them from the influence of continental waters. One important factor influencing nutrient distribution occurs where major currents move offshore to produce upwelling. This process was described in Chapter 24 related to offshore winds in Newfoundland. In the ocean, predominant wind patterns, the Coriolis Effect (due to different rotation speeds at different latitudes), water temperatures, and water density influence patterns of circulation. Temporary wind shifts do not alter the overall circulation pattern much, but changes in wind direction and magnitude can add significant variation to water current patterns. The distribution of primary production throughout the oceans indicates high primary production along the continental shelves (Figure 31-2). Upwelling areas occurring off the west coasts of northern Africa, North America, and South America have high primary production, as do coastal areas restricted from major currents, such as the North Sea, Gulf of Mexico, Bay of Biscayne, and the Mediterranean Sea. Upwelling results in cold and nutrient-rich water moving up to the surface; these are therefore areas of high primary production and are often areas of high fish production and catch as well. The largest fishery in the world exists off the coast of Peru, where pelagic fishes (anchoveta) in that upwelling zone are the major targets. There may be higher primary production in the middle of those currents where cross currents occur even though no local nutrient addition or upwelling occurs. The overall pattern is one of high primary production in coastal or some current areas, separated by zones with low primary production.

Although the oceans cover over 70% of the world's surface and may appear rather homogeneous, there is broad variation in factors influencing the abundance of animals throughout the

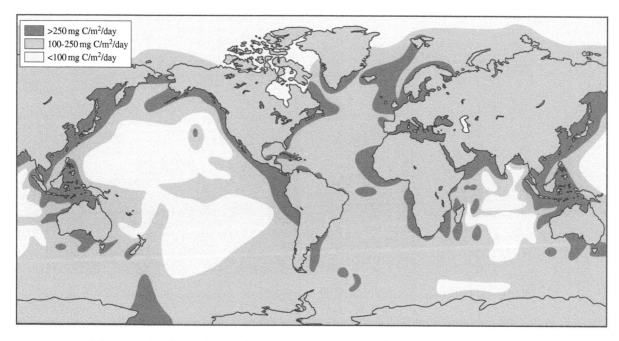

FIGURE 31-2   Primary production in the oceans. *Source: Adapted from Nybakken (1988).*

world's oceans. There are large differences in nutrients and primary production locally, as well as in currents, salinity, and temperature that influence dispersal. Therefore, zoogeographic patterns still occur in which animals are locally distributed in the north or south temperate zone because of water temperature or other conditions but may have difficulty crossing from one zone to the other because of limited nutrients, primary production, or temperature conditions between zones.

Vertical Profile

One other characteristic of the ocean, besides its surface area, is it is very deep. The average depth of the ocean is approximately 4000 m. The process of light penetration into lakes is described in Chapter 1, and enough light for photosynthesis is only transmitted to shallow depths in water, at most 150 m into the ocean. The vast majority of the ocean is deep and dark water. It is also a relatively continuous habitat throughout the world, that is, fishes may stay at depths of 3000 m and still migrate around the world without many geographical barriers. Even physical conditions at these depths, such as temperature, salinity, and oxygen content, are relatively constant.

Many characteristics of the ocean vary with depth, resulting in biological zonation of oceanic waters. The epipelagic zone – a lighted surface zone – is in the top 150 m or less of the ocean (Figure 31-3). While the pelagic zone is considered an important area in terms of fish production, it represents less than 5% of oceanic habitat available by volume. In this zone, most solar radiation is absorbed, infrared radiation in shallow depths (<1 m). Heat is distributed by wind mixing rather than by actual heat penetration in water. The epipelagic zone is the main area influenced by wind mixing; it is warm, has seasonal changes in temperature, and is well mixed. Virtually all phytoplankton production occurs in the epipelagic zone because it has sufficient light to allow photosynthesis. Most species commonly considered pelagic fishes occur in this zone.

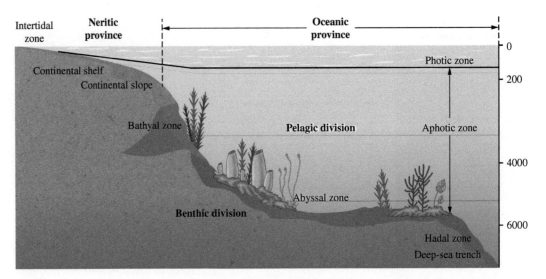

FIGURE 31-3 Diagrammatic vertical profile through the ocean. *Source: Redrawn from Morrissey et al. (2018).*

Waters deeper than 200 m represent the deep sea area and comprise over 65% of the Earth's surface area and 95% of the ocean volume. The mesopelagic zone has twilight conditions with some light penetrating from the surface, but light intensity is very limited. There is usually a permanent thermocline within this zone, so temperature conditions are much less seasonally influenced here than at the surface. This zone does not warm much in the summer, nor cool much in the winter; temperatures are usually in the range of 5–15 °C, and depths are in the range of 200–1000 m. Sixteen percent of the volume of the ocean is in the mesopelagic zone.

The final pelagic area is the bathypelagic zone, which occurs at depths below 1000 m. Virtual darkness is evident day and night in this area, cold temperature is consistent, no seasonal change in light occurs, and temperature is usually around 4 °C. Pressure in water increases with depth at the rate of 1 atmosphere of pressure (about 100 Pa or 760 mm of mercury) for every 10 m in depth. Because of great depth and lack of circulation, water pressure is extreme in this zone, and dissolved oxygen is limited. This zone is typically subdivided into the bathyal (200–3000 m), abyssal (3000–6000 m), and hadal (>6000 m) categories. The bathypelagic zone occupies over 75% of the world's ocean volume, which makes it the largest habitat in the world. Life in the bathypelagic zone is largely based on materials produced in the upper zones which die and sink. Phytoplankton that are either dead or unable to continue floating, as well as dead and dying animals from the upper zones, may sink eventually into this zone and become the source of energy in the area. The abundance of food in any given area of the bathypelagic zone is extremely low. Food may be much more abundant along the bottom where it can accumulate than in the area through which it falls. Since decomposition of many organisms occurs in the deep sea, nutrient concentrations are higher than in surface waters. However, stratification mostly keeps these nutrients locked up within this zone.

The benthic zone is that region under the influence of the ocean floor, and it can be subdivided into the same zones depending on depth. The depth of the benthic zone affects local species, but the distribution of food and other conditions there are much more similar among benthic zones than between the mesopelagic and bathypelagic zones. Productivity of animals may be quite high in the benthic zone, particularly in heterotrophic areas, such as thermal vents, where chemicals from Earth processes are liberated and form the basis of a food web.

Marine Biogeographic Realms

Given the extent of the oceans, their freshwater inputs, differences in climate and other physical gradients, and different evolutionary histories in oceanic basins, there should be strong differences in species richness among large oceanic areas. The differences in climate, physical histories – such as continental drift, sea-level rise, and glaciation – and biotic assemblages have been used to propose Marine Biogeographic Realms (MBRs) with boundaries covering areas reasonably similar. These marine boundaries are areas that may have similar histories of factors that cause the present distribution of marine species. Therefore, species residing in one MBR are believed to have had similar evolutionary pressures and reasons for adaptation, which would lead to endemic species within an MBR. Due to the open nature of the oceans, considerable overlap in species presence could also occur between adjacent realms. Plate tectonics and other geological processes have resulted in major changes in these MBRs over time, which may mix or split assemblages and influence speciation. Meta-analysis has recently been used to try to define boundaries of MBRs and relate them to distributions of fishes in the ocean.

Costello et al. (2017) conducted a meta-analysis of endemic species of marine organisms to better define MBRs. They used data on the distribution of 65 000 species of marine plants and animals to distinguish 30 distinct MBRs, which they considered similar to the number of realms per area as found on land. Their 30 MBRs included large areas in the open ocean as well as some smaller areas in coastal zones and inland seas (Figure 31-4). They described 18 continental shelf and 12 offshore realms with much wider geographic ranges for species found in the pelagic and deep sea environments compared to coastal areas. These large ranges are also reflected in the size of those MBRs. The top 100 widespread species included 46 fish

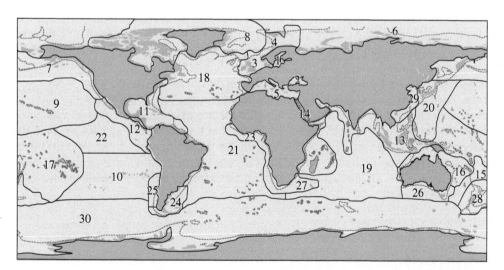

FIGURE 31-4   Map of global MBRs. 1, Inner Baltic Sea; 2, Black Sea; 3, NE Atlantic; 4, Norwegian Sea; 5, Mediterranean; 6, Arctic seas; 7, N Pacific; 8, N American Boreal; 9, Mid-tropical N Pacific Ocean; 10, SE Pacific; 11, Caribbean & Gulf of Mexico; 12, Gulf of California; 13, Indo-Pacific seas & Indian Ocean; 14, Gulfs of Aqaba, Aden, Suez, Red Sea; 15, Tasman Sea; 16, Coral Sea; 17, Mid S Tropical Pacific; 18, Offshore & NW North Atlantic; 19, Offshore Indian Ocean; 20, Offshore W Pacific; 21, Offshore S Atlantic; 22, Offshore mid-E Pacific; 23, Gulf of Guinea; 24, Rio de La Plata; 25, Chile; 26, S Australia; 27, S Africa; 28, New Zealand; 29, N W Pacific; 30, Southern Ocean. *Source: Costello et al. (2017)/ Springer Nature/Licensed under CC BY 4.0.*

species. Overall, about five times more pelagic species were widespread compared to benthic species. Species diversity varied from 192 in the Black Sea Realm to 16 508 in the tropical Indo West Pacific Realm. The unique species in each MBR varied from 3 to over 4000. This analysis provided strong evidence for regional factors influencing zoogeography of marine fishes and found a high degree of endemism (an average of 42% including all organisms) in a particular MBR.

Pelagic Zone

We have described various ecological processes influencing fish populations in the pelagic zone of the ocean throughout this book. MBRs have major significance in determining the fish assemblages. The major physical and chemical conditions within a realm, in addition to evolutionary history, have caused a large number of species to be locally endemic in spite of the open nature of the ocean. Pelagic fishes are among the most widely distributed of all oceanic groups and, as a result, have lower rates of endemism within an MBR. However, there are few cosmopolitan species, and most species of pelagic fishes reside within a few MBRs.

Pelagic fishes show higher species richness in tropical regions, and richness declines with latitude, much like the conditions described earlier for tropical rivers. These patterns have been believed to reflect the same ecological process such as stability of climate and of climatic history as described earlier. Maps of species "hotspots" have generally shown the species richness of areas is much higher in the tropics. Roesti et al. (2020) did a global meta-analysis of pelagic fish diversity from commercial catch data and indicated that species richness is much higher in the tropics than at higher latitudes – up to twice as high. However, they also found that predation pressure on pelagic fishes is most intense in temperate latitudes, counter to expectations that higher diversity is actually driven by higher predation rates in the tropics, which tend to favor coexistence of species. This brings up a crucial point about biodiversity: it not only includes species richness, but also evenness in abundance, phenotypic diversity, and functional trait diversity, which all help to define the diversity of ecological functions. Stuart-Smith et al. (2013) completed a meta-analysis including species abundances and functional traits, and this analysis demonstrated a much wider distribution of hotspots, which included many temperate areas. Their analysis indicated that abundance of marine fish populations was more even in temperate locations, meaning that more of the species have roughly equal abundance and fewer species are rare. For pelagic fishes, it appears that the number of species peaks in tropical regions, but other measures of biodiversity do not follow the same trend, and that pelagic fishes tend to largely be limited to one MBR. Overall, thermal conditions, geological history, and species interactions all interact to create boundaries for the largely endemic marine pelagic fishes in the open ocean.

Coral Reefs

Coral reefs are interesting ecosystems in many respects. They represent fish communities of high species diversity in oceanic regions that are otherwise unproductive. Coral reefs are among the most complex ecosystems in the world, with enormous diversity of fishes overall, and yet considerable variation in species diversity between reefs in local areas as well as in regions and oceans of the world. Thirty to forty percent of all species of marine fishes described to date are associated with coral reefs. One small reef in Australia includes 500 species of fish; up to 2000 species of fishes are associated with reefs in the central Indo-West Pacific region. The kinds of fishes that occur on reefs include a variety of types, specializing in different foods. However, species diversity is dominated by fishes of the suborder Percoidea (a group of spiny-rayed fishes).

Coral reefs occur in shallow water and are limited by the depth of light penetration. Coral species are colonial in nature, with adjacent individuals often genetically identical – due to asexual reproduction – and with the potential for sharing of resources, such as nutrition, among adjacent polyps. Corals are limited to tropical and subtropical seas from 25 °N to 25 °S with temperatures above 18 °C or areas where warm currents might move along a coast. The high diversity of fishes and the high productivity of coral reefs appear to be an anomaly because the distribution of nutrients and primary production in the marine system indicates that most tropical seas should be relatively unproductive (see Figure 31-2). However, corals can overcome these nutrient limitations because they are symbiotic with zooxanthellae – algal species that can fix nitrogen. Zooxanthellae use atmospheric nitrogen for primary production and produce carbohydrates by photosynthesis, which aid in coral nutrition. The region around a reef becomes considerably more productive over time than nearby waters without the coral. This productivity is due partly to the attraction of animals to the reef – resulting in deposition there of waste material, which are byproducts of food consumed elsewhere – and partly because of the primary production by symbiotic algae. Large areas of coral have undergone bleaching (loss of symbiotic algae), which in many cases can cause demise of the reef corals. Bleaching of reefs is a major concern as it occurs in a wide variety of locations. The causes are not clearly understood although human-induced influences, such as climate change – with its increasing water temperature and ocean acidification, are considered the main factor.

Fish Diversity on Coral Reefs

Coral reefs contain a diverse fish fauna, and it is tempting to link the diversity in coral reef communities to the stability of the reefs, which have been in place for many years and may have allowed adaptive radiation of fishes. Many environmental conditions on coral reefs are relatively stable, and much of the specialization that coral reef fishes develop is linked to feeding. This evolutionary diversification of feeding mechanisms is one of the processes resulting in high species diversity for coral reef fishes. A wide variety of coral-feeding fishes uses many mechanisms to consume coral-based resources. Parrot fishes bite off a piece of coral, extract the energy from it, and defecate the hard coral parts as bits of sand, which form the white sand beaches typical of many tropical areas. Some coral feeders have a long, thin snout that can be used to reach into crevices and pick off individual coral polyps. There are also specialized feeders, such as cleaner fishes, that remove parasites, as well as necrotic tissue from fishes. There are also many kinds of herbivores; some scrape off periphyton and some filter feed. These are only a few of the foraging types of fishes using coral reefs. The variety of feeding types may exceed those described earlier for tropical cichlids.

One interesting concept related to energetics of coral reef fishes is that they appear to be capable of existing on materials of lower energy density than fishes in other regions. Harmelin-Vivien (2002) demonstrated that herbivores and sessile invertebrate feeders are much more common in the tropics than in the temperate zone. These organisms consume food of much lower quality than temperate species, yet survive. Harmelin-Vivien has shown that both the percentage of species exhibiting herbivory and relative abundance of herbivores from all fish collected increase dramatically with decreasing latitude in oceanic areas. For example, herbivores are totally absent from cold temperate reefs, represent 2–8% of the species in warm temperate reefs, and include 12–22% of the species on coral reefs. This increase in herbivory as a foraging technique seemingly adds to the species diversity of fishes found on coral reefs.

Coral Reef Assemblages

Since only 3 of the 30 MBRs considered by Costello et al. (2017) include areas with abundance of coral reefs, the relationships between MBR and reef fish assemblages have not been investigated. Two major concepts can be related to species diversity on coral reefs: equilibrium and nonequilibrium theory. Nonequilibrium theory, proposed for coral reefs by Sale (1978), states that communities in coral reefs are at least partly regulated by chance dispersal – a nonequilibrium process. He proposed that dispersal of fishes causes a major destabilization to the coral reef ecosystem because nearly all fishes that live on coral reefs produce young that are pelagic, are carried in currents, and disperse widely. Eventually, some of these young colonize a coral reef but not necessarily reefs where their parents spawned. Because of this open ocean dispersal of larvae, Sale believed settlement is unpredictable, and he termed their settlement a "space lottery."

Contrasting Sale's "space lottery" concept is the equilibrium model for tropical species diversity. Equilibrium theory would hold that competition and other biotic processes have overriding importance in determining fish species diversity on coral reefs. The focus of studies on coral reef fishes has changed dramatically over recent years because of this controversy. In 1991, Peter Sale edited a book on coral reef fishes, which mainly consisted of work on adult or resident juvenile fishes. In 2002, a book on the same topic almost exclusively included work on larval dispersal and stochastic processes (Sale 2002). While some controversy still exists, much information is accumulating to demonstrate how both equilibrium and nonequilibrium processes are important in establishment of fish assemblages on coral reefs.

One interesting concept related to diversity on coral reefs is the concept of rarity. Jones et al. (2002) demonstrated the importance of rare species to overall species diversity. They showed that as many as 85% of the species found in reef fish assemblages have individuals that comprise less than 1% of all fish collected. Several questions come from such work: are these species always rare or only rare in places? How do rare species maintain themselves with pelagic larvae that disperse widely? Jones et al. defined rarity based on low local abundance and a number of ecological factors are obvious correlates of abundance in organisms. Large animals are usually less common than smaller ones. Animals with larger individual home ranges are also generally rarer. Reef fishes that have longer larval durations might be rarer due to longer exposure to predators and larger ranges over which the larvae settle. Jones et al. reviewed data that are related to Rapoport's rule, which states that organisms from higher latitudes have larger distribution ranges (Rapoport 1994). This rule is based on the observation – mainly from terrestrial systems – that organisms from higher latitudes tend to occupy larger ranges due to their generalist feeding habits and tolerance of variable environmental conditions. Jones et al. (2002) demonstrated that for coral reef fishes the opposite was true – that organisms from lower latitudes had larger ranges. They explained this on the basis of climate variability, that is, near the equator, these animals occur over a much wider range of latitudes and still experience similar climatic conditions compared to animals at the north or south extreme of this range. They also reviewed the literature on rare species and found that only one species that had been studied extensively was rare everywhere, while the majority were common in some areas and rare in most. Therefore, it appears that reef fishes generally have some selected habitats in which they are very abundant but then often remain as minor components of many other assemblages.

Other studies of coral reef fishes have focused more mechanistically on the larval dispersal stage. For example, Cowen (2002) evaluated larval behavior and produced evidence that both passive and active transport are needed for larvae to disperse from pelagic areas to reef settlement areas. While some larvae remain pelagic for months and may disperse hundreds of kilometers, most spend less time in the pelagic zone and may be dispersed only tens of kilometers. Larvae have the ability to partly control their drift pattern by vertical migrations, and most commonly, larvae are located from 10 to 60 m in depth and show vertical migrations on a diel basis. In one case, from 15 to 60% of the pomacentrid larvae settling near Lizard Island, Great Barrier Reef, were of local origin, indicating the ability to remain largely in the local area (Jones et al. 1999). Recent interest has focused on the impacts of local current gyres that may disperse or retain larvae near their natal site. Limer et al. (2020) demonstrated that gyres in the Gulf of Mexico would normally keep planktonic larvae near their natal reef, but given the instability of such gyres, considerable drifting would also occur in some larvae. This pattern would help explain both of the observations above, the space lottery for dispersing larvae – most likely the rare types – and local retention.

Deep Sea Fishes

The deep sea has been explored for a considerable period of time but remains the least known of all the Earth's regions. Imants Priede from the University of Aberdeen in Scotland has recently produced a comprehensive evaluation of deep sea fishes (Priede 2017). This work indicates an extremely diverse fish fauna in the deep sea with approximately 3500 described species and the expectation for 5000 total species inhabiting this region. Each zone within the deep sea has different conditions that drive different fish diversity. One interesting point is that recent research has shown approximately 284 circumglobal fish species – found in all oceans but not necessarily in all parts of each ocean – in the world, 80% of which occur in the deep sea. Circumglobal distributions would be expected due to the continuity of deep sea habitat and potential dispersal by deep currents, but it is rare for fishes to have a circumglobal distribution even in that ecosystem.

Unique oceanic habitats – and the largest ones on Earth by volume – are the mesopelagic and bathypelagic zones. The characteristics of fishes from the mesopelagic and bathypelagic zones considerably differ from typical pelagic fishes described earlier and from each other (Table 31-1). These differences include adaptations to the conditions of life in each zone. In the mesopelagic zone, twilight conditions do allow for limited vision. Coloration patterns including countershading frequently occur here, and the animals have developed eyes. Mesopelagic fishes are generally silvery colored with countershading, typically lighter ventrally and darker dorsally. This countershading may not be as distinct as it is in epipelagic fishes, but predators from underneath or above still have difficulty detecting fishes because their shading matches the background.

Fishes in the mesopelagic zone usually have very large eyes since vision is difficult but possible. Jaw development is relatively normal for mesopelagic fishes. Mesopelagic fishes are also accomplished swimmers, and they have developed locomotor muscles. These fishes often show vertical migration; they move deeper in the day when light is brighter, and shallower at night – often into the epipelagic zone – when light is dimmer, to optimize feeding or escape predation. Also correlated to swimming, most mesopelagic fishes have well-developed swim bladders. One final pattern for mesopelagic fishes is the presence of luminescent organs – i.e. photophores. These organs are very numerous, in terms of both numbers of fishes that have

TABLE 31-1 Some General Characteristics of Mesopelagic, Bathypelagic, and Benthic Fishes.

Character	Mesopelagic	Bathypelagic	Bathybenthic
Color	Silvery, countershaded	Black, dull	Variable
Eyes	Large, developed	Small, regressed	Large, developed
Jaws	Normal	Large, hinged	Normal, tearing teeth
Musculature	Well developed	Weak, >90% water	Robust
Skeleton	Scales and bones well developed	No scales, cartilaginous	Normal
Swim Bladder	Well developed	Absent or regressed	Variable
Gills	Well developed	Small	Developed
Photophores	Numerous, used in countershading	Less numerous, as bait	Common

photophores and the photophore patterns. Mesopelagic fishes demonstrate the kinds of distribution patterns that might be anticipated based on primary production in the oceans, that is, the greatest number of species have subtropical centers of abundance, and they are rarer under areas of low productivity.

Bathypelagic fishes have eyes that are small to nonexistent, and instead use olfactory, electroreception, or lateral line senses more than visual senses to locate prey or mates. Bathypelagic fishes often have very large, hinged jaws; many of these fishes can swallow food particles as large as or larger than themselves. Since food is relatively rare in this zone, the fishes may rarely eat, and the large jaws help them take advantage of opportunistic conditions. Bathypelagic fishes are generally very weak swimmers and have poorly developed muscle tissue; their muscles may be as much as 90% water, which also helps to counter massive pressures exerted at these depths. Concurrent with poor muscle development are limited or cartilaginous skeletons in bathypelagic fishes. Most bathypelagic fishes have no swim bladders as the intense water pressure also makes swim bladders difficult to fill with gas. Instead, they can regulate buoyancy using lipid reserves in their bodies.

Unlike mesopelagic fishes, those from the bathypelagic zone are usually red or black. They generally have no reflective surfaces because most of the light in this zone – if there is any – is bioluminescent light produced by animals. Any reflection of that light might make fishes obvious to other animals, including predators. Bathypelagic fishes, with limited activity, have very small gills and might even use skin surfaces as breathing organs.

Bathypelagic fishes are widely distributed, commonly under highly productive areas. However, the level and geographical distinction of surface productivity becomes mixed by currents before materials sink 4000 m, so the centers of distribution are not as strongly in high-productivity areas. Relatively few of the bathypelagic species show circumglobal distribution. Priede (2017) indicated that the distribution of most mesopelagic and bathypelagic assemblages is largely affected by biogeochemical processes in the ocean, which have been described as provinces with similar characteristics, like the MBR concept. However, there are far more provinces defined for them, and they are not influenced by evolutionary history like the MBR (Reygondeau et al. 2013).

Benthic fishes in the bathyal zone are most similar to mesopelagic fishes, as they are strong swimmers, with strong teeth and jaws. They have strong subtropical centers, and their distribution is mainly limited to a single ocean. Both pelagic and benthic fishes are limited to depths less than 8400 m, and distribution of benthic species may be disrupted by deep trenches and oceanic ridges. While there have been relatively few explorations into depths beyond 8400 m, the common belief is that extreme pressures at these depths limit fish distribution. Their distributions and assemblages follow other marine species with a number of endemic fishes within an MBR as well as a few species distributed more widely.

Deep hydrothermal vents were first discovered in 1977, and their conditions and abundance have been studied extensively since 2000. There are heated as well as cold water vents in the deep ocean, but the main interest in recent times has been the heated vents. Along with very hot water (often over 100 °C), the vents expel many reduced compounds that form the basis for chemosynthetic productivity of animals in the area. A high biomass of animals, including annelids, polychaetes, mussels, clams, and shrimp, obtains much of their energy needs from symbiotic bacteria that are chemoautotrophic – synthesize energy from chemicals rather than from photosynthesis. These vents tend to occur along deep sea ridges, and these areas have been little explored. Some benthic fishes also populate the vent regions, with up to 90 species of more common benthopelagic fishes found near some vents, mainly including rays, cods, grenadiers, and eelpouts. Conditions around the vents vary dramatically in temperature, and it is likely that the fishes occupy temperatures from 5 to 10 °C most commonly (Priede 2017).

Bioluminescence

For fishes that live in zones where there is no light from the Sun, bioluminescence has many important functions and as a result has independently evolved at least 27 times in different fishes (Davis et al. 2016). This luminescence is largely produced by symbiotic bacteria that inhabit the light organs of fishes. Haddock et al. (2010) detailed the various functions of bioluminescence as providing defense against other species, offense toward other species, or communication among conspecifics. Their defensive functions include using luminescence to startle predators, counter illuminate fishes, serve as a smoke screen to misdirect predators, distract predators to less sensitive body parts, serve as a burglar alarm by making the predator more obvious and vulnerable to higher level predators, provide a sacrificial tag by shedding a luminescent body part to attract predators away from the prey animal, and produce warning coloration to warn predators that the prey may be toxic. Offensive functions include luminescence serving as a lure to prey, a bright emission to stun the prey, and a lower level of emission to illuminate prey for easier capture. The widespread evolution and importance of bioluminescence to deep sea fishes has led to a higher diversity of species in the deep sea that have specialized to produce a variety of bioluminescent functions and patterns.

Estuaries

Estuaries represent the environment of interchange between freshwater and marine ecosystems. They are defined as constricted areas with brackish water caused by freshwater rivers flowing into the ocean. They are partially enclosed in most cases although free connection to the open sea is also a characteristic of an estuary. Estuaries receive sediments and other materials from the river system, as well as tidal exchange from the ocean, making them quite dynamic ecosystems. They often vary dramatically in salinity and temperature throughout the extent of one estuary. Many consider estuaries to be among the most important ecosystems in the world with high levels of productivity and importance in terms of processing nutrients and sediments before they reach the open ocean. Costanza et al. (2014) produced a monumental study evaluating the value to humans of ecosystem services for a variety of global ecosystems. Their analysis indicated that estuaries had the highest value per unit area based on their importance in processing nutrients, alleviating flooding, and production of organisms. On a global basis, these organisms form assemblages of mangrove forests, seagrass communities, oyster beds, and a variety of other productive habitats. Sedimentation also creates large mud flats within estuaries that are difficult to colonize and therefore somewhat limited in productivity. Of course, the type of organisms found in estuaries dramatically depends on their geographic location. Estuaries have suffered dramatic alteration from human impacts – including sedimentation and exploitation – that have destroyed oyster beds, cleared mangrove forests, and produced other forms of pollution damaging the organisms in estuaries.

While estuaries are considered very productive and important ecosystems, their fish communities are generally not that ddiverse, probably due to the variable conditions in an estuary. Vasconcelos et al. (2015) from Universidade de Lisboa in Portugal completed a meta-analysis of the global characteristics of estuaries and their fish assemblages. They found that the average estuary had 29 species of fish and maximum species richness was 214. These species involved several guilds, including migratory and resident fishes. However, freshwater fishes are far less abundant than marine species in estuaries. Vasconcelos et al. evaluated the fish species richness and various ecological and physical conditions of the estuaries. Globally, the main factors influencing species richness was MBR and continent with positive correlations of species diversity to mean sea surface temperature, terrestrial productivity, and stability of the connection to the marine system. Estuaries demonstrated similar patterns as described earlier, with tropical warm temperature regions having the highest species diversity and high latitudes having the lowest diversity. However, the overall number of species is far lower than the other ecosystems evaluated.

Elliott et al. (2007) evaluated the fish assemblages in estuaries and described 11 functional guilds of fishes utilizing estuaries. Their guilds included marine and freshwater stragglers that reside temporarily in the estuaries, anadromous fishes crossing the estuary to complete their life cycle, and estuarine residents that remain in the system throughout their life. Estuaries are widely known as nursery grounds for a number of marine fishes, including European plaice, which is one reason for their high economic values. Among the residents like mummichog, silversides, and anchovies are some species that are dependent on the estuary and cannot persist without using this habitat (obligative dependents), and others that are opportunistic in estuary use. These guilds help to determine the extent, purpose, and importance of estuaries to the species that utilize them as well as the variety of resources being accessed by estuarine fishes.

Summary

Vast oceanic environments contain over 17 000 fish species with many having the potential for global distribution. Yet the assemblages of fish in the ocean are generally limited geographically by biological, physicochemical, and geological conditions. The assemblages of fishes in the pelagic zone and in estuaries are mainly related to the Marine Biogeographical Realm, which combines many of the attributes listed above. About 40% of all marine fish species are endemic to one realm, and few are cosmopolitan in distribution. Fishes from coral reefs are limited to only three MBRs but demonstrate extreme species diversity. Their diversity is due in part to equilibrium processes, such as predation, feeding adaptation, and competition as well as nonequilibrium process like random dispersal of larvae. Deep sea fishes are also quite diverse with wide distributions and evolution of a variety of mechanisms to survive the challenging environment there. As climate change and oceanic conditions are altered by human interventions, many of these controlling conditions in the ocean will change, and the impact on marine fish species remains uncertain.

Literature Cited

Anderson, G.R.V., A.H. Ehrlich, P.R. Ehrlich, et al. 1981. The community structure of coral reef fishes. The American Naturalist 117:476–495.

Costanza, R., R. De Groot, P. Sutton, et al. 2014. Changes in the global value of ecosystem services. Global Environmental Change 26:152–158.

Costello, M.J., P. Tsai, P.S. Wong, et al. 2017. Marine biogeographic realms and species endemicity. Nature Communications 8:1–10.

Cowen, R.K. 2002. Larval dispersal and retention and consequences of population connectivity. Pages 149–170 in P.F. Sale, editor. Coral Reef Fishes: Dynamics and Diversity in a Complex Ecosystem. Academic Press, New York.

Davis, M.P., J.S. Sparks, and W.L. Smith. 2016. Repeated and widespread evolution of bioluminescence in marine fishes. PloS One 11:e0155154.

Elliott, M., A.K. Whitfield, I.C. Potter, et al. 2007. The guild approach to categorizing estuarine fish assemblages: A global review. Fish and fisheries 8:241–268.

Haddock, S.H., M.A. Moline, and J.F. Case. 2010. Bioluminescence in the sea. Annual Review of Marine Science 2:443–493.

Harmelin-Vivien, M.L. 2002. Energetics and fish diversity on coral reefs. Pages 265–274 in P.F. Sale, editor. Coral Reef Fishes: Dynamics and Diversity in a Complex Ecosystem. Academic Press, New York.

Jones, G.P., M.I. Milicich, M.J. Emslie, and C. Lunow. 1999. Self recruitment in a coral reef fish population. Nature (London) 402:802–804.

Jones, G.P., M.J. Caley, and P.L. Munday. 2002. Rarity in coral reef fish communities. Pages 81–102 in P.F. Sale, editor. Coral Reef Fishes: Dynamics and Diversity in a Complex Ecosystem. Academic Press, New York.

Limer, B.D., J. Bloomberg, and D.M. Holstein. 2020. The influence of eddies on coral larval retention in the Flower Garden Banks. Frontiers in Marine Science 7:1–16.

Morrissey, J., J.L. Sumich, and D.R. Pinkard-Meier. 2016. Introduction to the Biology of Marine Life. Jones & Bartlett Learning, Burlington, MA.

Nybakken, J.W. 1988. Marine Biology: An Ecological Approach. Harper and Row Publishers, New York.

Priede, I.G. 2017. Deep Sea Fishes: Biology, Diversity, Ecology, And Fisheries. Cambridge University Press, Cambridge, UK.

Rapoport, E.H. 1994. Remarks on marine and continental biogeography: An areographical viewpoint. Philosophical Transactions of the Royal Society of London, Series B 343:71–78.

Reygondeau, G., A. Longhurst, E. Martinez, et al. 2013. Dynamic biogeochemical provinces in the global ocean. Global Biogeochemical Cycles 27:1046–1058.

Roesti, M., D.N. Anstett, B.G. Freeman, et al. 2020. Pelagic fish predation is stronger at temperate latitudes than near the equator. Nature Communications 11:1–7.

Sale, P.F. 1978. Coexistence of coral reef fishes – a lottery for living space. Environmental Biology of Fishes 3:85–102.

Sale, P.F., editor. 2002. Coral reef fishes: Dynamics and Diversity in a Complex Ecosystem. Academic Press, New York.

Stuart-Smith, R.D., A.E. Bates, J.S. Lefcheck, et al. 2013. Integrating abundance and functional traits reveals new global hotspots of fish diversity. Nature 501:539–542.

Vasconcelos, R.P., S. Henriques, S. França, et al. 2015. Global patterns and predictors of fish species richness in estuaries. Journal of Animal Ecology 84:1331–1341.

PART 7

Human Influences on Fish and Fisheries

CHAPTER 32

Fisheries Harvest

H umans have used fishes and aquatic organisms for food throughout history. While there is growing importance of aquaculture (see Chapter 34), most aquatic system food production occurs through capture fisheries, which take wild fish from marine or freshwater systems for human consumption. Capture fisheries remain the only major food production system that relies on exploitation of wild animals. In addition, many marine and freshwater fisheries now involve capturing fish not for human consumption but for recreational purposes. The purpose of this chapter is first to overview the history and current status of capture fisheries and introduce approaches for managing fisheries. We then consider harvest by fisheries as a potential stressor and present some of the consequences of harvest on fish populations and communities.

Types of Fisheries

There are a multitude of types of fisheries. Fisheries are often differentiated as (a) commercial fisheries, in which fishes are caught for the primary purpose of selling; (b) recreational fisheries, in which fish are caught primarily for enjoyment and sport; or (c) subsistence fisheries, in which fish are caught for the primary purpose of feeding an associated community. Globally, the majority of wild fish harvest comes from commercial fisheries. More specifically, the majority of harvest comes from marine systems, and a large proportion of this harvest is by industrial commercial fisheries. Industrial commercial fisheries involve the use of highly powered fishing vessels and modern, technologically advanced gear to exploit fish stocks. Industrial fishing fleets generally include large vessels that may travel long distances throughout the world to capture fish. This is in contrast to artisanal commercial fisheries, which tend to use less advanced gear and smaller vessels with less power to exploit local fish stocks. Large-scale industrial fisheries harvest similar total biomass as small-scale, nearshore fisheries. However, there are many more fishers employed as artisanal-scale fishers than fishers employed as industrial-scale fishers; this is especially true in developing nations (Jacquet and Pauly 2008).

Harvest Patterns

Capture fisheries have been a major source of food for even prehistoric human populations. Studies on the remains of animals in aboriginal middens (depositories of waste materials) indicate that many tribes near water used fishes – particularly migratory fishes, as a major source

Biology and Ecology of Fishes, Third Edition. James S. Diana and Tomas O. Höök.
© 2023 John Wiley & Sons Ltd. Published 2023 by John Wiley & Sons Ltd.

of food. While fisheries have been exploited by humans throughout recorded history, major exploitation of fishery resources did not begin until humans were capable of pursuing fish stocks from ocean-going vessels. Some of the earliest marine fisheries occurred in the North Sea and were pursued by sailing vessels near their home port. Nowadays, modern trawlers move throughout the ocean to pursue fish, which are often processed and frozen on the ship right after capture. Because of this dramatic growth in technology, the exploitation of fishery resources is very high. The importance of fishes as a food source for humans led the United Nations, through the Food and Agriculture Organization (FAO), to collect statistics on fisheries of the world. These statistics have been compiled globally, mainly since 1950, but also extrapolated back into earlier times.

The earliest records of commercial fishing indicated that global harvest was approximately 1.5 million metric tons (MMT) per year in the mid-1800s. The harvest of fishes increased to nearly 100 million metric tons annually between 1850 and 2000 (Table 32-1). This dramatic increase in harvest mainly occurred since 1950 when the harvest was about 18.7 MMT. The harvest taken by commercial fisheries has largely been from marine sources. If one considers just capture of marine animals, global harvest has essentially remained at a constant level since 1995. Marine harvest has remained fairly stable despite a trend of increased global fishing activity since 1995. In other words, global marine fisheries harvest per equivalent level of fishing effort has declined since 1995. This strongly suggests that there has been an overall depletion of wild marine fish stocks throughout the world.

The total harvest of animals through fisheries and aquaculture increased to nearly 180 MMT by 2018. Much of this recent trend has been a result of increases in aquaculture throughout the world. When FAO first began collecting data on fish yields, aquaculture products were not

TABLE 32-1   Harvest estimates for various categories of fisheries, in millions of metric tons (MMT). Data from FAO (2020) and FAO statistics.

Year	Freshwater animals	Marine animals	Total capture	Culture	Total
1850			1.5		1.5
1900			4.0		4.0
1930			10.0		10.0
1950	2.5	16.2	18.7	0.6	19.3
1960	3.5	30.3	33.8	1.7	35.5
1970	5.3	57.6	62.9	2.6	65.5
1975	5.4	56.9	62.3	3.6	65.9
1980	5.5	62.2	67.7	4.7	72.4
1985	6.3	72.8	79.1	8.0	87.1
1990	6.9	78.6	85.5	13.1	98.6
1995	7.7	84.7	92.4	24.4	116.8
2000	8.6	86.0	94.6	32.4	127.0
2005	9.4	84.3	93.7	44.3	138.0
2010	10.9	77.3	88.2	57.8	146.0
2015	11.2	81.5	92.7	72.8	165.5
2018	12.0	85.4	97.4	82.1	179.5

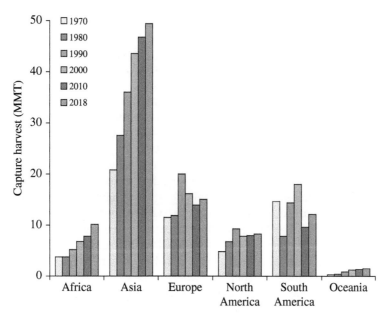

FIGURE 32-1 Capture harvest (millions of metric tons, MMT) for each continent since 1970. *Source: Data from FAO (2020) and FAO statistics.*

separately analyzed but included in both freshwater and marine capture fisheries. However, by the mid-1970s, it became apparent that aquaculture was an important component of fish harvest, and from that point forward, it has been evaluated separately from commercial fisheries. Culture harvest has increased dramatically since the 1950s and is now over 80 MMT per year (Table 32-1). The growth of aquaculture as a proportion of fish harvest has likewise increased dramatically over recent times (Table 32-1) with over 50% of global harvest now coming from aquaculture. However, unlike capture fisheries, global inland aquaculture production exceeds marine aquaculture production (see Chapter 34).

Global capture fisheries have experienced extreme changes in recorded history and show interesting patterns primarily related to changes in marine commercial harvest. The pattern of capture by continent (Figure 32-1) indicates successes, failures, and growth in fishery production. These continental data are reflective of continent of landing, not necessarily that of capture. A Korean boat fishing off the coast of the United States is listed here as Asian catch rather than North American. Asia clearly dominates the commercial fishery harvest, currently taking approximately 48% of the world's fish catch. Harvest by Asian countries has increased continually since 1970 at a rate higher than any other continent. An interesting contrast to Asia is South America, which has shown tremendous fluctuations in harvest. In the 1970s, South America harvested nearly as much of the world fish catch as Asia did, but the collapse of the anchoveta fishery in the 1970s resulted in a severe decline in yield. Europe and North America have seen declines in harvest since the 1990s, while wild harvest in Africa has increased.

Freshwater and Recreational Harvest Patterns

Based on FAO reports, freshwater organisms have only contributed a small fraction (generally around 10%) of the total yield. However, several analyses suggest that freshwater commercial harvest is likely underreported, and total harvest from freshwater systems may be particularly biased, as relatively high recreational and subsistence harvest in freshwater may not be accurately indexed in FAO reports. For example, Fluet-Chouinard (2018) used surveys of consumption of freshwater fishes by people in 42 countries to estimate that freshwater harvest is approximately 65% greater than generally reported. Even without accounting for potential underreporting, freshwater capture fisheries are by no means an insignificant contributor to global harvest (12 MMT of harvest in 2018; Table 32-1) and are very important for local ecosystems and economies. Similar to marine fisheries, freshwater fisheries include commercial, recreational, and subsistence fisheries. Subsistence and commercial fisheries continue to dominate freshwater harvest in some parts of the world. In several developing countries, freshwater fish consumption may constitute an important protein source. However, in several more economically developed nations, local freshwater recreational fishery harvest exceeds subsistence and commercial harvest. David Allan from the University of Michigan and colleagues (2005) presented some of the challenges facing global freshwater fisheries. They highlighted several similar patterns to those observed in marine fisheries. Reported global freshwater fisheries harvest has been steadily increasing since the 1950s, and harvest in Asia has consistently constituted at least half of the global freshwater harvest. Many of the patterns of overfishing and the biological consequences of fishing practices (discussed below) are also similar for marine and freshwater fisheries.

Recreational harvest – specifically angling – represents a large portion of freshwater fisheries harvest. While commercial harvest dominates in marine systems, recreational harvest can also account for the dominant source of mortality for some marine stocks and regions (Shertzer et al. 2019). The importance of recreational fishing is especially apparent in more industrialized countries (Arlinghaus et al. 2015), and globally recreational harvest may represent 12% of total harvest (Cooke and Cowx 2004). Laws regarding recreational fisheries and requirements for reporting of catches are highly variable from country to country, and even among jurisdictions within countries. In many cases, freshwater fisheries harvest must be estimated and extrapolated from creel surveys. The term creel refers to a traditional basket for carrying a catch of fish. While creel baskets are less common now, the term creel has come to more broadly represent the catch by individual recreational fishers, and creel surveys of recreational fishers are common practice to determine the types and numbers of fishes harvested. However, such surveys often reach only a small proportion of the total number of people who are angling. Individual recreational fishers generally catch a small number of fish (especially relative to individual commercial fishers). However, in many systems there are a rather large number of recreational fishers, and it is their cumulative harvest that is important to estimate. Given the distributed nature of recreational fisheries and the need to extrapolate from surveys of rather small subsets of fishers, there are often major challenges to accurately estimate the total harvest by recreational fishers. This has likely contributed to an earlier perspective that recreational fishing was unlikely to lead to the same levels of overfishing as commercial fishing. By bringing together diverse data and modeling recreational catches and fish population trajectories, various studies have now demonstrated the potential for recreational fisheries to overharvest – and even contribute to – the collapse of fish populations (see Post et al. 2002; Cooke and Cowx 2004).

An added challenge with appreciating the cumulative impact of recreational fisheries relates to the practice of catch and release. This practice of catching fishes and releasing them back into the water body of capture is intended to limit the actual number of fish deaths caused by recreational anglers. However, many fishes that are released back into water bodies do not actually survive under certain conditions. The frequency of such hooking mortality differs among species, environmental conditions, and how the fishes are handled during and after capture. For example, hooking mortality may vary between barbed versus barbless hooks, and between artificial lures versus live bait. If fish struggle for a very long time during the capture event, the risk of hooking mortality is greater. If upon capture, fish are confined for long periods in live wells with several other fish and poor water quality, they are more likely to not survive upon release. Multiple studies and symposia have now compiled information on the lethal and non-lethal effects of catch-and-release fishing including the conditions and fishing practices that may be most likely to lead to mortality (e.g. Arlinghaus et al. 2007; Cooke and Schramm 2007).

Overharvest

Unfortunately, a common feature of both marine and freshwater capture fisheries is over-harvest. In 2017, FAO estimated that 34.2% of global fish stocks were fished at biologically unsustainable levels, and of the 65.8% of stocks fished at a sustainable level, 59.6% were being maximally fished, and only 6.2% were underfished (FAO 2020). If a fish stock is harvested at a rate greater than it can naturally replenish itself, then it is harvested at an unsustainable level and may be considered overfished. More specifically, to allow for some variability, FAO has previously defined stocks below 80% of their target biomass under maximum sustainable exploitation as being overfished (see surplus production models in Chapter 11). While some stocks that were previously overfished at an unsustainable level have shifted to now being exploited sustainably, the general trend over the last 50 years has been an ever-increasing proportion of the world's fish stocks being fished in an unstainable manner. Even after overharvested fisheries are closed and no harvest is legally allowed, it can take many years for fish stocks to recover and ecosystem level changes may make it impossible for some fish stocks to ever truly recover.

Overharvest of fish stocks has followed a repeated pattern. As unexploited or underexploited stocks begin to be fished, the initial effort (f) directed to catch fish is low, total catch (C) is low, and catch per effort (C/f) is high. As effort increases, so does catch. However, if fishing practices remain the same, C/f will decrease with increasing f because of harvest leading to decreased abundance. If f continues to increase over time, C may eventually start to decrease despite increasing fishing effort. And, if f continues to increase beyond this point, C may eventually decline to 0 as the stock is extirpated (Figure 32-2). This pattern was demonstrated quite clearly in Allan et al. (2005) when depicting the catch of Mekong giant catfish in Chiang Khong, Thailand, from 1986 to 1997. During this relatively brief period, fishing effort directed toward the Mekong giant catfish increased dramatically (increasing to be approximately four times greater over ten years). The increase in fishing effort was initially accompanied by an increase in catch, but catch declined eventually, despite high fishing effort; presumably because stock abundance was greatly depleted (Figure 32-3).

Given repeated temporal patterns of f, C, and C/f, it may seem rather simple to curtail over-fishing by not allowing fishing effort to grow too high. While this is an attractive proposition, it has proven quite challenging for many fisheries for a variety of reasons. It can be difficult to accurately track fishing effort. Effort is often reported in units, such as time spent fishing, area fished, or number of nets in the water. Over time, however, fishing has become much more effective. Fishers are now able to use sophisticated methods (e.g. spotting planes and sonar fish finders) to locate fish, and fishing gear and methods have become more effective as nets are made less visible to fish. Gear can be towed more quickly, and fishing lures are designed to be more attractive to fish. As fish stocks decline, individual fish may congregate in smaller areas, and as fishers are now able to find and effectively target such concentrated areas, they are able to maintain high C/f even when overall abundances are low. Several studies have described the technology creep of both commercial (e.g. Eigaard et al. 2014) and recreational (e.g. Cooke et al. 2021) fisheries, where gradual increases in fishing efficiencies may be difficult to notice, but over time may strongly influence catchability. For example, Eigaard et al. (2014) considered a number of different studies that estimated change in catchability by marine commercial fisheries due to technology creep and found an overall mean increase in catchability of 3.2% per year. One hour spent bottom trawling or angling may now represent a very different level of

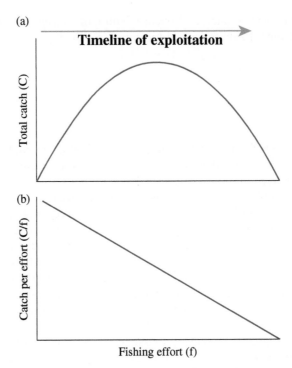

FIGURE 32-2 General relationships among total catch (C), effort (f), and catch per effort (C/f) in exploited fisheries. As a previously unexploited stock begins to be exploited, one might initially expect (a) low total effort and catch, but (b) high catch per effort. As the fishery grows and effort increases, total catch increases, but C/f decreases. Then as fishing effort becomes sufficiently high as to deplete the stock, C begins to decrease, and C/f decreases even further.

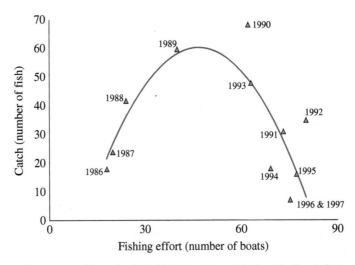

FIGURE 32-3 Annual catch of Mekong giant catfish in Chiang Khong, Thailand, from 1986 to 1997.
Source: Figure generated from data presented in Allan et al. (2005).

fishing effort than in the past. It may appear as if C/f remains relatively high, even if fish stocks are depleted, if this is not accounted for.

Another consideration related to fisheries and overharvest is the efficiency of harvest and especially fuel use. Parker et al. (2018) considered the greenhouse gas emissions associated with marine fisheries. They estimated that in 2011 marine fishing activities generated 179 million metric tons of carbon dioxide equivalent greenhouse gases. Based on their data compilation, they estimated that global fisheries emissions increased by 28% from 1990 to 2011. However, during this period there was no appreciable change in total fisheries harvest. Fishers were using more fuel and emitting more greenhouse gases to catch approximately the same biomass of fish. Parker et al. (2018) reported that from 1990 to 2011, the greenhouse gas emissions necessary to capture a given unit of fish mass increased by 21%.

In addition to the issue of fisheries catch per effort not necessarily reflecting actual abundance, tracking fisheries harvest is confounded by challenges related to accurate reporting and the potential for underreported and illegal fishing activities. Reporting fishing effort and catch is highly variable among fisheries and jurisdiction. In many cases, some level of self-reporting of fishing effort and catch is required, and this can lead to inaccurate reports. If fishing regulations limit the number or mass of fish harvested, fishers may be tempted to underreport their catch so as to be allowed to catch more fish. If regulations limit the size of fish than can be harvested, fishers may be tempted to harvest undersized fish. There are various forms of illegal fishing with fishers targeting species not allowed to be harvested, harvesting more fish than allowed, harvesting sizes of fish not allowed to be harvested, and capturing fish in areas or during seasons when harvest is not allowed.

In many fisheries, individual fishers or corporations compete to catch a limited number of fish. This sets up a dynamic that can exacerbate the risk for overharvest as individuals may feel pressure to capture fishes that will otherwise be taken by other fishers. Fisheries have often been held up as an example of *tragedy of the commons*, where individuals compete to exploit a common resource and, in doing so, over-exploit the resource to the detriment of all. The risk for such an outcome has been exacerbated by certain governmental practices. As fisheries harvest has declined, some governments have facilitated continued fishing by providing subsidies to fishers or low-interest loans for fishing equipment. Such practices can lead fishers to become overinvested or overcapitalized, where they have so much funding invested in fishing equipment they feel they have to keep fishing to be able to fund their loan payments. According to Schuhbauer et al. (2020), of the $35.4 billion provided in 2018 by public entities to subsidize global fisheries, the vast majority (80%) went to support industrial fisheries, and $7.7 billion was in the form of fuel subsidies. Therefore, subsidies have especially exacerbated intensive fishing operations and likely served to further increase greenhouse gas emissions attributed to commercial fishery operations.

Fishing Regulations

Given the common issue of overharvest, it is important to be able to limit harvest through regulations. Fishing regulations involve promoting inefficiencies and therefore forcing fishers to capture a lower number of fishes – and a limited size range of fishes – than they feasibly could. Common fishing regulations include seasonal closures of fisheries, fishing gear limitations, limited entry fisheries, quotas on how many fish may be captured, and size regulations. Many of these regulations are not mutually exclusive, and commercial, recreational, and subsistence fisheries may be managed by some combination of regulations.

In many jurisdictions, a first step in regulating fisheries harvest often includes licensing of fishers. Recreational fishers are often required to purchase a daily, monthly, or annual license to be allowed to fish in a certain jurisdiction or target certain species. This can be important not only to generate funding for a management agency but also to be able to track fishing pressure. The information provided as part of the licensing process may allow agencies to track the demographics of fishers and improve the ability to communicate directly with licensed fishers.

Fisheries quotas can take many forms and be applied at the level of the entire fishery or at the level of the individual fisher. Total allowable catch (TAC) quotas are often used to limit the overall harvest from fish stocks. TACs are generally set each year, and through reporting and monitoring, fishery harvest is tracked such that once overall harvest for that year exceeds the TAC level, fishing is closed until the next year. Individual transferable quotas (ITQs) are used in commercial fisheries to limit the harvest of a single fisher or corporation. As the name suggests, ITQs can be sold and transferred among fishers. In limited entry fisheries, ITQs can cap the total number of fishers and require a fisher to purchase someone else's ITQ if they want to enter the fishery. ITQs have also been purchased by government entities and retired to reduce overall pressure on a fishery. Daily or annual bag limits are often used for recreational fisheries to constrain the total number of legal-sized fish an individual angler captures in a day.

Size regulations are common in both recreational and commercial fisheries and are based on the notion that fish can only be retained after capture if they are of appropriate size. This type of regulation is potentially appropriate if capture methods do not lead to the mortality of fish, as it only makes sense to return live fish to the water body of capture. The most common type of size regulation by far is a minimum size limit, where fish are only allowed to be harvested if they are above some minimum length. The idea behind such limits is to allow fish to grow sufficiently large and have a chance to reproduce before they are captured. However, other consequences of this type of regulation are possible, such as reduced growth due to compensatory density-dependent effects and stockpiling of fish below the size limit of harvest. This type of regulation may also have evolutionary effects (see below). Other types of size limits have also been implemented for these reasons. These include maximum size limits, where relatively large fish are not allowed to be harvested, as well as slot limits and inverse slot limits, where only fish of some intermediate size are allowed to be harvested or only rather small and very large fish are allowed to be harvested (i.e. protecting intermediate sized fish; see Figure 32-4).

Another option for influencing size of fish harvested – and for affecting risk of mortality during capture – is to use gear regulations. By restricting mesh size used in gillnets, one can target fish of a certain size, as small fish can swim through the mesh and very large fish will not be tangled in the gear. Anglers may be restricted to unbarbed hooks or only artificial lures to minimize the likelihood of hooking mortality. The practice of snagging – where fish are hooked in their side rather than relying on biting a lure or bait – is often not allowed. Various gears have also been designed to limit the likelihood of bycatch or habitat destruction (see below).

Several fishing regulations involve closing fisheries during a particular period or location. A spawning closure is a common temporal closure intended to protect fishes from harvest while they are actively spawning and providing protection for their offspring. The intent of spawning closures is to protect fishes when they may be particularly vulnerable to capture and when the removal of adults may compromise survival of young fish. Closures of whole fisheries for prolonged periods constitute more severe actions but have been implemented when fisheries are sufficiently over-exploited to the point that all harvest pressure must be removed so the stocks can recover.

Spatial fishery closures restrict fisheries harvest in certain areas. These types of regulations have been in place in freshwater and marine systems for many years and have been referred to as refuges, sanctuaries, or protected areas. Support for expansion of marine-protected areas (MPAs) as a tool to conserve fish stocks has gained a great deal of support in recent years. An attractive aspect of MPAs is that they are a quite simple concept. Protecting large areas of ocean would ideally allow fish stocks to thrive in these areas and serve as a source of recruits to surrounding, unprotected areas. In order for such effects to come to fruition, the movement

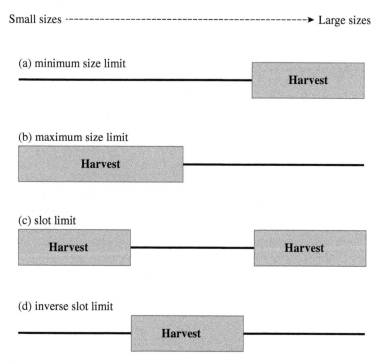

FIGURE 32-4 Illustration of typical size limits used to regulate fisheries harvest, including (a) minimum size limits (most common), (b) maximum size limits, (c) slot limits, and (d) inverse slot limits.

patterns of young and adult fishes are critical. On one hand, if adult fishes move frequently between protected and unprotected areas, high harvest outside of protected areas could lead to stock declines and undermine the benefits of an MPA. On the other hand, if many fish remain inside of MPAs and other fish move out of MPAs where they can be harvested, an MPA may be able to simultaneously protect fish stocks and provide for sustainable fisheries in surrounding areas. In 2016, the IUCN World Conservation Congress set a goal of protecting 30% of the world's oceans. Indexing of global coverage of MPAs is difficult in part because of inconsistencies in defining what constitutes an MPA and summarizing area or volume. However, various indices generally demonstrate that global coverage is well short of the 30% target. For example, in 2015 global coverage may have been as low as 3.5% (Lubchenco and Grorud-Colvert 2015).

The types of regulations described above are key tools for effective fisheries management. Scientifically sound regulations and active fisheries management can serve to protect fisheries from overharvest and may help rebuild previously over-exploited fish stocks. Hilborn et al. (2020) compiled fisheries assessment data from throughout the world, covering 822 marine stocks from the early 1970s to late 2010s (**ramlegacy.org**). These compiled data represented roughly half of the world's marine fishery catch and include estimates of stock biomass, catch, and fishing pressure (or mortality). Hilborn et al. (2020) found that abundance of many stocks had increased in recent years. As part of this analysis, they also considered whether fisheries were intensively managed (using a Fisheries Management Index from Melnychuk et al. 2017). Hilborn et al. (2020) reached the conclusion that fisheries in regions with more intensive fisheries management were in better shape in terms of higher relative biomass and more sustainable catch and fishing pressure. It is worth noting that this analysis only included information for stocks with reliable stock assessments. Hilborn et al. (2020) pointed out that for the other half of marine fish stocks, countries and organizations lack reliable information to assess stock status. They suggest that most countries with limited reliable information are also characterized by low levels of fisheries management, and stocks in such regions likely remain depleted and over-exploited. Globally, it is important to continue to expand the number of stocks reliably assessed and actively managed.

Biological and Ecological Consequence of Fisheries Harvest

The management of fisheries involves simultaneous consideration of fish biology and ecological interactions, biotic and abiotic habitat characteristics, and socio-economic anthropogenic dynamics. Much of fisheries management focuses on management of human actions toward the aim of providing humans with sustainable food sources. However, from the perspective of fish stocks, fisheries harvest can have a variety of ecological – and even evolutionary – consequences. Fisheries harvest can directly influence the abundances and demographics of fish stocks and can affect specific vital rates through indirect pathways. Fisheries harvest can also alter habitats and have food web and ecosystem level effects on exploited systems.

Population Depletion and Demographic Effects

The most obvious and direct effect of fisheries harvest is reduced abundances of target populations. By introducing fishing mortality in addition to natural mortality, exploited fish stocks experience greater total mortality and will not be as abundant. Fisheries-related population depletion is evident across the world in both freshwater and marine systems and as a consequence of exploitation by commercial, recreational, and artisanal fisheries. Abundances of countless fish populations have declined due to fisheries harvest, and some species have even been extirpated.

While fisheries harvest is obviously expected to reduce abundances, the effects may not always be as severe as expected due to tempering effects of compensatory mechanisms. If a specific fish stock is limited by compensatory density-dependent effects, harvest may serve to thin out the stock and release surviving individuals from some density-dependent controls. Surviving individuals may experience lower natural mortality, for example, if reduced intra-population competition decreases probability of starvation or if lower densities reduce disease transmission. Surviving individuals may experience greater per capita growth rates and higher per capita reproductive rates. While release from density-dependent control may somewhat offset harvest-related mortality, per capita compensatory effects will generally not overcome the overall declines in population abundance and biomass associated with fisheries harvest.

In addition to effects on mortality, fisheries harvest will strongly affect the demographics of fish stocks. With increased mortality, fish will on average not survive to as old of an age. Therefore, exploited stocks will tend to be dominated more by younger fish than unexploited stocks. Since individual size tends to increase with age, exploited stocks will also tend to be characterized by relatively small fish. These types of demographic effects can be exacerbated by the reality that increases in mortality will not tend to be equivalent across a fish stock. All fisheries harvest methods are essentially selective, and certain individuals are more likely to be harvested than others. Larger fish (and therefore older fish) are often more likely to be harvested than smaller fish. Many fisheries gears allow small fish to escape catch, while large fish are vulnerable. In addition, minimum size limits are one of the most common types of fisheries regulations. Even if small and large fish are as likely to be caught, only large fish can be legally harvested under such regulations. Together with increased total mortality, the common phe-nomenon of size-biased harvesting can greatly alter demographics of stocks toward younger ages and smaller sizes.

In addition to size and age, other attributes of fishes may make them more likely to be har-vested. Fish that occupy certain habitats may be more likely to be harvested than fish occupying other habitats. The specific timing of migrations by fish stocks may make some individuals more vulnerable to harvest than others. Sex is another attribute that often affects vulnerability to harvest. Fish roe (eggs) is targeted by many fisheries, including fisheries for many species of sturgeon, herring, salmon, vendace, and cod. For many of these fisheries, other parts of the fish – in addition to the roe – are consumed by humans. However, the roe is the only sought-after tissue for some fisheries. In such cases, the risk of harvest is much greater for females than males. Many species of fish display sexual dimorphism, and it is common for female fish to grow to a larger size than male fish. Therefore, many size-selective fisheries that target larger fish are simultaneously sex-selective. Both roe fisheries and size-selective non-roe fisheries can contribute to sex-biased fish populations, with an overrepresentation of males in the popula-tion, after females have been harvested.

The southern Lake Michigan yellow perch fishery experienced prolonged size-selective harvest by both recreational and commercial fishers until the commercial fishery closed in 1997. Anglers were more interested in harvesting larger yellow perch, and commercial fishers used selective gillnets to catch larger perch. These harvest dynamics over time contributed to a truncated size distribution with an overrepresentation of smaller, younger fish in the population. Also, since yellow perch display strong sexual size dimorphism, with females growing larger than males, size-selective harvest contributed to a sexually skewed population. After years of intensive commercial fishing, Lauer et al. (2008) found that in 1994 females constituted less than 25% of individual perch collected in southern Lake Michigan during assessment trawling.

Fisheries Harvest and Recruitment Variability

By altering demographics of fish populations, fisheries harvest can also affect the recruitment potential and level of recruitment variation of fish populations. Fisheries harvest can lead to a strong decrease in a population's egg production potential by reducing the number of adult fish and specifically reducing the number of large, old, and highly fecund female fish, caus-ing overall egg production by a fish stock to decline precipitously. Also, as described in Chap-ter 21, for many fish species, larger, older female fish not only produce a particularly large number of eggs, but they also appear to produce larger offspring with a better chance to survive through early life. Therefore, these spawners may be particularly important for recruitment potential, and their loss may have disproportionate effects on population production. This was highlighted in a review by Hixon and colleagues (2014), where they called for the protection of BOFFFFs (big, old, fat, fecund, female fish).

Recruitment is highly variable, and while relationships between spawner abundance and recruitment tend to be rather weak, they also tend to be nonlinear (see Chapter 12). For example, the Ricker stock-recruitment model suggests that recruitment would be greatest at an intermediate level of spawner abundance. Therefore, even if fisheries harvest reduces the abundance of spawning fish, and even if the demographics of populations shift toward younger fish, it is possible that average recruitment levels may actually increase, and it is highly likely that populations will still be able to produce occasional strong recruitment year classes. How-ever, year-to-year variability in recruitment – and therefore variability in overall population size – may increase in response to a truncated population. A truncated adult fish population will consist of a smaller number of year classes, and therefore population abundance will be dependent on a lower number of annual recruitment events. So, there will be less potential for population size to even out across a larger number of recruitment events. Smaller offspring (i.e. those produced by relatively young and small spawners) may also be more susceptible to environmental effects and may therefore experience more variable survival and contribute to variable recruitment success. Some studies have demonstrated that the timing of spawning will differ among adult fish of different ages. During a spawning season, older spawners often spawn earlier in the season, while young, first-time spawners tend to spawn later. Due to year-to-year differences in seasonal conditions, the offspring of early spawners may survive better during some years, while in other years, later spawners' offspring may fare better. A population with a diversity of spawners of different ages is more likely to produce at least some offspring each year that encounter favorable conditions during early life. In contrast, an exploited, truncated fish population may only spawn during a rather narrow time window and therefore the mean environmental conditions experienced by offspring may be quite different from year to year, leading to high annual variation in early life survival and recruitment success.

This possibility of fisheries exploitation exacerbating recruitment variation was evaluated by Chih-hao Hsieh and colleagues (2006). These authors evaluated catches of different species of larval fish in the Pacific Ocean as part of the California Cooperative Oceanic Fisheries Investigation (CalCoFI). This survey had sampled larval fishes in the Pacific each year over a 50-year period (1951–2002). The authors assumed that species-specific catches of larval fishes during a year were indicative of the annual abundances of adults of particular species. They then examined how variable species-specific catches of larval fishes were from year to year based on the idea that high variation in the catch of larval fishes from year to year would be indicative of more variable population abundances. They examined temporal variation of 29 different species. These species differed in terms of various life-history traits, habitat preferences, and phylogenetic associations. The species also differed in terms of whether or not they were intensely harvested by fisheries. Hsieh and colleagues accounted for the effects of life-history and ecological traits on populations' variability and therefore attempted to separately consider the effects of fishing on population variability. They found a pattern consistent with expectations. The 13 species directly targeted by fisheries displayed a significantly greater degree of population variability (as indexed by the coefficient of variation of larval fish catches) compared to the 16 species not directly targeted by fisheries.

Fisheries-Induced Evolution

In addition to short-term effects on population abundances and vital rates, fisheries harvest can have longer term evolutionary effects on fish populations. Fisheries harvest can exert very strong selective pressures on fish populations and can select against various behaviors and traits. If such traits and behaviors have a genetic component and are partially inherited, they may, over time, become less prevalent in a population through the artificial selection introduced by fisheries harvest. Essentially, any trait that could lead some individuals to be more vulnerable to capture or may interact with harvest risk to influence individual fitness could be under selective control. One can imagine a whole suite of traits and behaviors that may be susceptible to fisheries-induced evolution (FIE), including habitat occupancy, movement and migration behavior, reproductive investment, and body shape. However, the traits and behaviors that have most frequently been considered to be under FIE influence are growth rates, maturation schedules, and traits – such as aggression – which may influence vulnerability to harvest. The potential for FIE may allow populations to adapt when faced with the reality of fisheries harvest. However, such selection is simultaneously concerning in that it may alter a population's genetic variability and – in extreme cases – may lead populations to lose genetic variability. Such loss of genetic variability may limit a population's ability to adapt in the future and leads to the expectation that, while populations may evolve under fisheries harvest, if fisheries harvest is subsequently ceased, populations may not rapidly evolve back to pre-harvest attributes.

Evolution of Growth Rates

In the absence of size-selective fisheries, individual fish are generally expected to fare better if they grow faster. Larger fish are generally less susceptible to many predators; they can swim faster, capture and consume a larger variety of potential prey, and they have greater energy stores and lower mass-specific metabolic rates, and are therefore not as vulnerable to starvation or other negative effects during periods of resource scarcity. Larger, faster growing fish are also more likely to mature at an earlier age, and relatively large fish generally produce larger gonads and have greater reproductive potential. Therefore, several natural processes should favor faster growth and larger size, and many fish populations have evolved to grow relatively fast and reach a relatively large size. However, with the introduction of size-selective fishing and targeted harvest of large fish, the selective environment favoring rapid growth and large size may be severely altered. There may instead be limited benefit for fish to grow rapidly to a large size, and if size-selective harvest is sufficiently high, individual fitness may actually decrease at larger sizes. It follows that FIE may lead genetically determined growth rates to decrease over time.

A relatively simple study demonstrating the potential for FIE of fish growth rates was conducted by Conover and Munch (2002), who were both with the State University of New York, Stony Brook, at the time. Conover and Munch established a captive population of Atlantic silverside. They then divided the population into various groups and exposed the different groups to distinct harvest regimes across multiple generations. The Atlantic silverside is an annual species, meaning that generation time is short (one year), and one can therefore examine multi-generational effects over just a few years. The harvesting imposed by Conover and Munch was extreme in that in one treatment they harvested the largest 90% of individuals, and in another treatment they harvested the smallest 90% of individuals. In yet another treatment, they harvested 90% of individuals, independent of size. The individuals in each group not harvested were instead allowed to spawn. The offspring of these spawners were then exposed to the same harvest treatments as their parents, and this was repeated for four generations. Conover and Munch then examined growth and size of individuals in these three different groups and found that the groups exposed to harvesting of large individuals were smaller (mean mass 1.05 g) compared to neutral-harvested groups (3.17 g) and small harvested groups (6.47 g). The large harvested groups had evolved to grow more slowly, while neutral-harvested groups had seen no change in growth rates and small harvested groups had evolved to grow more quickly. Some have criticized the experiment by Conover and Munch (2002) because it was too severe (see Hilborn 2006). While size-selective fisheries harvest can be extreme, it is highly unlikely that the largest 90% of individual fish would be captured in a single year or during a single generation. Most harvested fish species are not annual species, and multiple overlapping generations may be expected to slow the pace of FIE. However, Conover and Munch demonstrated the potential for such effects even if considering on a somewhat simplified system.

Another complication in examining FIE of growth rates relates to realized growth rates. While a fish population may evolve genetically to grow more slowly, environmental conditions may somewhat counteract such effects. For example, as described above, fisheries harvest may release populations from compensatory density-dependent controls and lead to faster individual growth rates. It is therefore possible that a population may simultaneously experience FIE toward slower growth and compensatory effects favoring more rapid growth, and therefore realized growth may ultimately not change a great deal. Such effects have been demonstrated via modeling analyses, including an individual-based ecogenetic model developed by Wang and Höök

(2009). In this study, the researchers developed a model with simulated fish whose growth was partially influenced by their environment – including the overall density of the population – and partially influenced by genetic parameters that affected the growth of each individual. Those individuals that survived and grew were able to reproduce and pass on their genetic parameters to the next generation of simulated fish. When size-selective harvesting of the simulated population was introduced, the distribution of genetic parameters in the population shifted toward slower growth. However, average realized growth of individuals in the population actually shifted toward faster growth because, while they were genetically pre-disposed to grow slowly, release from density-dependent control actually led them to grow more rapidly. This example demonstrates the difficulty in being able to detect FIE of growth rates in natural systems.

Evolution of Maturation Schedules

The potential for size-dependent fisheries harvest to lead to selection on genetically determined maturation schedules has long been recognized. However, it is only recently that methods have allowed for more explicit evaluation of this phenomenon. Many populations of fish have evolved to mature at relatively large sizes and old ages in the absence of intense fisheries harvest. As mentioned in previous chapters, there are several potential benefits for individuals to delay maturation. By dedicating energy to somatic growth during early life, individuals may grow larger and have a greater chance of survival. Once they do reproduce, they are larger and likely more fecund. Larger size is also related to being able to hold better spawning locations and attract more and potentially preferred mates. Early reproduction may not only limit somatic growth, but reproduction is stressful and can lead to mortality if an individual's energy reserves are insufficient. At the same time, if individuals delay maturation until they reach a very large size and old age, their lifetime fitness may decrease simply because this would require a longer survival duration before initial reproduction and – for iteroparous fish – could lead to reduced numbers of lifetime spawning events. There is a natural balance between reproducing at an early versus a late age.

Fisheries harvest and size-selective fishing mortality targeting larger fish is expected to alter the balance between delayed and early maturation toward individual fish reproducing at younger ages and smaller sizes. Increased risk of mortality is expected to select for maturation at an earlier age, and selective mortality of relatively large individuals is expected to select for maturation at both earlier ages and smaller sizes. In fact, several monitoring programs have documented that age at maturation (e.g. indexed as age at 50% maturity, A_{50}) shifts toward younger ages with increased fishing pressure. While such a shift may be suggestive of FIE, it does not prove the phenomenon. Removal of individuals through fisheries harvest can release surviving individuals from compensatory density-dependent controls and lead to more rapid growth, favoring earlier maturation, even in the absence of genetic selection. Maturation schedules are controlled both genetically and through environmental experiences and vary related to both age and size. These realities have complicated the ability to unequivocally demonstrate FIE of maturation schedules.

Researchers have developed tools to try to tease apart the influence of genetic controls of maturation schedules versus environmental and growth effects. One such tool is the estimation of probabilistic maturation reaction norms (PMRNs). A reaction norm is a type of genetically defined model of how an organism will respond phenotypically to environmental conditions. This type of model is often used in considering plastic expression of phenotypes. Various researchers – including Ulf Dieckmann from the International Institute for Applied Systems Analysis in Austria and Mikko Heino from the University of Bergen – developed the PMRN approach based on the reaction norm concept but expanded to use statistical approaches to consider and fit population level PMRNs (see Box 32-1). The PMRN approach is somewhat

Box 32-1 Probabilistic Maturation Reaction Norms

Probabilistic maturation reaction norms (PMRNs) are a tool to describe maturation schedules. They are intended to index maturation schedules, while accounting for the effects of individual growth and mortality on maturation schedules. PMRNs attempt to control for the influence of phenomena, such as release from compensatory density-dependent limitations on growth or differences in growth and mortality environments on maturation schedules. Therefore, PMRNs facilitate evaluation of how intrinsic maturation schedules differ among populations or change within a population over time. It is attractive to interpret PMRNs as controlling for environmental effects and revealing the genetic component of maturations schedules, but this would be over-interpreting PMRNs, as these statistical models cannot fully account for all environmental effects. However, they come much closer to depicting adaptive, intrinsic maturation schedules than several other maturation indices (e.g. A_{50} and L_{50}).

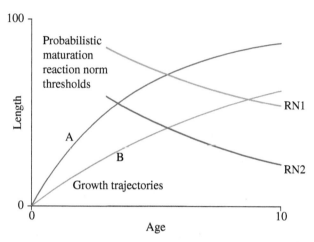

Most PMRNs model the probability of fish becoming mature as a function of age and length. In the figure associated with this box, the bold curves depict the PMRN 50% thresholds for two different populations (1 and 2). The thin curves represent growth trajectories for two female individuals (A and B). These individuals could be genetically identical, and their different growth trajectories could come about due to different environmental experiences. As individuals grow (increase in age and length), their probability for maturing increases, and in particular, when the growth trajectory crosses the PMRN 50% threshold, the probability of maturing is 50%. If maturation of individuals A and B followed the PMRN for population 1 (RN1), individual A is expected to mature at a younger age and larger size than individual B. If their maturation is related to PMRN for population 2 (RN2), individual A would be expected to mature at a younger age and larger size than individual B, although at an younger age and smaller size than under RN1. If maturations of the two individuals are influenced by different PMRNs, they may be expected to mature at similar lengths or ages despite having different growth trajectories. For example, if maturation of individual A is depicted by RN1 and maturation of individual B is depicted by RN2, they may be expected to mature at similar ages although their size threshold for maturation may be different. Similarly, if maturation of individual A is depicted by RN2 and maturation of individual B is depicted by RN1, they may be expected to mature at similar length, but their age threshold for maturation would be different, with individual B maturing much later than individual A. This demonstrates the added insight PMRNs provide beyond what traditional maturation indices – such as age at 50% maturation (A_{50}) or length at 50% maturation (L_{50}) – provide on their own.

Statistical methods are used to estimate PMRNs for different populations. These generally require a large number of records on individual fish size, age, and maturation status. Age-specific probabilities of maturation as a function of length can only be estimated for ages with sufficient records of both mature and immature fish. Nonetheless, for many populations it has been possible to estimate sex-specific PMRNs across a number of ages. This has allowed for comparisons of how these intrinsic maturation indices change over time within a population as the population is exposed to varying levels of size-selective fisheries harvest.

different than the traditional reaction norm approach in that it does not depict phenotypic expression relative to a specific environmental factor. Rather, the approach aims to account for growth and survival influences on age-specific population maturation schedules and therefore allow for comparison of maturation patterns while controlling for potential compensatory and plastic effects.

Esben Olsen worked with Dieckman, Heino, and other collaborators (2004) to estimate PMRNs for Atlantic cod from the northwest Atlantic. After decades to centuries of intensive harvest, the fishery for cod in this region collapsed during the late 1980s and early 1990s. Olsen and colleagues examined size, age, and maturity data on cod, before, during, and after this collapse (1977–2002). They demonstrated that age at 50% maturity (A_{50}) declined from the 1980s to 1990s. However, they also acknowledged that this could be related to release from compensatory density-dependent control. The PMRN approach allowed them to account for such effects and showed that intrinsic maturation schedules of cod had shifted over time. In a particular spatial district in the Atlantic (2J), they showed that, while in the early 1980s, 50% of age-5, female cod were likely to become mature at approximately 58 cm in length, by the early 1990s, 50% of age-5, female cod in the same district were likely to become mature at approximately 42 cm in length. This represents a dramatic shift in age-specific maturation size threshold and appears to be related to intensive size-selective harvest of Atlantic cod.

Several authors have now used the PMRN approach to examine changes in maturation schedules of both marine and freshwater fish populations in response to size-selective fishing. Many of these studies have found similar results as Olsen et al. (2004), i.e. a downward shift of PMRNs over time consistent with harvest of larger fish. Many of these studies have also speculated that with the cessation of harvest, it would be difficult for populations to evolve back to their pre-harvest maturation schedules because (a) the selective pressure to evolve in the opposite direction may not be sufficiently strong and (b) the genetic variation in the population may not be sufficient to allow for such evolution. For example, Pinsky and Palumbi (2014) examined genetic variation between fish populations either overfished or not overfished and found there was less genetic diversity in overfished populations. In contrast to this expectation, Feiner et al. (2015) examined PMRNs of yellow perch in southern Lake Michigan during 1979–2006. In the beginning of this period, there was an intensive size-selective commercial fishery in place, but stricter regulations were put into place during the early 1990s and the commercial fishery was closed in 1997. These authors found an unexpected rapid recovery of perch PMRNs following this fishery closure. After the commercial fishery was closed, the age-specific sizes at maturation for female yellow perch shifted back to greater lengths after only 10 years. It is unclear why PMRNs were able to seemingly recover so quickly after fishing ceased, but the authors suggested it may be related to the relatively short generation time of yellow perch, other environmental changes in Lake Michigan that could have influenced maturation schedules over time, and migration of yellow perch from other locations in the Lake Michigan system that could have introduced novel genetic variation and facilitated the rapid recovery of PMRNs.

Vulnerability to Harvest and Selection for Behavior

The potential for FIE has generally been examined more for commercial fisheries than for recreational fisheries. However, similar to commercial fisheries, many recreational fisheries target the largest individuals, in particular through minimum size regulations that only allow for harvesting of larger individuals. Therefore, if recreational fishing pressure is sufficiently intense, it could have similar evolutionary effects on maturation schedules and growth rates as size-selective commercial fisheries. Recreational fisheries may also select for additional individual traits. In particular, recreational angling is based on individual fish being sufficiently aggressive to strike an artificial lure or bait. Individual fish are likely to vary in terms of aggression and may differ in terms of their likelihood to strike lures or bait, and their susceptibility to harvest. If aggression is a heritable trait, this type of effect may select against aggression in a population of fish targeted by recreational angling.

Various studies have now examined how selective harvesting can select for different heritable behavioral traits in fishes. As an example, David Sutter and colleagues (2012) examined recreational angling effects on not only behavior but also fitness. They compared two lines of largemouth bass – one that had been selected to be very susceptible to angling harvest over three generations and another that had been selected to be much less susceptible to angling harvest over three generations. They stocked four mature males in each of six different ponds from the highly susceptible line and four mature males of the low susceptibility line. Each male was tagged to allow for easy identification of the individual and the selection line they belonged to. They also stocked 42 females from a wild population across the ponds and then observed behavior of the largemouth bass and documented reproductive success of the different males. Recall that largemouth bass are nest breeders and males build nests and then guard nests after females spawn in them. The authors found an effect of male size on reproductive output and behavior. Among relatively large males, individual males from the highly susceptible to angling line had more eggs deposited into their nests and spent longer amount of time guarding nests as eggs developed. The authors also found, through observations, that males from the highly susceptible to angling line were more engaged in egg care (fanning eggs with caudal fin to allow for oxygenation, actively chasing off potential egg predators), and they were more aggressive (more willing to strike at hookless fishing lures presented to them while guarding nests). In fall, the ponds were drained, and surviving juvenile fish were genetically assigned to different parents in each pond. The authors found that larger males tend to have sired a larger number of surviving offspring, and in particular, large males from the highly susceptible to angling line sired the greatest number of offspring. The results of this study demonstrate that angling can remove more aggressive fish from a population and lead to differences in behaviors across generations. Since more aggressive male bass also appeared to be more reproductively fit, angling may not only select for behavior, but may affect the reproductive potential of targeted fish populations.

Habitat Effects and Bycatch

Fisheries harvest can be detrimental, not only for the targeted fish stock but also for other, non-targeted fish stocks, as well as habitat conditions in the exploited system. Several fishery practices directly affect habitat features. In particular, bottom trawling has received a lot of attention based on the potential damage of this practice on bottom habitats. In the process of catching fish, heavy bottom trawls may scoop up aquatic vegetation, corals, rocks, and other structures that support biotic productivity. This can transform the bottom to a much less structurally complex environment and may be compared to repeatedly ploughing an agricultural field, therefore preventing establishment of plants or other structures. Such fisheries-related habitat alterations can subsequently affect the performance of fish stocks as primary production and prey availability may be reduced and refuges from predation for young and small fish may be lost. Bottom trawling is not the only fisheries practice that may be destructive for habitats. Dynamite has been used to capture fish in both marine and freshwater systems and can clearly be destructive. Divers apply cyanide in crevices in some coral reefs to stun and capture fish, especially for the ornamental aquarium market. Such chemical poisoning can not only lead to unintended mortality of fish, but can also be quite destructive to living corals.

In addition to effects on habitat conditions, many fishing practices harvest taxa other than the intended target species. Bycatch has been defined, summarized, and reported in various ways, which leads to inconsistent perspectives of how important bycatch is on a global basis. However, bycatch is clearly a big concern in a variety of fisheries and has been estimated conservatively to equate to about 10% of total global fisheries harvest in 2015 (Zeller et al. 2017). Bycatch is difficult to track because many of the fish caught as bycatch are not reported. They may be released alive back into the system of capture but may not survive long term. In some cases, dead fish are thrown back into the water simply because regulations do not allow harvest of such non-target species. In others, fishes caught as bycatch may be retained and secretly harvested even though such practices are illegal. The result is that, in addition to reported global fisheries harvest, there is a large amount of unreported bycatch.

The issues of fishing-related habitat destruction and bycatch are ongoing management challenges. The extent of these issues may not be fully appreciated, and therefore a first step in addressing these issues is synthesizing their effects. Oliver et al. (2015) synthesized global information on bycatch of sharks and rays. They considered different fisheries and the ratios of shark or ray bycatch versus catch of target species. They found that elasmobranch bycatch was strongly related to fishing practices. Sharks were particularly vulnerable to being harvested as bycatch by offshore, longline fisheries, while rays were susceptible to trawling. Oliver et al. (2015) also demonstrated that many fisheries with high elasmobranch bycatch operated in international waters, where regulations may be less straightforward to implement.

Some fisheries gear and regulations are now designed to limit deleterious effects of habitat destruction and bycatch. Bottom trawls have been modified in various ways to have less contact with the bottom and therefore reduce habitat destruction. A variety of bycatch reduction devices have been developed. Several of these have been designed for trawls, such that when nontarget species are caught in a trawl, they are able to escape. An example is a turtle exclusion device that involves some sort of deflector in the trawl that allows target organisms to pass through, while large organisms – like turtles – are deflected out and escape through a hatch in the trawl net. There are also hatches placed toward the end of trawl nets targeting shrimp, which allow some strongly swimming fishes to escape while shrimp are retained. The most effective approach for limiting fishing-related habitat destruction and bycatch involves simply not fishing in certain areas and time periods. Therefore, spatial closures and MPAs (described above) are often used, not only to constrain harvest of target species but also to limit habitat destruction and bycatch.

Food Web Effects of Fisheries Harvest

Even in the absence of bycatch, fisheries harvest can have effects on species other than the primary target species. Each fish population maintains a variety of interactions with other populations, including predator–prey, competitive, parasitic, and mutualistic interactions. When the abundance of a population is reduced through fisheries harvest, this affects interactions with other populations and may contribute to decreases or increases in abundances of other populations. If fisheries harvest alters the size, age, or sex distribution of a population, this may alter ecological interactions. For example, smaller individuals often consume smaller prey and are vulnerable to more gape-limited predators. Thereby, an exploited fish population that becomes dominated by relatively small individuals may be characterized by very different ecological interactions than a population that is characterized by more large individuals. Through such altered ecological interactions, fisheries harvest can have a variety of cascading effects on non-target species and alter the structure and composition of a food web. Such effects have been directly observed in several systems and have been modeled in other systems using tools such as EcoPath and EcoSim described in Chapter 26.

Fisheries harvest of top predators has been hypothesized to trigger top-down trophic cascades as described in Chapter 30. That is, by removing top predators, their prey populations are released from predatory control. The increased abundance of populations one trophic step down from top predators allows them to exert increased predatory control on populations two trophic steps down from top predators, which leads these populations to decline in abundance. For example, Ken Frank and colleagues from Canada's Department of Fisheries and Ocean, Bedford Institute of Oceanography, examined data collected over several decades in the northwest Atlantic – specifically the Scotian Shelf near Nova Scotia (Frank et al. 2005). This is a system previously dominated by Atlantic cod and other groundfish. Prolonged intensive fisheries harvest led to the dramatic decline of cod, haddock, two species of hake and various other benthic species. After the decline of these species, their prey – including pelagic fish and large benthic invertebrates (crab and shrimp) – increased in abundance. Planktivorous fish and invertebrates often select for relatively large zooplankters, and – consistent with this tendency – as small pelagic fish and benthic invertebrates increased, the abundance of large herbivorous zooplankton (>2 mm in length) decreased, while the abundance of small zooplankton (<2 mm in length) remained fairly constant. Overall decreases in herbivorous zooplankton were accompanied by increased phytoplankton densities and decreased concentrations of nitrate – the limiting nutrient for photosynthesis in such marine systems. These patterns are very much consistent with the expectations of a trophic cascade when top predators are removed. Although, it is difficult to unequivocally prove that all of these observed temporal patterns are entirely driven by trophic cascade dynamics as there are so many processes operating in large natural systems.

As described in Chapter 26, regime shifts are sudden, dramatic changes in the composition, structure, and function of ecosystems, which may be quite difficult to reverse. Regime shifts have been demonstrated to occur in aquatic systems due to gradual changes in nutrient loading and less gradual shifts in climate and ocean circulation patterns. Regime shifts may also occur in response to fisheries harvest as depletion of key populations has sweeping effects on the entire ecosystem. For example, the various changes on the Scotian Shelf, in response to cod harvest as described by Frank et al. (2005), could be considered a fisheries-induced regime shift. Christian Möllman and Rabea Diekmann from the University of Hamburg examined data from a number of Northern Hemisphere systems, including the Baltic Sea, Black Sea, North Pacific, North Sea, and Scotian Shelf (Möllman and Diekmann 2012). They used multivariate methods (principal components analyses) to evaluate changes in systems and found evidence across systems for sudden changes in the structure of communities. Many of these shifts occurred coincidently across the Northern Hemisphere during the late 1980s and early 1990s. The authors suggested that these shifts were likely driven by a combination of factors, including overfishing and climatic effects. Overfishing may alter trophic connections in a system and render it susceptible to dramatic shifts in structure, while changes in climate and ocean circulation may be the dramatic trigger for the sudden shift in community structure.

In addition to fisheries exploitation affecting non-target organisms through ecological interactions, past fishing practices have frequently involved the sequential harvest of populations at decreasingly lower trophic levels. This phenomenon, known as *Fishing Down the Food Web*, was highlighted in a study by Daniel Pauly and colleagues (Pauly et al. 1998). These authors examined fisheries harvest data compiled by the FAO for different regions of the world's oceans from 1950 to 1994. Various applications of the EcoPath model allowed them to estimate the mean trophic level of different species and groups of fish. By using EcoPath estimates, each species or group could potentially have non-integer trophic levels as these estimates represent the mean cumulative trophic level of a group consuming a diversity of prey and encapsulating different prey consumed during different stages of ontogeny, during different seasons, and across diverse habitats. Pauly and colleagues then integrated information on species groups harvested during a year in a particular region together with estimates of each species group's trophic level to estimate mean trophic level harvested in a region during a year. They were then able to examine temporal patterns of the trophic levels harvested in marine fisheries. They consistently found that, over time, lower and lower trophic levels were being harvested. In particular, this pattern was evident in the Northern Hemisphere, including the Atlantic and Pacific Oceans and Mediterranean Sea. Essentially, this phenomenon of *Fishing Down the Food Web* seems to come about because of a preference of fishers to catch large fish. Since larger fish are often top predators, targeting larger fish generally implies targeting higher trophic level fish. So, initially higher trophic level fish are targeted. As these larger-bodied, higher trophic level species are depleted, fishers shift to target the next size and lower trophic levels. As these slightly smaller species are subsequently depleted, fishers switch to even smaller and lower trophic levels, and so on.

Summary

Fisheries harvest has served as a major consumptive use of fish stocks over human history. Dramatic increases in capture fisheries occurred globally from the 1950s to 1990s, and many stocks now appear to be overexploited. Concerns regarding overexploitation are widespread and relevant to both marine and freshwater fisheries, as well as commercial, subsistence, and recreational fisheries. While fisheries harvest can lead to depletion of fish stocks, fisheries exploitation can have a variety of additional demographic, ecological, and evolutionary effects. Exploitation tends to lead to fish stocks dominated by small, young fish due to the size selectivity of fisheries harvest. The selective nature of fisheries can also lead to the evolution of fish stocks in terms of growth potential, maturation schedules, and behavior. Fishing activities and harvest can have impacts beyond the specific species targeted, by potentially altering habitat conditions and affecting non-target species, either through mortality via bycatch or through modified ecological interactions and food web effects. Therefore, fisheries simultaneously constitute a key sector for feeding the global human population as well as perhaps the greatest single stressor facing many wild fish populations.

Literature Cited

Allan, J.D., R. Abell, Z. Hogan, et al. 2005. Overfishing of inland waters. Bioscience 55:1041–1051.

Arlinghaus, R., S.J. Cooke, J. Lyman, et al. 2007. Understanding the complexity of catch-and-release in recreational fishing: An integrative synthesis of global knowledge from historical, ethical, social and biological perspectives. Reviews in Fishery Science 15:75–167.

Arlinghaus, R., R. Tillner, and M. Bork. 2015. Explaining participation rates in recreational fishing across industrialized countries. Fisheries Management and Ecology 22:45–55.

Conover, D.O., and S.B. Munch. 2002. Sustaining fisheries yields over evolutionary scales. Science 297: 94–96.

Cooke, S.J., and I.G. Cowx. 2004. The role of recreational fishing in global fish crises. Bioscience 54:857–859.

Cooke, S.J., and H.L. Schramm. 2007. Catch-and-release science and its application to conservation and management of recreational fisheries. Fisheries Management and Ecology 14:73–79.

Cooke, S.J., P. Venturelli, W.M. Twardek, et al. 2021. Technological innovations in the recreational fishing sector: Implications for fisheries management and policy. Reviews in Fish Biology and Fisheries 31:253–288.

Eigaard, O.R., P. Marchal, H. Gislason, and A.D. Rijnsdorp. 2014. Technological Development and Fisheries Management. Reviews in Fisheries Science & Aquaculture 22:156–174.

FAO. 2020. The State of World Fisheries and Aquaculture 2020. Sustainability in action. Rome. http://www.fao.org/3/ca9229en/CA9229EN.pdf

Feiner, Z.S., S.C. Chong, C.T. Knight, et al. 2015. Rapidly shifting maturation schedules following reduced commercial harvest in a freshwater fish. Evolutionary Applications 8:724–737.

Fluet-Chouinard, E., S. Funge-Smith and P.B. McIntyre. 2018. Global hidden harvest of freshwater fish revealed by household surveys. Proceedings of the National Academy of Sciences 115:7623–7628.

Frank, K.T., B. Petrie, J.S. Choi, and W.C. Leggett. 2005. Trophic cascades in a formerly cod-dominated ecosystem. Science 308:1621–1623.

Hilborn, R. 2006. Faith-based fisheries. Fisheries 31:554–555.

Hilborn, R., and 22 co-authors. 2020. Effective fisheries management instrumental in improving fish stock status. Proceedings of the National Academy of Sciences 117:2218–2224.

Hixon, M.A., D.W. Johnson, and S.M. Sogard. 2014. BOFFFFs: On the importance of conserving old-growth age structure in fishery populations. ICES Journal of Marine Science 71:2171–2185.

Hsieh, C.H., C.S. Reiss, J.R. Hunter, et al. 2006. Fishing elevates variability in the abundance of exploited species. Nature 443:859–862.

Jacquet, J., and D. Pauly. 2008. Funding priorities: Big barriers to small-scale-fisheries. Conservation Biology 22: 832–835.

Lauer, T.E., J.C. Doll, P.J. Allen, B. Breidert, and J. Palla. 2008. Changes in yellow perch length frequencies and sex ratios following closure of the commercial fishery and reduction in sport bag limits in southern Lake Michigan. Fisheries Management and Ecology 15:39–47.

Lubchenco, J., and K. Grorud-Colvert. 2015. Making waves: The science and politics of ocean protection. Science 350:382–383.

Melnychuk, M.C., E. Peterson, M. Elliott, and R. Hilborn. 2017. Fisheries management impacts on target species status. Proceedings of the National Academy of Sciences 114:178–183.

Möllman, C., and R. Diekmann. 2012. Marine ecosystem regime shifts induced by climate and overfishing: A review for the northern hemisphere. Advances in Ecological Research 47:303–347.

Oliver, S., M. Braccini, S.J. Newman, and E.S. Harvey. 2015. Global patterns in the bycatch of sharks and rays. Marine Policy 54:86–97.

Olsen, E.M., M. Heino, G.R. Lilly, et al. 2004. Maturation trends indicative of rapid evolution preceded the collapse of northern cod. Nature 428:932–935.

Pinsky, M.L., and S.R. Palumbi. 2014. Meta-analysis reveals lower genetic diversity in overfished populations. Molecular Ecology 23:29–39.

Pauly, D., V. Christensen, J. Dalsgaard, R. Froese, and F. Torres Jr. 1998. Fishing down marine food webs. Science.279:860–863.

Parker, R.W.R., J.L. Blanchard, C. Gardner, et al. 2018. Fuel use and greenhouse gas emissions of world fisheries. Nature Climate Change 8:333–337.

Post, J.R., M. Sullivan, S. Cox, et al. 2002. Canada's recreational fisheries: The invisible collapse? Fisheries 27:6–17.

Schuhbauer, A., D.J. Skerritt, N. Ebrahim, F. Le Manach, and U.R. Sumaila. 2020. The global fisheries subsidies divide between small- and large-scale fisheries. Frontiers in Marine Science 7:539214

Shertzer, K.W., E.H. Williams, J.K. Craig, et al. 2019. Recreational sector is the dominant source of fishing mortality for oceanic fishes in the Southeast United States Atlantic Ocean. Fisheries Management and Ecology 26:621–629.

Sutter, D.A., C.D. Suski, D.P. Philipp, et al. 2012. Recreational fishing selectively captures individuals with the highest fitness potential. Proceedings of the National Academy of Sciences 109:20960–20965.

Wang, H.Y., and T.O. Höök. 2009. Eco-genetic model to explore fishing-induced ecological and evolutionary effects on growth and maturation schedules. Evolutionary Applications 2:438–455.

Zeller, D., T. Cashion, M. Palomares, and D. Pauly. 2017. Global marine fisheries discards: A synthesis of reconstructed data. Fish and Fisheries 19:30–39.

CHAPTER 33

Invasive Species

This chapter deals with non-native species and problems they can cause to fish populations. An important priority in preserving our natural heritage is ensuring continuance of biotic diversity. Freshwater ecosystems are among the most damaged in this regard as uses of water for irrigation, human consumption, or industry – as well as discharge of sewage into natural waters – often result in degradation of these waters and damage to aquatic communities. This damage may be largely invisible to people nearby as it occurs underwater with species rarely or never seen. Yet among vertebrates, fishes represent over 50% of all known species, and this percentage will likely increase if all existing species are described. Thus, there is potential for large loss of biodiversity through human influences on fish populations. Coupled with this, other aquatic organisms also rely on the quality of aquatic ecosystems and may suffer even more dramatically than fish in losses of biodiversity.

The situation with non-native species covers a variety of different impacts as well as types of invasions (Copp et al. 2005). Clearly, non-native simply means a species that was not historically found in the watershed or location. Such a species could have moved there on its own accord (often aided by human modifications of the watershed), been intentionally transplanted there, or now be expanding to that location after introduction from a remote source such as another country. Non-native fishes, also termed nonindigenous or exotic species, may have beneficial, neutral, or negative impacts in a given ecosystem. Nuisance species are those that have negative impacts and may include native as well as non-native species. Invasive species are a special category defined by the US National Aquatic Invasive Species Act of 2003 as a nonindigenous species whose introduction may cause harm to the economy, environment, human health, recreation, or public welfare. These terms intertwine in many ways and a particular species may be classified in several of these groups in different locations. The purpose of this chapter is to overview non-native species and their effect on freshwater fishes in the United States as a means to understand future issues in maintenance of aquatic biodiversity.

Non-native Species and Introductions

One of the most important factors threatening biodiversity of native fishes is the introduction of species. This seems like a contradiction because addition of new species should increase the local biodiversity. However, immigration and extinction processes in aquatic ecosystems have often resulted in loss of native fauna after introduction of a non-native fish. The management of fishes by stocking new species into natural waters has been a very popular activity in human history. This intentional translocation of organisms has occurred far more commonly

Biology and Ecology of Fishes, Third Edition. James S. Diana and Tomas O. Höök.
© 2023 John Wiley & Sons Ltd. Published 2023 by John Wiley & Sons Ltd.

in fishes than in other animals. Intentional introductions to improve fishing often result in new genotypes or entirely new species being introduced into a watershed, sometimes with damaging effects on resident fishes. In many ecosystems today, the percentage of non-native species found in a given location exceeds 40% of total species richness. Not all fish introductions have been intentional. On occasion, modification of habitat – such as building canals to connect different watersheds or circumvent waterfalls – has resulted unintentionally in fishes gaining access to new areas as well. Regardless of the means of distribution, the spread of invasive species has caused much damage to native biodiversity and is a major concern today.

There are currently strong regulations against intentional introduction of non-native species by private citizens. However, state and federal agencies charged with managing sport fishing still have considerable opportunity to introduce non-native fishes. While such management is coming under increasing scrutiny, it is still a potential threat to biodiversity.

Governments have recently become very interested in nuisance aquatic species, invasive species, and other categories of organisms that have damaging impacts. This interest has evolved as the damage done by invasive organisms has increased and knowledge of that damage has spread. For example, the United States passed the Nonindigenous Aquatic Nuisance Prevention and Control Act in 1990 and later changed it to the US National Aquatic Invasive Species Act of 2003. The purpose of these acts was to federally control unintentional and intentional introductions of invasive species and to develop control methodology to deal with current problems related to invasive species. However, the recognition of problems with invasive species is not a recent development. One can examine the history of Australia and see the devastating effects caused by introductions of rabbit, prickly pear cactus, cane toad, and many other species. While such introductions may have had positive intents of increasing food production, controlling pests, or simply adding a familiar and enjoyable species to the area, many have turned out to be disasters. In many cases, these invasive species dominate the ecosystem there, often resulting in major habitat alteration and declines or extirpation of native species. In spite of these histories, we have been slow to accept that problems with ecological balance cannot often be solved by simply introducing new species to fill a "vacant niche" in damaged ecosystems. Indeed, the opposite effect is often achieved.

The introduction of non-native sport fishes is a particularly pervasive problem that can generate a downward cycle of demand and degradation. It can also generate considerable economic and recreational value by increasing tourism and fishing in the area. This trade-off can be very difficult to manage. Improving fishing by stocking a favored non-native species like brown trout can result in higher fishing pressure, increased public demand for more stocking, and more fishing pressure on other native species in the same ecosystem. One example is that the great popularity of trout fishing has resulted in rainbow trout being introduced into most areas of the world with temperature conditions that allow it to persist. Rainbow trout are introduced to increase fishing for an immensely popular gamefish and has resulted in increasing angler recreation and enjoyment. However, to provide better trout survival, such introductions have often been coupled with programs to eliminate native "coarse" or "trash" fishes, species that are not as interesting to the angling public but are native and mainstays of our aquatic biodiversity. Additionally, trout have been introduced into a number of fishless lakes, particularly in the high mountains, and have had strong effects on native species there such as frogs, salamanders, and many invertebrates.

Rainbow trout was native to the US and Canada in Pacific coast drainages, extending inland to the Rocky Mountains, but has been widely transplanted and probably represents the most widely introduced fish globally. In western North America, introduction of rainbow trout to improve fishing opportunity has led to hybridization and the subsequent loss of unique genetic lineages of cutthroat trout. The native trout fauna of intermountain western North America is a variety of local species or subspecies of cutthroat trout. Thermal and physical habitat features are believed to be factors that allowed coexistence of the species in native areas, but that isolation is lacking in areas where rainbow trout have been introduced. Muhlfeld et al. (2009b) found that hybridization between rainbow and westslope cutthroat trout was positively correlated with warmer summer water temperatures and more frequent road crossings. Hybridization was negatively correlated with distance to a source population of rainbow trout. Human influences on habitat and water temperature have resulted in increased hybridization in recent years, and climate change over the last 30 years also seems associated with increased hybridization in the Flathead River system (Muhlfeld et al. 2014). Currently, pure cutthroat strains seem to be in isolated in cold headwater areas or locations with physical barriers. Hybridization of the westslope cutthroat trout has led some to consider its listing as an endangered species. Hybrid trout showed up to 50% lower reproductive output than pure species, further threatening persistence (Muhlfeld et al. 2009a). Land-use disturbance, increasing water temperature, and proximity to a source of rainbow trout all threaten the westslope cutthroat trout with genomic extinction, and may require removal or reduction of populations with high frequency of admixture (hybridization) and high density to reduce the rate of hybridization.

Distributions of Non-native Fishes

The importance of non-native fish introductions cannot be overstated. Williams and Meffe (1998) summarized the status and trends in nonindigenous species in the US and showed that individual states had between 12 and 114 species of non-native fish introduced into inland waters (Figure 33-1). On the basis of the raw number of introductions, the leading states included California, Texas, Florida, North Carolina, and Nevada. Some of these introductions have been done intentionally to "improve" biodiversity or fishing, and others have been unintentional. Williams and Meffe reviewed the history of impacts for various non-native species and proposed that as many as 30% of fish species introductions had known harmful effects, and another 25% had an uncertain effect (Figure 33-2). In general, the introduction of non-native species tended to cause more harmful than beneficial effects to natural ecosystems.

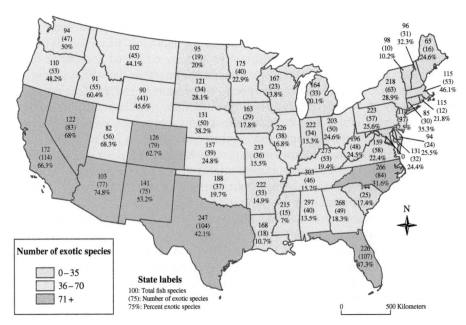

FIGURE 33-1 Total native fauna, number of non-native species, and percent non-native species for inland waters of each US state. *Source: Data from Williams and Meffe (1998) and Master et al. (1998).*

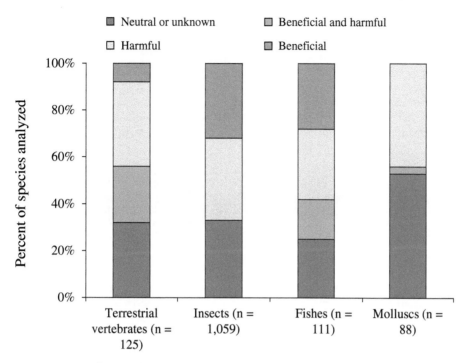

FIGURE 33-2   Percent of non-native species in each taxonomic group that have caused harmful, beneficial, both harmful and beneficial, and neutral or unknown effects on other organisms. *Source: Redrawn from Williams and Meffe (1998).*

How important is the change in biodiversity due to local extinction compared to the change due to introduction? Frank Rahel (2000) evaluated change in species numbers in the 48 contiguous states. His first analysis evaluated changes in the number of shared species among the 1128 pairwise combinations of these states. This analysis (Figure 33-3) indicated that, on average, states now share 15 more species of fish than they did historically. Very few states have fewer shared species, while the majority have more, up to an addition of 40 shared species between two states. This indicates that most states have more similar faunas than they did before human settlement. While extirpations may be more permanent, introductions have been more common (Figure 33-4). In this analysis, an extirpation would be loss of a species from a state, while introduction would be the establishment of a new species in a state. In all, there were 196 extirpation and 901 introduction events. Most states had few extirpations, and no states had more than 10. In contrast, only 50% of the states had small numbers of introductions (mainly states with already diverse fish faunas so fewer species could be added), and some had as many as 50 introductions. The result of these two changes is homogenization of the fauna or reduction of local differences in species composition.

While intentional introductions cause homogenization, so do changes that occur due to indirect influences of humans. One major effect of humans is changing the condition of natural habitats, and the subtle effects of habitat change may also influence persistence of endemic or non-native species. Scott and Helfman (2001) demonstrated correlations between the abundance of endemic (native) or invader (non-native) fish species and the intensity of land use in a watershed (Figure 33-5). In their analysis, they demonstrated a negative correlation between the number of endemic species and land-use intensity, and the correlation explained 50% of the variance in the number of endemic species in the Little Tennessee and French Broad rivers. Land-use intensity was quantified by accounting for the proportion of land deforested, the number

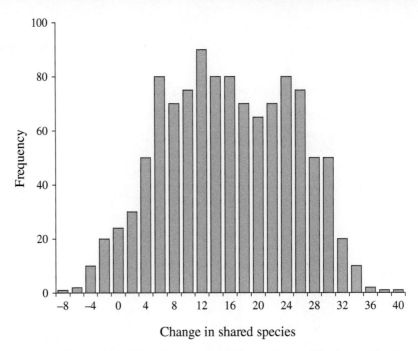

FIGURE 33-3 Changes in number of shared species for 1128 pairwise combinations of the 48 coterminous states. *Source: Redrawn from Rahel (2000).*

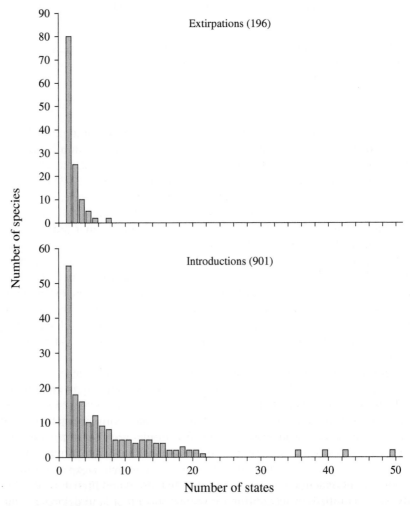

FIGURE 33-4 Number of species extirpated from (top) or introduced to (bottom) a given number of US states. *Source: Redrawn from Rahel (2000).*

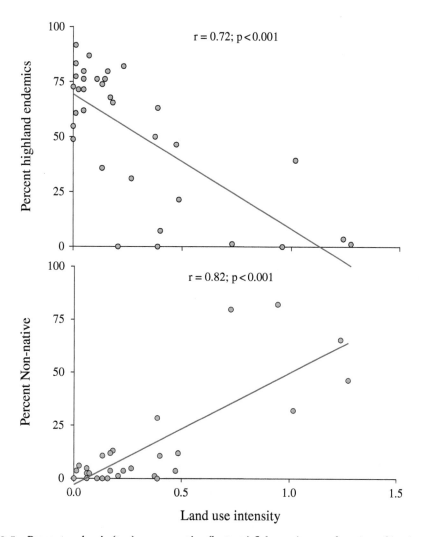

FIGURE 33-5 Percent endemic (top) or non-native (bottom) fish species as a function of land-use intensity for 36 sites in Little Tennessee and French Broad river basins. *Source: Redrawn from Scott and Helfman (2001).*

of buildings per hectare, and the kilometers of roads per hectare, which is a wide spectrum of land-use change but not directly impacting aquatic systems. Scott and Helfman also showed a positive relationship between the abundance of non-native species (considered nuisances) and land-use intensity, and the correlation accounted for 64% of the variance in number of non-native species. Such an analysis does not prove cause and effect but shows strong correlation between the human development of landscapes and homogenization of the fauna.

Changes in the ranges and abundances of fishes have occurred as a result of alteration of habitat, land-use change, climate change, and new species intentionally or inadvertently introduced to an area. A study by Buckwalter et al. (2018) evaluated changes in the size of ranges for fishes in the New River system of North Carolina from historical and current collections. They found 77 species in the watershed with 55% of them being non-native. Spreaders (fish with expanding ranges) included 31 species, which consisted of 31% of native fishes and 52% of introduced species. In contrast, all four decliner species were native. In a related study, Peoples and Midway (2018) evaluated the functional traits of 272 fish species from historical and current collections in the eastern United States to relate range size changes with ecological and biological conditions. Functional traits, such as diet breadth, temperature tolerance, and serial spawning, were correlated with native range size for the fishes. Fishing pressure, life-history characteristics (greater size, fecundity, age at maturity, and life span), and native range size were most related to establishment of expanded ranges. For both studies, there was clear evidence of expanding ranges in a large number of native and introduced fishes, and biological traits of the fish, as well as human use and habitat alteration, were also important in invasions.

Invasive Carps

It is quite possible that no ecosystem in the United States has been damaged as much by invasive species as the Laurentian Great Lakes. We have covered much of this history in a previous chapter, but at this point, the recent problems related to "Asian" carps (we will use the term carps because there are multiple species involved, and either refer to their common names or the generic term carps) movement throughout the Mississippi River drainage and possibly into the Great Lakes have been very controversial. That history and attempts to manage the species invasion are covered here.

There are seven species of carps now in the United States, and five of them are common to the Mississippi River drainage. The common carp was brought to the United States in 1831 and intentionally introduced as a potentially major food fish. It has become a destructive species throughout the United States, causing erosion, plant removal, and turbidity increases throughout wetlands and shallow lakes. Its abundance has also had severe impacts on survival of other native species. Grass carp were imported into the United States in 1963 to use in aquatic vegetation control in aquaculture ponds and lakes. They escaped to become established in the Mississippi River in 1971 and were introduced intentionally into other states to be used in vegetation control in lakes and ponds. Silver carp and bighead carp were imported from China to Arkansas in 1973 for use in improving the water quality and production of aquaculture ponds. They subsequently invaded Midwestern rivers by pond escapement or unknown deliberate introductions. They became established in the Mississippi River system in 1982. Black carp exist in aquaculture ponds and have also spread but are not a major problem at present.

The life history of the invasive carps has a lot to do with their future impacts and distributions. Silver carp filter feed on phytoplankton, while bighead carp mainly consume zooplankton. Grass carp consume large rooted vegetation. Silver, bighead, and grass carp all spawn in large river systems, and their buoyant eggs disperse downstream while developing. As a result, eggs must remain buoyant in moving water for several days until hatching, during which time eggs and larvae can disperse to distances of 80 km or more downstream (Zhu et al. 2018). Their prime spawning habitat is large rivers with a continuous distribution of acceptable currents. While the Mississippi River works well in this regard, dammed rivers are less likely to be successful spawning habitat. The young-of-year disperse into shallow flats and floodplain areas (Collins et al. 2017) and develop there. Therefore, the invasion is largely from adults moving upstream to locations to spawn rather than the young invading new habitats in upstream areas. Also, while the species can survive in lakes, they cannot breed unless they have access to flowing waters for many kilometers.

Silver, bighead, and grass carp were introduced into the Mississippi River system by 1982 or earlier, and their abundance increased dramatically in subsequent years (Figure 33-6). Early surveys showed a dramatic increase through 2000 (Koel et al. 2000), while later surveys indicate even more dramatic increases, as well as recent declines between 2000 and 2020 (Upper Midwest Environmental Sciences Center 2021). Silver carp seem to be the dominant species in most of

(a)

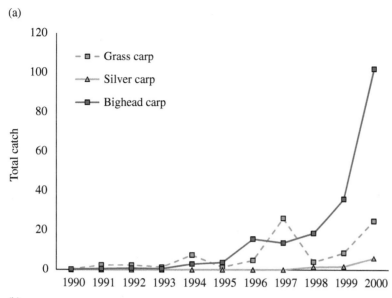

(b)

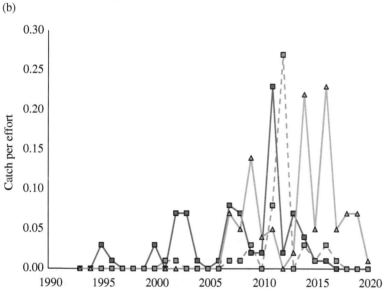

FIGURE 33-6   (a) Annual catch of carps collected by the Long Term Resource Monitoring Program from pool 26 of the Mississippi River, 1990–2000. *Source: Redrawn from Koel et al. (2000).* (b) Annual catch per net night for carps from Pool 26 in large hoop nets from 1993 to 2020. *Source: Data from Upper Midwest Environmental Sciences Center (2021).*

the Mississippi River system, but all three are commonly found. A graphical indication of the status of carps in 2017 for the Mississippi River system is shown in Figure 33-7. They have spread throughout the Mississippi and its tributaries but have not yet become established in the Great Lakes. The silver carp has become an internet sensation because it has reached high abundance in the Mississippi River system and has the tendency to jump after passage of speeding boats. Images of jumping fish have become a major symbol of invasive species issues in the United States.

One common issue related to non-native species is that they are often fairly well established in a new habitat before their presence is detected. This has been true for most invaders in the Great Lakes, including zebra mussels, spiny water fleas, sea lamprey, and alewife. It is

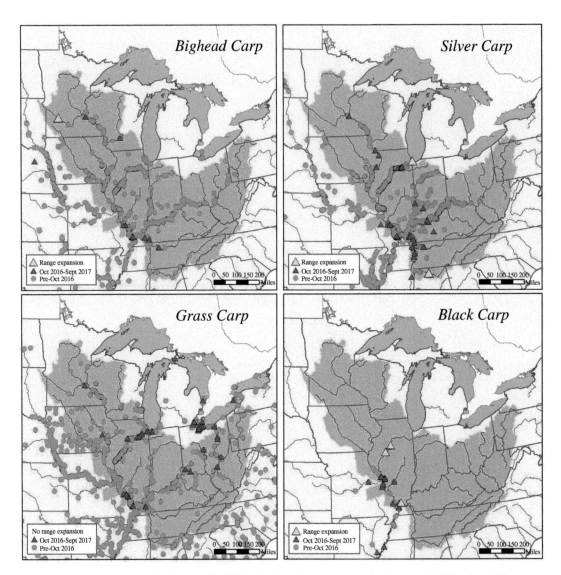

FIGURE 33-7 Range expansion maps of all four species of carp: green circles represent the data points of occurrences before October 2016; red triangles identify the data points collected from October 2016 to September 2017; and yellow triangles indicate occurrences that expanded the range of that species.
Source: Reproduced with permission from ACRCC (2019)/Asian Carp Regional Coordinating Committee.

very difficult to eliminate a non-native species once it has become well established. However, preventing their spread to a new area can be more effective in maintaining the ecological integrity of the region.

This process is clearly true for the carps. The species have spread throughout the Mississippi River system and have become extremely abundant in many areas. As a result of their establishment, native species, such as the gizzard shad and bighead buffalo, have declined in abundance, biomass, and body condition (Pendleton et al. 2017). The silver carp in particular seems to be competing with native species of planktivores (and possibly young-of-year of other species) for their diet. In addition to this impact, native sport fishes have also shown a decline in response to invasion of the carps. Chick et al. (2020) compared locations that have been invaded to areas without carp populations and found abundance of silver carp had a direct negative relationship to abundance of adult sport fish in the upper Mississippi River system. While extirpations have not been apparent, reductions in abundance of native species can clearly be attributed to these carps.

The Mississippi River is a continuous aquatic system, with dams that may disrupt movements but locks for ship passage which provide passage over these dams, so there are no barriers to fish dispersal throughout most of its length. As a result, once established, carps have been able to expand into upstream areas over time. Adult carps are now common in Minnesota and North Dakota, but young-of-year have been limited to farther downstream areas around Illinois.

There is a major concern about introduction of these carps into the Great Lakes. This is possible because the Chicago River system (originally in the Great Lakes drainage) was reversed in flow from Lake Michigan into the Mississippi River. Originally, wastewater from Chicago flowed into Lake Michigan and contaminated the drinking water supply for the city. As a result, diseases like cholera became common and killed 2–3% of the population annually in the 1850s. The engineering solution was to reverse the flow of the Chicago River into the Mississippi and disperse that sewage downstream. However, this connection also opened up an avenue for exchange of fauna between the Great Lakes and the Mississippi River watersheds. One example is that this connection allowed zebra mussels – which had become common in the Great Lakes – to disperse downstream into the Mississippi River system.

A key in stopping the spread of an invasive species is to be able to detect when that species first occurs in a new locale and then mount a response to eliminate it before the population expands. Detection is a difficult problem, as normal methods such as netting or electrofishing have a low probability of collecting rare species, so much effort would be necessary for detection. In response to this, a number of scientists have proposed using environmental DNA (eDNA) as an indicator of presence (Lodge et al. 2012). eDNA is simply DNA that is eliminated from an animal by excretory or other processes and can then be magnified in a water sample using polymerase chain reaction (PCR) and identified to species. Using this method, chemical analysis of a water sample could show whether a particular species is present and may even show the diversity of species present and their abundance. This would be a much simpler method of detection if the testing system is reliable, but the reliability of eDNA testing has been questioned.

eDNA testing is used to detect if carps are present in Great Lakes waters and has been widely applied (ACRCC 2019). Current surveillance of the carps includes extensive testing of eDNA as well as netting and electrofishing. There have been vocal disagreements about the precision of eDNA for species detection; undocumented claims of bias include dead animals being displaced by predators or boats into a new area, fecal matter being deposited by birds in water where animals are not present, and difficulties with detection where water blanks have indicated presence of an organism by eDNA. However, there has been extensive genetic evaluation of best methods for eDNA use, and at present, protocols for species detection in the Great Lakes include the coupling of initial detection of an organism by eDNA with extensive physical collections to validate detection.

Many different methods have been proposed to prevent the expansion of carps into the Great Lakes. The largest fear for their introduction is that they will compete for food with young sport fishes and, as a result, have a dramatic impact on the valuable sport fishery of the Great Lakes – estimated at over $1 billion annually. An electrical barrier has been placed on the Des Plaines River near Brandon Road (Figure 33-8) in order to kill carps that might try to pass that point. Commercial and research fishing has been applied to the area near the barrier to reduce the density of invaders near the structure and reduce chance of passage. A simple (but expensive) solution would be to block off access for all organisms by some sort of physical barrier, but this would also block shipping. So far, there has been no agreement on separation of waters between the Great Lakes and Mississippi drainage. The plan proposed by the Army Corps of Engineers in 2017 was a $275 million effort to control expansion. That has recently been expanded to a proposed $778 million activity, which would involve upgrading the complex at Brandon Road Lock and Dam to include an additional electrical barrier, a bubble barrier, a loud noise deterrent, and a flushing lock that would send dead organisms back to the Mississippi system. While these actions may result in protection against carps, a number of other species like zebra mussel would not be deterred by such devices. In addition, it has been nine years since the initial concept of blocking carps passage was proposed by the Army Corps of Engineers, with little action beyond planning so far. The projected date to complete this newly proposed barrier works would be 2027. It is entirely possible that the carps could be established in the Great Lakes before this operation is even completed. There have been collections of eDNA from carps in Lake Michigan (Figure 33-8) and Lake Erie, giving some credence to the idea that they may already be in the lakes in a propagule large enough for successful reproduction. In reality, only physical separation of the waters would produce a lasting result (Rasmussen et al. 2011), and the cost of it would likely not exceed the $778 million price tag for the currently proposed works.

The invasive carps issue also brings up the importance of non-native species as a societal concern. In the case of the Great Lakes, one of the main arguments for preventing the invasion is the potential damage to sport fishing for salmon, which are also non-native species. Salmon fishing provides a major economic input into the communities around the lakes, so it is not simply a case of desire but also of economics. It is relevant to realize that on land there have been major manipulations of biodiversity for farming, logging, and urban development, and

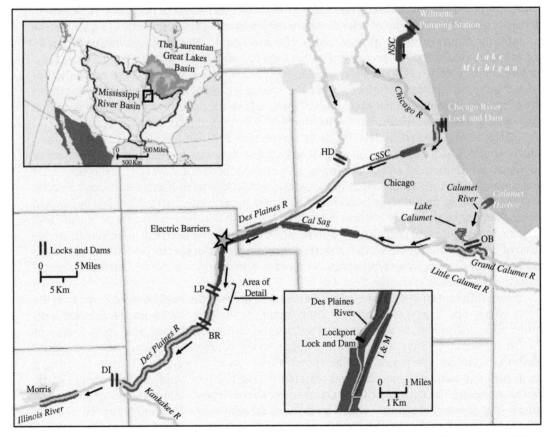

FIGURE 33-8 eDNA detections for carps (red) in the Chicago waterways upstream and downstream of the electric fish barriers in 2009 and 2010. *Source: Reproduced from Rasmussen et al. (2011)/Elsevier.*

these have generally been considered acceptable losses to human populations. The same could be said for aquatic habitats, which have been heavily manipulated in many cases for the benefit of human development. The important point in the expansion of non-native fishes is that we understand and accept that the changes brought upon natural ecosystems are large and mainly irreversible, so it is important that we clearly recognize the costs and benefits of changes in aquatic ecosystems in present times as well as in the future.

Summary

The entire world faces an enormously important issue regarding native diversity of fishes. Introductions have had large effects on the biota, and these changes result in more similar faunal compositions between locations and higher abundance of non-native species. Extinction and critical endangerment have received the most attention and resulted in major expenditures of funding for management. The introduction and spread of non-native species have received far less attention until recent times even though it also has major significance on our biodiversity. Reversing the homogenization of our biota is a much more difficult issue than dealing with single species and critical habitats. The conservation, restoration, or rehabilitation of habitat, as well as physical reintroduction of native species and removal of invasive species, are all necessary components of restoring aquatic biodiversity. The example of the carps in the Mississippi drainage demonstrates that even with advanced warning about possible expansion, intervention is a costly and uncertain prospect. It is not at all clear that such interventions will succeed. Whether improved environmental conditions can result in natural reclamation of biotic diversity is also yet to be seen. Clearly, the challenge is ahead of us – one of living more in balance with the world than living in domination of it.

Literature Cited

ACRCC (Asian Carp Regional Coordinating Committee. 2019. Asian Carp Monitoring and Response Plan. Asian Carp Regional Coordinating Committee, Council on Environmental Quality, Washington, D.C.

Buckwalter, J.D., E.A. Frimpong, P.L. Angermeier, and J.N. Barney. 2018. Seventy years of stream-fish collections reveal invasions and native range contractions in an Appalachian (USA) watershed. Diversity and Distributions 24:219–232.

Chick, J.H., D.K. Gibson-Reinemer, L. Soeken-Gittinger, and A.F. Casper. 2020. Invasive silver carp is empirically linked to declines of native sport fish in the Upper Mississippi River System. Biological Invasions 22:723–734.

Collins, S.F., M.J. Diana, S.E. Butler, and D.H. Wahl. 2017. A comparison of sampling gears for capturing juvenile silver carp in river-floodplain ecosystems. North American Journal of Fisheries Management 37:94–100.

Copp, G.H., P.G. Bianco, N.G. Bogutskaya, et al. 2005. To be, or not to be, a non-native freshwater fish? Journal of Applied Ichthyology 21:242–262.

Koel, T.M., K.S. Irons, and E.N. Ratcliff. 2000. Asian Carp Invasion of the Upper Mississippi River system. U.S. Department of the Interior, U.S. Geological Survey, Upper Midwest Environmental Sciences Center, La Crosse, Wisconsin.

Lodge, D.M., C.R. Turner, and 7 coauthors. 2012. Conservation in a cup of water: Estimating biodiversity and population abundance from environmental DNA. Molecular Ecology 21:2555–2558.

Master, L.L., S.R. Flack, and B.A. Stein, editors. 1998. Rivers of Life; Critical Watersheds for Protecting Freshwater Biodiversity. The Nature Conservancy, Arlington, Virginia.

Muhlfeld, C.C., S.T. Kalinowski, T.E. McMahon, et al. 2009a. Hybridization rapidly reduces fitness of a native trout in the wild. Biology Letters 5:328–331.

Muhlfeld, C.C., T.E. McMahon, M.C. Boyer, and R.E. Gresswell. 2009b. Local habitat, watershed, and biotic factors influencing the spread of hybridization between native westslope cutthroat trout and introduced rainbow trout. Transactions of the American Fisheries Society 138:1036–1051.

Muhlfeld, C.C., R.P. Kovach, L.A. Jones, et al. 2014. Invasive hybridization in a threatened species is accelerated by climate change. Nature Climate Change 4:620–624.

Pendleton, R.M., C. Schwinghamer, L.E. Solomon, and A.F. Casper. 2017. Competition among river planktivores: Are native planktivores still fewer and skinnier in response to the Silver Carp invasion? Environmental Biology of Fishes 100:1213–1222.

Peoples, B.K., and S.R. Midway. 2018. Fishing pressure and species traits affect stream fish invasions both directly and indirectly. Diversity and Distributions 24:1158–1168.

Rahel, F.J. 2000. Homogenization of fish faunas across the United States. Science 288:854–856.

Rasmussen, J.L., H.A. Regier, R.E. Sparks, and W.W. Taylor. 2011. Dividing the waters: The case for hydrologic separation of the North American Great Lakes and Mississippi River Basins. Journal of Great Lakes Research 37:588–592.

Scott, M.C., and G.S. Helfman. 2001. Native invasions, homogenization, and the mismeasure of integrity of fish assemblages. Fisheries 26:6–15.

Upper Midwest Environmental Sciences Center. 2021. Graphical Fisheries Database Browser. U.S. Geological Survey, LaCrosse, Wisconsin. Accessed April 15, 2021 at www.umesc.usgs.gov.

Williams, J.D., and G.K. Meffe. 1998. Nonindigenous species. Pages 117–130 *in* M.J. Mac and P.A. Opler, editors. Status and Trends of the Nation's Biological Resources. U.S. Geological Survey, Washington, D.C.

Zhu, Z., D.T. Soong, T. Garcia, et al. 2018. Using reverse-time egg transport analysis for predicting Asian carp spawning grounds in the Illinois River. Ecological Modelling 384:53–62.

CHAPTER 34

Aquaculture

Introduction

Aquaculture is a long practiced method of procuring fish and other seafood for human consumption, much like capture fisheries. As mentioned earlier for integrated carp culture, aquaculture existed at least 2000 years ago in China and the Roman Empire. Much of early aquaculture focused on maintaining fish alive so they would be available fresh for consumption, and eventually growing those fish to a larger size. Aquaculture has dramatically changed in recent years as intensification and high rates of production have been common since about 1970. Aquaculture can simply be defined as the controlled production of an aquatic crop. Aquaculture mainly focuses on aquatic animals, in contrast to agriculture, which is dominated by plant production. This controlled production can vary from simply rearing young fish from eggs and releasing them to grow in the wild, to completion of the entire life cycle under controlled conditions. Aquaculture species include a wide variety of fish, shellfish, and algae, but in this chapter, the focus is mainly on finfish production and its influence on the ecology of both cultured fish and the nearby ecosystem.

Culture Harvest

Culture fisheries have helped fill the growing global demand for seafood because of dramatic growth in recent years with little evidence of future decline. While capture fisheries have remained fairly constant since 1990, with harvests around 90 MMT, culture fisheries have steadily increased in harvest from 13 MMT in 1990 to 82 MMT in 2018 (Figure 34-1). Since 1970, growth in aquaculture has been about 7% per year, and the fraction of aquaculture products in global seafood harvest is now nearly 50%. Culture fisheries are dominated by Asian countries, with 89% of the aquatic animal yield coming from Asia (Figure 34-2). Europe has a considerable yield, followed by Africa and South America. However, no other continents are near the level of production of Asia. About 59% of all aquaculture occurs in freshwater. Much of the marine culture harvest consists of molluscs, crustaceans, and seaweeds, with only 12% of finfish culture harvest coming from marine sources. Much of the freshwater aquaculture is done in ponds or cages in natural waters with recent attention on development of indoor recirculating aquaculture systems for fish production.

    The species cultured are considerably different than those taken in capture fisheries. One beneficial aspect of aquaculture is that species can be selected based on biological and food characteristics, so they are largely used as human food, in contrast to some capture fisheries.

Biology and Ecology of Fishes, Third Edition. James S. Diana and Tomas O. Höök.
© 2023 John Wiley & Sons Ltd. Published 2023 by John Wiley & Sons Ltd.

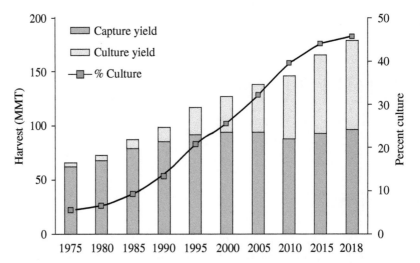

FIGURE 34-1 Changes in capture and culture harvests (in MMT) and percent of the harvest from aquaculture from 1975 to 2018. *Source: Data from FAO (2021b).*

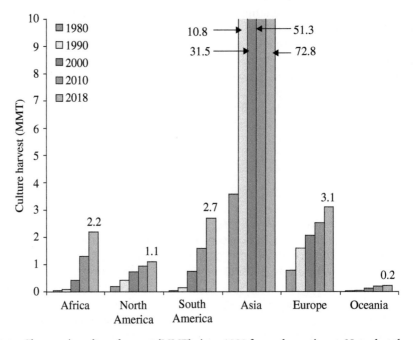

FIGURE 34-2 Changes in culture harvest (MMT) since 1980 for each continent. Note that the values for Asia are truncated. *Source: Data from FAO (2021b).*

Many of the species taken in capture fisheries are used for production of other products such as fishmeal and fish oil. Since the capture harvest cannot be precisely selective, many non-target animals are taken, and even many of the target species are not consumed directly by humans. The dominant group cultured is carp, with about 36% of all production in 2018 being carp species, while finfish overall comprise 67% of production (Figure 34-3). Molluscs are also important, while other groups make small contributions in weight (although sometimes not small in monetary value). Tilapia culture has expanded dramatically in recent years, while salmonids have relatively low production, as do shrimp and other crustaceans. One interesting

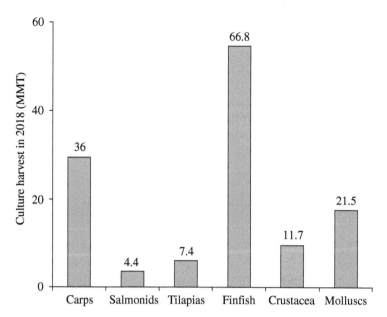

FIGURE 34-3   Culture harvest (MMT) in 2018 for various species and groups. Figure at the top of each bar is percent of the total culture harvest. *Source: Data from FAO (2021b).*

situation with aquaculture is that most of the dominant species grown are low trophic herbivores and detritivores, which tend to be cheaper to grow due to their lower requirements of costly dietary protein (e.g. carps, tilapia, and molluscs).

Management of Aquaculture

Aquaculture is in some ways much simpler to manage than capture fisheries although it does require considerable governmental regulation. In its simplest form, each owner manages one to several ponds, cages, or pens by stocking them with some species, adding feed or fertilizer, possibly exchanging water or aerating to improve water quality, and harvesting fish at market size when local market conditions are good. Since the fish are alive up to the time of harvest, culturists can delay harvest until market conditions are good. With this enterprise, management is simply a matter of when to stock, what to add, and when to harvest. Of course, from a larger perspective, the effects of effluents, diseases, and drugs on local waters are concerns that extend to all aquaculture practices and require a higher level of regulation to deal with the industry overall.

Aquaculture increases in intensity from extensive (stocking of fish at low density with input of fertilizers to increase productivity of water and later harvesting them), to semi-intensive (stocking of fish at moderate density with addition of fertilizers and some feeds), and ultimately to intensive (provision of complete feeds, aeration, flushing, and other water quality management). Obviously, the environmental impact of aquaculture depends a lot upon intensity of culture. Most concerns to date focus on aquaculture systems with intense use of feeds, fertilizers, antibiotics, and other chemicals, and the release of those materials into natural waters. One other large concern related to aquaculture involves the release of exotic fish into natural waters.

Unlike terrestrial production of domestic crops, aquaculture is still conducted utilizing largely natural species. A large number of species are commonly grown, and little domestication has occurred for most cultured species, particularly in less developed countries. This has both positive and negative aspects. On the positive side, undomesticated fish that escape – if they are from the local area – may not cause damage to natural communities of fish. However, the lack of domestication means less predictability in growth characteristics and performance of the crop as well as lack of efficiency in growing target species. Many culturists today call for further domestication of aquaculture lines and use of these improved genotypes for higher rates of production. However, domesticated fish could then be considered an exotic species when they are inadvertently released into natural waterways. They may well compete with native stocks of the same species or affect other species in different ways than natural genotypes of the same species.

Pond Aquaculture

Ponds have been used for fish culture for millennia, and the management of ponds ranges from extensive to intensive depending on the species grown. Chinese aquaculture historically used an Integrated Multi-Trophic Aquaculture (IMTA) system, where a number of species at different trophic levels were grown together to help control water quality and utilize pond resources fully. These ponds were integrated with farms so that farm wastes were added as food or fertilizer. As mentioned earlier, some of the species like grass carp functioned to consume large vegetation, partially digest it, and excrete it as nutrients or decomposed material, which was then utilized by other species to increase productivity in the pond. Silver carp and bighead carp were also used in these ponds to consume phytoplankton and zooplankton, which were stimulated to grow on the waste material from grass carp. All three species were harvested and consumed by the farmers, and production levels were quite high – up to several metric tons of fish produced per hectare of water annually. While these ponds were managed semi-intensively, water quality largely remained high because the fish species controlled plankton and nutrient contents, and water was rarely exchanged. This integrated system serves as the historic background for much of aquaculture in Asia. IMTA has also been restructured for marine cage culture systems today, with fish cultured in cages, kelp nearby on ropes to absorb waste nutrients, and oysters – also nearby – to consume solid particles excreted in the cages. Like the Chinese system, the intent is to utilize all of the materials produced in the culture system to maintain high water quality and reduce effluent damage to the ecosystem while producing the added marketable crops of kelp and oysters.

Pond systems can vary in size from 100 m² to several hectares (Figure 34-4). Common management is to drain a pond at harvest and collect remaining fish following initial seining of the pond and then dry the pond for a time to improve bottom conditions. Draining a pond is the one practice that tends to release most of the nutrient and sediment loads into natural waters although ponds can be drained into other reservoirs and the water reused in another aquaculture system after some treatment. Fish are usually stocked in ponds as small fingerlings and reared to a marketable size. Since the individual size of fish may increase from 20 g to 2 kg during a grow-out, the biomass of fish in the pond also changes over time along with the amount of feed needed to continue to grow these fish. This can result in ponds stocked well below their capacity until late in the grow-out. Longer grow-out species may be moved into larger ponds as they grow, resulting in a multi-stage grading of the fish, to allow for better use of the available fish ponds. Faster growing species like tilapia are commonly grown from fry to market size without grading.

Pond culture continues to be practiced today with up to half of all aquaculture occurring in ponds. In addition to the increase in total production that has occurred since 1970, there has been a strong increase in intensity of production systems used. For example, catfish ponds in the United States are highly intensive, with high stocking density of fish, high feeding rates,

FIGURE 34-4 Pond aquaculture system for intensive fish production. *Source: Personal photo from J. Diana.*

often continual mechanical aeration of the pond to increase oxygen content, and periodic flushing of the water to reduce the concentration of waste products. Such a system can produce much more yield but also at a much higher cost than lower intensity systems. These intensive pond systems are also much more demanding in terms of water use, energy use, and waste discharge into local environments. Depending on the fish culture system used, ponds can have water exchange rates over 20% per day with the need to treat discharge water to reduce nutrient and solids concentration.

Raceway Systems

As intensity of aquaculture increases, it becomes difficult to maintain adequate water quality in stagnant ponds. One solution for this is to use a raceway system, where water is run through a channel and fish are held at very high density, and water is continually flushed through the system to maintain water quality. Raceways are the most common systems used in hatchery production of many sport fishes – particularly salmonids – which also prefer water with some current. Water exchange rates can be more than 5–10 turnovers per hour, and the systems can then be used to collect solid waste before discharge. High densities of fish – up to 100 kg per m² – can be maintained and fed in these systems. Traditional raceways used multi-pass systems, where water went through one raceway at a rate that allowed the fish to use oxygen to a concentration where fish growth began to be affected, and then the water was aerated by passing over a falls and used in another raceway (Figure 34-5). After two or three raceway passes, fish waste caused the water to become concentrated with nitrogen or phosphorus, which further limited fish growth, so the water was discharged. At the end of each raceway, sumps were designed to collect solid wastes from the system, which could be used as fertilizer or landfilled. Depending on the size of system and the water body used for culture and discharge, nutrients could be directly discharged into natural waters, into settling and treatment ponds, or through wetlands to reduce the nutrient concentration before reaching a stream. These systems have high costs as they might require discharge treatment, much more labor, high energy costs if groundwater

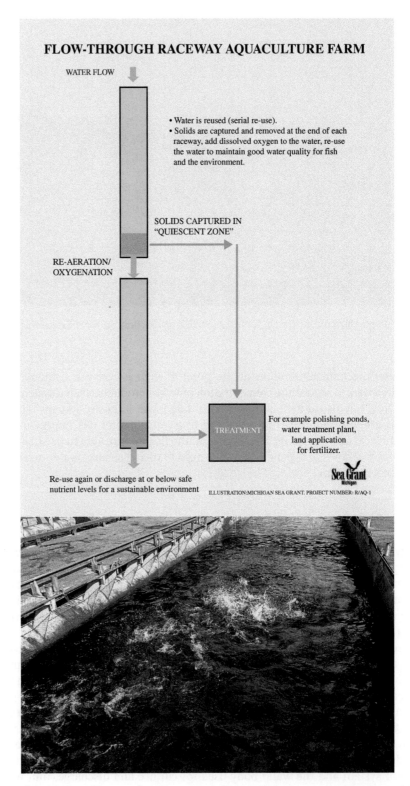

FIGURE 34-5 Schematic and image of raceway aquaculture system. *Source: Maskim/Adobe Stock Photos.*.

was used as a source, and more elaborate facilities. They also produce much higher yields per unit area. Raceway systems have had significant environmental impacts in some regions with phosphorus discharge to lakes causing algal blooms. Canale et al. (2016) produced a mass balance and bioenergetics model to minimize the discharge of phosphorus from state owned fish hatcheries and reduce the pollution burden to receiving waters by managing phosphorus inputs in feed and managing discharge of solids and feces.

Modern raceways have used oxygen generation to replace the three-pass system with two alternatives. Oxygen can be added to levels above saturation, and added sequentially into the raceway, so all of the nitrogen and phosphorus capacity can be used in one raceway. Such a system provides better overall water quality for fish growth. Another alternative is the use of round tanks, which are flushed, injected with oxygen, and can be maintained at higher densities and production levels than traditional raceways. These tanks can maintain optimal current velocity for fish with high oxygen conditions and have waste collection at a basin near the drain. They essentially function as a raceway turned in on itself. They have an added advantage of better water quality conditions throughout the tank, so more consistent growth conditions for the fish. They can also better utilize the space of an indoor facility than raceways.

Cage Culture

Another historic aquaculture system is the use of cages to grow fish. Cages or pens made of net were placed in natural waters, fish confined to the cages and fed, and the water quality maintained by dilution from natural water, as well as sinking of solid waste. Cage culture systems were historically used in countries like Cambodia where many houses floated over suspended cages to simplify management. The grow-out system is intensive due to the need for feeding, but other kinds of water quality manipulation are not necessary. Of course, cages disperse waste products into the natural waters and can contaminate those waters when there are excessive numbers of cages involved. Currently, cage culture is conducted both in fresh and marine waters, and is really the only marine system to culture finfish nearshore or offshore.

Cages can vary from framed squares 1 m³ in volume submerged into rivers, lakes, or ponds to extremely large systems up to 25 m in diameter or more, either completely enclosed as a sphere or floated on the surface with a circular collar (Figure 34-6). Concerns about cage culture have arisen mainly due to the dispersal of waste into natural waters. This is particularly acute in freshwater systems where simple cages can be added and, if unregulated, can result in a very high density of cages, high oxygen demand, and damage done to the natural ecosystem. Oceanic cages have the same potential, but due to the higher cost of managing cages in the open ocean due to turbulence as well as distance from land, most culture has been done in fjords or bays. Fjords and bays provide a benefit of having regular flushing by tidal action, resulting in the dispersal of waste products away from the culture site. Many freshwater fish are commonly grown in cages, as well as cobia, yellowtails, and salmon, which are grown in cages and net pens in the ocean. There are a variety of cage types, from open cages with a floating collar to maintain the dimensions of the net, to completely closed and submersible cages that can be moved in the water column to avoid waves and current conditions.

Recirculating Systems

In recent times, aquaculture technology has moved toward recirculating systems, which utilize the same water, treat for waste products, and recirculate it to the cultured fish. These systems have been developed to reduce the amount of water used in aquaculture, to more efficiently utilize nutrients, as well as to reduce the impacts of waste materials on receiving waters.

FIGURE 34-6 Schematic and image of cage culture system. *Source: Reproduced with permission from Michigan Sea Grant;rh2010/Adobe Stock Photos.*

Recirculating systems can be indoors or outdoors and may do all of the recirculation in one small or large system or use attached ponds and other facilities to treat the waste. Indoor recirculating systems use a biofilter, which removes nitrogen waste from the water – similar to a sewage treatment system – as well as collectors to reduce solid waste before the water recirculates to fish tanks (Figure 34-7). In comparison, outdoor systems might utilize separate ponds to sustain the target fish and to circulate the water to plants, which consume the nutrients produced and detritivorous animals to consume the solid waste.

An interesting elaboration of recirculating aquaculture is the development of aquaponics. Similar to IMTA, aquaponics is a system where the wastewater from fish grow-out

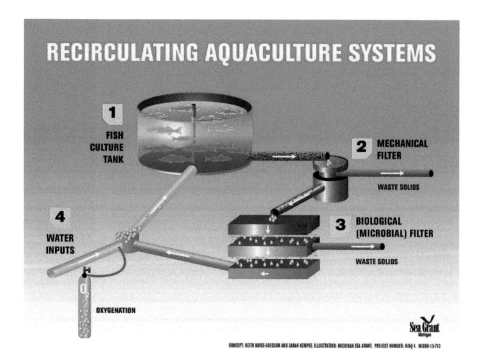

FIGURE 34-7   Schematic of recirculating aquaculture system. *Source: Reproduced with permission from Michigan Sea Grant.*

is used to fertilize plants hydroponically – with roots suspended in the water – so plants like lettuce remove surplus phosphorus and nitrogen from the water, bind it up in their tissue as another marketable crop, and then the water is returned to use for aquaculture. Aquaponics has received much interest lately, and some major facilities have been built, although it is still a relatively experimental type of aquaculture. Overall, indoor recirculating systems are commonly used for species of high value, grown in indoor facilities where escapement is eliminated and effluents are not generated. This is often the most expensive type of aquaculture. Outdoor aquaponics facilities can just be a way of using plants and animals to generate another marketable crop while also treating waste material. Some current pond systems are using partitioned aquaculture to intensively feed fish in some areas of ponds and grow herbivorous or detritivorous fish in other parts of the ponds to utilize the production stimulated by waste from the partitioned area. In this way, it is very similar to IMTA but may also use biofilters and other mechanisms to control water quality. It is also similar to integrated pond culture, where pond water and mud was often used to grow crops on the banks of a pond.

Positive and Negative Effects of Aquaculture

Since aquaculture is often seen as a new and experimental agricultural system, it is sometimes viewed with suspicion, namely with regard to its impacts to the ecosystem. Some of these have already been described, such as eutrophication of receiving waters by aquaculture wastes and introduction of exotic species or genotypes by fish escapement from the system. There are a number of aspects of aquaculture that can negatively impact natural ecosystems and species, as well as a number of positive impacts aquaculture can have on biodiversity. These were summarized and evaluated by Diana (2009). Negative effects included:

1. Escapement of aquatic crops and their potential hazard as invasive species.
2. The relationships among effluents, eutrophication of water bodies, and changes in the fauna of receiving waters.
3. Conversion of sensitive land areas, such as mangroves and wetlands, as well as water use.
4. Other resource use, such as fish meal and its concomitant overexploitation of fish stocks.
5. Disease or parasite transfer from captive to wild stocks.
6. Genetic alteration of existing stocks from escaped hatchery products.
7. Predator mortality caused, for example, killing birds near aquaculture facilities.
8. Antibiotic and hormone use, which may influence aquatic species near aquaculture facilities.

Positive impacts included:

1. Production of fish can reduce pressure on wild stocks that are already overexploited.
2. Stocking organisms from aquaculture systems may help enhance depleted stocks with limited reproductive success.
3. Effluents and waste from aquaculture can increase local production, abundance, and diversity of species.
4. Destructive land-use patterns, such as slash-and-burn agriculture, may be replaced by more sustainable patterns, such as aquaculture in ponds, which may also generate income, reduce poverty, and improve human health.

Aquaculture has expanded rapidly in recent years, which is important because food production – particularly protein production – is of great importance to human populations, and fish are a healthy food source. What is interesting in aquaculture is that as environmental impacts have been identified, aquaculture systems have been modified to deal with those environmental issues (Naylor et al. 2021). For example, the use of fish meal has been a concern in aquaculture because of its effect on marine biodiversity and on other organisms that utilize fish harvested for fish meal. However, research on feeds has resulted in replacement of fish meal with plant-based proteins or other food sources. This has occurred not only because of environmental impacts of aquaculture on fish meal, but also the cost. Alternative protein sources in feed have resulted in significant savings in aquaculture, as well as significant reductions in the use of fish meal, a win–win situation. It has also reduced the carbon footprint of fish feeds. There are a number of examples such as this, where aquaculture has adapted practices to accommodate environmental concerns in a significant manner. In this regard, being a new industry helps because a lot of development can still occur. The following examples are included to more fully explain the aquaculture systems used, their environmental impacts, and the changes in management over time.

Atlantic Salmon

The first example is Atlantic salmon. They were a minor commercial species, averaging around 12 000 MT of harvest between 1950 and 1990. They began a serious decline in harvest and abundance in the 1990s, with commercial harvest about 2000 MT from 2010 to 2018 (Figure 34-8). This decline was largely due to changes in their spawning habitat, where streams were impounded, polluted, or modified, resulting in poor spawning conditions. Harvest, although

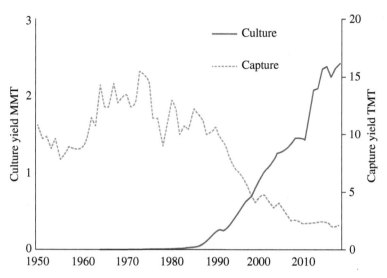

FIGURE 34-8   Yield of Atlantic salmon from aquaculture and capture fisheries from 1950 to 2018. *Source: Data from FAO (2021b).*

relatively small, also played a part in this decline. By 2000, Atlantic salmon were considered an endangered species in the United States, which they remain today and are prohibited from commercial or recreational exploitation. There is a very limited recreational and subsistence fishery in Canada and European countries, with commercial fisheries banned in rivers and severely limited in the ocean.

Atlantic salmon aquaculture began around 1980 as a response to the limited availability of wild caught fish. Atlantic salmon are mainly grown by pen culture in fjords or nearshore waters, where they are fed intensively. Norway led the development of culture techniques and grows over half of global production today. Production has increased steadily since 1980, reaching 2.4 MMT in 2018 (Figure 34-8). Recent increases in production have been moderate, with annual growth of 5.2% between 2001 and 2018. The leading producers are Norway (1.3 MMT), Chile (0.6 MMT), United Kingdom (0.15 MMT), and Canada (0.12 MMT). Atlantic salmon culture has had dramatic effects on Atlantic salmon harvest in the wild because farmed salmon have largely replaced commercially caught salmon as a product. This has reduced pressure on Atlantic salmon stocks in nature and resulted in increased natural salmon populations in many areas. Combined with the more intensive conservation of wild Atlantic salmon, the reduced market pressure has helped increase wild salmon returns to many rivers. However, several concerns exist regarding Atlantic salmon culture. These include sedimentation near pens, as well as releases of nutrients, chemicals, and farmed salmon into natural waters.

In the 1970s, Norway began an aquaculture system for salmon based on net pens in their fjords. Fish were fed intensively, with mainly protein-based feeds from fish meal, and had limited impact on natural ecosystems because of low density of pens in the fjords. Fish were initially reared from eggs and sperm of wild caught salmon, but soon a domestication process was begun to produce faster growth, which resulted in an isolated brood stock of fish. As the industry grew, pollution problems were first recognized from the nutrients and solid waste generated in net pens. This began a series of activities to evaluate the risk produced by salmon aquaculture as well as reduce that risk in the industry. Taranger et al. (2015) indicated the Norwegian Atlantic salmon farming industry had major potential risks in genetic integration of farmed salmon into wild populations, extension of salmon lice and viral diseases to wild fish, nutrient and organic discharge polluting fjords, and use of fish meal resulting in declines of oceanic biodiversity. In 2009, the Norwegian government established five primary goals to reduce aquaculture impacts focusing on improving the industry's impacts. The risk assessment for Atlantic salmon showed a significant number of wild salmon populations had moderate to high risk for genetic introgression, and some stations along the Norwegian coast had salmon lice problems, but there was low prevalence of viruses caused by salmon culture, and only 2% of fish farms displayed unacceptable conditions for organic loading. In addition, feed development has resulted in reductions of marine protein and oil in salmon feeds, with 89% marine source material in feeds from 1990 declining to 25% by 2016 (Aas et al. 2019). Norway and other countries have responded to a large suite of culture problems, but improvement is still needed mainly in genetic and disease impacts. However, there have been far more serious issues in Chile, where antibiotic use and escapement have been major problems for the industry. These resulted in a serious decline in the industry in 2007, but production is now increasing to new highs. Regulatory issues make Chilean production questionable today to a number of environmentally sensitive markets. While conditions in Norway and elsewhere have improved, the Monterey Bay Aquarium's (MBA) Seafood Watch program still lists most net pen aquaculture for Atlantic salmon as unsustainable (MBA 2021).

One recent issue with Atlantic salmon is the production of genetically modified fish for aquaculture. Genetic modification is not like normal domestication, where fish are selected based on their biological traits and then crossed in breeding programs to artificially improve their performance in a selected trait, such as growth rate. Genetic modification is the insertion of specific RNA or DNA from other species to produce desired traits in the organism. For plants, these traits have spanned from resistance to herbicides or pesticides to allow chemical use to prevent pests, to increased growth or drought resistance. Atlantic salmon have undergone GMO development to dramatically increase their growth rate and reduce feed consumption compared to other domestic salmon stocks (Cebeci et al. 2020). AquAdvantage salmon are now being produced for human consumption, and the main concerns the Food and Drug Administration had in approving this GMO were related to the effects of escaping salmon. GMO salmon are currently grown only in indoor recirculating systems to reduce escapement possibilities and are reared as triploid fish to reduce genetic contamination issues if any fish do escape the system. Much of the concerns for agricultural GMOs include human perception of the risks imposed by consuming these plants either by humans or by other animals. While such concerns might also exist for the GMO salmon, it seems less likely because the only manipulations have been for changes in growth, not in chemical resistance. GMOs remain controversial, with some scientists believing they have great potential to dramatically enhance human food production for expanding populations, while others remain skeptical about their impact on humans and natural ecosystems.

Nile Tilapia

The second species reviewed here is Nile tilapia (henceforth called tilapia in this section). Tilapia were commonly captured and cultured at low levels of production in the 1900s and generally served as a low-priced and common fish for local consumption in Africa and Asia. They were caught by artisanal fishers in rivers and lakes as well as cultured in ponds. Most aquaculture was done by extensive to semi-intensive techniques in small ponds. Tilapia are hardy fish, which makes them easy to culture. Culture methods were extended to rural poor populations by a number of aid agencies like the US Peace Corps.

Tilapia were "discovered" in America and Europe as a valuable and tasty food fish in the 1990s. As a result, their culture expanded to accommodate the export market (Figure 34-9). Tilapia production in 2018 was 4.5 MMT or approximately 7.4% of all production (Figure 34-3). Tilapia production had dramatic increases in recent years with an 8.7% annual growth rate between 2001 and 2018. The leading countries in aquaculture production in 2018 were China

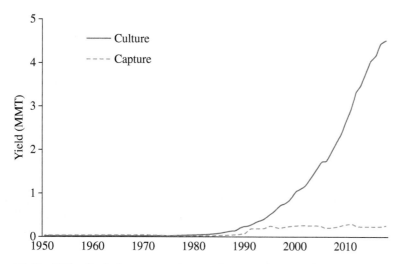

FIGURE 34-9 Yield of Nile tilapia from aquaculture and capture fisheries from 1950 to 2018. *Source: Data from FAO (2021b).*

(1.2 MMT), then Indonesia (1.17 MMT), Egypt (1.1 MMT), Brazil (0.3 MMT), and Thailand (0.2 MMT). Unlike Atlantic salmon, tilapia production is still done in ponds using semi-intensive to intensive technology. Tilapia ponds vary from about 0.1 to several hectares in size. Throughout the world, tilapia culture is mainly enhanced by fertilization (Diana et al. 2013), but intensive production is commonly done today with complete feeding as well. In many developing countries, the level of intensity depends on the target market. Fish for local consumption may be about 150 g in size, which are eaten whole and can be grown using fertilization techniques in about five months. In comparison, markets for export – usually as frozen fillets – require fish of 300–500 g and use supplemental or complete feeding. However, supplemental feeding of tilapia is not as intensive as in Atlantic salmon. Since tilapia in fed systems can still find considerable food from organisms in the ponds – and this results in more efficient growth and higher profits – ponds can be managed as natural ecosystems rather than intensive feed lots (Diana 1997). Water quality may remain reasonable through grow-out (commonly about six months), and aeration may be applied toward the end of grow-out when fish and feed biomass is maximum.

Early tilapia production systems used fry taken from brood ponds, and the mixed sex culture that resulted often caused breeding in the ponds, overpopulation, and stunted fish growth. A number of innovations have improved this issue. The most common is use of sex reversal to produce all-male fish fry for culture as males grow substantially faster and can reach larger market sizes than females. Sex control is done by providing early fry stages with feed containing methyltestosterone for a short period, which causes female fish to become phenotypic males. Systems for sex reversal have been adapted to locations with access to higher technology or more rural settings and have been widely successful (Belton et al. 2009). Other sex manipulations done for tilapia include producing YY males or polyploid individuals. Another selection program for tilapia is the GIFT tilapia developed by WorldFish, which is essentially rearing of an improved genetic strain by normal domestication, resulting in fry that have up to 85% faster growth (Eknath and Hulata 2009). While sex reversal can be done on existing brood stocks of tilapia anywhere, use of GIFT fish requires managing a brood stock of these selected individuals to produce fry for culture. This has been a limiting factor to the widespread adoption of the GIFT fish.

Ecological concerns regarding tilapia mostly involve the release of tilapia into natural waters. Of all species in aquaculture, tilapia has probably had the largest impact on native fishes throughout the world. Releases of tilapia have not only occurred because of aquaculture, but tilapia species have taken over many freshwater and brackishwater environments when they were released or when they escaped. Their outdoor culture in a location outside their range is problematic, given their competitive effects on natural fish communities, and even in warm water locations where they are already present, expanded culture can result in further spread of tilapia.

Summary

Aquaculture is predicted to continue its dramatic growth through 2030 (FAO 2021a). Almost all species demonstrate continuing increases in production, so this increase will likely occur across a wide range of species. FAO believes the major growth in aquaculture production will occur in China with high amounts throughout Asia and Latin America. The main issues on sustainability for aquaculture will continue to be the impact on natural waters, the introduction of exotic species, and the introduction of diseases and chemicals to natural waters. As production technology improves, these concerns are likely to be handled much better than they are currently. Examples of government regulations regarding the placement of farms, the willingness of governments to enact regulations regarding use of chemicals in aquaculture, and other activities today indicate clear trends that aquaculture will be much more heavily regulated in the future.

Literature Cited

Aas T.S., T. Ytrestøyl, and T. Åsgård. 2019. Utilization of feed resources in the production of Atlantic salmon (*Salmo salar*) in Norway: An update for 2016. Aquaculture Reports 15:100216.

Belton, B., D. Turongruang, R Bhujel, and D.C. Little. 2009. The history, status, and future prospects of monosex tilapia culture in Thailand. Aquaculture Asia 14(2):16–19.

Canale, R.P., G. Whelan, A. Switzer, and E. Eisch. 2016. A bioenergetic approach to manage production and control phosphorus discharges from a salmonid hatchery. Aquaculture 451:137–146.

Cebeci, A., I. Aydin, and A. Goddard. 2020. Bigger, stronger, better: Fish transgenesis applications and methods. Biotech Studies 29(2):85–97.

Diana, J.S. 1997. Feeding strategies. Pages 245–263 *in* H. Egna and C. Boyd, editors. Dynamics of Pond Aquaculture. CRC Press, Boca Raton, Florida.

Diana, J.S. 2009. Aquaculture production and biodiversity conservation. BioScience 59:27–38.

Diana J.S., H.S. Egna, T. Chopin, et al. 2013. Responsible aquaculture in 2050: Valuing local conditions and human innovations will be key to success. BioScience 63:255–262.

Eknath, A.E., and G. Hulata. 2009. Use and exchange of genetic resources of Nile tilapia (*Oreochromis niloticus*). Reviews in Aquaculture 1:197–213.

Food and Agriculture Organization, United Nations (FAO). 2021a. The State of the World Fisheries and Aquaculture. FAO, Rome, Italy.

Food and Agriculture Organization, United Nations (FAO). 2021b. Fisheries Division, Global Statistics Collection. Accessed at http://www.fao.org/fishery/statistics/en on June 19, 2021.

Monterey Bay Aquarium (MBA). 2021. Monterey Bay Aquarium Seafood Watch. Accessed at https://www.seafoodwatch.org/ on June 19, 2021.

Naylor, R.L., R.W. Hardy, A.H. Buschmann, et al. 2021. A 20-year retrospective review of global aquaculture. Nature 591:551–563.

Taranger, G.L., Ø. Karlsen, R.J. Bannister, et al. 2015. Risk assessment of the environmental impact of Norwegian Atlantic salmon farming. ICES Journal of Marine Science 72:997–1021.

CHAPTER 35

Climate Change and Consequences for Fish

Climate change is arguably the most profound environmental stressor of our time. The temperature of Earth has clearly increased over the past several decades. While human activities have the potential to influence the degree of future warming, various future projections imply that temperatures will continue to increase. Moreover, temperature is not the only climatic feature projected to change. Climate models also point to complex altered precipitation patterns. Unlike temperature – which is broadly expected to increase – mean precipitation may increase in certain regions and decrease in other regions. The seasonality and variability of precipitation are expected to change. For example, some regions may experience increased precipitation during the spring but decreased precipitation in the fall. Droughts may become more common in some regions, but short, very intense precipitation events may become more frequent. Various other climate features – such as cloud cover, solar radiation, and wind patterns – have received less attention than temperature and precipitation, but these features may also change in the future. From a fish perspective, all of these climatic conditions will directly and indirectly affect physical, chemical, and biological attributes of aquatic systems. Therefore, the habitats occupied by fish populations and assemblages are expected to undergo fundamental changes, which may have strong effects on fish vital rates, population trajectories, and community dynamics. The species that thrive in a given system may change, some species may alter their latitudinal distributions, and the species composition of many aquatic systems may look very different in the future.

We describe below some of the effects of climatic change on aquatic systems, including responses that have already occurred and expected responses in the future. We describe some of the ways in which physical, chemical, and biological changes may affect fish at the individual, population, and assemblage levels, and we present some potential interactive effects of climate change with other stressors. We also point to some considerations and challenges related to managing fish populations in the face of climatic change.

Biology and Ecology of Fishes, Third Edition. James S. Diana and Tomas O. Höök.
© 2023 John Wiley & Sons Ltd. Published 2023 by John Wiley & Sons Ltd.

Climate-Induced Changes to Aquatic Systems

Thermal Patterns

With increased atmospheric temperatures, mean temperatures in essentially all water bodies are expected to increase. However, the magnitude of increases in mean temperatures will likely vary depending on latitude, region, and system type. Large marine systems are expected to respond differently from small freshwater systems, and responses in freshwater systems are likely to vary between lotic and lentic systems, small, shallow systems versus large, deep systems, and systems with different drainage characteristics (e.g. relative contributions of surface runoff versus groundwater inputs). Importantly, thermal characteristics of aquatic systems are much broader than simple mean temperatures. Along various temporal and spatial scales, water bodies can be characterized by their thermal variability, e.g. diurnal temperature changes, vertical stratification period, ice cover, and upwellings. All of these features are expected to respond to climatic change.

Ice cover of bodies of water provides an interesting index of climate as it reflects an integration of atmospheric conditions over relatively long time scales. Annual measures of ice cover are therefore less sensitive to day-to-day variation in air temperatures, and such measures have been used widely to index long-term changes in thermal conditions (e.g. Magnuson et al. 2000; Sharma et al. 2019). In fact, several records of ice cover of lakes and bays are suggestive of atmospheric warming over the past 150 years (Magnuson et al. 2000). With warmer mean atmospheric temperatures, periods of ice cover are expected to decrease further in the future with ice formation occurring later in the winter and ice melting occurring earlier. Many aquatic systems that have previously been covered by ice during most winters may remain ice free in the future (Sharma et al. 2019).

As described in previous chapters, the density of water responds to temperature, which can lead to thermal stratification in sufficiently deep water bodies. During spring warming in temperate regions of the Northern Hemisphere, a water body may shift from being essentially isothermal (~4 °C throughout the water column) to being slightly warmer in surface waters. The continued heating of surface water over time can lead to increased differences in temperatures and densities of water near the surface of the water body (epilimnion) versus water near the bottom of the water body (hypolimnion). With cooling in the fall, surface water temperatures subsequently decrease, and eventually thermal stratification breaks down as the water column becomes isothermal again. The initiation, duration, and breakdown of growing season thermal stratification all vary annually in response to atmospheric conditions. While the timing of these events will continue to vary in the future, on average the initiation of stratification will shift earlier, stratification break down will occur later, and the total duration of the stratified period will increase. The duration and depth of stratification may also change in response to not only air temperatures, but also wind and storms, which influence the mixing depth of the epilimnion, and solar radiation and water turbidity, which influence depth-specific warming.

Many lentic systems experience reverse stratification during winter with dense 4 °C water near bottom and less dense, cooler water higher in the water column. With warmer temperatures, many such dimictic lakes may not become sufficiently cold to stratify during winter. They will instead shift to become monomictic lakes, meaning that they will only stratify during a single period each year – growing season – and only experience one period of mixing each year.

Precipitation, Runoff, and Hypoxia

Changes in precipitation patterns with increased variability leading to more intense storms and more frequent drought periods will have important implications for various aquatic systems. The quantity of water in aquatic systems is expected to become more variable. For example, many fluvial systems will become flashier with greater swings between peak and low flows. This will be especially true for fluvial systems receiving relatively high surface runoff and low groundwater contribution. Due to periods of limited precipitation and increased evaporation potential, small shallow water bodies – including small streams and wetlands – may experience more frequent periods of drying.

The timing of river discharges may also change. In temperate regions, it is common for river discharges to be especially high during the spring. Meltwater runoff of snow and ice that accumulated during winter can combine with high precipitation to lead to very high river flows and discharges during spring. Spring peak discharges may occur much earlier with warmer temperatures. Snow and ice may also not accumulate over winter, therefore the spring discharge peak may be less pronounced.

Many of the nutrients that enter aquatic systems do so through direct precipitation over aquatic systems or precipitation onto land and subsequent runoff. Altered precipitation patterns in particular may have strong effects on runoff and non-point source nutrient loading into aquatic systems. Decreased overall precipitation may lead to decreased nutrient loading in some regions, but changes in the timing and magnitude of precipitation – coupled with higher temperatures – may lead to increased nutrient loading in many regions. Increased spring precipitation may in particular contribute to increased nutrient loading in temperate regions. Manure and fertilizers are broadly applied to agricultural fields and residential lawns during this season, and frequent precipitation could contribute to large amounts of these nutrients running off and loading to aquatic systems.

Increased runoff of nutrients and sediment are expected to exacerbate eutrophic conditions in many water bodies. Higher sediment loading may contribute to increased turbidity in the water column and sedimentation over bottom substrates. Higher concentrations of phosphorous has the potential to enhance primary productivity of phytoplankton, epiphyton, benthic algae, and macrophytes. Higher phytoplankton production may contribute to decreased water clarity similar to sedimentation; although there are important differences between sediment turbidity and algal turbidity. Increased primary productivity may also directly influence oxygen dynamics of various water bodies and likely exacerbate hypoxic conditions.

Several processes may contribute to increased hypoxic conditions under climate change. First, the solubility of oxygen in water decreases with increasing temperatures. Second, stratification is expected to establish earlier in the year and break down later in the year for sufficiently deep systems. A longer overall period of stratification implies a longer period of separation between bottom waters and the atmosphere. Therefore, bottom oxygen depletion will progress over a longer period and is more likely to lead to hypoxia or anoxia. Finally, greater nutrient loading in the spring will fertilize larger plankton blooms, which will settle to the bottom where their subsequent decomposition will only exacerbate the rate of oxygen depletion.

Water Levels

The phenomenon of rising sea levels due to climate change has received a great deal of attention. There is evidence that ocean levels have already substantially risen (Nichols and Cazenave 2010). One of the processes contributing to sea level rise is the thermal expansion of water. As water temperatures increase liquid expands, leading to increased volume and higher ocean levels. A large portion (~1/3) of the sea level rise that has occurred since the industrial revolution is attributable to thermal expansion of water. Warmer atmospheric temperatures have also contributed to melting of glaciers, leading more global water to be in liquid form in the world's oceans and not sequestered as ice on land or above ocean surfaces. The consequences of sea level rise will be most apparent in coastal zones, where previously terrestrial areas will become permanently inundated or experience more frequent flooding. From a fish perspective, this will lead to changes in the location and access of important coastal habitats, many of which serve important spawning and nursery functions.

Water levels of enclosed water bodies may not respond similar to oceans. While thermal expansion will still affect such water bodies, many enclosed systems will be less directly influenced by glacial melt. Water levels will instead respond to the regional balance between altered precipitation and evapotranspiration along with potential human regulation of inflows and outflows. Some enclosed systems may experience an overall increase in water levels, while others will experience a decrease. With increased frequency of droughts, in addition to increased periods of intense precipitation, a common feature of many enclosed water bodies will likely be increased variability of water levels, with the temporal scale of water level variation differing in part based upon system size and volume.

Wind Effects

In contrast to temperature and precipitation, future wind patterns have received relatively limited attention. However, retrospective analysis and future projections indicate that wind patterns may change over annual and decadal time scales. Increased frequencies of intense storm events may be accompanied by greater frequency of extreme wind events. Average winds speeds may also change, and the prevailing direction of winds may evolve. For example, Waples and Klump (2002) analyzed prevailing summer wind patterns over the Laurentian Great Lakes from 1980 to 1999 and found a general pattern of winds shifting to come more from the south over this decadal time frame.

Wind patterns are especially influential on enclosed aquatic systems, including many freshwater lakes. Water currents in such systems are largely wind-driven. Altered wind patterns can influence how water circulates in a system and may have implications for how organisms – including fish eggs and larvae – are transported throughout the system. Sustained winds can contribute to seiches and upwellings. If the prevailing direction and intensity of wind changes, frequencies and locations of seiches and upwellings may also change. Wind patterns can also have strong influence on how water bodies stratify. Intense winds in the spring and summer can increase physical mixing in a water body and delay when a system stratifies or lead to stratification at a greater depth. Intense storms and winds can contribute to the breakup of stratification in fall. Frequent storms and intense short duration winds are predicted to become more frequent or severe in different regions and based on different climate models. If such phenomena occur during critical life stages of fish, they could have negative effects on performance. For example, in studying smallmouth bass in western Lake Erie, Steinhart et al. (2005) found that the frequency of storms during the nest guarding period negatively affected survival success of eggs and larvae in nests. Future wind patterns, including direction, intensity, and variability, will likely interact with other climatic factors to influence physical, chemical, and biological conditions in various water bodies.

Ocean Acidification

In addition to temperature change, increased anthropogenic emission of carbon dioxide has contributed to ocean acidification. Estimates suggest that oceans absorb about 1/4–1/3 of CO_2 emitted by humans. Elevated levels of CO_2 in marine systems lead to increased concentrations of carbonic acid and, in turn, increased concentrations of hydrogen and bicarbonate ions. Therefore, the pH levels of oceans have become more acidic – a 0.1 decrease in pH by 2020 (NOAA 2020). Further, decreases in oceanic pH levels are projected to continue in the future as elevated atmospheric CO_2 continues to be absorbed by marine systems.

The most obvious biological effects of ocean acidification relate to calcifying organisms, including various shell-building mollusks. Elevated concentrations of hydrogen ions will bind with carbonate ions – carbonate ions are less available to form calcium carbonate as shells develop. Developing animals therefore may not be able to effectively secrete calcium carbonate structures – such as shells – and extant calcium carbonate structures may dissolve. The widespread phenomenon of coral bleaching has been related to ocean acidification and may have immense impacts on coral-dominated ecosystems.

Effects of ocean acidification on fishes are less obvious, but studies demonstrate that altered pH levels have the potential to impact fish in various direct and indirect ways. Acidification impacts on other species that serve as food or provide habitat for fish may have individual-, population-, and community-level effects on fish. Altered pH may directly influence olfaction of fishes, and altered sensing ability may have behavioral consequences. The overall potential influence of ocean acidification on fishes is very much an ongoing line of inquiry, and there is not broad consensus as to how impactful ocean acidification will be on fish population and communities. A meta-analysis by Clements and colleagues (2022) integrated findings from 91 studies that had examined the influence of acidification on behavior of coral reef fishes. They attempted to synthesize seemingly contradictory findings with some studies reporting extreme effects of acidification on behavior and other studies pointing to very weak or no effects. In particular, they found that early studies that initially examined effects of acidification tended to find strong effects on fish behavior, while more recent studies found weak effects. They also documented that the studies that had found large effects of acidification tended to have rather small sample sizes, while the more involved and larger sample size studies generally did not detect an effect. The authors pointed to the overall pattern as an example of a "decline effect," where effects detected in initial studies within a field of inquiry decline over time as more studies – including more robust studies – are conducted. However, when it comes to effects of ocean acidification, Clements et al.'s (2022) meta-analysis serves to demonstrate that researchers have not reached consensus on effects on fishes, and future studies will likely expand our understanding of acidification effects.

Effects on Fish Populations and Communities

A large number of studies have evaluated how aquatic ecosystems – as well as fish communities and populations – may respond to climate change. Some studies have taken a retrospective approach and evaluated the influence of past climatic variation on fish populations and communities and then extrapolated such influences based on projected future climates. Other studies have exposed fish to potential future conditions in the laboratory and evaluated responses. Still other studies have used a simulation approach and modeled how fish populations and communities may respond to future conditions. Studies have collectively pointed to a variety of potential influences of altered thermal conditions, oxygen availability, water currents, and system productivity. However, the ability to make very precise, accurate predictions of how fish populations and communities will respond to climate change is likely limited. The magnitude of climate change will likely vary – depending on future human activities – and differentially affect physicochemical conditions in aquatic systems. Responses of fish may be influenced by processes, such as contemporary evolution, where selective pressures may alter the genetic composition of populations and mediate responses to future conditions. Responses of fish populations will also likely be strongly influenced by trajectories of other populations. The trajectories of prey, predators, competitors, and positively interacting populations will collectively have strong influence on an individual population's response. Nonetheless, studies of climate change effects on fish are useful for appreciating plausible responses under certain scenarios and assumptions, and such studies can ultimately help us qualitatively understand likely climatic effects.

Thermal Effects

Temperature is incredibly influential on a variety of aspects of fish performance, affecting foraging, metabolism, growth, reproduction, and other processes. The influence of temperature on foraging, metabolism, and growth in particular was demonstrated through the presentation of bioenergetics in previous chapters. Temperature effects on the balance between metabolic energy expenditures and consumptive capacity differ based on species-specific physiology, leading some species to experience greatest growth potential at relatively cold temperatures and other species to experience greatest growth potential at relatively warm temperatures.

Bioenergetics models lend themselves to explore how fish consumption and growth patterns may respond to warmer temperatures under climate change. Several authors have used a bioenergetics approach to consider how warming temperatures due to climate change may affect fish growth and habitat quality. One of the earliest such analyses was conducted by Hill and Magnuson (1990) from the University of Wisconsin's Center for Limnology. They used a bioenergetics analysis to model potential climate change effects of warmer temperatures in three of the Laurentian Great Lakes on three fish species. They focused specifically on Lake Superior (the coldest of the three lakes), Lake Michigan, and Lake Erie (the warmest of the lakes), and they modeled a representative cold water, cool water, and warm water species (lake trout, yellow perch, and largemouth bass, respectively). The authors considered a scenario in which atmospheric CO_2 doubled from baseline conditions (pre 1980) and relied on predictions from three different climate models to assess how air temperatures may change under this scenario. They then used past relationships between air temperatures and water temperatures in lakes Superior, Michigan, and Erie to develop regression equations, and used these equations to simulate water temperatures in the upper part of the water column, based on potential future air temperatures.

Using baseline and potential future water temperatures in lakes Superior, Michigan, and Erie, Hill and Magnuson (1990) modeled one year of growth for the three representative species. They initially assumed that all three species could behaviorally thermoregulate. They assumed lake trout would consistently be able to find water temperatures of 10 °C if near-surface temperatures exceeded this level, and they assumed yellow perch and largemouth bass would be able to thermoregulate and never experience temperatures greater than 23 and 27.5 °C, respectively. They also assumed that fish would be able to find food and feed at a constant proportion of maximum consumption (*P*-value) during both baseline and future scenarios. Since maximum consumption rate varies with temperature, this latter assumption implies that, while the *P*-value is fixed, the actual amount of food consumed varies with temperature. Under these assumptions, Hill and Magnuson's (1990) analyses suggested that all three species have the potential to increase their growth under warmer conditions. This conclusion may seem counterintuitive, given that warmer temperatures should lead to greater metabolic costs. However, with the ability to thermoregulate, the simulated fish were never exposed to exceedingly warm temperatures, and by assuming a constant *P*-value, their simulated food consumption increased. The Great Lakes – especially Lake Superior – are relatively cold systems that limit growth during many months of the year. Warmer temperatures actually implied that fish were able to experience positive growth for a longer period of the year. The one exception was largemouth bass in Lake Superior. It was too cold for the warm water bass to grow under baseline temperatures in this system, and even under climate change, temperatures did not increase sufficiently to allow for positive growth (Figure 35-1).

The assumptions of perfect thermoregulation and elevated consumption rates under climate change may be unrealistic. For example, the central basin of Lake Erie experiences hypolimnetic hypoxic conditions most years, potentially limiting access to cold waters during the summer and early fall. Such hypoxic conditions are also likely to become more common and

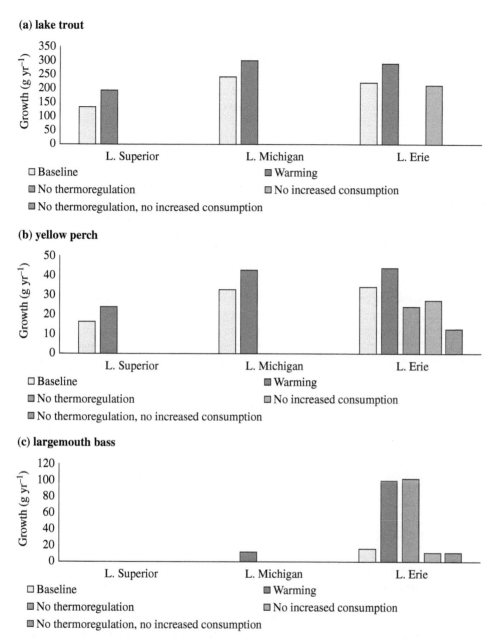

FIGURE 35-1   Simulated annual growth of (a) lake trout, (b) yellow perch, and (c) largemouth bass in lakes Superior, Michigan, and Erie (from Hill and Magnuson 1990). For all three lakes, growth was simulated under baseline temperatures (yellow bars) and potential future elevated temperatures (blue bars), while assuming species could thermoregulate and maintain (i.e., increase) their proportional consumption rate. Note that under these assumptions, largemouth bass were unable to grow in lakes Superior and Michigan under baseline temperatures and were even unable to grow in Lake Superior under elevated temperatures. For Lake Erie, growth was additionally simulated under future temperatures, while assuming fish were not able to thermoregulate (orange bars), unable to increase their consumption (green bars), and both unable to thermoregulate and maintain consumption rate. (pink bars) Note that for lake trout, if thermoregulation was not possible, the species could not maintain positive growth in Lake Erie. *Source: Adapted from Hill and Magnuson (1990).*

severe under climate change as described above. Therefore, fish may not be able to effectively thermoregulate in this system. Hill and Magnuson (1990) also simulated growth in Lake Erie without allowing fish to thermoregulate. Feeding rates of fish may also be more dependent on prey availability than temperature-dependent maximum consumption rate. That is, fish may not be able to increase the amount of food they eat as temperatures increase. Hill and Magnuson (1990) also considered scenarios where the amount of food consumed by fish under climate change was equal to the amount of food consumed under baseline conditions (note that this implies that P-values decrease under climate change).

When the assumptions of thermoregulation and increased prey consumption were relaxed, Hill and Magnuson's (1990) predictions regarding growth generally decreased. If they assumed prey consumptions rates remained equal under baseline and climate change conditions – i.e. P-values decreased under warmer climate change conditions – growth rates of lake trout, yellow perch, and largemouth bass in Lake Erie would all be expected to decrease under climate change. These predicted decreases in growth under climate warming were apparent regardless of whether fish were allowed to thermoregulate. The inability of fish to thermoregulate and instead be exposed to epilimnetic temperatures throughout the year leads to qualitatively different responses among species. If lake trout were unable to thermoregulate under climate change, they would be expected to be exposed to temperatures so high they would be lethal. While yellow perch would not be expected to be exposed to lethal temperatures, without thermoregulation they would nonetheless experience temperatures sufficiently high to experience reductions in growth relative to baseline conditions. Finally, largemouth bass growth in Lake Erie was largely unaffected by thermoregulation. Even under climate change, simulated water temperatures only very briefly reached a level (>27.5 °C), which would have triggered bass to thermoregulate (Figure 35-1).

It should be noted that the scientific community's abilities to model future climate dynamics and thermal conditions in aquatic systems have dramatically improved since Hill and Magnusson's (1990) analysis, some of which is summarized for the Laurentian Great Lakes in Collingsworth et al. (2017). Hill and Magnusson's (1990) analysis is nonetheless a useful demonstration for considering thermal effects via bioenergetics, including the differential responses among species and the influence of behavioral thermoregulation and prey consumption rates on predicted effects on growth.

Latitudinal Shifts

As climate change contributes to altered conditions, the suitability of environments for species may change such that previously suitable locations are no longer suitable and previously unsuitable locations become suitable. Such effects can lead to range shifts of species distributions. Latitudinal range shift is an active area of research and monitoring with ongoing efforts to describe species' (a) leading edges – the portion of species' distributional ranges that are expanding, often toward higher latitudes, and (b) trailing edges – the portion of species' distributional ranges that are contracting, often away from lower latitudes. Shifts in geographical ranges can be viewed both positively and negatively. On the one hand, species' abilities to shift ranges may allow them to persist as portions of their current ranges become unsuitable. On the other hand, species moving into new ranges are analogous to invasive species and can lead to various novel biotic interactions, including negative effects on extant species. The ability of species to shift their ranges will also depend on their traits and the connectivity of different environments they may move between.

As an example of how changing climate conditions could contribute to range shifts, consider an analysis of the distributions of yellow perch, Eurasian perch, and smallmouth bass (Shuter and Post 1990). These authors proposed that a key factor limiting the northerly range of these species is the ability of young perch and bass to survive winter. Winter is a period of resource scarcity when fish lose energy. Because small fish have higher mass-specific metabolic rates and lower overall energy stores, fish may not be able to survive winter if they are not sufficiently large. This constraint would be especially severe in higher latitudes, where the duration of resource scarcity and starvation is relatively long. Young fish must grow sufficiently large before winter to be able to survive, and Shuter and Post (1990) suggested that it is largely growth potential and duration from hatch to winter that actually limits the northern extent of yellow perch, Eurasian perch, and smallmouth bass. They demonstrated through growth models that this limitation was able to explain the current latitudinal distribution of these species. They then used projections of future temperatures (+4 °C) to estimate the potential future northerly extents of yellow perch and smallmouth bass in North America. They showed how warmer temperatures could lead to a dramatic northern shift of the range limit of these two species.

Many freshwater systems are relatively unconnected, and therefore it is not always obvious how freshwater fish may shift their latitudinal range in response to changing climate. In contrast, many marine species occupy large, relatively open systems where movement may appear to be less constrained. However, even in marine systems, geographic barriers and habitat conditions may limit species' abilities to shift latitudinal ranges, and other stressors – such as fisheries harvest – may strongly mediate realized ranges. Many studies have evaluated climatic change effects on marine fish distributions either by examining range shifts in the recent past or simulating how species' ranges may shift in the future in response to warmer temperatures. As an example of the former, consider a study by Bell and colleagues (2015) from the National Marine Fisheries Service's Northeast Fisheries Science Center (NEFSC). These researchers considered how the ranges of four fish species had shifted from 1972 to 2008 in the Atlantic Ocean, on the shelf off the Northeast United States. Specifically, they considered black sea bass, scup, summer flounder, and winter flounder. All of these species are harvested by both recreational and commercial fisheries, and all four species have been actively monitored – including by an NEFSC fishery independent trawling program. This trawling program indexed fish distributions along the Atlantic shelf off Virginia in the south to Maine in the north during both spring and fall of each year. Each year and season, Bell and colleagues calculated the geographic center of biomass for each species. They then analyzed how the center of biomass of each species shifted each year, and developed statistical models to evaluate what factors might have contributed to observed shifts in the centers of biomass. They not only considered temperature as a potential driver of shifts, but also considered two factors related to fisheries harvest, abundance and mean length of the population. Fisheries harvest contributes to smaller population sizes and truncated length distributions with lower mean lengths. If populations become lower in abundance, they may be characterized by a restricted overall range. There is also evidence that for these species their latitudinal range along the Atlantic shelf is related to size, and therefore truncated size distributions may contribute to range shifts.

Bell et al. (2015) observed that the center of biomass for three of the four species shifted toward the North Pole from 1972 to 2008. Specifically, black sea bass and scup centers of biomass shifted northward in the spring, summer flounder center of biomass shifted northward in the fall, and winter flounder center of biomass did not trend. Based on statistical models (generalized additive models), Bell et al. (2015) concluded that the northward shift of summer flounder appeared to be related to changes in fisheries harvest. Larger summer flounder seemingly were more likely to be distributed farther north. There was a general reduction in fisheries harvest and an increase in mean size of summer flounder during this study. The northward shift of summer flounder center of biomass in the fall was simply better explained by this increase in relatively large individuals than climate-related trends. In contrast, the northward shifts of black sea bass and scup in the spring were most strongly associated with a trend of warming temperatures. Therefore, climatic changes appeared to be contributing to the northward shift of these two species (Figure 35-2).

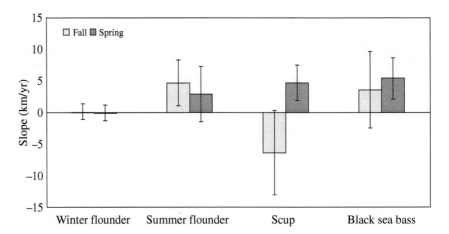

FIGURE 35-2 Estimated shifts in centers of biomass for winter flounder, summer flounder, scup, and black sea bass in the Atlantic Ocean along the east coast of the Unites States. Bars represent slopes of shifts (km yr^{-1}) with 95% confidence intervals and positive values, indicating a northward shift in biomass over 1972–2009. Summer flounder in the fall along with both scup and black sea bass in the spring exhibited significant northward shifts in biomass. *Source: Adapted from Bell et al. (2015).*

In addition to retrospective studies, such as Bell et al. (2015), researchers have also projected the future distributions of marine fish stocks. Cheung and colleagues (2009), from the University of British Columbia, projected future range shifts of 1066 marine fish (836 species) and invertebrates (230 species) at a broad, global scale during 2003–2050. They specifically considered distributions of these exploited species across a global grid (cell area: 30′ latitude × 30′ longitude). Within each of these cells, they used an algorithm from a previous study (Close et al. 2006) to estimate baseline distributions. These baseline distributions were based on relative abundances observed from 1980 to 2000 with the understanding of species-specific habitat preferences. They then generated predictions of future distributions using dynamic bioclimate envelope models. These models simulated relative abundance of a particular species in a grid cell based on habitat suitability, which included thermal suitability, in addition to the likely cumulative input of settling larvae and migrating adults. Cheung et al. (2009) considered three different climate change scenarios (high, mid, and low) in applying bioclimate envelope models, representing different assumptions regarding future carbon emissions. They estimated that the median poleward shift of the leading edge of modeled species would be 291 km under the high emission scenario and 223 km under the low emission scenario. They also estimated that distributional centroids of species would have median shifts of 79 and 44 km under high and low emission scenarios, respectively. However, these predicted range shifts were not equivalent across all species groups. For example, pelagic species were estimated to shift poleward to a much greater degree than demersal species.

Cheung et al. summarized range shifts across all species to determine the extent of local species invasions and local species extinctions within individual grid cells. They predicted local species invasion intensity – number of invading species relative to the number of species predicted to be in a grid cell before climate change – and extinction intensity – proportion of species becoming locally extinct in a grid cell. Species invasion intensity was predicted to be especially high in the Arctic and Southern oceans. This prediction reflected both (a) that species richness in these regions was rather low during their baseline period and (b) that a number of species distributions shifted poleward into these regions. Local extinction intensity was predicted to be high in various regions, including the tropics, Southern Ocean, and semi-enclosed seas, such as the Mediterranean. As temperatures increase, extant species may be expected to shift away from these regions.

Phenology

Climatic changes may have strong influence on fish phenology – the seasonal timing of processes, such as migration, reproduction, and larval migration. Such processes vary annually, even in the absence of climate change. People have a long history of noting the annual timing of terrestrial phenomena, such as the arrival of migrating birds, flowering of plants, and emergence of insects. While it may be less straightforward to document phenological events in aquatic systems, researchers have been interested in aquatic phenology for a long time. Phenology in particular has been of interest as it relates to recruitment success in fish populations. For example, Cushing's (1969) match–mismatch hypothesis builds from annual variation in the timing of larval emergence relative to the timing of plankton blooms.

It may seem obvious that the timing of biological events will shift as annual thermal conditions vary and that the timing of multiple events will shift in a similar manner. However, this is not often the case. Many seasonal biological events – such as fish migration or spawning – are not simply a function of temperature but may respond to processes, such as fluvial discharge rates and plankton blooms. Many events are dependent on seasonal changes in daylight duration. Unlike many other factors that influence phenology (e.g. thermal conditions and discharges), seasonal progression of daylight length is essentially set and will not respond to climate change. Therefore, daylight length serves to constrain the phenological variation of many events. Different biological events may be more or less sensitive to variation in daylight length compared to other phenological drivers. With warming temperatures and other consequences of climate change, this can lead some biological events to continue to be approximately synchronized, while others may come out of phase. This can have important consequences for various biological interactions, population trajectories, and community dynamics.

A study by Winder and Schindler (2004) of the University of Washington effectively demonstrated the potential consequences of phenological events becoming out of phase and uncoupled. The authors examined a long-time series of physical and biological data collected from Lake Washington from 1962 to 2002. They documented that during these four decades, water temperatures increased. The onset of stratification in the spring also tended to occur earlier, shifting forward on average by 21 days during the 40-year period. They also examined dynamics of diatoms – phytoplankton that are important food for herbaceous zooplankton and the dominant component of the spring phytoplankton bloom. They found that spring diatom blooms shifted earlier from 1962 to 2002, and blooms occurred 27 days earlier by the end of the time series. They also considered the timing of two herbaceous zooplankters *Keratella* spp. (a rotifer) and *Daphnia* spp. (a cladoceran). Both of these zooplankton taxa express spring peaks that tend to occur coincident – or just after – spring diatom blooms. They found that *Keratella* spring peaks occurred earlier during the study period, essentially tracking the change in water temperatures, onset of stratification, and diatom blooms. In contrast, the timing of *Daphnia* spring peaks did not change in a consistent manner during the study period.

The differential responses of *Keratella* and *Daphnia* to changing physical and lower trophic level conditions appeared to have an effect on their performance in Lake Washington. Winder and Schindler (2004) found that *Karetella* densities were positively associated with temperatures. When water temperatures became warmer in March and April, *Karetella* peaked earlier. *Daphnia* densities, on the other hand, were not linked to water temperatures. Photoperiod instead appeared to trigger *Daphnia* to hatch from resting eggs and subsequently emerge and peak. Therefore, *Keratella* were seemingly able to take advantage of the earlier diatom bloom, while *Daphnia* became out of phase with this important food source. Winder and Schindler (2004) documented that overall *Keratella* densities (April–June) remained at fairly similar levels during the study period, while *Daphnia* densities declined.

The Winder and Schindler (2004) study demonstrated how phenological responses can have strong effects on population trajectories of plankton. Differential phenological responses of fish populations have also been observed and may have similar effects on population trajectories.

The CalCOFI program (California Cooperative Oceanic Fisheries Investigation) has surveyed larval fish assemblages off the coast of California in the Pacific Ocean multiple times per year since 1951. Asch (2015), of Scripps Institute of Oceanography and Princeton University, analyzed CalCOFI larval fish collection data to determine how the timing of different species of larval fish caught in the California Current Ecosystem (CCE) may have changed over time. In total, they considered the phenology of the 43 most common species from 1951 to 2008. For each species, they calculated the seasonal mean central tendency, an index of when a species was on average collected during the year.

Sea surface temperatures (SST) increased in the CCE during the study period (1951–2008), with earlier annual warming. In contrast, physical models suggest that seasonal upwelling in the CCE will occur later with climate change. Both temperature and upwelling dynamics influence the timing and magnitude of phytoplankton and zooplankton blooms, and therefore the seasonal availability of prey for larval fish. Warmer temperatures may contribute to fish spawning earlier and eggs developing faster, leading to the expectation that larval fish may show a tendency to emerge earlier over time. However, delays in upwelling and associated effects on prey availability and larval transport would lead to the opposite tendency, with larval fish emerging later. Asch (2015) found evidence of both tendencies. A total of 39% of seasonal peaks of larval fish occurred earlier, while 18% occurred later. The remaining species did not demonstrate a strong trend across the 57-year study period. Many of the species that displayed earlier seasonal peaks in larval abundance were characterized by epipelagic, offshore distributions. Such species may be less influenced by delayed coastal upwelling and instead respond to warmer offshore temperatures. In contrast, many of the species that showed delayed peak larval densities used coastal, near bottom habitats and may have been particularly responsive to upwelling dynamics. This study by Asch (2015) demonstrates the complexity of some fish phenological responses. It is unlikely that all populations will shift their reproductive timing and offspring development in a similar manner. Some populations of larval fish may consequentially continue to experience frequent temporal matching with suitable early life environmental conditions, while other populations may experience frequent mismatches.

Interactive Effects

While climate change is a profound stressor of aquatic ecosystems, it is by no means the only stressor of such systems. Climatic effects will interact with various other stressors in influencing fish populations and communities. Some potential interactive effects are highlighted in examples above. For example, altered temperatures and precipitation patterns may interact with anthropogenic land-use practices to contribute to greater nutrient runoff and exacerbation of hypoxic conditions. Warming of marine systems can also lead to range shifts and invasion by some species into novel areas as highlighted by Cheung et al. (2009). Therefore, fish populations and communities will not only respond to climate-driven changes in physical and chemical conditions, but will also simultaneously face stressors, such as overfishing, chemical pollution, invasive species, and habitat destruction. Such interacting stressors and uncertainty related to their trajectories and effects serve to further complicate abilities to appreciate climate change effects.

The potential interactive effects of climate change and invasive species have been highlighted by various researchers. Rahel and Olden (2008) reviewed some of the ways by which climatic change can facilitate the spread of invasive species. They point out that climatic changes can alter environmental conditions such that different species are able to reach and establish in new environments. They describe how, past conditions, such as cold temperatures, flow regimes and salinity levels may have limited pathways for invasive species to reach new environments and prevented establishment of certain species. However, with climate change, there may be new pathways for introductions and possibilities for establishment. They also highlight that in areas of the world where precipitation and runoff are likely to decrease. There may be an increase in projects to manage and deliver water, including development of new reservoirs and construction of canals. Canals can create new connections among aquatic systems – providing a new pathway for introductions – and reservoirs create an artificial lentic environment within a lotic system and can favor non-native species.

Bell et al. (2021) provide a more specific example of how climate change can have interactive effects with invasive species to determine performance of native species. They considered the distributions of five species of trout in streams and rivers of the northern Rocky Mountain region of Montana: two native species (bull trout and cutthroat trout) and three non-native species (brook trout, brown trout, and rainbow trout). They compiled data from 21 917 surveys to consider how these species distributions had changed over the period from 1993 to 2018. They then modeled potential species distributions under future climate change by 2080 – based on the A1B scenario, which predicts an intermediate level of climatic change. Both Bell et al.'s (2021) retrospective and forecast analyses focused on whether a species was present or not in a given stream. Specifically, they developed a multispecies dynamic occupancy model based on past occupancy data and various environmental data. They applied this model to consider how future conditions may influence occupancy patterns.

By modeling past occupancy over multiple decades, Bell and colleagues (2021) considered the type of conditions that allowed the five different species to persist in different environments. They found that bull trout were likely to persist in streams with relatively cold temperatures with high flow, while cutthroat trout were likely to persist across a range of temperatures and flow rates. Non-native brown trout and rainbow trout persisted in streams with relatively warm temperatures and high flow, while native brook trout persisted in streams with cool temperatures and low flow. They found that non-native species affected the persistence of both native species, with brown trout negatively affecting persistence of bull trout, and brook trout and rainbow trout negatively affecting the persistence of cutthroat trout. Bell et al.'s (2021) analysis indicated that the percent of occupied habitat decreased for bull trout, cutthroat trout, and brook trout in the past, and would likely decrease further in the future. In contrast, brown trout percent of occupied habitat had decreased very little in the past and was predicted to remain stable in the future, while rainbow trout percent of occupied habitat had increased in the past and was predicted to increase even further in the future (Figure 35-3). Bell et al.'s (2021) analysis demonstrated that native species distributions are

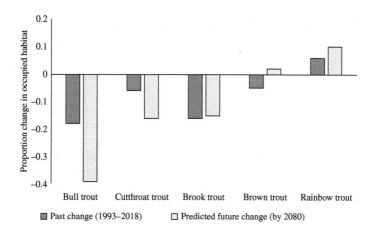

FIGURE 35-3 Estimated past changes and predicted future changes in stream occupancy in the Rocky Mountains region of Montana, USA, for native (bull and cutthroat) and non-native (brook, brown, and rainbow) trout. The decreased proportional occupancy of native trout was related to climatic change, both directly (response to warmer temperatures and lower stream flow) and indirectly (through climate-facilitated spread of non-native species). *Source: Adapted from Bell et al. (2021).*

not only dependent on their own response to changes in abiotic conditions – such as temperature and stream flow rates – but also the responses of non-native species to changes in abiotic conditions.

In addition to invasive species, other stressors will have strong interactive effects with climate change, and in some cases, climate change can interact with multiple other stressors. For example, Schartup et al. (2019) considered how climate change, overfishing, and chemical pollution can have interactive effects. They focused on methylmercury (MeHg), a widespread contaminant in the world's oceans known to have various negative effects on fish and acts as a neurotoxin for humans that consume contaminated fish. Some piscivorous, higher trophic level fish can amass very high MeHg concentrations through biomagnification. Schartup et al. (2019) specifically focused on piscivorous Atlantic cod and spiny dogfish in the Gulf of Maine, Atlantic Ocean. They examined MeHg concentrations of these two species from the 1970s and 2000s, and used food web, predator–prey, and bioenergetics models to simulate potential future MeHg under increased ocean temperatures (+1 °C) and changes in fishing patterns and mercury emission patterns. They specifically considered how overfishing of large herring could affect cod and dogfish MeHg concentrations, and the effect of a 20% reduction in seawater MeHg concentrations – a plausible future reduction based on decreased loadings.

Schartup et al.'s (2019) simulations indicated that with warmer temperatures MeHg concentrations of both Atlantic cod and spiny dogfish should increase. With higher temperatures, fish are able to eat more, while they also expend greater energy through metabolism. Therefore, with warmer temperatures, a fish will need to consume more prey over its lifetime to reach a certain size, and there is greater potential for biomagnification and accumulation of contaminants, such as MeHg. Analyses of past Atlantic cod and spiny dogfish diets – in addition to modeled prey preferences – indicated that they would likely respond differently to overfishing of preferred prey. As large herring are depleted, Atlantic cod switch to consume a greater proportion of small clupeids, which are lower in MeHg. In contrast, as large herring are depleted, spiny dogfish switch to consume a greater proportion of cephalopods (i.e. squid), which have relatively high MeHg concentrations. Therefore, with overharvest of large clupeids Atlantic cod express decreased MeHg concentrations, while spiny dogfish express increased MeHg concentrations. If temperature and diet were unchanged, both Atlantic cod and spiny dogfish respond to a 20% reduction in seawater MeHg concentrations with decreased tissue MeHg concentrations. However, responses to reduced seawater MeHg concentrations are quite different, if temperature increases by 1 °C and large herring are overharvested. Schartup et al. predicted that even if seawater MeHg decreases by 20%, MeHg concentrations for a 5 kg spiny dogfish could increase by >60% due to the combined effects of warmer temperature and overharvest of preferred prey (Figure 35-4).

Summary

There is broad evidence that climatic changes have affected physical, chemical, and biological conditions in aquatic ecosystems in the recent past, and projections indicate that future climatic conditions will continue to have profound influence on aquatic systems and the fish populations and communities that inhabit them. Altered temperatures can directly affect processes, such as fish feeding and growth rates, reproductive timing, and spatial distributions. Various studies have demonstrated such effects through retrospective analyses, and a growing number of studies have projected how future climatic changes may further influence fish populations. While it is important to consider climatic effects on fish, precise predictions of fish population responses to future climatic conditions are challenging because of the large variety of

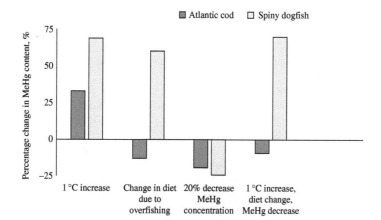

FIGURE 35-4 Estimated changes in methylmercury (MeHg) concentrations of Atlantic cod (blue bars) and spiny dogfish (yellow bars) in the Gulf of Maine under potential future scenarios: a 1 °C increase in temperature, a change in diet due to overfishing of their herring prey, a 20% decrease in water MeHg concentrations, and the combined effect of a 1 °C increase in temperatures, a change in diet, and a 20% decrease in water MeHg. *Source: Adapted from Schartup et al. (2019).*

interacting factors and mediating effects. The near-term evolution of fish populations may imply that their resilience to climatic changes may be much different from what could be anticipated based on their current genetic composition. Climatic changes are by no means the only stressors fish populations will face, and the effects of multiple interacting stressors may be quite complex. Most fish populations interact directly and indirectly with a multitude of other populations (e.g. as predators or prey) that strongly influence their trajectories. Therefore, anticipating how populations respond to climatic change will not only depend on how individual populations respond to changes in physical conditions, but how a potentially huge number of other interacting populations also respond. However, through various approaches, researchers are improving our abilities to anticipate climatic effects and therefore inform the future management of fish populations and communities.

Literature Cited

Asch, R.G. 2015. Climate change and decadal shifts in the phenology of larval fish in the California Current ecosystem. Proceedings of the National Academy of Sciences 112:E4065–E4074.

Bell, D.A., R.P. Kovach, C.C. Muhlfeld, et al. 2021. Climate change and expanding invasive species drive widespread declines of native trout in the northern Rocky Mountains, USA. Science Advances 7:eabj5471.

Bell, R.J., D.E. Richardson, J.A. Hare, P.D. Lynch, and P.S. Fratantoni. 2015. Disentangling the effects of climate, abundance, and size on the distribution of marine fish: An example based on four stocks from the Northeast US shelf ICES. Journal of Marine Science 72:1311–1322.

Cheung, W.W.L., W.Y.V. Lam, J.L. Sarmiento, et al. 2009. Projecting global marine biodiversity impacts under climate change scenarios. Fish and Fisheries 10:235–251.

Clements, J.C., J. Sundin, T.D. Clark, and F. Jutfelt. 2022. Meta-analysis reveals an extreme "decline effect" in the impacts of ocean acidification on fish behavior. PLoS Biology 20:e3001511.

Close, C., W.W.L. Cheung, S. Hodgson, et al. 2006. Distribution ranges of commercial fishes and invertebrates. Pages 27–37 in D. Palomares, K.I. Stergiou, and D. Pauly, editors. Fishes in Databases and Ecosystems. Fisheries Centre Research Report 14. University of British Columbia, Vancouver.

Collingsworth, P.D., D.B. Bunnell, M.W. Murray, et al. 2017. Climate change as a long-term stressor for the fisheries of the Laurentian Great Lakes of North America. Reviews in Fish Biology and Fisheries 27:363–391.

Cushing, D.H. 1969. The regularity of the spawning season of some fishes. ICES Journal of Marine Science 33:81–92.

Hill, D.K., and J.J. Magnuson. 1990. Potential effects of global climate warming on the growth and prey consumption of Great Lakes fish. Transactions of the American Fisheries Society 119:265–275.

Magnuson, J.J., D.M. Robertson, B.J. Benson, et al. 2000. Historical trends in lake and river ice cover in the Northern Hemisphere. Nature 289:1743–1746.

NOAA (National Oceanic and Atmospheric Administration). 2020. Ocean acidification. www.noaa.gov/education/resource-collections/ocean-coasts/ocean-acidification. (Retrieved 10 May 2022)

Nichols, R.J., and A. Cazenave. 2010. Sea-Level rise and its impact on coastal zones. Science 328:1517–1520.

Rahel, F.J., and J.D. Olden. 2008. Assessing the effects of climate change on aquatic invasive species. Conservation Biology 22:521–233.

Schartup, A.T., C.P. Thackray, A. Qureshi, et al. 2019. Climate change and overfishing increase neurotoxicant in marine predators. Nature 572:648–650.

Sharma, S., K. Blagrave, J.J. Magnuson, et al. 2019. Widespread loss of lake ice around the Northern Hemisphere in a warming world. Nature Climate Change 9:227–231.

Shuter, B.J., and J.R. Post. 1990. Climate, population viability, and the zoogeography of temperate fishes. Transactions of the American Fisheries Society 119:314–336.

Steinhart, G.B., N.J. Leonard, R.A. Stein, and E.A. Marschall. 2005. Effects of storms, angling, and nest predation during angling on smallmouth bass (*Micropterus dolomieu*) nest success. Canadian Journal of Fisheries and Aquatic Sciences 62:2649–2660.

Waples, J.T., and J.V. Klump. 2002. Biophysical effects of a decadal shift in summer wind direction over the Laurentian Great Lakes. Geophysical Research Letters 29:43–4.

Winder, M., and D.E. Schindler. 2004. Climate change uncouples trophic interactions in an aquatic ecosystem. Ecology 85:2100–2106.

CHAPTER 36

Conservation of Freshwater Fishes

Introduction

Conservation biology arose from the concern of biologists protecting endangered and threatened species, who initially developed techniques involved in recovery of these critically endangered organisms. It has since expanded to involve the application of biological principles toward preserving biotic diversity of organisms. It applies many principles described earlier in this book and, in particular, relies on some key areas of study, such as the theory of island biogeography, especially the nature of species–area relationships. The relationship between the number of species present and the area of habitat they have to occupy (termed a species-area curve) is useful in defining what minimum area might be necessary to maintain a certain level of biotic diversity as well as to evaluate changes that might occur in persistence of species as area of suitable habitat declines.

A second field critical to conservation biology is population genetics. Conservation biologists are interested in genetic change of a population due to conditions including low population density, domestication, or competition with exotic organisms. Conservation genetics also focuses on determining minimum viable populations, which are levels of populations of a species or stock necessary to maintain its biotic diversity in specific habitats or regions. Minimum viable populations include the number of organisms necessary to effectively breed and maintain the level of genetic diversity that exists in a species.

Endangered Species and Extinctions

The concern for persistence of vulnerable species was first formally expressed in the United States in management of endangered species. Laws on endangered species have evolved through a series of legislative acts, starting with the Endangered Species Preservation Act (ESPA) in 1966 and terminating in the current Endangered Species Act (ESA), originally passed in 1973. These acts were established to protect and manage species considered vulnerable to extinction. Currently, there are two legal categories of endangerment in the ESA: endangered species (under immediate threat of extinction) and threatened species (under threat of becoming endangered). Some states and other organizations have additional

Biology and Ecology of Fishes, Third Edition. James S. Diana and Tomas O. Höök.
© 2023 John Wiley & Sons Ltd. Published 2023 by John Wiley & Sons Ltd.

categories such as species of special concern, not yet threatened but likely to become so. For example, The Nature Conservancy (TNC) uses a 1–5 ranking system: GX, presumed extinct; GH, possibly extinct; G1, critically imperiled; G2, imperiled; G3, vulnerable; G4 and 5, secure (Master et al. 1998). These categories are defined with numerical and stability rankings for each species. Internationally, the Convention on International Trade in Endangered Species (CITES) regulates sale and transport of vulnerable organisms and lists these organisms for protection. Protection categories vary within other countries as well. Legal definition and listing of endangered species under ESA allows for protection of these organisms, protection of their critical habitat, and establishment of plans to recover the population to a non-threatened status. Endangered species management has generally focused on means to protect single species.

Conservation of fish populations were some of the issues that led to the development of the ESA and were involved in the evolution of the act. The first lawsuit under the ESPA focused on pupfish. The Devil's Hole pupfish was under threat of extinction by removal of groundwater, which reduced the water levels in the small spring habitat where they exist. This damage began prior to the existence of a specific act, and ultimately the passage of ESPA facilitated management and survival of this species (Pister 1990). Early legal protection focused on damage caused by federally approved projects, and issues with fish came to the forefront since federal approval is often required for development of irrigation, hydropower, dams, and large-scale removal of groundwater. Thus, federal projects with such direct impacts on species of fish and aquatic organisms were often litigated under ESPA and ESA.

The ESA was developed to protect any genetically unique form of organism. While it was originally defined to protect species, some unique subspecies also came under its protection, and eventually, even unique populations. Many of the current problems with endangered species in the United States have to do with specific spawning populations of species such as sockeye salmon in the Pacific Northwest. While some populations of this species are not considered threatened or endangered, the loss of other populations is considered a threat to maintenance of the species and regulated under ESA.

The targeting of certain unique or important species for management has particular importance in maintenance of biodiversity. Although all species are precious, organisms that may be the only living member of a family, order, or even higher level of taxonomy may have much more importance in preservation of biotic diversity than subpopulations of a widely distributed species. Such importance is not legally recognized by ESA or CITES, but it definitely influences activities of practitioners. A good example of such a critical group of fish is lungfish from the order Ceratodontiformes. There are currently only six living species of this entire order, and most populations of these species are rare. The Australian lungfish is the sole remnant in what were seven species in the family Ceratodontidae. It is believed to be the most primitive of the existing lungfish species, and it does not have the ability to aestivate (become dormant during drought conditions), as do the five South American and African species in the remaining two families.

The Australian lungfish currently persists in several river systems in southeastern Queensland, including the Burnett, Mary, and Brisbane rivers. Some transplantation has also occurred. This limited distribution is of some concern as changes in the rivers may result in loss of this very unique fish. Recent studies have shown that the population in the Burnett River is quite healthy and viable although future options of adding dams or conducting other manipulations on the river still make this species vulnerable (Brooks and Kind 2002, Kind 2003). For this and many others species, dedicated fishery biologists have led efforts to understand and preserve unique fish, often in opposition to the desires of their employer or the public that may want to develop critical habitat for irrigation or other purposes. The global significance of an organism that is the only living member of its family and one of very few members of an entire order of vertebrates means that management for species like Australian lungfish is extremely important in global preservation of the biotic diversity. Similar statements could be made for the coelacanth *Latimeria* and a score of other unique fishes. While the significance of rare orders or families of animals does not legally enter into management of endangered species, these animals are indeed important to our understanding of biodiversity and evolution.

Conservation Genetics

Maintaining biodiversity is a key concept of conservation biology, and biodiversity is defined as the variety of life in a given habitat or on the planet, so it includes the diversity of species, populations, and genetic types of organisms. The interest in genetics reflects the belief that a genotype with much variability has a greater potential to persist under stress than one more homozygous. Because of various human activities, many species now show restricted genetic variation compared to the pattern prior to human disturbance. For many populations that have undergone a major bottleneck, when they recover from that bottleneck, genetic variability remaining is often far less than what existed before. Stocking of sport fishes has also affected genetic diversity as fishes in hatchery systems are often selected for domestication traits. These traits may include rapid growth rate, lack of aggressive behavior, willingness to consume pelleted feed, etc. While these traits have strong significance for the ability to rear fish in hatcheries, they may have strong negative effects on survival of fish in the wild. When hatchery fish are introduced into areas where native fishes of the same species exist, interbreeding can result in changing of the gene pool, from what naturally existed there to a mix or domination of genetic material from stocked fish.

The recognition of the importance of knowing population genetics and local adaptations of species have caused many scientists to question practices of rearing and stocking. In particular, introduction of reared fish from a domesticated gene pool – even within the native range of that species – may influence existing populations. Domesticated traits of that species can result in genetic change in the natural population, and many have called for the cessation of hatcheries or the change of hatchery practice to account for natural patterns of genetic diversity. Changes now being tested include only stocking fish reared from eggs in natural waters, using egg sources from the same watershed into which the fish are stocked, or intentionally reintroducing natural genetic variation into stocked fish. All of these methods attempt to reverse the degradation of genetics by hatchery programs. However, stocking of sport fish remains a popular management technique today and may even be the mainstay of many fishery management agencies. Controversies on the genetic impacts of stocking are likely to persist in the future.

Conservation Status of Freshwater Fishes in the United States

Interest in conservation of biodiversity has led scientists from many countries to create conservation programs and evaluate biotic diversity of their fauna and flora. Many nonprofit groups have organized to focus on management of natural heritage and protection of biotic resources that have often been ignored in the past.

Threatened Fish Fauna of the United States

The number of species of freshwater fishes in various regions was reviewed earlier. North America has a relatively diverse fauna for a temperate area with over 1000 species of freshwater fish. This is four times the diversity of Europe, but far less diversity than in Asia, Africa, or South America. TNC has evaluated the global significance of freshwater species in the United States as part of their strategy to conserve natural areas that have significance in maintenance of biodiversity (Master et al. 1998). They found that the diversity of described species of freshwater organisms in the United States is extremely high for many taxonomic groups compared to those described in other regions (Table 36-1). For example, the United States has approximately 10% of all known freshwater fish species worldwide, 61% of known crayfishes, and 40% of known stoneflies. This is an overestimate of the fractions of fishes in the United States as there are over 15 000 species of freshwater fish recognized globally (Nelson 2006), not the 8400 included in the TNC analysis. Still, the overall biotic diversity of freshwaters in the United States is relatively high. High levels of biodiversity of freshwater organisms across the United States are mainly influenced by two major patterns: distribution of precipitation and glacial history. There is high and relatively consistent precipitation in the eastern half of the country, with less total and more sporadic precipitation in the west, except the Pacific Northwest. Rainfall provides water – the basis for diverse aquatic ecosystems – and produces more stable conditions in the eastern half of the country. There are obviously exceptions to this, such as precipitation patterns in western mountain ranges, but these areas have not developed conditions for high diversity of aquatic organisms. Also, in northern areas such as the states bordering Canada or in mountainous regions, continental glaciation resulted in major perturbation of freshwater communities. The result of this differential precipitation and glacial history is that the states

TABLE 36-1   The Number of Described Species in Various Taxonomic Groups for the United States and the World, and the US Ranking Among Countries in Species Diversity for Each Group.

Taxonomic group	US described species	Worldwide described species	Percent found in the United States	US ranking in species diversity
Freshwater fishes	801	8400[1]	10	7
Crayfishes	322	525	61	1
Freshwater mussels	300	1000	30	1
Freshwater snails	600	4000	15	1
Stoneflies	600	1550	40	1
Mayflies	590	2000	30	1
Caddisflies	1400	10564	13	1
Dragonflies	452	5756	8	Uncertain
Stygobites	327	2000	16	1

Source: Master et al. (1998)/The Nature Conservancy.
[1] This is an underestimate, as Nelson (2006) indicates a much larger number of freshwater fish species (over 14 000), which is now widely accepted.

in the southeast have the highest number of aquatic species, while the arid western states have the lowest number (Figure 36-1).

Somewhere between 15 and 70% of aquatic species in the United States are currently at risk depending on taxonomic group (Figure 36-2). Over 70% of mussel, 50% of crayfish, and 35% of freshwater fish species were in categories GX to G3 (at risk) in 1998. The geographic distribution

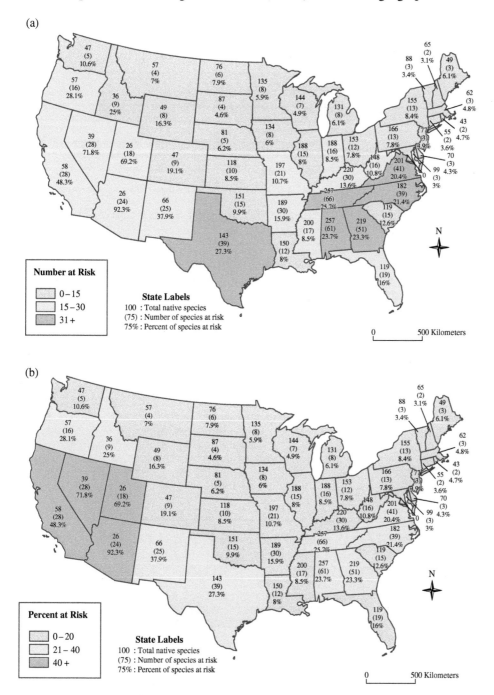

FIGURE 36-1   Total number of fish species, the number at risk, and the percent at risk for each state in the coterminous US: (a) categorized by total number at risk and (b) categorized by percent at risk.
Source: Drawn from data in Master et al. (1998) and Williams and Meffe (1998).

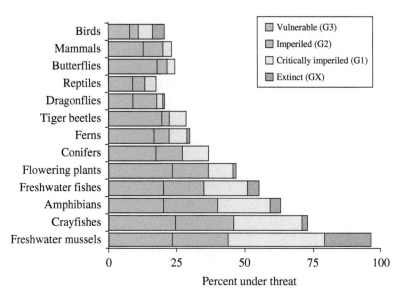

FIGURE 36-2 Percent of all US species in each risk category for various groups of aquatic organisms. *Source: Redrawn from Master et al. (1998).*

of these species shows a strong concentration within the southeastern portion of the country with lesser but important concentrations in the arid west (Figure 36-3). There is high diversity of fish and mussel species and a high degree of endemism (existence of species with very local distributions) in the southeastern United States as many of these species evolved and persist in relatively small, isolated watersheds. On the other hand, the arid west is an area where the fauna is far less diverse or endemic, but far more impacted by human activity. A much higher

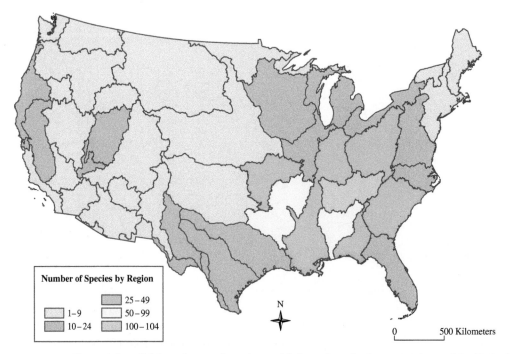

FIGURE 36-3 The number of fish and mussel species at risk in various freshwater regions of the United States. *Source: Redrawn from Master et al. (1998).*

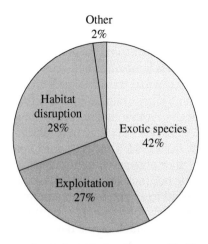

FIGURE 36-4 Major known causes of vertebrate extinctions worldwide. Cases included 90 mammals, 92 birds, 100 reptiles or amphibians, and 100 fishes, and did not include 134 cases where cause was unknown. *Source: Data from Cox (1993).*

fraction of the aquatic fauna in western states is under threat even though the total number of species is less than in the southeast (Figure 36-1).

Many scientists have been concerned about the degree of human influence on our planet and its impact on future biodiversity. For example, some estimates indicate that humans consume an unsustainable fraction of the total primary production (directly or through its use in animal feeds and other products) that occurs on Earth today. Similarly, human events such as population increase, agricultural and industrial water use, and water pollution have resulted in major challenges to the persistence of aquatic biota. Cox (1993) reviewed 384 cases of extinction for vertebrates in the world (including 100 fish species) and determined that the most important cause of extinction was introduction of exotic species, causing about 42% of all vertebrate extinctions (Figure 36-4). Except for 35% of the cases that could not be attributed to a specific cause, all of the causes of extinction were human interventions, with exploitation being a minor factor, while habitat disruption was also a major cause.

Case Histories in Conservation and Management

As a means of understanding the situation with sustaining fish populations, we describe three case histories: the Devil's Hole pupfish, lake sturgeon, and northern pike. All three are North American species, with the pupfish having a very small body size and very limited geographic distribution, while the pike and sturgeon have a very broad distribution and larger body size, the lake sturgeon so large it is considered a member of the charismatic megafauna. Charismatic megafauna are large animals with high public appeal, and as such, they have not only received a much more con-centrated effort at their conservation, but they have even been used as a means to protect whole ecosystems (Thompson and Rog 2019). The first two cases are for threatened and endangered species and will cover the issues causing threats to their persistence, as well as the management used to overcome those threats. In addition, we include one more widely distributed and harvested species – the northern pike – to describe the history of their management and actions being done to allow a sustainable harvest for the future. Through these cases, we hope to give examples of the challenges and methods to conserve fish species under varying human-related stresses.

The Devil's Hole Pupfish

Death Valley is a basin in eastern California and western Nevada with dry and hot conditions, which few fish species can endure. Its temperature has reached 134 F (56.7 °C) – the warmest temperature recorded on the planet. While this location is challenging for aquatic biodiversity, a number of springs connected to a large underground aquifer have resulted in isolated pools that have served as locations for the evolution of a number of different fish and other aquatic species. The unique nature of Death Valley resulted in its designation as a national monument in 1933 and later as a national park in 1994. This was done to protect this unique environment and its organisms from unguarded or unsupervised human activity.

Devil's Hole is a small opening in a rock structure (surface area 77 m²) with access to a pool of very warm (33 °C) water attached to a subterranean aquifer over 150 m deep. There is only a 17.5 m² shelf shallow enough to allow for some photosynthesis, and the steep terrain around the hole eliminates direct sunlight for much of the day. Within this harsh environment, the Devil's Hole pupfish has evolved from other cyprinodont species also found throughout the arid southwest. Some unique evolved characteristics of the Devil's Hole pupfish include their ability to tolerate very warm water (over 40 °C) and high salinity (over 60 ppt). To protect its unique biology, as well as geology, Devil's Hole was included as a separate part of Death Valley National Monument in 1952. Due to low numbers (total population counts less than 200 individuals) and unique characteristics, the Devil's Hole pupfish was listed as endangered in 1967 under the ESPA although no clear protections were included in that designation.

Unfortunately, receiving endangered status was not sufficient to protect the pupfish. In the 1960s, inadvertent stocking of mosquitofish, as well as development of farms and housing in nearby Ash Meadows, challenged persistence of the pupfish. In 1969, wells were drilled to use groundwater to irrigate fields for growing cattle feed in Ash Meadows. Pumping of groundwater resulted in a decline in the surface water level in the pool to almost a meter below its reference level by 1972 (Deacon and Williams 1991). This threatened the submerged shelf where pupfish foraged and bred with desiccation. Several public groups, especially the Desert Fishes Council (a group of scientists instrumental in keeping challenges to persistence of the pupfish before the public as well as management organizations), petitioned the US Fish and Wildlife Service, which then took the case of groundwater use to court. In 1972, the Supreme Court ruled that the fish have a right to exist and that groundwater use for agriculture had to cease in the area. The land was then sold to a developer who planned to build a housing subdivision and use groundwater for domestic purposes, which was allowed under the Supreme Court decision. This led to an emotional 14-year court battle about endangered species and water rights with automobiles in the area displaying bumper stickers that read "Save the pupfish" or "Kill the pupfish," and an editorial in the Elko Nevada Daily Free Press suggesting that someone use rotenone to solve the problem. After considerable effort in the courts, funding was arranged by The Nature Conservancy to purchase the Ash Meadows area and establish a National Wildlife Refuge (NWR) that would protect the pupfish and other species from groundwater extraction.

Establishing an NWR did not end the challenges to persistence for pupfish. Population numbers of pupfish declined dramatically in the early 2000s, reaching a low of 35 fish in 2013. Habitat manipulations, such as adding another shelf to increase food production and using artificial light to increase photosynthesis, were attempted to increase pupfish numbers. In another effort to save the species, 27 pupfish were translocated in 1972 to a refuge at Hoover Dam, where they were reared in captivity until 1986 when the facility failed. Additional fish had been moved to several other refuges to safeguard the stock but failed due to low numbers of the founding population and other challenges. Recent research shows that Devil's Hole fish held at other locations have developed physical differences from the original pupfish population, casting question on the genetic integrity of the artificially reared fish (Lema and Nevitt 2006). Population numbers at Devil's Hole have varied from highs of 500 individuals in 1980 and 1995 to below 200 individuals since 2005. Recent surveys of the pupfish population have shown an increase of more than 100 fish since 2016. In spite of these changes, the Devil's Hole pupfish is still under threat and considered critically endangered by the International Union for the Conservation of Nature (IUCN).

The battle for existence of this species has been long, expensive, and tested the Endangered Species legislation that existed over this time frame. A Recovery Plan and listing of Critical Habitat have since been developed, and intensive management of Devil's Hole, as well as Ash Meadows, have allowed the fish to persist. There have been a number of groups supporting this process, especially the Desert Fishes Council, the US Fish and Wildlife Service, and even the Supreme Court.

The Lake Sturgeon

Lake sturgeon present almost an opposite case to that of the Devil's Hole pupfish as far as a threatened species is concerned. Lake sturgeon are widely distributed with populations existing in major river systems and the Great Lakes – from Alberta to the St. Lawrence Seaway and from Hudson's Bay to Tennessee. Within this area, lake sturgeon were once quite abundant with commercial fisheries harvesting their populations for caviar and other products until closure of fisheries in late 1920s. The fish are large-bodied and long-lived with adults exceeding 200 cm in length and 150 kg in weight. It is difficult to age sturgeon because of their longevity, but maximum age is at least 100 years. Lake sturgeon do not spawn until they reach 20 years or more, and then the energy drain of spawning may cause females to forgo reproduction for another 2–5 years. These characteristics mean that lake sturgeon cannot sustain populations unless overall mortality rates are fairly low. These native fish have now become commonly managed for conservation purposes throughout a much of their range. Their current status varies with location, with 19 of 20 states in their range considering them threatened, while different Canadian populations are considered threatened, endangered, or of special concern. There is a current effort underway to list them as a threatened species under the ESA in the United States.

Overfishing became a problem for lake sturgeon in the late 1800s (see Figure 36-5 for harvest of sturgeon in Lake Huron). In the Great Lakes, as well as Lake of the Woods and many

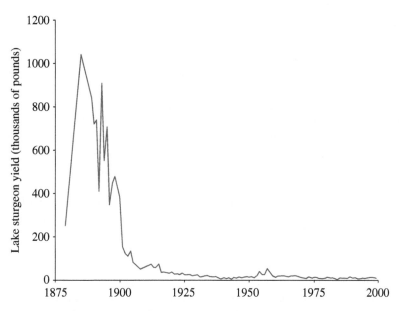

FIGURE 36-5   Landings of lake sturgeon from commercial fisheries in Lake Huron, 1867–2000.
Source: Drawn from data in Baldwin et al. (2002).

rivers, lake sturgeon were harvested by commercial fisheries since early times. Peak lake sturgeon landings in Lake Huron occurred around 1890, and these declined dramatically by 1900. Lake sturgeon were originally considered a nuisance species, and they were intentionally eradicated by commercial operations because these large fish damaged nets. Fish were caught and either discarded in the water, fed to pigs, dried on shore to be burned, or used to extract oils from their carcasses (Scott and Crossman 1973). After about 1860, they were more commonly used for human food and other products such as isinglass (a clarifying agent for beer and wine, as well as a thickener used in pottery and jellies). By the early 1900s, lake sturgeon were rare throughout the Great Lakes, and they remain so today. A commercial fishery with limited harvest still exists in the St. Lawrence River. Some sport fishing harvest occurs in the Great Lakes as well as some inland lakes.

Besides overfishing, most lake sturgeon populations were also affected by human-induced changes to their habitat. Lake sturgeon exist in rivers, as well as large lakes, and migrate to spawn in the rapids of rivers in the spring. Dams built on river systems blocked access to many spawning habitats, and pollution damaged egg survival in degraded rivers. Spawning populations were decimated, and alternative spawning locations, such as rocky substrate immediately below a dam, were often chosen by sturgeon. However, these locations had variable flow depending on use of the dam, and downstream reaches did not necessarily have adequate nursery habitat for young sturgeon. Based on the existing literature, Auer (1996) proposed that sturgeon species need access to 250–300 km of open river and lake to allow for normal behavior and population restoration. Fragmentation of habitat into smaller units by dams or habitat change results in lower overall abundance of sturgeon, as well as potential for population collapse if conditions in that smaller range deteriorate. However, other research has shown variation in sturgeon behavior in different natural and impacted habitats, which led McDougall et al. (2017) to propose that differing "spawn–drift–settle–establish" habitats occurred in different river systems. Their results indicate that self-sustaining sturgeon populations could occur in locations with as little as 10 km of open river habitat, if conditions for all life-history stages were included in that area. While this is possible, it is not necessarily optimal for sturgeon populations to be constricted in such small areas. There are a variety of environmental conditions where sturgeon exist and lake sturgeon populations have adapted to persist under those differing conditions. Nonetheless, historic overfishing and habitat alteration caused a huge decline in the sturgeon populations throughout their range and continue to limit recovery of stocks. For example, the catch on Lake of the Woods in 1957 was only 0.005% of the level from 1893 (Scott and Crossman 1973).

Lake sturgeon are now popularized by public media throughout their range, where management is applied to protect and increase their populations, even though they are still exploited in some areas. Commercial fisheries exist only in the St. Lawrence River, where harvest is regulated by small total allowable catch, gear restrictions, and seasons. Sport fishing for lake sturgeon is allowed in several states and provinces, although most seasons are for catch-and-release fishing only, while a few locations allow very limited winter spear fisheries or very short seasons for harvest with a single fish possession limit. Overall, sport and commercial fishing regulations have been set to restore sturgeon populations rather than to allow much fishing or harvest.

Restoration has been attempted by stocking lake sturgeon into locations with poor or non-existent populations in areas once known to support the fish, allowing access to upstream spawning locations through fish passage over dams and the creation of spawning reefs in existing habitat. Stocking was originally done by taking gametes from ripe fish in locations where fish could be predictably collected and spreading those young fish reared to fingerlings in a hatchery system to new locations, but concerns arose about genetic integrity of recipient populations. It is now common for eggs to be collected from a river, or adult fish to be collected and artificially spawned, and then these eggs are reared in streamside facilities using water from the native river (Osborne et al. 2020). Such hatcheries have allowed specific spawning groups to be enhanced within their native locations but at greater cost than traditional hatchery systems. This is partly done in an effort to preserve the unique genetic composition of each population although the unique genetic nature of each stock is not well known. Another advantage for considering streamside hatcheries is maintaining local life-history types evolved to the specific river conditions, whether these adaptations are genetic or environmental.

Another restoration technique has been construction of artificial spawning reefs in sturgeon habitats affected by habitat loss. Such reefs are rock and boulder substrates where sturgeon can deposit eggs. Reefs have been added extensively to the St. Clair and Detroit River systems and have resulted in new spawning aggregations and egg deposition (Fisher et al. 2018). Whether these reefs contribute to increases in juvenile and adult stocks are uncertain although such contributions will take considerable time.

An additional tool to improve sturgeon spawning in some systems is the passage of sturgeon over or around dams, allowing migration to historic spawning sites. This has been hampered because sturgeon do not readily use fish ladders, the most common fish passage method historically built for salmon or walleye. On the Menominee River – a Lake Michigan tributary on the border between Michigan and Wisconsin – a fish lift has been installed, and fish passed over two dams to open up an additional 4–112 km of river for spawning. Fish bypass areas have also been installed to allow downstream migration of adults and juveniles. While this system is passing adults that migrate upstream and appear to spawn, it will take time to determine if this results in an increase of young fish in the population.

An extensive amount of research and management has been done on lake sturgeon populations in the last 20 years. Differences in behavior have been shown among land locked, open river, and Great Lake populations that make characterization of sturgeon populations more difficult. Funding of recovery programs has been extensive and done by state, provincial, federal, and tribal programs. Private groups such as Sturgeon for Tomorrow have encouraged citizens to adopt a sturgeon population and monitor their safety during the spawning interval, when sturgeon are often found in very shallow and exposed waters where poaching can be a major problem. All of these efforts have been successful in rehabilitating populations of lake sturgeon, and many stocks have recovered to be above threat of extinction. While it is unlikely the fish will reach historic levels of abundance, many populations can support limited recreational and commercial fisheries.

Northern Pike

The third case history deals with changing perspectives for harvest, exemplified by regulations and management of northern pike populations. Some populations of pike need and receive considerable amounts of management effort in order to sustain populations in a given aquatic system. Pike – more widespread than lake sturgeon – have a circumpolar distribution in the northern hemisphere and with distribution in North America from Alaska south into Missouri and Nebraska. Within this range, they are very common in inland lakes, rivers, and the Great Lakes. Pike are a relatively large apical predator and generally fairly low in abundance in any particular system. Their history and management has varied based on human perceptions of our role in nature and of the fish.

While northern pike are widespread and common in a large number of systems, there have been activities that have damaged pike populations. Northern pike spawn in flooded vegetation, often wetlands attached to lakes and rivers in spring. These areas have often been drained, filled, or cleared for riparian development, damaging pike spawning and nursery habitats. Commercial fishing commonly harvested pike populations although never at a high level. Still today, commercial fisheries for pike exist in waters of Canada, Europe, and Asia. In addition, pike are very popular sport fish that sustain harvest rather than catch and release, which has become more common for many other species. As a result, overfishing for northern pike has become an increasing problem in some locations.

The philosophy of harvest for northern pike has changed dramatically over time as have regulations restricting harvest. Diana and Smith (2008) reviewed the evolution of pike management in Michigan, and we use Michigan as an example. The Michigan State Board of Fish Commissioners originally believed pike were voracious predators that "mutilated and killed their more peaceful neighbors." By 1876, the commissioners reported that pike were "freshwater devil fish," and there should be only one policy for their management, that is "the policy of extermination." Such beliefs were common for predatory animals in those days, which were considered to devastate populations of more favorable sport fish and game animals. Thankfully, attitudes changed over time, and conservation was fostered by regulations to protect pike from overharvest. In 1929, the Sportfishing Law first established a size limit at 35 cm for northern pike and a closed season from 1 March to 1 May to protect spawning fish. Creel limits were also set at five. A size limit of 50 cm was passed in 1959 to allow pike to spawn at least once before they would be harvested. More recently, there have been major changes in the philosophy of management for northern pike with the realization that pike populations are important in controlling populations of stunted perch and panfish. These changes have recognized pike as having a broad potential for a fishery within the state. Various other states and provinces now regulate pike with different sizes and harvest regulations to allow for harvest-based fisheries, trophy fisheries, and fisheries to reverse stunting of pike in certain lakes.

Human alterations of riparian areas have resulted in loss of spawning habitat for pike. As a result, many populations are now supported by stocking. Stocking of pike has varied over time with early hatchery operations producing numerous small fish that were stocked into recipient lakes in spring or early summer and generally did not survive well. Later, growing pike to advanced sizes and stocking in colder temperatures of late fall – or even the following spring – was shown to enhance survival of stocked fish. However, such enhanced hatchery operations are far more costly than ones stocking numerous small fish. The trade-off between number of fish stocked, size of fish stocked, and survival has been evaluated in many states, and the best result depends on the ecological conditions of recipient waters. Many jurisdictions have no stocking programs to support pike populations in most lakes, while stocking continues in areas at the extreme of their range or areas with very high fishing pressure.

Another supplementation of reproduction for pike is enhancement of spawning habitat. In the 1950s, it was realized that access to spawning marshes, either natural or constructed wetlands, could allow pike to spawn in such areas and enhance reproduction. Connections were constructed between distant wetlands and a lake or river, and young fish produced in the wetland could either naturally migrate back to the lake in summer or be seined and released in the lake. Spawning marshes are still constructed in areas depleted of wetland vegetation with resulting increases in pike populations. Design of spawning habitat for pike has proceeded in shoreline areas of lakes as well with corresponding increases of pike populations in areas such as the Toronto Harbor.

Conservation is important for threatened or endangered species as well as for populations influenced by habitat loss or heavy harvest pressure. Sport fishing regulations are intended to reduce harvest or direct it to certain times of year or sizes of fish that will allow for successful reproduction. It is important to realize that producing more restrictive regulations does not necessarily result in changes in fish mortality. Heavily fished populations can benefit from stricter regulations, while lightly fished populations will most likely show no response to regulation changes. Most anglers do not catch their limit of fish, so changes in bag limits do not directly alter fish populations to similar degree. Another issue in fishing regulations is the willingness of the public to follow the rules. In most areas, conservation officers are rarely encountered so voluntary compliance is important. Unpopular regulations, or ones which are overly complicated, may not have the intended effect on the fishery.

Summary

As human populations increase, our influence on fish populations becomes more severe, and our need to manage populations to sustain abundance or even survival also increases. A number of laws have been enacted to identify and sustain at risk species, and the science of their conservation is developing. All too often, such cases end up in courts with controversial results. Human alteration of landscapes and harvest of fishes also cause a need to provide laws to sustain even common species as well as methods to manage their populations. This chapter provides several examples of efforts used to rehabilitate fish populations, which is an important role that fishery managers have played for many years and will continue to play in the future as more challenges occur and more conservation science is developed.

Literature Cited

Auer, N.A. 1996. Importance of habitat and migration to sturgeons with emphasis on lake sturgeon. Canadian Journal of Fisheries and Aquatic Sciences 53(S1):152–160.

Baldwin, N.A., R.W. Saalfeld, M.R. Dochoda, H.J. Buettner, and R.L. Eshenroder. 2002. Commercial fish production in the Great Lakes 1867–2000. http://www.glfc.org/databases/commercial/commerc.asp

Brooks, S.G., and P.K. Kind. 2002. Ecology and demography of the Queensland lungfish (*Neoceratodus forsteri*) in the Burnett River, Queensland with Reference to the Impacts of Walla Weir and Future Water Infrastructure Development. Final Project Report QO02004. Department of Primary Industries, Brisbane, Queensland, Australia.

Cox, G.W. 1993. Conservation Ecology. William C. Brown Publishers, Dubuque, Iowa.

Deacon, J.E., and C.D. Williams. 1991. Ash Meadows and the legacy of the Devil's Hole pupfish. Pages 69–87 *in* W.L. Minckley and J.E. Deacon, editors. Battle Against Extinction: Native Fish Management in the American West. The University of Arizona Press, Tucson.

Diana, J.S., and K. Smith. 2008. Combining ecology, human demands, and philosophy into the management of northern pike in Michigan. Hydrobiologia 601:125–135.

Fisher, J.L., J.J. Pritt, E.F. Roseman, et al. 2018. Lake sturgeon, lake whitefish, and walleye egg deposition patterns with response to fish spawning substrate restoration in the St. Clair-Detroit River system. Transactions of the American Fisheries Society 147:79–93.

Kind, P.K. 2003. Movement Patterns and Habitat Use in the Queensland Lungfish *Neoceratodus Forsteri* (Krefft 1870). Unpublished Ph.D. Dissertation. University of Queensland, Brisbane, Australia.

Lema, S.C., and G.A. Nevitt. 2006. Testing an ecophysiological mechanism of morphological plasticity in pupfish and its relevance to conservation efforts for endangered Devils Hole pupfish. The Journal of Experimental Biology 209:3499–3509.

Master, L.L., S.R. Flack, and B.A. Stein, editors. 1998. Rivers of Life; Critical Watersheds for Protecting Freshwater Biodiversity. The Nature Conservancy, Arlington, Virginia.

McDougall, C.A., P.A. Nelson, D. Macdonald, et al. 2017. Habitat quantity required to support self-sustaining lake sturgeon populations: An alternative hypothesis. Transactions of the American Fisheries Society 146:1137–1155.

Nelson J.S. 2006. Fishes of the World, 4th edition. John Wiley & Sons, New York.

Osborne, M.J., T.E. Dowling, K.T. Scribner, and T.F. Turner. 2020. Wild at heart: Programs to diminish negative ecological and evolutionary effects of conservation hatcheries. Biological Conservation 251:108768.

Pister, E.P. 1990. Desert pupfishes: An interdisciplinary approach to endangered species conservation in North America. Journal of Fish Biology 37A:183–187.

Scott, W.B., and E.J. Crossman. 1973. Freshwater Fishes of Canada. Fisheries Research Board of Canada, Bulletin 184, Ottawa.

Thompson, B.S., and S.M. Rog. 2019. Beyond ecosystem services: Using charismatic megafauna as flagship species for mangrove forest conservation. Environmental Science and Policy 102:9–17.

Williams, J.D., and G.K. Meffe. 1998. Nonindigenous species. Pages 117–130 *in* M.J. Mac and P.A. Opler, editors. Status and Trends of the Nation's Biological Resources. U.S. Geological Survey, Washington, D.C.

Index

Note: Page numbers in *italics* refer to figures; page numbers in **bold** refer to tables.

Biology and Ecology of Fishes, Third Edition. James S. Diana and Tomas O. Höök.
© 2023 John Wiley & Sons Ltd. Published 2023 by John Wiley & Sons Ltd.